DIGITAL ELECTRONICS

Concepts and
Applications for
Digital Design

Saunders College Publishing

A division of Holt, Rinehart and Winston, Inc.

Philadelphia   Fort Worth   Chicago   San Francisco   Montreal   Toronto   London   Sydney   Tokyo

Richard J. Prestopnik

Fulton-Montgomery Community College

DIGITAL ELECTRONICS

Concepts and Applications for Digital Design

Requests for permission to make copies of any part of
the work should be mailed to: Copyrights and
Permissions Department, Holt, Rinehart and Winston,
Inc., Orlando, Florida 32887.

Text Typeface: ITC Cheltenham Light
Compositor: General Graphic Services, Inc.
Acquisitions Editor: Barbara Gingery
Managing Editor: Carol Field
Project Editor: Anne Gibby
Copy Editor: Elaine Honig
Manager of Art and Design: Carol Bleistine
Art Director: Christine Schueler
Art and Design Coordinator: Doris Bruey
Text Designer: Tracy Baldwin
Cover Designer: Lawrence Didona
Cover Photo Researcher: Teri Stratford
Text Artwork: GRAFACON
Director of EDP: Tim Frelick
Production Manager: Bob Butler

Cover Credit: Photography by Weinberg & Clark.
Provided by Advanced Micro Devices.

Printed in the United States of America

Digital Electronics: Concepts and Applications
for Digital Design

ISBN: 0-03-026757-9

Library of Congress Catalog Card Number: 89-043499

3   032   98765432

To my wife, Jan, who makes every day worth living.

One constant in the digital logic industry upon which we can depend is that the technology will always be changing. Over the brief life history of digital design we have witnessed an evolution as component design has progressed from bulky vacuum tubes to slick VLSI integrated circuits. Naturally, these technological changes have influenced how digital logic textbooks present their subject matter.

In more recent times, logic families such as TTL have seemed like the all-encompassing technologies of the future due to their preponderance of integrated parts. Variations in TTL as well as the introduction of CMOS and ECL parts ushered in the world of vendor logic: predesigned parts that could be interconnected in a building block approach. However, therein lies the rub. The logic family influence caused the digital design course of study to rapidly migrate toward teaching how to interconnect these parts to create digital systems, but tended to shy away from teaching the knowledge base necessary to design basic elemental functions. A whole generation of digital logic students were taught not how to design various synchronous counters, but taught how to wire specific part numbers to function as synchronous counters. The details of logic design progressed no further than an understanding of pin function for the logic family of choice. This has created a disparity between academics and industry that becomes increasingly evident as programmable logic devices and other large-scale integrated circuits dominate the design industry. Systems are no longer designed by interconnecting basic parts. Complete systems are now integrated on a single chip—but, what a chip! With the likelihood that several thousand logic devices can be crammed onto one piece of silicon, it is imperative that future logic designers know how to design—*not* just interconnect. It is no longer sufficient to serve up a helping of TTL part numbers and expect students to grow into logic designers.

Text Philosophy

Though technology has changed, one thing hasn't. Students who understand how to design logic systems—from basic gate structures to complex systems using the fundamental aspect of logic design rather than prepackaged building blocks—become innovative creative logic designers. This basic philosophy underscores the presentation of all material in this text. The students are trained to comprehend and apply the basic concepts of digital design. These basic tenets are then used as a springboard to the understanding of more advanced ideas. A topic is not simply developed in the text to enable students to wire a predesigned part to function; the students are taught how to approach a design task and develop a logic circuit appropriate to the design objective at hand. If students then wish to implement the function using predesigned parts, the task will be easy. If students wish to create the function using more advanced logic technologies, the task will still be easy because the design approach is still based on fundamentals. Students are not hindered in their design efforts if they find they cannot use familiar part numbers; the students are taught how to design using basic logic design concepts as their guide.

Text Structure

Thirteen chapters take students from the most basic of digital electronic topics to an advanced understanding of digital design. Along the way, students are supplied with an interesting, balanced, and measured progression of circuits and ideas to transform what appears as simply another electronic subject into an area of high interest worthy of their pursuit.

Chapter 1 introduces the concepts of digital signals and binary numbers providing a base set of knowledge for all to follow. Digital gates are introduced in Chapter 2 and then supported with in-depth discussions on important analytical elements such as timing diagrams and truth tables. Chapters 3 and 4 delve into the mathematical side of Boolean Algebra and introduce many simplification techniques from algebraic reduction through Karnaugh Mapping and the Quine–McCluskey method. Computer-aided design is also covered in this section as well as in other key areas of the book.

Mathematical circuitry ranging from simple half adders through carry look-ahead circuitry are detailed in Chapter 5. Important mathematical ideas such as complement math are also investigated thoroughly in this chapter. Students learn how to design and use fundamental digital circuitry such as decoders and multiplexers based on the material covered in Chapter 6. A significant portion of the chapter material also covers binary codes and error-checking techniques.

Chapter 7 comprises a comprehensive look at latches and flip-flop devices. Topics range from basic device operation to flip-flop timing, hazards, and race conditions. Timing circuitry such as one-shots are also investigated. Chapter 8 builds on the knowledge gained in Chapter 7 to enable the student to design and understand many common sequential circuits including asynchronous counters, synchronous counters, and shift registers. It is worth emphasizing that the material is presented to allow students to design a variety of circuits regardless of the future technologies they may encounter in industry.

Chapter 9 treats students to memory devices including important RAM and ROM technologies. Memory expansion is also covered. Chapter 10 opens up a new horizon in the student's design insight by providing a detailed look at the operation and design considerations involved with the use of programmable array logic (PAL) devices. This material is supplemented by two complete appendix sections on the software development aspects of programmable logic, including information on how to obtain free software for this purpose.

Chapters 11 and 12 cover a wide range of practical topics including A/D and D/A conversion, interfacing, logic technologies, and noise in digital systems. Chapter 12, in particular, covers many ideas rarely mentioned in textbooks, but which are critical to successful digital design. Chapter 13 introduces students to the powerful design world of state machines.

Throughout each chapter, attention has been paid to providing a consistent approach to the presentation of material and to linking together successive topics and chapters to provide continuity in the learning process.

Chapter Structure

Each chapter includes many pedagogical features designed to enhance the learning process and to maintain reader motivation and interest. These elements include

- Chapter outline.
- Chapter objectives.
- Chapter preview setting the tone and goals for the concepts to follow.
- Figures supporting the ideas brought out in the text.
- Text written in an easy-to-follow and understandable style.
- Design examples exploring topics in further detail to enable students to apply and reinforce theory and concepts. These are excellent problem-solving exercises for the instructor as well.
- "For Your Information" sections are scattered throughout the text, providing useful and interesting information related to or enhancing the topic under discussion.
- Chapter summary itemizes and reviews key chapter information.
- An extensive set of end-of-chapter questions supports the chapter material.
- Use of two colors to emphasize the important aspects of figures and illustrations.

End-of-Text Materials

Provided at the end of the book are appendices containing the following useful materials:

- Manufacturers' data sheets on integrated circuit parts used in the text.
- Precautions for the handling of MOS devices.
- Extensive information regarding the use of PALASM2 for programmable logic development.
- Designing state machines using PALASM2.
- IEEE/ANSI Std. 91–1984 logic symbol usage.
- Answers to selected end-of-chapter questions.
- Glossary.

Ancillaries

A complete set of associated material is available to support this text and the students' learning process. These include

- A laboratory manual containing 43 laboratory experiments and several longer projects.
- An instructor's laboratory manual with completed designs and typical test data.
- An instructor's manual with solutions to all end-of-chapter problems and transparency masters of 50 key figures from the text.

All ancillary material was written by the textbook author to provide continuity and consistency across the spectrum of teaching material.

Acknowledgments

Much of my effort involved in the writing and production of this textbook was backed by those who gave their time, knowledge, or support to this project. I gratefully acknowledge the advice, dedication, and professionalism exhibited by the people at Saunders College Publishing/Holt, Rinehart and Winston, Inc., including Barbara Gingery, Senior Acquisitions Editor; Anne Gibby, Project Editor; and Laura Shur, Editorial Assistant.

Many companies that have changed the face of the digital electronics industry supplied information and products critical to the completion of this project. These companies include Advanced Micro Devices/Monolithic Memories, Altera Corp., Analog Devices, Cypress Semiconductor, Data I/O Corp., Ford Microelectronics, General Instruments, Integrated Device Technologies, Intel Corp., Logical Devices, LSI Logic Corp., Maxim Integrated Products Inc., Motorola, National Semiconductor, Precision Monolithics, Raytheon, RCA Semiconductor, Samsung Semiconductor, Signetics Corp., Siliconix, Sprague Electric Company, Standard Microsystems, Teledyne, Texas Instruments, SGS-Thomson, VTC Inc., Xicor Inc., Xilinx. Several of these companies also provided datasheets and other information printed in this text as noted by accompanying credit lines. I am grateful for their support.

I appreciate the comments and helpful suggestions provided by the following technical reviewers for this book: Steve Yelton, Cincinnati Technical College; Gerald Schickman, Miami Dade Community College, North Campus; Andrew Fioretti, formerly of Suburban Tech; George Mason, Indiana Vocational Technical College, South Bend; John Morgan, DeVry Institute of Technology, Irving; Norman Crowder, Illinois Technical College; Thomas Bingham, Flourissant Valley Community College; John Harrell, Colorado Technical College; Walter Buchanan, Indiana University/Purdue University, Indianapolis; Stephen Kator, formerly of Metropolitan State College; Jim DeLoach, DeVry Institute of Technology, Atlanta; David Hata, Portland Community College; Bradley Jenkins, St. Petersburg Junior College; Herbert Daugherty, Indiana Vocational Technical College, Muncie; Marybelle Beigh, Chemeketa Community College.

Creative efforts such as book writing occur easily in an environment where new ideas are welcomed and the stimulus for innovation occurs daily. I am fortunate that this environment flourishes at Fulton-Montgomery Community College. The combination of a superb teaching faculty and an encouraging academic administration makes the teaching profession an ongoing pleasurable experience. May it always remain this way.

This section would not be complete without acknowledging the love and support given to me by my children Nathan, Emily, and Adam. My biggest thank you is reserved for my wife Jan who has acted as editor, proofreader, and source of common sense throughout my writing projects. I'm extremely blessed to have her love and friendship.

Richard J. Prestopnik
Johnstown, New York

Contents

DIGITAL ELECTRONICS

Concepts and
Applications for
Digital Design

CHAPTER 1

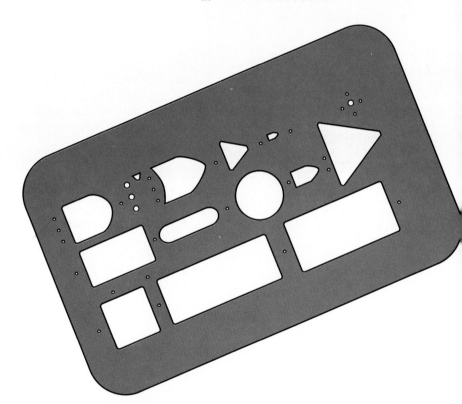

INTRODUCTION TO DIGITAL ELECTRONICS

OBJECTIVES

To explain the meaning of logic levels and how they are used in logic circuit design.

To discuss the binary number system and its relationship to digital logic. The binary and decimal systems will be compared.

To introduce integrated circuits, integrated circuit logic families, and their application to digital design.

To define the nanosecond.

PREVIEW

The term "digital electronics" encompasses many technological fields that may be familiar, such as integrated circuits, computer systems, and consumer electronics. Because of its wide-ranging influence, digital electronics is a necessary field of study for the electrical technician, electrical engineer, computer engineer, industrial engineer, and many others. Unlike other electronic fields of study, digital electronics deals with logic design, the binary number system, and electronic circuitry that is already designed and proven. The digital engineer, designing at a higher functional level, designs "logic circuits" rather than "electronic circuits," thereby avoiding many of the electrical distractions accompanying electronic circuit design. This does not suggest that digital design is approached in a simplified cookbook fashion, for complexity is evident in many digital applications. But it does suggest to the novice that some of the mathematical rigor and circuit complexity is less intimidating than in electronic design, making digital electronic fundamentals relatively easy to master. This chapter begins our exploration of digital electronics.

1.1 ■ DIGITAL ELECTRONICS—A HISTORICAL PERSPECTIVE

The development of the digital electronics field parallels that of computer development. Although many related historical incidents preceded the modern digital computer, today's digital electronics field reflects mainly the development of computer systems over the past 50 years. Computer system development encouraged the growth of digital concepts and the invention of digital technology. This stimulus fostered improvements in circuitry and components whose benefits spilled over into other areas. For instance, early computers were designed around the vacuum tube and relay circuits. These components were relatively slow (electrically) and very power-hungry. In addition, the physical size of these devices resulted in enormous machines that demanded expensive ventilation and space requirements. During this time the availability of the newly invented and much smaller transistor came to the rescue. Many design problems were alleviated as transistors became the backbone for the newer generations of computing machinery. As the computer industry advanced, the need for smaller, faster, and better devices led to the development of the integrated circuit. An extremely small device that could hold many digital circuits, the integrated circuit created an explosion in the electronics industry that is still felt today. The microprocessor chip, the heart of almost everything electronic from personal computers to microwave ovens, was a result of this rapid and fascinating development. Digital electronics development, predominately in the computer and integrated circuit marketplace, raced ahead at full steam, bringing new devices and ideas to the industry on a daily basis. We begin our investigation of the field by examining many of the fundamental concepts dictating how digital electronic circuits are designed and developed.

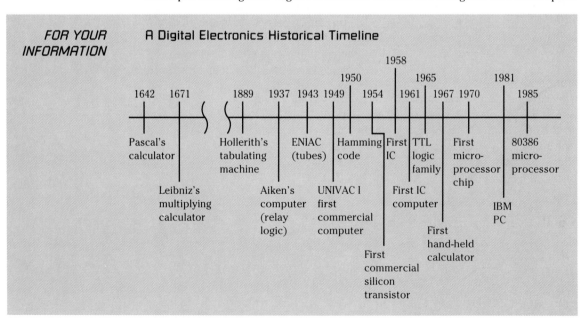

FOR YOUR INFORMATION

A Digital Electronics Historical Timeline

1.2 ■ DIGITAL?

Before proceeding, we need to acquire an appreciation for the term "digital." If the term digital were one side of a coin, then the term "**analog**" would be the other side. We are very well aware of analog signals even if we are not familiar with the term itself. Analog signals manifest themselves as voltages, currents, temperature, and many other common measurements. In fact, most physical phenomena are analog in nature. The distinguishing characteristic of analog signals is that they can possess an unlimited number of values; that is, the values are **continuous** (although in practice, analog data are limited to a range of continuous values). For example, a temperature reading can be expressed as 74.2°F, or 74.24°F, or 74.241°F, and so on. The number of digits representing an analog measurement is limited only by the accuracy of the equipment; analog signal measurements can be refined by increasing the number of digits.

Digital signals are represented by only two possible conditions or **logic levels**. Therefore, digital signals are **discrete** measurements. For instance, as the seconds tick by on a digital timer, they occur in incremental steps, that is, 48 seconds, 47 seconds, 46 seconds, and so on. Certainly each increment of time lasts for a second, but the interval between the displayed time does not exist in a digital representation. We know very precisely what time is on the timer just by looking at it, but 48.3 seconds or 48.34 seconds is not represented. The continuous measurement of time or any other parameter cannot occur with a digital representation. We can increase the accuracy of the reading by including more digital displays or devices (i.e., subdivide the seconds into parts), but the readings will always be incremental.

Fortunately, the incremental nature of digital information processing is not a problem. Rather, it is an advantage in many applications. The use of digital electronics in computer systems stems from the fact that digital circuits are very easy to design, that they are easily fabricated on an integrated circuit to take up little space, and that they operate at very high rates of speed. Furthermore, as the remaining chapters in this text will explain, creating digital systems from relatively simple digital circuits is not altogether that difficult. Very often we will find that larger and more accurate digital systems can be obtained simply by adding or cascading smaller circuits together, a decided advantage to the designer. The continuous nature of analog quantities can also be accommodated by using digital circuits. Converting analog information into digital information, and vice versa, is a common application. The following sections show how digital information is defined.

1.3 ■ HOW LOGIC LEVELS ARE DEFINED

The notion of a **logic level** is important in digital design because it is the basis of circuit design, circuit testing, and circuit operation. Electronically, a logic level is equal to a voltage, or more specifically, a range of voltages. Digitally, only two logic levels exist. This means that the digital designer deals with signals in the digital logic world where only two distinct levels are possible as opposed to the electronic engineer's concern with voltages, currents, and other electronic phenomena. The digital designer doesn't ignore electronic circuits, but rather, uses circuits already designed and proven to function properly from an electronic point of view. The digital designer treats these circuits as logic building blocks that can be interconnected to produce logic circuits and logic systems. Since the electronic circuits are designed to be interconnected, the digital designer is not so concerned with matters of voltage levels and current flow. How these electronic circuits are viewed as logic building blocks is the subject of discussion in the next chapter.

The concept of electronic circuits versus digital circuits is based on the notion of the logic level. In most circuitry the levels are created by switching circuits that can be either on or off. Thus, the two logic levels can be designated simply as "on" and "off." From this definition a basic light switch functions as a digital device; it can be on or off. In this example the physical motion of moving the switch is the device **input** signal, whereas the on or off condition is described as the device's **output** signal. In practical digital circuits both input and output signals are electronic in nature. However, this example shows how a continuous analog quantity (motion) is converted into a digital quantity (on-off).

From a digital design standpoint other terms are also used to describe the two logic levels. **High** and **low, one** and **zero, true** and **complement,** and **true** and **false** are the most frequently used terms. Furthermore, a logical device whose output is high is said to be in the high "state." Likewise, when the output is low, the device is said to be in the low "state." The **state** of a logic device or circuit is also a notation that is frequently used.

Assigning the levels one and zero to the output of a digital device is simply a matter of convention. For instance, a switch could be defined as being in the one state when closed and in the zero state when open. However, it is just as easy to define the closed switch as the low state and the open switch as the high state. Conventionally speaking, most digital devices are designed so that the higher output voltage is deemed as a one, high, or true level, whereas the lower voltage is deemed as a zero, low, or complement level. This system is referred to as a **positive logic** system.

DESIGN
EXAMPLE 1.1

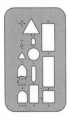

Using the logic level terms one (1) and zero (0), determine the digital quantities for the following device outputs:

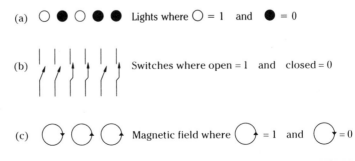

(a) ○ ● ○ ● ●   Lights where ○ = 1   and   ● = 0

(b)   Switches where open = 1   and   closed = 0

(c)   Magnetic field where ( ) = 1   and   ( ) = 0

(d) Pulse train where $\overline{} = 1$ and $\underline{} = 0$

| t0 | t1 | t2 | t3 | t4 |

(e) 0.1v 2.6v 3.1v

Transistors where an output voltage
from 0v to 0.8v = logic 0 and
2.4v to 5v = logic 1

Solution   Each device may have only one logic level at a time. The exact
levels depend on how the levels are defined for each device. Using the definitions given, we have the following logical representations for the preceding
devices:
(a) 10100   (using on as logic level one and off as logic level zero)
(b) 110010
(c) 011
(d) 01001
(e) 011

1.4 ■ THE BINARY NUMBER SYSTEM

As human beings we are well aware of the decimal system. The simple fact
that humans have ten fingers and ten toes had a significant impact on the
development of the decimal system. You might say that the decimal system
was technology driven, meaning that it evolved along with the technology that
uses it—us. The digital systems and computer systems that we design operate
on the binary number system for the same basic reason—they are technology
driven. The electronic circuits that comprise our digital logic building blocks
are constructed using transistors. You may not be familiar with the transistor,
but a transistor can function as a simple on/off switch, having the capability
to possess two distinct states. Since this switching action is fundamental to
digital circuit design, technology dictates that we use the binary number system
in digital design. (We can design transistor circuits to have more than two
states if desired, but it is much easier and more reliable to work with only
two.)

The binary number system contains only two numerals to represent all
numbers—zero (0) and one (1). It should be apparent why logic levels are
also often referred to as ones and zeros. (However, binary zeros and ones do
not necessarily equate with logic level zeros and ones, because, as mentioned,
logic levels may be arbitrarily assigned.) Whereas decimal numbers are comprised of digits, binary numbers made from groups of ones and zeros are said
to be comprised of **bits**. Since a single bit can only represent a one or zero,
multiple bits are required to represent most numbers. In order to understand
binary numbers fully, we need first to reexamine some characteristics of decimal numbers.

Terminology Defining the Grouping of Binary Bits

Bit—a single binary number (0 or 1)
Nibble—four binary bits (i.e., 0101)
Byte—eight binary bits (i.e., 10110011)
Word—16 binary bits
Double word—32 binary bits

Figure 1.1
Decimal Digits

0 1 2 3 4 5 6 7 8 9

Decimal numbers are formed from the ten decimal digits listed in Figure 1.1. When grouped together, they form decimal numbers. Grouping decimal digits into decimal numbers is done in accordance with a **positional weighting system**, which means that the position of a digit within a group of digits determines its value. Figure 1.2 illustrates this well-known fact.

Figure 1.2
Decimal Number
System Positional
Weighting

128

— units position
— tens position
— hundreds position

$128 = (1 \times 100) + (2 \times 10) + (8 \times 1)$

The 2 in the number 128 has a greater positional value than the 8 because the 2 resides in the "tens" position, whereas the 8 resides in the "units" or "ones" position. Therefore, the 2 represents 2 tens or 20, whereas the 8 represents 8 units or 8. The complete number is determined by summing the individual digit values. Thus, $128 = (1 \times 100) + (2 \times 10) + (8 \times 1)$. Notice that the positional values are related to powers of ten (Figure 1.3). That is, the units position is equivalent to 10^0, the tens position is equivalent to 10^1, the hundreds position is equivalent to 10^2, and so on. The positional value in-

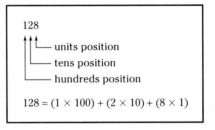

$$\ldots 10^4 \quad 10^3 \quad 10^2 \quad 10^1 \quad 10^0 \; . \; 10^{-1} \quad 10^{-2} \quad 10^{-3} \; \ldots$$

10000 1000 100 10 1 ⦙ 0.1 0.01 0.001

Decimal point

That is, $10^4 = 10 \times 10 \times 10 \times 10 = 10000$

Figure 1.3
Decimal Powers of 10

$$10^{-2} = \frac{1}{10^2} = \frac{1}{100} = 0.01$$

creases by a factor of ten for each position moving from right to left. Understanding the basic principles of a system used every day may seem trivial, but
it lays the groundwork for an understanding and appreciation of the binary
system, a system with which you are probably not quite so familiar.

Binary numbers are also formed according to a positional weighted system
that is based on the **powers of 2**. This is illustrated in Figure 1.4. As we can
see, the positional value doubles for each bit position as we proceed from
right to left. Knowledge of these positional values allows us to translate binary
numbers into decimal number equivalents and also points out the fact that
any decimal number can be expressed in the binary system. Thus, digital
computer systems have no trouble dealing with decimal numbers; we simply
translate and retranslate numbers from decimal form to binary form.

$$\ldots 2^4 \quad 2^3 \quad 2^2 \quad 2^1 \quad 2^0 \; . \; 2^{-1} \quad 2^{-2} \quad 2^{-3} \quad 2^{-4} \ldots$$

$$16 \quad\quad 8 \quad\quad 4 \quad\quad 2 \quad\quad 1 \;\; 0.5 \quad\quad 0.125$$

$$0.25 \quad\quad 0.0625$$

Binary point

That is, $2^3 = 2 \times 2 \times 2 = 8$

$$2^{-3} = \frac{1}{2^3} = \frac{1}{8} = 0.125$$

Figure 1.4
Binary Powers of 2

Figure 1.5 shows how binary numbers are related to decimal numbers.
Each binary position has a weighted value equivalent to a decimal value. All
positions containing ones are converted to their equivalent decimal value and
then the sum of the positional decimal values forms the equivalent decimal
number. In general, representing equivalent numbers requires more binary bits
than decimal digits.

Decimal numbers can be converted to binary equivalents following the
simple procedure illustrated in Figure 1.6. The decimal number is divided
repetitively by 2 until further division is impossible. The remainder from each

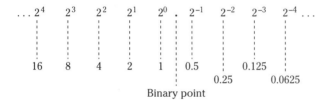

Positional weight ⟶ $2^3 \; 2^2 \; 2^1 \; 2^0$
Binary number ⟶ 1 1 0 1

$2^0 \times 1 = 1 \times 1 = 1$ — Value of
$2^1 \times 0 = 2 \times 0 = 0$ each binary
$2^2 \times 1 = 4 \times 1 = 4$ bit position
$2^3 \times 1 = 8 \times 1 = \underline{8}$
13 ⟵ Decimal
value
of 1101

Figure 1.5
Binary to Decimal
Conversion

Convert 16 to its binary equivalent number

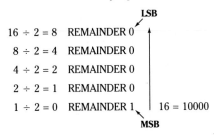

$16 \div 2 = 8$ REMAINDER 0

$8 \div 2 = 4$ REMAINDER 0

$4 \div 2 = 2$ REMAINDER 0

$2 \div 2 = 1$ REMAINDER 0

$1 \div 2 = 0$ REMAINDER 1   16 = 10000

Convert 28 to its binary equivalent number

Figure 1.6
Converting Decimal
Numbers to Binary
Using Repetitive
Division by 2

$28 \div 2 = 14$ REMAINDER 0

$14 \div 2 = 7$ REMAINDER 0

$7 \div 2 = 3$ REMAINDER 1

$3 \div 2 = 1$ REMAINDER 1

$1 \div 2 = 0$ REMAINDER 1   28 = 11100

step of the division process is recorded. Since the remainder will be either one or zero, the remainders become part of the binary number. The division process will end with a division of 1 by 2. The result is zero with a remainder of 1. This last remainder of 1 becomes the **most significant bit** (MSB) of the result. This MSB is the bit with the largest positional weight. The other bits in the answer are recorded in order from the MSB until the last bit (the first remainder in the process) is recorded as the **least significant bit** (LSB). The LSB is the bit with the smallest positional value.

Fractional numbers are converted using repetitive multiplication by 2 as Figure 1.7 illustrates. Each multiplication generates a carry of one or zero from the fractional portion of the number to the whole portion of the number. Each carry becomes part of the binary result. The multiplication process is carried out until the fractional result equals zero.

Decimal numbers can also be converted to binary by determining the positional weights whose sum equals the desired decimal number. For instance, decimal 9 is equal to the positional weights of 8 + 1. The binary number

Convert 0.125 to binary

Figure 1.7
Converting Fractional
Decimal Numbers to
Binary Using
Repetitive
Multiplication by 2

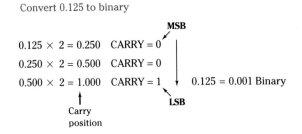

$0.125 \times 2 = 0.250$ CARRY = 0

$0.250 \times 2 = 0.500$ CARRY = 0

$0.500 \times 2 = 1.000$ CARRY = 1   0.125 = 0.001 Binary

Carry
position

9 is composed of ones representing these weights placed in their proper positions. Thus, decimal 9 = binary 1001. (Zeros are assigned to positions 2 and
4 since these positional weights cannot be summed together to create 9, but
they are necessary as place holders in the binary number.)

Convert to decimal:
(a) 110110   (b) 11001.01   (c) 10000000

**DESIGN
EXAMPLE 1.2**

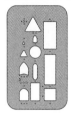

Solution   Determine the decimal equivalent number by summing the individual binary bit values. The binary bit value is equal to the binary bit times its
positional weight.

(a) 110110

$$(1 \times 32) + (1 \times 16) + (0 \times 8) + (1 \times 4) + (1 \times 2) + (0 \times 1) = 54$$

(b) 11001.01

$$(1 \times 16) + (1 \times 8) + (0 \times 4) + (0 \times 2) + (1 \times 1)$$
$$+ (0 \times 0.5) + (1 \times 0.25) = 25.25$$

(c) 10000000

$$(1 \times 128) + (\text{all other bit values} \times 0) = 128$$

Convert to binary using the binary system's positional weights:
(a) 18   (b) 52   (c) 512

**DESIGN
EXAMPLE 1.3**

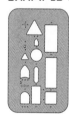

Solution   Determine the binary number by finding the positional values that
add together to produce the decimal number. Assign these numbers to their
proper bit positions. Use zeros for all other bit positions.

	512	256	128	64	32	16	8	4	2	1	
(a)	0	0	0	0	0	1	0	0	1	0	= 18
(b)	0	0	0	0	1	1	0	1	0	0	= 52
(c)	1	0	0	0	0	0	0	0	0	0	= 512

1.5 ■ COUNTING IN THE BINARY SYSTEM

Using the information on binary positional weights, we can create a binary
counting sequence, as shown in Figure 1.8. Each number in succession is one
greater than the number before it. For instance, binary 9—1001—is one greater
than binary 8—1000. This can be confirmed by converting from binary to

Binary	Decimal
0000	0
0001	1
0010	2
0011	3
0100	4
0101	5
0110	6
0111	7
1000	8
1001	9
1010	10
1011	11
1100	12
1101	13
1110	14
1111	15

Numbers are in sequence, so each number is one greater than the preceding number.

Figure 1.8
Binary Counting
Sequence

That is, $8 = (2^3 \times 1) + (2^2 \times 0) + (2^1 \times 0) + (2^0 \times 0)$
$9 = (2^3 \times 1) + (2^2 \times 0) + (2^1 \times 0) + (2^0 \times 1)$

decimal. This "**pure or natural binary**" counting sequence is used extensively in digital design. Notice the counting pattern. The LSB position varies in a
0
1
0
1 pattern. The next most significant position varies in a
0
0
1
1
0
0
1
1 pattern. Examine the other columns for similar patterns. Notice how each pattern is related to powers of 2 and make every effort to become familiar with the patterns. Also notice that the number of binary combinations is related to the number of binary bits. For instance, four binary bits produce $2^4 = 16$ combinations of ones and zeros—0000 through 1111. Five binary bits produce $2^5 = 32$ combinations—00000 through 11111. Note that the all zeros combination is just as valid as any other combination. It doesn't mean "nothing"; the all zeros combination implies a condition where all binary bits happen to be zero. In general, the number of binary combinations possible from a group of bits equals 2^n, where n represents the number of bits. Furthermore, for any

counting sequence the largest number possible is $2^n - 1$. Therefore, with 4-bits, the largest number that can be represented is $2^4 - 1 = 16 - 1 = 15$; with 5-bits the largest number is $2^5 - 1 = 31$.

Additional uses of binary numbers, such as addition, subtraction, and other applications will follow in chapters to come. This preliminary introduction to binary numbers sets the stage for their use in digital systems design.

How many binary bits are required to represent:
(a) 68 combinations   (b) 512 combinations?

Solution   The number of combinations possible for any number of binary bits is equal to 2^n, where n equals the number of bits.

(a) $2^6 = 64$ and $2^7 = 128$
Since 68 is greater than 64, 6-bits is insufficient. Seven bits are necessary to represent 68 combinations (or 69, 70, 71, . . . , 128).
(b) $2^9 = 512$; 9-bits represent 512 combinations exactly.

How many combinations are possible with:
(a) 8-bits   (b) 16-bits?

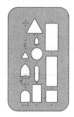

What is the range of binary numbers possible? What is the largest equivalent decimal number possible?

Solution   The number of binary combinations possible is equal to 2^n. The range of binary numbers is from all zeros to all ones where all ones equal $2^n - 1$. Converting this number to decimal yields the largest decimal number possible for the number of bits allotted.

(a) 8-bits: $2^8 = 256$ combinations
    The range = 00000000 through 11111111; the largest number is 11111111
    or:

$$128 + 64 + 32 + 16 + 8 + 4 + 2 + 1 = 255$$

(b) 16-bits: $2^{16} = 65,536$ combinations
    The range = 0000000000000000 through 1111111111111111; the largest number is 1111111111111111 or:

$$32,768 + 16,384 + 8,192 + 4,096 + 2,048 + 1,024 + 512$$
$$+ 256 + 128 + 64 + 32 + 16 + 8 + 4 + 2 + 1 = 65,535$$

DESIGN
EXAMPLE 1.6

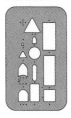

Write out the counting sequence for:
(a) three binary bits   (b) five binary bits

Solution   The counting sequence begins with a count of all zeros and ends with the largest number possible using the number of bits specified ($2^n - 1$). Each number is numerically one greater than the number before it.

3-bits MSB LSB	5-bits MSB LSB
⋮ ⋮	⋮ ⋮
000	00000
001	00001
010	00010
011	00011
100	00100
101	00101
110	00110
111	00111
	01000
	01001
	01010
	01011
	01100
	01101
	01110
	01111
	10000
	10001
	10010
	10011
	10100
	10101
	10110
	10111
	11000
	11001
	11010
	11011
	11100
	11101
	11110
	11111

1.6 ■ DIGITAL INTEGRATED CIRCUITS

Many of the major advances in digital system design have come about because of the **integrated circuit (IC)** or "chip." No doubt, you will be using some of these devices in laboratory experiments to strengthen your understanding of digital logic. Digital integrated circuits are constructed to hold digital circuitry ranging from a very few simple devices to complex microprocessors. These chips are easy to interconnect, so relatively complex "logic circuits" can be assembled in a short time. Before you begin work with these chips or proceed further in this book, a few points of information are in order.

Pin Outs

A chip **pin out** is a diagram defining the purpose of each pin on an integrated circuit. When you connect several chips together, you do so by attaching a wire from one chip's output pin to another chip's input pin. Naturally, you must know which pin is which for this to be possible. Common digital ICs range in size from 14-pin chips through 68 pins. The chips also come in a variety of shapes and sizes. In all cases a manufacturer's data book containing all pertinent chip specifications is essential in order to utilize these devices.

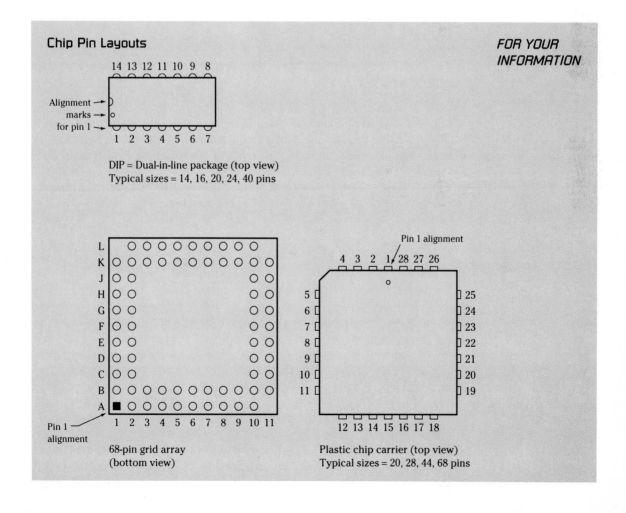

Chip Pin Layouts

FOR YOUR INFORMATION

DIP = Dual-in-line package (top view)
Typical sizes = 14, 16, 20, 24, 40 pins

68-pin grid array
(bottom view)

Plastic chip carrier (top view)
Typical sizes = 20, 28, 44, 68 pins

A Logical Look Inside a Chip

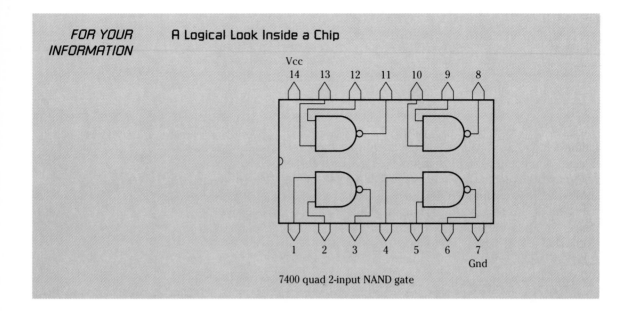

7400 quad 2-input NAND gate

Logic Families

Most digital ICs belong to a **logic family**. A logic family contains many chips, each with a different digital function. The chips within the logic family are all fabricated using the same semiconductor technology and are designed to be easily interconnected, and, as you might expect, each family has its own advantages and disadvantages. The three major families you are likely to encounter are TTL—**Transistor-transistor logic**, CMOS—**complementary metal oxide semiconductor**, and ECL—**emitter coupled logic**. These families will be discussed in more detail later in this book.

Power

All electronic devices consume power as they operate. The source of power is provided by a system power supply that must be sufficiently large to provide current and voltage for all devices attached to it. A logic device demanding large amounts of power requires a correspondingly large power supply, a costly item. In addition, devices with excessive power consumption generate greater quantities of heat than devices that consume little power. This may necessitate special cooling systems to draw off the extra heat and to maintain safe operating temperatures. Cooling systems can range from simple fans to chilled water circulating systems, all of which add cost and bulk to the overall digital system. Heat also stresses components and tends to reduce reliability. Furthermore, large power consumption makes battery-operated systems impractical because the batteries will not last long. At this point in your studies the most important function of the power supply is delivering the proper voltage to your chips. Every chip you use requires a power and ground connection from the power supply. The manufacturer's data sheet specifies which pins on the chip are set aside for this purpose.

Speed

When referring to logic device speed, we are specifically referring to **propagation delay time**. Propagation delay time is the time it takes an input signal change to appear at the device output. Devices with fast propagation delay times generally produce digital systems that operate at high rates of speed. For instance, fast computers are fast partially because they are constructed with fast logic chips. Other factors, such as rise and fall times, also affect circuit speed. These considerations are also discussed in later chapters. But since propagation delay time affects circuit performance and the way that we approach some circuit designs, a preliminary understanding of this delay is important.

1.7 ■ THE NANOSECOND

When discussing integrated circuit speed, we are talking about devices that are blazingly fast, with speeds in the **nanosecond** range. A nanosecond is one-billionth of a second, and many logic families have delay times in the 5–10 nanosecond range. Such speeds have a strong impact on the way logic circuits are designed and constructed.

Several analogies may help you gain an appreciation for nanosecond speeds. Assume that you can snap your finger once a nanosecond. At this rate you can snap your finger 1,000,000,000 times a second, or 60,000,000,000 times a minute, or 3,600,000,000,000 times an hour. You get the picture—a lot can happen at nanosecond rates. Computers and digital systems operate near these speeds, accounting for their fast performance.

Finally, consider a jet traveling at a speed of 750 miles per hour. This is equivalent to 12.5 miles per minute, or 0.208 miles per second, or 2.08×10^{-10} miles per nanosecond. This equates to only 0.0000132 inches per nanosecond. Even compared to some of humans' fastest accomplishments, the brevity of a nanosecond is obvious.

Time and Powers of 10		FOR YOUR INFORMATION
millisecond	10^{-3}	
microsecond	10^{-6}	
nanosecond	10^{-9}	
picosecond	10^{-12}	

SUMMARY

■ The development of digital electronics as a distinct electronics field has strong links to the growth of the computer industry.

■ The term digital describes signals that are discrete, whereas the term analog defines signals that are continuous.

■ Practical digital signals are represented electronically as two distinctive logic levels.

■ Typically, logic levels are defined as high and low, true and complement, true and false, or one and zero.

■ The relationship between the two digital logic levels and the binary number system (base 2) influences digital circuit design.

■ Conversion between the binary and decimal number systems is a necessity if digital systems are to function in a decimal world.

■ Most digital circuits are fabricated as integrated circuits.

■ Integrated circuits provide many benefits for digital design, including small size and high operating speeds.

PROBLEMS

Section 1.2

1. Which of the following are analog quantities? Which are digital quantities?
*(a) volume of water flowing over a dam
 (b) body temperature
*(c) dates on a calendar
 (d) steps on a staircase
 (e) voltage at an electrical outlet
*(f) music information on a compact disc

Section 1.4

2. Convert the following binary numbers into equivalent decimal numbers:
*(a) 100110   (b) 0011001   (c) 10000000   (d) 10000000.001
*(e) 1010101.101   (f) 1000001   (g) 10001100   (h) 111111111110

3. Convert the following decimal numbers into equivalent binary numbers using positional weighting:
 (a) 64   (b) 65   *(c) 12.5   (d) 129   (e) 1023   *(f) 26.125   (g) 9

4. Convert the following decimal numbers into binary numbers by the methods of dividing by 2 or multiplying by 2:
 (a) 256   (b) 255   *(c) 16.125

---

* See Appendix F: Answers to Selected Problems.

Section 1.5

5. Show how to represent the decimal 7 as a:
   (a) 3-bit binary number
   *(b) 4-bit binary number
   (c) 8-bit binary number
6. Write down the binary counting sequence for the numbers 0 through 63.
7. How many combinations of ones and zeros are possible with:
   (a) 2-bits   (b) 5-bits   (c) 8-bits   *(d) 10-bits   (e) 12-bits   (f) 16-bits
   *(g) 20-bits
*8. What is the largest number that can be represented using the number of bits listed in problem 7(a)–(g)?

Section 1.6

9. Using a TTL data book, determine the function and package styles for the following TTL part numbers:
   (a) 74LS32   (b) 7486   (c) 74AS195   (d) 54280
10. Using CMOS data books, if available, determine the function and package styles for the following CMOS parts:
    (a) 74HC74   (b) 4073   (c) 4518   (d) 74HCT393
11. Using an ECL data book, if available, determine the function and packaging styles for the following ECL parts:
    (a) 10101   (b) 100125   (c) 10130   (d) 10164
12. Using a programmable logic data book, if available, determine the function and packaging styles for the following parts:
    (a) PAL16L8   (b) PAL16R4   (c) PAL20R6

Section 1.7

13. Determine how many nanoseconds elapse during:
    *(a) 20 minutes   (b) one day   (c) one week

CHAPTER 2

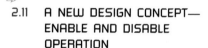
DIGITAL LOGIC FUNDAMENTALS

OBJECTIVES

To introduce the fundamental building blocks—AND, OR, INVERT, NOR, NAND—used in digital design.

To demonstrate several techniques used to represent logical operations.

To show methods by which basic logic gates may be combined into more complex and useful circuitry.

To demonstrate methods that correctly verify the operation of logic circuits.

To illustrate the differences and similarities between truth tables and timing diagrams.

To show the usefulness of the Exclusive-OR function.

To enable the student to move from a simple understanding of digital logic concepts toward a higher level of proficiency in the application of digital circuits.

PREVIEW

Every digital design consists of fundamental building blocks that interact to form a logical structure. This means that we begin our study of digital logic design by using simple digital devices called gates and extend our knowledge of digital design based on an understanding of fundamental logic gate concepts. This chapter introduces the basic gates used in digital design and describes various methods that represent the gate functions in circuits and equations.

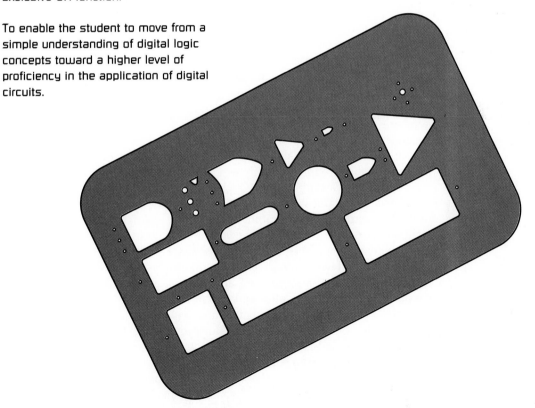

2.1 ■ AN INTRODUCTION TO DIGITAL LOGIC GATES

Five fundamental logic circuits—AND, OR, INVERT, NAND, and NOR gates—form the basic building blocks for digital logic design. The term "gate" is used to describe any of the basic logic circuits. The application of the term gate will become evident as your knowledge of their application unfolds.

Electronically, these gates have been built over the years around relay, vacuum tube, diode, and transistor circuits. Some well-proven designs have evolved into integrated circuit logic families that greatly reduce the design complexities facing the digital logic designer. The basic gates have also been constructed to operate using water, light, or air pressure rather than electricity, and the future possibility of organic logic gates is being investigated. In any case, regardless of the technology used to implement the logic itself, the concept of what an AND, OR, INVERT, NAND, or NOR gate is intended to do remains the same. Each one of these logic functions will be investigated thoroughly in the following sections to lay the foundation on which we can progress to more complex logic designs. We can be confident that the material learned now is perfectly applicable to any past, present, or future logic technology.

2.2 ■ HOW LOGIC GATES ARE DESCRIBED

Each logic gate performs a logical function. In fact, each gate's name states the particular function the gate is capable of performing. In the following sections a verbal description corresponding to the logic gate name will begin the discussion of each gate, providing a natural introduction and description of device behavior.

There are other useful ways to describe logic gates as well. Logical functions are similar to the standard mathematical functions used every day. For instance, the AND gate performs the AND function, which is similar to multiplication. Therefore, just as we need a mathematical way to describe common mathematical operations, we will need a mathematical way to describe these logic functions. Describing these logical functions in a mathematical expression provides us with logic design advantages. The mathematical expression, or **Boolean expression**, describing the logical function can be mathematically manipulated to give the logic designer many benefits in terms of reduced circuit complexity, faster circuits, and less expensive circuits.

Since we will work extensively with circuit schematics, schematic symbols representing the various gates are also necessary as design aids. Each gate has its own unique symbolic representation, or device identification. Logic symbol usage becomes second nature after a short while. Throughout this text traditional logic symbols are used. However, in the appendix an alternate method is discussed for describing logic gate and circuit functions.

Finally, a **truth table** will depict all possible gate input and output combinations. A truth table is a chart showing the logic gate binary output conditions for every set of binary input combinations. We use the truth table often to illustrate not only gate operation, but also the operation of the digital circuits that we eventually design. The truth table is also a useful tool aiding in solving or **debugging** circuit problems. (The term "debugging," or working the "bugs" or errors out of a circuit design, stems from the fact that one problem in an early computer system was actually caused by a moth caught between relay contacts. This truly was a bug in the circuit.)

2.3 ■ THE AND FUNCTION

A verbal definition for the AND function can be stated as follows: **The output of an AND gate is a binary one if and only if all gate inputs are also at the binary one level. If any other binary input combination is present, then the output is a binary zero.** Many students beginning to study digital logic assume that the significance of the AND function is only that all one levels at the gate inputs result in a one level at the output. This incorrect assumption ignores the fact that any zero, at any input, provides a zero output condition. Although not immediately evident, the zero output condition is just as beneficial as the one output condition.

With the AND definition in mind, let's examine the AND logic symbol (Figure 2.1). The symbol shown is for a 2-input AND gate. The letters A, B, and C represent variables. In this case A and B are inputs, whereas C is an output. These variable letters are merely chosen for convenience. Since we are working with the binary number system, the variables can only represent a binary one or zero. Therefore, A can be either a one or zero; the same is true for variable B. Variable C is also one or zero, but the specific value of C at any time depends on the ANDing of A with B.

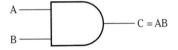

Figure 2.1
AND Logic Symbol

Applying the AND definition to the symbol leads us to understand that output C is equal to one when inputs A AND B are also one. This **unique condition** for the AND gate is the only input condition that produces a one output. Any other combination of A and B (there are three others) produces a zero output since either A or B, or both, will be zero.

The truth table concisely illustrates all the potential input combinations. Recall that the number of combinations based on a grouping of binary bits is equal to 2^n. Therefore, two inputs provide $2^2 = 4$ input combinations. The 2-input AND truth table is shown in Figure 2.2. Notice that inputs A and B have four combinations of zeros and ones listed below them. Notice also that the combinations are listed in a strict binary count progression. All truth tables you create should list the input combinations following this progression in order to prevent errors that may occur by accidentally omitting combinations. The C column lists the output conditions for the AND function. The only combination indicating a one at the C output occurs when both inputs are at one levels. The remaining combinations, all with at least one zero level input,

A	B	C
0	0	0
0	1	0
1	0	0
1	1	1

Figure 2.2
2-Input AND Truth
Table

will produce a zero level at the output. At a glance, the truth table quickly shows all possible AND gate conditions.

The AND function is not limited to 2-input devices. Theoretically, an infinite number of inputs are possible, in which case the AND gate output still only equals one when all inputs are one. If any of the infinite number of inputs is at a zero level, then the output is zero. (From a technology perspective, however, the number of inputs is restricted. Logic families such as TTL and CMOS generally limit inputs to a maximum of eight because of the physical packaging constraints of the devices. Some programmable logic devices may have AND gates with up to 32 inputs.)

Figure 2.3 illustrates a 4-input AND gate. The truth table for this gate is shown in Figure 2.4. With 4 inputs present, 16 possible input combinations

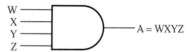

Figure 2.3
4-Input AND Gate

W	X	Y	Z	A
0	0	0	0	0
0	0	0	1	0
0	0	1	0	0
0	0	1	1	0
0	1	0	0	0
0	1	0	1	0
0	1	1	0	0
0	1	1	1	0
1	0	0	0	0
1	0	0	1	0
1	0	1	0	0
1	0	1	1	0
1	1	0	0	0
1	1	0	1	0
1	1	1	0	0
1	1	1	1	1

Figure 2.4
4-Input AND Truth
Table

exist, but only the combination where WXYZ = 1111 creates an output of one. The 15 remaining input combinations will all result in a zero output level. You can see that the AND definition can be extended over any number of inputs.

Referring back to Figures 2.1 and 2.3, we notice the Boolean expression representing the AND function. In Figure 2.1, C = AB. This is read as: C is equal to A AND B. The AND operator is generally implied in these expressions much the same way that the multiply operation is implied in Y = 2T. In this case we would read the expression as: Y is equal to 2 times T. For the AND operation an expression such as C = AB is read as: C is equal to A ANDed with B. Sometimes for clarity, a multiplication sign or a dot is used to represent the AND operation. Therefore, C = A*B or C = A·B; both are equivalent to C = AB. Throughout this book the C = AB notation is usually used. We now see that the equation in Figure 2.3, A = WXYZ, means that input variables W, X, Y, and Z are ANDed together to produce A (A is equal to W AND X AND Y AND Z). Thus, the Boolean expression mathematically conveys the same information presented by symbols and truth tables. This is true for all logic functions.

A designer wishes to AND circuit variables R, Q, and W to create output Y. Draw the symbol, Boolean equation, and truth table for this function.

DESIGN
EXAMPLE 2.1

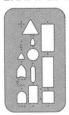

Logic symbol:

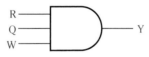

Logic equation:

$$Y = RQW$$

Truth table:

R	Q	W	Y
0	0	0	0
0	0	1	0
0	1	0	0
0	1	1	0
1	0	0	0
1	0	1	0
1	1	0	0
1	1	1	1

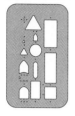

Write the Boolean equation and draw the logic symbol corresponding to the following truth table:

X_3	X_2	X_1	X_0	OUT
0	0	0	0	0
0	0	0	1	0
0	0	1	0	0
0	0	1	1	0
0	1	0	0	0
0	1	0	1	0
0	1	1	0	0
0	1	1	1	0
1	0	0	0	0
1	0	0	1	0
1	0	1	0	0
1	0	1	1	0
1	1	0	0	0
1	1	0	1	0
1	1	1	0	0
1	1	1	1	1

Solution   The truth table is that of a 4-input AND gate—a high output condition exists only for the input combination where all inputs are high.

4-input AND symbol:

X_0
X_1
X_2 ── OUT
X_3

Equation:

$$OUT = X_0 \, X_1 \, X_2 \, X_3$$

2.4 ■ THE OR FUNCTION

The OR function can be defined as follows: The OR gate output will be at a binary one level when any OR gate input is also at the binary one level. The OR gate output will be at the binary zero level only when all OR gate inputs are at the binary zero level.

Once again, as with the AND gate, there are two broad ways to consider OR operation. Most input combinations to an OR gate produce a one level output. This **nonunique condition** occurs whenever any input is at the one level. The unique zero level output condition occurs only for the all zeros input combination, meaning that for any number of input variables only one combination produces the zero output condition. As you will notice, unique and nonunique output combinations are two useful ways to consider logical device operation.

The symbol for the OR gate is given in Figure 2.5. As you can see, this 2-input OR gate provides an output equal to the ORing of input A with B. The expression listed at the output, $C = A + B$, is the Boolean expression representing the OR operation. Notice that the plus sign $(+)$ is used to represent OR. This expression is read as: C is equal to A OR B. The truth table in Figure 2.6 provides us with complete information on the 2-input OR gate. From this table we can see exactly when to expect a high or low output signal.

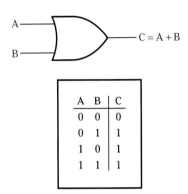

Figure 2.5
2-Input OR Gate
Symbol

A	B	C
0	0	0
0	1	1
1	0	1
1	1	1

Figure 2.6
2-Input OR Truth Table

Naturally, the OR function can be extended to include more than two inputs. For example, if a 3-input gate were available, as illustrated in Figure 2.7, then eight input combinations are possible ($2^3 = 8$). The expression $Z = F + G + H$ indicates that the three inputs F, G, and H are ORed together to provide output Z. Remember, even though we have several input variables and an output variable to contend with (which may initially be confusing), these variables represent only a binary one or zero level. The truth table in Figure 2.8 confirms this. All combinations of F, G, and H, including at least one input

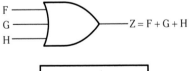

Figure 2.7
3-Input OR Gate
Symbol

F	G	H	Z
0	0	0	0
0	0	1	1
0	1	0	1
0	1	1	1
1	0	0	1
1	0	1	1
1	1	0	1
1	1	1	1

Figure 2.8
3-Input OR Truth Table

variable at the one level, will produce an output level also equal to one. The only combination that results in a zero output level is the combination wherein F, G, and H are all zero. We could have any number of inputs, and as long as all inputs are at the zero level, the output will be zero.

DESIGN EXAMPLE 2.3

Draw the logic symbol, Boolean expression, and truth table resulting from the ORing of variables C, D, and E:

Logic symbol:

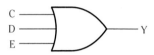

Boolean expression:

$$Y = C + D + E$$

Truth table:

C	D	E	Y
0	0	0	0
0	0	1	1
0	1	0	1
0	1	1	1
1	0	0	1
1	0	1	1
1	1	0	1
1	1	1	1

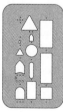

DESIGN EXAMPLE 2.4

A light goes on if at least one of four switches—S1, S2, S3, or S4—are also on. Write the logic equation describing this circuit.

Solution   Since the light can be lit by either S1 or S2 or S3 or S4 in any combination, the OR equation describes circuit operation. Thus:

$$\text{Light} = S1 + S2 + S3 + S4$$

2.5 ■ THE INVERT FUNCTION

Inversion is a very simple, easy to understand function that is nonetheless an extremely powerful logic building block. The **INVERT**, or **complement** operation, is electronically carried out using an **inverter gate** whose function is described as follows: **The output of an inverter is binary one if the inverter**

FOR YOUR
INFORMATION

Use of the Inversion Representation

- $\overline{A}\,\overline{B} \longrightarrow$ A is inverted; B is inverted; $\overline{A}$ and $\overline{B}$ are ANDed.
- $\overline{AB} \longrightarrow$ A and B are ANDed, then inverted (NAND).
- $\overline{A} + \overline{B} \longrightarrow$ A is inverted; B is inverted; $\overline{A}$ and $\overline{B}$ are ORed.
- $\overline{A + B} \longrightarrow$ A and B are ORed, then inverted (NOR).

Alternate Inversion Representations

$$\overline{A} = A' = !A = /A$$

input is binary zero. If the inverter input is binary one, then the inverter output is binary zero.

The inverter symbol in Figure 2.9 confirms an implication in the preceding definition: The inverter gate has only one input. The sole logical purpose of this gate is to reverse the logic level of the input. Notice that in Figure 2.9 the input is designated as A and the output is designated as $\overline{A}$. The bar placed over the variable indicates that the variable has been inverted. We generally describe this as "complementing a variable." The word complement (not compliment) also means that the variable has been inverted.

Figure 2.10 is the truth table for the inverter. The basic nature of the invert function is evident in the simplicity exhibited by its truth table.

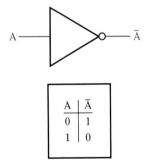

Figure 2.9
Inverter Symbol

Figure 2.10
Inverter Truth Table

2.6 ■ TRUE AND COMPLEMENT NOTATION

Often we refer to the initial value of a variable as its true value and the inverted value as its complement value. True and complement values of a variable are frequently generated as a matter of course in digital systems. For instance, if variable X is present, then usually $\overline{X}$ is also present. Sometimes the word true is related to the one level of a variable and the word complement is related to the zero level of a variable, so care must be taken when using this nomenclature. For example, a glance at the inverter truth table clearly shows that variable A can be at the one or zero level just as can variable $\overline{A}$. Thus, unless additional information is specified, the overbar designation indicates only that a variable has been inverted in a digital design. When specifically using the terms true and complement, common practice is to refer to A as the true value and $\overline{A}$ as the complement value. This is a practice that we will follow throughout this text.

As simple as they are, the inverter, AND, and OR gates can theoretically be used to build any digital structure. We will soon study other gates that make this task even easier, but at this point we need to understand that these fundamental logic functions do form the underlying operation of all logical designs, an important fact to appreciate. However, this does not call for an overnight session memorizing truth tables. **Understand** the purpose of the various gates and functions and you will soon have a command of the material to follow.

DESIGN EXAMPLE 2.5

If the stream of binary information shown here is applied to the input of an inverter, what will be the resulting output information?

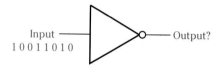

Input — 1 0 0 1 1 0 1 0 ——▷o— Output?

Solution   Each input bit will be available at the output in complement form. Therefore, 10011010 becomes 01100101.

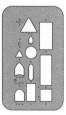

DESIGN EXAMPLE 2.6

An alarm is purchased for a burglar alarm circuit. The alarm is activated when a low logic level is presented to its input. However, the burglar alarm circuitry produces a high logic level output indicating that the alarm should sound. How can the alarm be connected for proper operation?

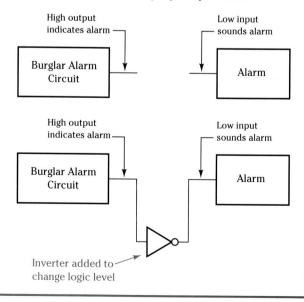

Inverter added to change logic level

FOR YOUR
INFORMATION

Common use for the terms true and complement:

$$A = 1 \longrightarrow \text{true level of A}$$
$$A = 0 \longrightarrow \text{complement level of A} \longrightarrow \overline{A}$$

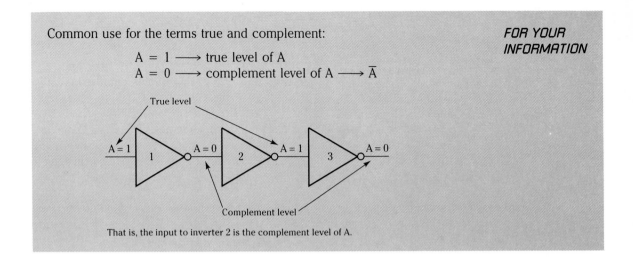

That is, the input to inverter 2 is the complement level of A.

2.7 ■ HOW TO USE LOGIC GATES

Now that we have several primitive or basic logic gates under our belts, you may be wondering how we use them. Certainly, you will not go down to a computer store, buy an AND gate, and expect to do any significant computing. However, the various gates we have begun to study are the building blocks to bigger and better things. In fact, it is truly remarkable that such simple functions can be combined to produce the computers, control systems, and other sophisticated systems to which we have become accustomed. We now examine how gates are combined to produce more extensive logical functions.

In general, logic circuits are designed to be connected so that the output of one gate feeds into the input of another gate. Figure 2.11 shows an AND gate output feeding an OR gate input. From an electronic (voltages, currents) point of view the manufacturers have already designed the gates' internal circuitry to accept this kind of connection easily. This makes design life for logic designers easier. They need not consider all the electronic details involved in connecting the circuits together—and there are plenty of details to consider—the designers simply deal with the logical considerations involved. Figure 2.12 shows how one output may feed the inputs of several other gates. There may be limits to the number of inputs that can be connected to a single output, but those technology constraints will be considered in a later chapter. For now, we just note that a logic gate output can feed more than one logic gate input and that the connections are made simply by attaching a wire between outputs and inputs.

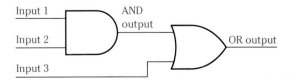

Figure 2.11
Connecting Logic
Gates

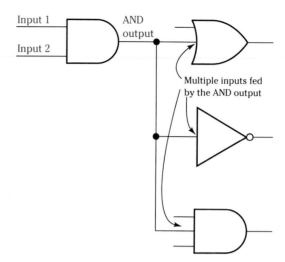

Input 1

AND
output

Multiple inputs fed
by the AND output

Figure 2.12
Gate Output Feeding
Several Inputs

Now a method is needed to describe any logic circuit we may wish to design or examine. Fortunately, the Boolean expression serves us well, since every digital circuit can be described with a Boolean equation. Figure 2.13 is an example. The variables assigned to the circuit inputs flow logically through the circuit to the output. That is, input variables are logically combined by logic gates to produce an output that reflects the circuitry through which the input signals have passed. This is essentially what we expect from any circuit; input conditions should produce a desired output condition. Therefore, it makes sense to start at the inputs to see how the various logic gates influence the variables on their way to the output. A Boolean expression is a mathematical way to summarize the effects of a logic circuit on input variables, but attempting to determine the output Boolean expression for a complex circuit in just one step is not an easy task nor a good idea. Rather, the best procedure is to examine how the input variables are logically modified on a gate-by-gate basis. For example, input variables A and B are immediately ANDed in the circuit shown in Figure 2.13. Therefore, the output of the AND gate is AB. Writing this down on the schematic near the output of the gate will help you to keep track of the circuit data flow. The output created by the AND gate now becomes an input to the 2-input OR gate. Keep in mind that the term AB still has only two possible binary values—one or zero. The fact that two letters now represent the OR gate input implies that the input is obtained from some previous logical operation; it does not mean that more than two binary levels are possible. The second input to the OR gate comes directly from input C. Thus, input C is ORed with input AB, resulting in the output Y = AB + C.

Figure 2.13
Finding a Circuit
Boolean Expression

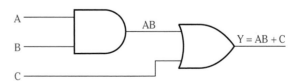

A

B

AB

Y = AB + C

C

Examine the expression $Y = AB + C$. Notice that the Boolean representation for the OR function $(+)$ is clearly evident. The AND operation is also indicated. Can you see why the AND of A with B is one of the inputs to the OR gate? Refer to Figure 2.13 again. The OR gate is only a 2-input gate. You can easily understand how the C input feeds the OR gate since C is a single input variable, but you should also notice that the other OR gate input must be written as a more complex term since it comes from a logic gate output rather than from a direct input variable.

Figure 2.14 illustrates another circuit and the steps necessary to obtain the logic expression for the circuit. The outputs of each AND gate are determined from the input variables. Note that one AND gate has the complement of B as an input and that this variable is ANDed with input variable D. The AND gate output is obtained directly from these two variables to yield $\overline{B}D$ since $\overline{B}$ and D are the two variables being ANDed together. The outputs of all three AND gates are applied to the inputs of a 3-input OR gate to create the final expression $Y = AB + \overline{B}D + ACD$.

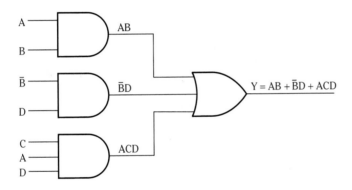

Figure 2.14
Circuit Boolean
Expression

Figure 2.15 is a more complex circuit. Notice that the output of the AND gate is $(A + B)C$. This is read as: "A OR B ANDed with C." Again, the inputs to this AND gate are variable C and the output of an OR gate $(A + B)$. The parentheses are used to indicate that the OR operation involves variables A and B. Without parentheses the equation would read "$A + BC$," implying that variable A is ORed with the results of the ANDing of B and C. Removing the parentheses has changed the meaning of the expression. A word to the wise— use as many parentheses as necessary to differentiate clearly logic circuit operations. Also keep in mind that there is a mathematical order involving arithmetic functions. In Boolean algebra the AND operation takes precedence over the OR operation. In other words, all AND operations appearing in a Boolean expression occur within the logic circuit before OR operations. The

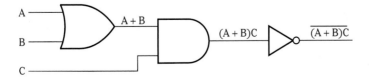

Figure 2.15
Using Parentheses
With a Circuit
Expression

only exception to this rule occurs for operations appearing within parentheses. Operations within parentheses take precedence over all others. Therefore, the parentheses are required in Figure 2.15 to identify correctly that the OR operation occurs within the logic structure before the AND operation.

In addition, Figure 2.15 also shows the effects of inversion on an output function. The AND gate output is connected to the inverter input. The inverter simply complements the logic level presented to it. Recall that when using Boolean algebra, the overbar indicates inversion. Since the input to the inverter is $(A + B)C$, the output is simply the complement of this input, or $\overline{(A + B)C}$. The overbar is placed across the entire expression because the variables comprising the expression, as they move from inverter input to output, are inverted as a function rather than as distinct variables. Therefore, an overbar placed over any group of variables implies that the logical function has passed through an inverter. Much more overbar notation is evident when we study NAND and NOR gates.

DESIGN EXAMPLE 2.7

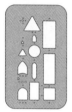

Determine the expression for the following circuit:

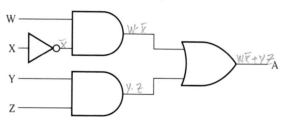

Solution Determine each gate equation; work from inputs to output.

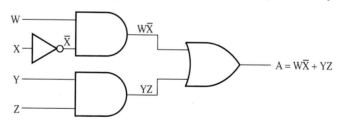

DESIGN EXAMPLE 2.8

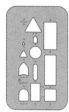

Determine the expression for the following circuit:

Solution Starting at the inputs, determine each gate equation. Use these to determine Y.

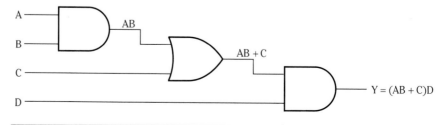

Determine the expression for the following circuit:

Solution Starting at the inputs, determine each gate equation. Use these to determine T.

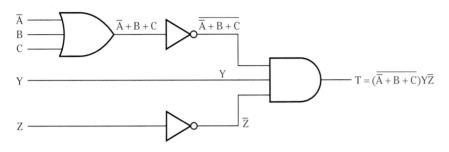

Draw the circuits for:
(a) $Y = AB + \overline{C}D + E$
(b) $Y = (A + B)CD$
(c) $Y = \overline{T + (U\overline{R} + S)Q}$

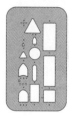

Solution Following mathematical precedence, draw functions within parentheses first; then AND functions; and finally, OR functions.

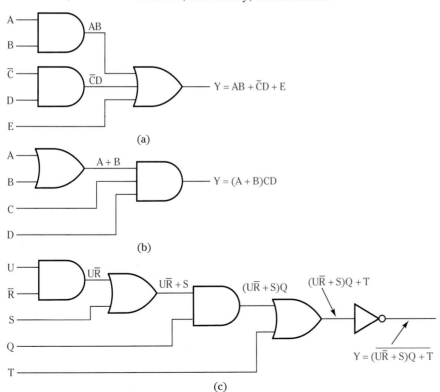

Using Binary Numbers with Logic Circuits

The capability to describe a circuit logically can be extended to predict the output level for any input combination. This is particularly helpful in testing a circuit.

Figure 2.16 shows a logic circuit with specific input logic levels. By working these logic levels through the circuit, we can succesfully predict the correct output level. This process is similar to creating the logic equation from input variables except that actual logic levels are used. Initially, you may find it convenient to place gate truth tables on the circuit schematic diagram as indicated in Figure 2.17. As you become more proficient using digital logic, this becomes unnecessary, although jotting down an equation or truth table for clarification purposes can never hurt. The gate truth tables in Figure 2.17 serve as a reminder of the basic functions. Also, by using the variable letters assigned to the circuit, you can clearly see how each gate contributes to the final output. For example, the AND gate ANDing A with B has four possible output conditions, but since the gate inputs are defined as A = 1 and B = 1, the only possible output for this AND gate is one. A quick look at the AND truth table confirms this. Marking a one at the output of this gate (Figure 2.18)

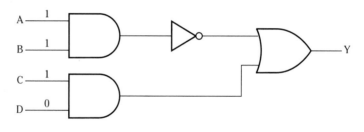

Figure 2.16
Circuit with Input
Levels

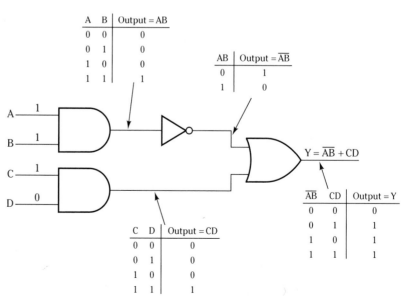

Figure 2.17
Determining Gate
Output Levels

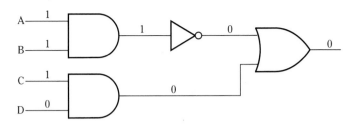

Figure 2.18
Determining Circuit
Levels

indicates not only that the output is one, but that the input to the next gate in line, the inverter, is also a one. Naturally, the output of the inverter is zero since the inverter complements whatever is present on its input. Also note that, from a Boolean equation point of view, the output of the AND gate is AB. Therefore, the inverted output becomes $\overline{AB}$. Here the overbar indicates that the AND operation of A with B has been inverted. Even though $\overline{AB}$ looks a bit complex, we can see that $\overline{AB}$ is simply equal to the binary number zero. It is advantageous to keep in mind that each wire (or line on a schematic) is only capable of carrying a single logic level, regardless of how complex the equation representating that wire may look.

To complete our discussion of the circuit illustrated in Figures 2.16, 2.17, and 2.18, let's look at the AND gate ANDing C with D. Since D is equal to zero, the output of the gate must also equal zero. (Any zero AND gate input forces the output to zero.) This output zero and the zero from the inverter output are then ORed together. Can you determine the output at this point? The truth table for the OR gate in Figure 2.17 reminds us that the output of the OR is zero when both inputs are zero. Thus, for input conditions A = 1, B = 1, C = 1, and D = 0 the circuit results in an output of $\overline{AB}$ + CD = 0.

Changing any of the previous input levels might possibly change the output levels as well. We would, of course, reanalyze the circuit to determine the new output level. Since the circuit has 4 inputs, there are 16 possible input combinations, resulting in 16 possible outputs. In actual practice, we may need to know how the circuit operates for many of these combinations, and the technique just described could become tedious for more thorough analysis. There is an easier way, with the truth table coming to our rescue again. This time, however, we write a truth table for a circuit rather than for just a gate. Since a truth table simply shows all output conditions for all input conditions, and no stipulation exists that the truth table is only accurate for gates, deriving a truth table for a circuit makes good sense, and, in fact, has several advantages for the diligent digital designer/troubleshooter. The table indicates not only the output for every input combination, but it also shows the output of every gate within the circuit, which is very useful for circuit testing. To utilize all this information, we need a complete truth table showing every gate output. Generally, constructing a complete truth table, including each inverter output, is an easy and fast way to determine circuit operation. Submitting to the temptation to leave gates off the table usually results in errors and time wasted. Furthermore, a complete truth table is essential for circuit testing and debugging.

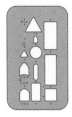

(a) A technician preparing to test the following circuit needs to know the output level for the input combination A, B, C, D = 1101. Determine the output level.
(b) Repeat the analysis for A, B, C, D = 1100.

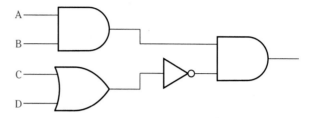

Solution Apply input levels to the schematic. Determine the output level at each gate, working from inputs to outputs.

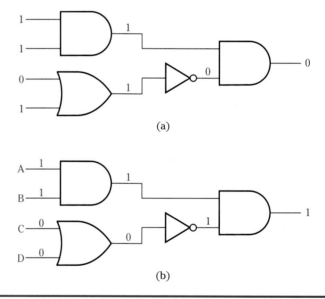

Developing Complete Circuit Truth Tables

A complete circuit truth table diagram is drawn in the same manner as a gate truth table diagram. For clarity, we will use the same circuit example discussed in the last section. First, all input variables are recorded in the first row of the table (Figure 2.19). The four inputs A, B, C, and D have 16 possible one/zero combinations, all of which must be recorded on the truth table since they all produce a corresponding output state.

We could assign the 16 binary combinations to the table in random order, but we run a significant risk of missing or repeating combinations if we take this approach. Assigning the input combinations by following a strict binary

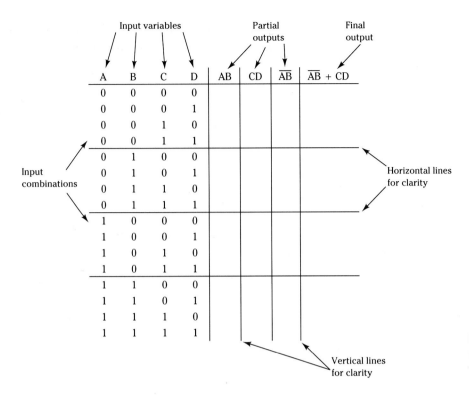

A	B	C	D	AB	CD	$\overline{AB}$	$\overline{AB}$ + CD
0	0	0	0				
0	0	0	1				
0	0	1	0				
0	0	1	1				
0	1	0	0				
0	1	0	1				
0	1	1	0				
0	1	1	1				
1	0	0	0				
1	0	0	1				
1	0	1	0				
1	0	1	1				
1	1	0	0				
1	1	0	1				
1	1	1	0				
1	1	1	1				

Input variables · Partial outputs · Final output · Input combinations · Horizontal lines for clarity · Vertical lines for clarity

Figure 2.19
Drawing a Circuit Truth Table

count allows us to account easily for all combinations both mathematically and visually, as indicated in Figure 2.19. Also notice that input A is listed as the most significant bit (MSB) and input D as the least significant bit (LSB) in this truth table. We could easily have assigned D as the MSB; as long as all combinations of the input variables are accounted for, the table will be correct. (*Caution:* If the variables were assigned so that the MSB was D and the LSB was A, the circuit would function the same, but the relationship between input variables and output levels would differ. Always make sure that the significance of the input variables is understood because the notion of the LSBs and MSBs is essential for mathematical representations and for binary codes.)

After listing the input variable combinations, we write down the output equations for each gate in the circuit. We refer to these as partial circuit outputs. The outputs of each gate, other than the last gate in the circuit, are considered as partial outputs since they contribute to the final logic circuit output. The best procedure is to write down the partial outputs in the order they are drawn on the schematic from input to output. Notice in Figure 2.19 that each partial output has been recorded along the top row of the table and all of them are separated by a vertical line to align each in a distinct column. Each column will eventually contain all the logical information for each partial output. Horizontal lines can also be used to align the rows of information. This will keep the lists of data neat and easy to read. The final output ($\overline{AB}$ + CD) is shown as the last column in this truth table. In a truth table that is correctly laid out, each partial output column will contribute the necessary input information to succeeding columns. All the items included in Figure 2.19 are typical of any complete circuit truth table. Now we will fill in the truth table values that correspond to this circuit's operation.

Each column representing a partial output is filled in with binary information for every input condition as shown in Figure 2.20. This process will be relatively easy to master if you understand how a logical function creates a partial output. For example, column AB represents the AND of inputs A with B and every entry in this column is the result of the ANDing of the A variable value with the B variable value. You may be bothered at this step by the fact that the 2-input AND gate under analysis has 16 truth table entries. Shouldn't a 2-input gate have only four possible combinations? Notice that the 16 entries are really just the four possible combinations repeated four times. That is, the input combination 00 occurs four times as do the combinations for 01, 10, and 11. These extra groupings occur because we are dealing with a 4-input circuit. The gate under study uses only two of the four possible input signals, but the truth table accounts for all 16 input combinations.

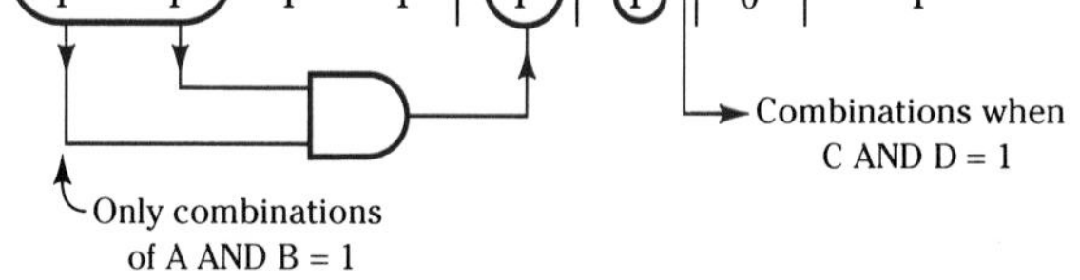

Figure 2.20
Completing a Circuit
Truth Table

Let's continue with the AB column and recall that the only time an AND gate can produce a high output is when all inputs are high. Notice that column AB is high only for the last four entries in the truth table. This corresponds to the times when inputs A and B are both high. Noting the AND operation in this way—looking for unique gate outputs—speeds analysis and completion of the truth table, and also promotes accuracy.

The column for CD is produced in a similar fashion. Like the preceding column, this column shows only four ones, but they do not occur in the same pattern as in the AB column. This is to be expected since the inputs for column CD are not the same as for column AB. The CD column is obtained by determining for each input combination the outcome of ANDing C with D.

The column for $\overline{AB}$ is not derived from the input listing since $\overline{AB}$ is not created directly from the input variables. Rather, since AB feeds the inverter input, $\overline{AB}$ is simply the opposite of column AB. You can see how easy it is to complete this column; each entry in the $\overline{AB}$ column is the complement of the entries in the AB column. The final output is created by ORing columns $\overline{AB}$ and CD. Since the OR gate inputs are already listed on the truth table (columns $\overline{AB}$ and CD), you can easily obtain the final circuit output levels by ORing each $\overline{AB}$ column entry with each CD column entry.

Develop a complete truth table for the following circuit:

DESIGN EXAMPLE 2.12

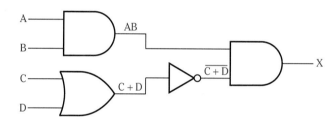

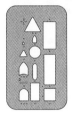

Solution Determine each gate output for every input combination. Use this information to determine the final output values.

A	B	C	D	AB	C + D	$\overline{C + D}$	X = AB $\overline{(C + D)}$
0	0	0	0	0	0	1	0
0	0	0	1	0	1	0	0
0	0	1	0	0	1	0	0
0	0	1	1	0	1	0	0
0	1	0	0	0	0	1	0
0	1	0	1	0	1	0	0
0	1	1	0	0	1	0	0
0	1	1	1	0	1	0	0
1	0	0	0	0	0	1	0
1	0	0	1	0	1	0	0
1	0	1	0	0	1	0	0
1	0	1	1	0	1	0	0
1	1	0	0	1	0	1	1
1	1	0	1	1	1	0	0
1	1	1	0	1	1	0	0
1	1	1	1	1	1	0	0

Circuit Troubleshooting with the Truth Table

The truth table developed in the previous section is now a convenient debugging aid for circuit troubleshooting. If, for instance, the circuit is producing an incorrect output level for input combination 1100, you can easily trace to the root of the problem using a logic probe, meter, or oscilloscope. The truth table guides you through the debugging process since it clearly indicates all potential circuit logic levels. Let's set up a troubleshooting problem using Figure 2.21 and the previously developed truth table to analyze a circuit debugging situation.

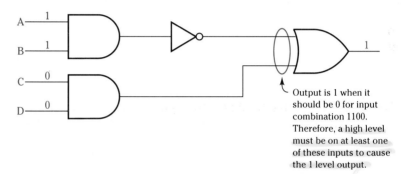

Figure 2.21
Circuit
Troubleshooting

Output is 1 when it should be 0 for input combination 1100. Therefore, a high level must be on at least one of these inputs to cause the 1 level output.

The circuit in Figure 2.21 shows that the output is high for input combination ABCD = 1100. By using the truth table and your knowledge of gate operation, we can logically deduce what is wrong with the circuit. Since the truth table indicates all correct circuit responses, we are able to isolate errors by finding discrepancies between actual circuit levels and levels indicated in the truth table. In fact, we know there is an error in the first place because the truth table final output and the circuit output do not agree. (The truth table in Figure 2.20 shows that input combination ABCD = 1100 should produce a low level output signal.)

You may recall that we determine a circuit's output level by tracing signals from input to output. When debugging a circuit, we often find it more convenient and expedient to trace from the output(s) back to inputs. In this manner portions of the circuit that operate correctly are easily noticed and need not be tested. This saves time and allows us to isolate a fault quickly. In this example the output is one when it should be zero. What could possibly cause this? Examining the gate controlling the output should give us a clue. The gate is an OR gate, and since the output is one, this indicates that at least one of the two gate inputs is probably high. We deduce that this is also incorrect since our truth table shows that both OR inputs are zeros in this instance (columns CD and $\overline{AB}$ in Figure 2.20). If only the OR input from the inverter is high and the other input is low, then we know that the problem resides in the portion of the circuit fed by the inverter. The other portion of the circuit need not be checked because the truth table confirms the correct operation.

Conversely, the AND gate feeding the OR would be in error if it were providing a one level. A glance at the truth table shows that neither the inverter nor AND outputs should be high for circuit input combination 1100, so this portion of the circuitry is checked. This is how circuit discrepancies are isolated. The troubleshooter traces back through the circuit following the signal path with the incorrect logic level. Eventually, a point will be reached where the cause of the error will be detected.

There are many reasons for a circuit error. Design rules that must be adhered to for reliable performance are associated with every logic technology. For instance, some logic families respond to a disconnected input as high logic levels (referred to as floating highs). A miswired circuit can create a floating high and cause the problem we are simulating. Other problems occur because of improper voltage levels, malfunctioning devices, or bent IC pins, to name a few. As you gain experience in digital design and troubleshooting, these common problems become obvious and easy to track down.

FOR YOUR
INFORMATION

Guidelines for Circuit Troubleshooting

1. Set inputs to levels that create an incorrect output response.
2. Trace back into the circuit from outputs toward the inputs with a logic probe, oscilloscope, or meter to determine the path of the error. A logic probe may be preferred since it can clearly indicate valid logic high and low levels. Oscilloscopes and meters do not easily distinguish between floating levels and low levels. However, an oscilloscope or logic analyzer is essential for timing analysis. (Timing is covered in later sections of the text.)
3. Compare measured levels with truth table levels at each gate input and output along the error path.
4. If or when a correct logic level is found along the error path, the point of the error is uncovered. The exact cause of the error can now be determined.

Some Troubleshooting Insights:

1. If, while tracing an error, valid but incorrect logic levels are found, the error is most likely due to improper wiring.
2. If floating levels (no response on a logic probe) are encountered, then disconnected inputs or broken wires are likely.
3. Incorrect levels on power supply pins indicate shorted power supply leads, outputs incorrectly tied together, or chips that are not properly supplied with power.
4. Check actual IC pins with the logic probe, not just the wiring. Bent pins can be uncovered in this way.
5. If a defective chip is suspected, substitute it with a known good device and recheck the circuit operation.

**DESIGN
EXAMPLE 2.13**

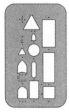

The following diagram shows the wiring layout for the circuit depicted in Figure 2.18 (using TTL parts):

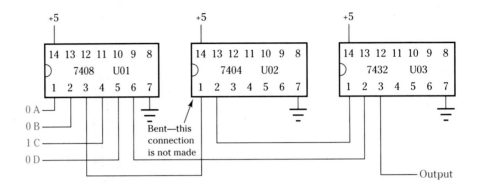

(a) For the input conditions shown (0010), what will be the circuit output level if pin 1 of IC U02 is bent under the chip?
(b) What levels would be found in the circuit as the technician traced signals from outputs to inputs?
(c) How will the technician actually spot the error?

Solution   Use a schematic and circuit truth table to record and compare measured levels to expected levels.

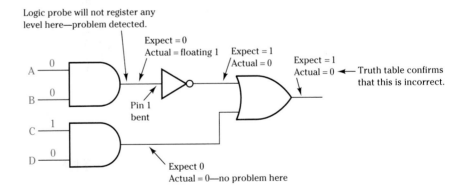

Since pin 1 is bent, a floating high exists at the inverter input.
(a) The output level resulting is low.
(b) Levels resulting from the input levels and the bent pin are indicated on the schematic as "actual" values.
(c) The error would be detected when pin 1 of U02 measured as a floating one on a logic probe or an invalid logic voltage level on an oscilloscope.

2.8 ■ USING TIMING DIAGRAMS

All the gates and circuits we discussed so far have been examined from a static operation point of view. Static operation means that specific input levels are provided to the devices, resulting in specific output levels. The input levels are maintained so that output levels never change during our analysis. Many circuits operate in a static fashion. This doesn't mean that they never change level, but, compared to the switching speed of the logic devices, the changes are few and far between. Logic probes and meters easily test circuits of this nature. However, many circuits are designed to operate at high rates of speed. This is referred to as dynamic operation. Oscilloscopes and logic analyzers are the test instruments of choice for dynamic circuit testing and debugging. **Timing diagrams** are the designers' analytical tool of choice for predicting and examining dynamic circuit operation.

2.9 ■ UNDERSTANDING THE TIMING DIAGRAM

The timing diagram pictorially presents the input and output activity occurring at a gate or within a logic circuit. Unlike the truth table, the timing diagram does not usually show every possible input combination, but rather, shows the actual input and output level changes occurring over a specified period of time. This useful analytical tool conveys an enormous amount of information about specific circuit behavior. To begin timing diagram study, we examine the 2-input AND gate illustrated in Figure 2.22. Waveform A, which accompanies this diagram, is associated with input A and waveform B is associated with input B. A and B are still variables just as they always have been. The difference is that we can see exactly what levels occur on A and B over a period of time and exactly when the changes in level occur. This figure clearly points out the differences between the static 1 and 0 levels used on truth tables and the dynamic, changing behavior of timing diagrams. The timing diagram generally portrays a more realistic picture of actual circuit operation. Variable C is the AND gate output, which we will examine during the following discussion.

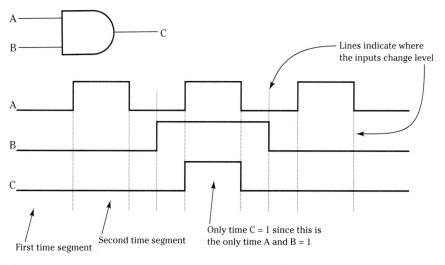

Lines indicate where the inputs change level

First time segment

Second time segment

Only time C = 1 since this is the only time A and B = 1

In a timing diagram of this sort time is assumed to begin on the diagram's left side. A horizontal time scale is not shown on the diagram. This is typical, but how does the rate of change of the inputs relate to normal time measurements? At this point the variables' rate of change is less important than the logical behavior of the output as compared to the inputs. When questions of time arise, we generally refer to the manufacturer's data books for specific device parameter values. Many specifications for device operation can be found in the data books, providing a complete device performance picture. We defer our study of device parameters until later. Also, we note that a gate switching levels at a 100,000 Hz rate will have the same timing diagram when operated at 200,000 Hz. In other words, the timing diagrams we are drawing are generic in nature. They simply show how the logic responds to ones and zeros without regard to how fast the levels are changing. This is a simplistic way to look at circuit operation, but very useful for circuit analysis and typical for preliminary design. Fear not, we will add the other details when needed.

Beginning at the input waveforms' left, we find it advantageous to sketch in vertical lines that indicate when an input variable changes (which is also illustrated in Figure 2.22). This gives us the flexibility to view individually the segments of time we have created, and also simplifies the timing analysis.

During the first segment of time, inputs A and B are both zero. The AND operation tells us that the output must also be zero for this condition, which is indicated as the level on C in Figure 2.22. During the next segment of time, A = 1 and B = 0. According to our understanding of the AND function, C should equal zero in this instance as is appropriately marked on the timing diagram. As we proceed, we will find that the only time the output is high is when both A and B are ones. (This is indicated on the diagram for our reference.) Each segment of time relates back to the simple truth table representation of the gate operation. By dividing up the diagram according to input changes, we reduce the timing diagram analysis to simple truth table analysis. Naturally, this becomes easier, quicker, and more obvious as we become proficient with the technique.

DESIGN EXAMPLE 2.14

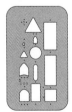

Complete the timing diagram for the following gate:

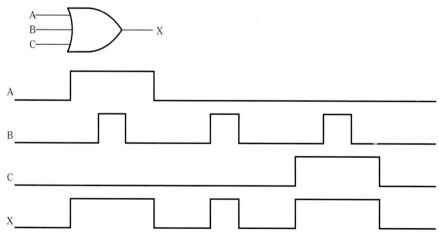

Solution   X will be high whenever A OR B OR C are high.

Let's now analyze circuit operation using a timing diagram. The output equation for Figure 2.23 is Y = AB + C$\bar{B}$. Our intent is to produce a timing diagram for output Y. It is very helpful to include the partial outputs for the two AND gates on the timing diagram, since the Y output is controlled by these two functions. Labeling these outputs W and X will assist us here. Lightly drawing vertical lines to indicate input changes also helps to depict accurately the timing relationship between input and output.

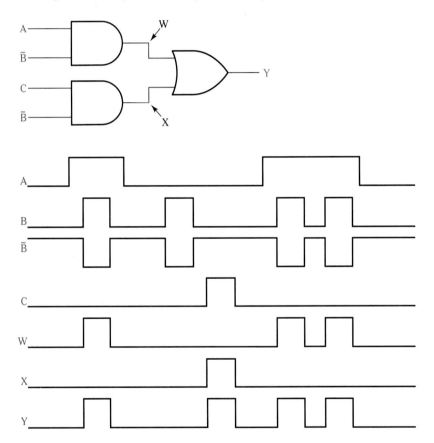

Figure 2.23
Circuit Timing Diagram

Partial output W is the AND of inputs A and B. Rather than viewing the complete circuit, we focus on only the AND operation and the results will be easy to obtain. Once the output for this section has been determined and noted on the timing diagram, we can proceed to another portion of the circuit without the results of the AND of A and B cluttering our thinking process. Next, we place the result (X) of ANDing C with $\bar{B}$ on the diagram. Of course, we notice that X is high only when C and $\bar{B}$ are both high. Y is obtained by determining how W and X respond when ORed. This is trivial since we expect Y to equal one whenever W or X is one. Notice how we often predict output levels before drawing them on the diagram. Making the effort to anticipate probable results seems like a minor point, but it can greatly assist us in better understanding the circuit and gate operation as well as help us to isolate errors quickly and to increase our design skills.

***DESIGN
EXAMPLE 2.15***

Complete the timing diagram for the circuit $Y = (A + B)(\overline{B} + C)$.

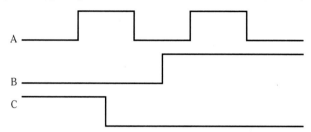

Solution   Draw the circuit, mark partial outputs, and draw waveforms for each gate.

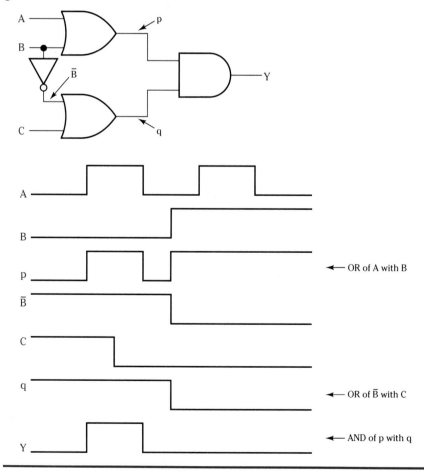

FOR YOUR
INFORMATION

Digital signals are characterized by a change in level. The folowing terminology identifies the components of a digital signal:

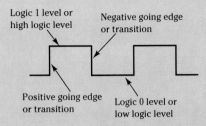

In digital circuitry both levels and edges are useful as controlling signals.
Some devices, for instance, will only respond to a negative going edge.
Therefore, each digital waveform cycle contains four logical indicators that
can be used for many design applications.

How Propagation Delay Time Affects Timing

Before progressing further, we ought to consider the accuracy of timing diagrams. We are assuming that output levels change as soon as controlling input
levels change. This presumes that the input signal is not delayed as it passes
through the gate. This, of course, is not true since the electronic circuitry within
the gate cannot switch instantaneously. The resulting delay time, as signals
move from input to output, is called **propagation delay time.** Therefore, a more
accurate representation of output activity includes this delay. It is not included
in our initial analysis for several reasons. For preliminary design, functional
circuit operation is of primary interest and delay times only add unnecessary
detail. Also, many circuits operate at relatively slow speeds. Considering a 15
nanosecond (nsec) delay time for circuitry operating at a 25 millisecond (msec)
rate is not very useful since this difference in time is roughly 1.6 million to
one. We do not ignore propagation delay time or other signal delays, however.
In good designs these delays are ultimately accounted for to prevent circuit
problems and to ensure reliable operation. These timing details are added in
after design work has progressed to a usable state. During this point in the
design cycle computer simulation techniques are usually employed to aid
timing analysis. Figure 2.24 sheds some light on timing delays with respect to
propagation delay times.

The AND gate in Figure 2.24 has a propagation delay time equal to 12 nsec,
which means that the output level changes 12 nsec after the input level changes.
The timing diagram shows a constant one level for input B, while input A varies
over time. By examining the changes in input A, you can see that the output
follows input A. The output is high when the A input is high and low when
the A input is low. In fact, except for the difference in delay time, input A and
the output are identical. As the diagram indicates, the output response to an

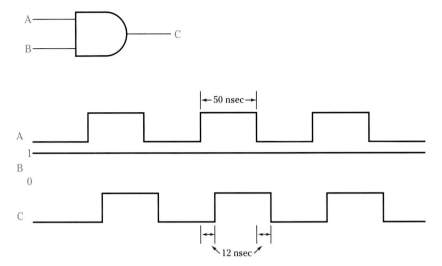

Figure 2.24
Gate Propagation
Delay Time

input level change is delayed by 12 nsec. For example, whenever input A makes a transition from a low level to a high level, the output makes the same transition, but 12 nsec later. It is the logic designer's responsibility to understand the effects and consequences of this delay. The delay values for specific gates are available in data books and vary from logic family to logic family since delay times depend on the technology used.

The cumulative effect of propagation delay or **skew** is evident in Figure 2.25. In this figure four inverters, connected in a cascade, are attached to the AND gate output. Each inverter's propagation delay time is 10 nsec, as indicated

**FOR YOUR
INFORMATION**

A closer look at propagation delay time shows that the delay is a combination of several factors and varies depending on the change in level.

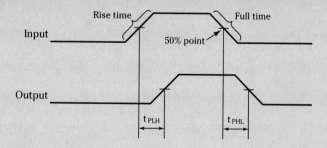

t_{PLH} = propagation delay time for a low to high transition
t_{PHL} = propagation delay time for a high to low transition

Delay times are measured between the 50% points of the input and output waveforms. The delay includes the circuit delay and rise/fall times. Factors controlling delay times are capacitance, temperature, and supply voltage.

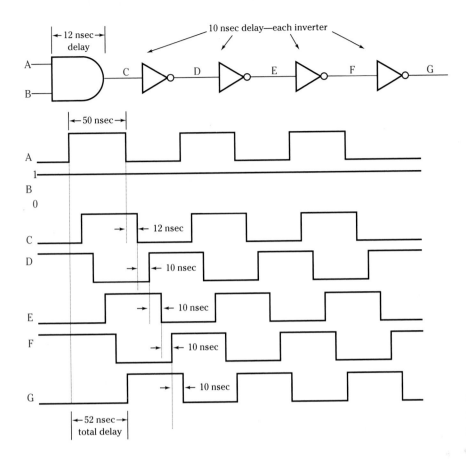

Figure 2.25
Cumulative Effects of
Delay Time

by Figure 2.25. Therefore, the signal arriving at point G is delayed not only by the AND gate, but also by the delay of four inverters. The output signal arrives at point G 52 nsec (12 + 10 + 10 + 10 + 10) after leaving point A. This process illustrates how delay time becomes troublesome. As more and more gates are connected together, the additive effects of individual gate delays increase significantly. This delay, combined with other timing delays (rise times and fall times), can seriously affect high speed circuit operation. Since our designs are basic and introductory at this point, we will generally not delve into delay analysis unless the specific circuits we design are directly affected.

2.10 ■ TWO POWERFUL FUNCTIONS—NAND AND NOR

AND, OR, and INVERT functions are three of five fundamental logic building blocks. The remaining two—NAND and NOR gates—significantly enhance the designer's ability to create sophisticated circuitry. NAND and NOR gates combine both logic capability and the inversion property in one simple circuit, providing tremendous design power. The following sections detail these two important logic devices.

The NAND Function

A verbal definition for the NAND function reads as: **The NAND gate output is zero if and only if all inputs are ones. When any other input combination is present, the output is a one.** Examining this statement carefully, you will notice that the NAND definition is the direct opposite of the AND definition. This means you can expect each NAND gate input combination to produce an output value that is the complement of an equivalent AND gate output.

There are also alternative ways to interpret this statement (as we did with the AND and OR functions). Unique gate output conditions occur when all inputs are at the one level. Under this circumstance the output is zero. Conversely, any NAND input placed at the zero level forces the output to the one state. These are nonunique conditions, since most input combinations place at least one input at the zero level. This is easily understood by examining the NAND gate symbol and equivalent circuit in Figure 2.26.

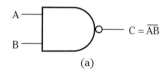

(a)

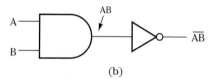

Figure 2.26
(a) 2-Input NAND Gate
(b) NAND Equivalent

(b)

Figure 2.26a illustrates the 2-input NAND gate logic symbol. Notice that the basic symbol is the AND function. The small circle on the output indicates inversion, suggesting further that the value resulting from the AND of inputs A and B is complemented. The small circle indicating inversion was introduced during discussion of the inverter, and, as we recall, the inverter symbol also has a small circle on its output. The small circle is sometimes called a **berry** or a **bubble**. Noting that the bubble represents inversion will prove useful when we introduce other circuit analysis techniques. Figure 2.26b is an equivalent representation for the NAND function, confirming the point that the NAND operation is an AND operation followed by inversion.

Figure 2.27 compares 2-input AND and NAND truth tables. You can easily see that the output conditions are complements of one another and that the NAND output is zero only when both inputs are one. Furthermore, the three input combinations driving the NAND gate output high are also evident (this occurs when any NAND input is low).

Naturally, a Boolean expression for the NAND function is required. Figure 2.28 shows the expression for a 3-input device. Perhaps you have derived this Boolean equation already (a pat on the back if you did). Initially, A, B, and C

A	B	AND	NAND
0	0	0	1
0	1	0	1
1	0	0	1
1	1	1	0

Figure 2.27
2-Input NAND Truth
Table

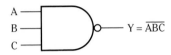

$Y = \overline{ABC}$

Figure 2.28
3-Input NAND Gate

are ANDed together, producing ABC. ABC is then complemented to produce $Y = \overline{ABC}$. The overbar extends over all three variables because A, B, and C were ANDed before the complement operation occurred. It is very important that you completely understand the use of the overbar. As we proceed through this text, we will encounter many complex equations with many overbars, which requires confidence and understanding of this notation on your part to sift through them all. Remember this helpful hint: The overbar always indicates that the variables or logical operation listed below the overbar were inverted. This means that you can expect to see overbars after an inversion operation has taken place. $\overline{A}$, for instance, means that input A has been inverted; $X = \overline{(AB + C)}$ means that A and B are ANDed, then ORed with C, and the final OR output is inverted. Note, A is not inverted individually, nor is B or C; it is the final OR function that is subject to inversion.

Of course, the NAND function is not limited to two inputs since no theoretical limit exists for the number of inputs, but the input limit is technology restricted. Fabricating 4096-input NAND gates with present semiconductor and integrated circuit technology is extremely difficult. Nonetheless, it is interesting to realize that a 4096-input NAND gate output still is zero only when all 4096 inputs are ones. If any of the 4096 inputs are zero, then the output will be at a one level, consistent with the NAND definition. The truth table for the 3-input device in Figure 2.28 is given in Figure 2.29.

A	B	C	Y
0	0	0	1
0	0	1	1
0	1	0	1
0	1	1	1
1	0	0	1
1	0	1	1
1	1	0	1
1	1	1	0

Figure 2.29
3-Input NAND Truth
Table

DESIGN
EXAMPLE 2.16

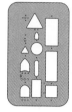

A designer wishes to NAND variables W, X, Y, and Z together to create output T. Draw the circuit for this function. Also write the Boolean equation and the truth table.

Circuit:

Equation:

$$T = \overline{WXYZ}$$

Truth table:

W	X	Y	Z	T
0	0	0	0	1
0	0	0	1	1
0	0	1	0	1
0	0	1	1	1
0	1	0	0	1
0	1	0	1	1
0	1	1	0	1
0	1	1	1	1
1	0	0	0	1
1	0	0	1	1
1	0	1	0	1
1	0	1	1	1
1	1	0	0	1
1	1	0	1	1
1	1	1	0	1
1	1	1	1	0

The NOR Function

The NOR function is defined as follows: The NOR gate output is zero when any input is one. The NOR gate output is a one level only when all inputs are zeros. Close inspection of this statement reveals that the NOR function is the direct opposite of the OR function. This means we can expect all NOR gate input combinations to produce the complement output values obtained from an equivalent OR gate.

Naturally, the NOR gate possesses the same unique input combinations as the OR gate. For instance, the only time a high NOR gate output occurs is when all inputs are low. The all zero input condition is also the unique input condition for the OR gate, with the exception that the OR gate output level is the complement of the NOR gate. Also similar to the OR function are the nonunique NOR input combinations. Any combination of ones on NOR gate inputs produces a zero level output condition (one output for an OR). In other words, the majority of input combinations to either NOR or OR gates forces the output to zero for a NOR or to one for an OR.

A glance at Figures 2.30a and 2.30b illustrates how the NOR function symbol is obtained from an OR function followed by inversion. The NOR symbol utilizes the bubble notation on the output, the same as does the NAND gate, to represent the device's inversion capability. Figure 2.31 compares the OR and NOR truth tables and further reinforces our knowledge that OR and NOR output conditions are complements of one another. All unique and nonunique input conditions are marked on this truth table.

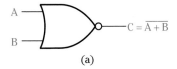

(a)

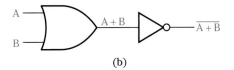

(b)

Figure 2.30
(a) 2-Input NOR Gate
(b) NOR Equivalent

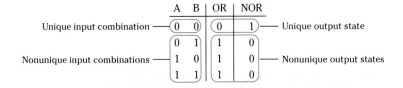

A	B	OR	NOR	
Unique input combination — (0	0)	(0	1) — Unique output state	
	0	1	1	0
Nonunique input combinations — 1	0	1	0 — Nonunique output states	
	1	1	1	0

Figure 2.31
NOR Truth Table

A Boolean expression identifying the NOR function is given for a 4-input NOR gate in Figure 2.32. As you may already suspect, the output expression consists of a complemented OR function, $T = \overline{(W + X + Y + Z)}$. The overbar is drawn over the entire OR function because the OR function output is inverted. This 4-input device also implies that NOR gates can theoretically have any number of inputs and still produce an output consistent with the NOR definition. The truth table for this multi-input NOR gate is given in Figure 2.33.

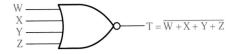

$T = \overline{W + X + Y + Z}$

Figure 2.32
4-Input NOR Gate

W	X	Y	Z	$T = \overline{W + X + Y + Z}$
0	0	0	0	1
0	0	0	1	0
0	0	1	0	0
0	0	1	1	0
0	1	0	0	0
0	1	0	1	0
0	1	1	0	0
0	1	1	1	0
1	0	0	0	0
1	0	0	1	0
1	0	1	0	0
1	0	1	1	0
1	1	0	0	0
1	1	0	1	0
1	1	1	0	0
1	1	1	1	0

Figure 2.33
4-Input NOR Truth Table

DESIGN EXAMPLE 2.17

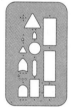

A solenoid is activated with a low logic level. Write the equation and draw the logic symbol that produces the activating signal when at least one of the switches—S1, S2, S3, S4, or S5—is high.

Solution   Since any switch can provide the activating signal, an OR function will suffice. However, the active output level required is low; therefore, the complement of the OR function—NOR—is necessary.

$$\text{Solenoid} = \overline{S_1 + S_2 + S_3 + S_4 + S_5}$$

NAND and NOR Timing Diagrams

Timing diagrams for devices or circuits exhibiting inversion require a bit more care than the basic diagrams previously discussed. The fact that digital devices can complement logic levels often is troublesome to those new to digital logic concepts, and nothing shows this up more clearly than an attempt at a timing diagram. Don't let this scare you. Just rely on a real understanding (not memorization) of logical device properties, and the correct concepts will soon fall into place. This idea deserves to be stressed. Memorizing truth tables is next to useless when you proceed to more complex design concepts. It takes a firm understanding of fundamental digital logic ideas, such as unique and non-unique input combinations, before the steps to serious logic design can begin. Make sure you gain that understanding now.

Figure 2.34 illustrates the timing diagram for a 2-input NAND gate. As usual,

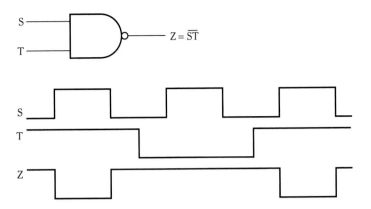

Figure 2.34
NAND Timing Diagram

we investigate the timing relationships according to changes in input logic levels, then analyze each segment of time for the proper logical output. A quicker way to complete the diagram is by looking for unique input variable combinations. In this example the NAND gate unique input combination occurs when all inputs are at the logic one level. We expect the output to equal zero when this happens. Understanding this, we examine the input waveforms from beginning to end and find the segments of time when both S and T are ones. The output is zero only during these times; a one level results for all other combinations of the inputs. The timing diagram is completed quickly once this is understood. The designer also increases his or her design skills following this approach because knowledge of digital logic is applied to a problem using insight beyond that of simple truth table lookup.

Figure 2.35 shows the timing diagram for a 3-input NOR gate. The gate indicated also includes an input variable naming convention typically used in industry. Rather than assign letters to the input variables, real digital systems have descriptive signal names. This is necessary in large systems with many inputs and outputs. During design the signals are constantly being cross-referenced, not only by the designer, but by many others in manufacturing, quality control, and other design support services as well. You can imagine the confusion that would occur if all signals were identified only with letters

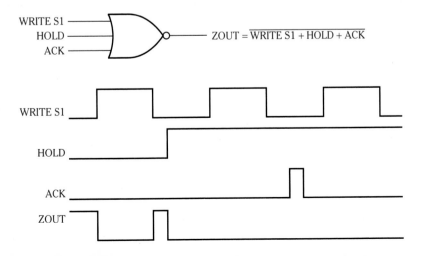

Figure 2.35
3-Input NOR Timing
Diagram

of the alphabet. A realistic name conveys information about the purpose of the signal to the person reading the schematic. Keep in mind the following: Each signal name still represents only a single binary level at any point in time. The signal name could be a mile long, but the level on that line is still either zero or one. Also notice that the Boolean equation example, ZOUT = $\overline{\text{WRITES1} + \text{HOLD} + \text{ACK}}$, still represents the NORing of three variables, even though the variable names are rather complex. Furthermore, the output is one only when all three inputs are zero (unique input combination). In all other cases (when any input is a one) the output is zero.

**DESIGN
EXAMPLE 2.18**

Complete the timing diagram for the following NAND gate:

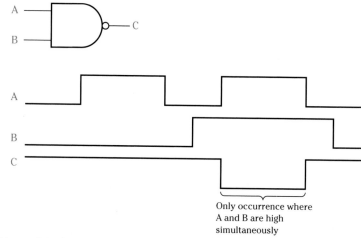

Only occurrence where
A and B are high
simultaneously

Solution   C is formed from the NAND of A with B.

**FOR YOUR
INFORMATION**

Names or descriptions are often used to identify the input and output signals on logic gates. Although the name assigned to a line may be complex, the signal level it represents is still a logic one or logic zero.

For example:

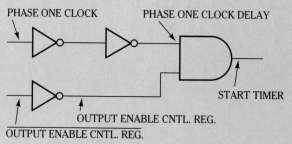

The output labeled "START TIMER" is one only when its inputs are both high. "PHASE ONE CLOCK DELAY" may seem complex, but its level can only be one or zero.

Complete the timing diagram for the following circuit:

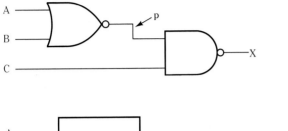

A

B

C

p   ← NOR of A with B

X   ← NAND of p with C

2.11 ■ A NEW DESIGN CONCEPT—ENABLE AND DISABLE OPERATION

Having discussed the five basic functions (AND, NAND, OR, NOR, and INVERT) that are the foundation for further digital logic study, we can now apply our understanding of these functions in a more advanced manner. Ideally, we strive to develop an intuitive sense of logic design, giving us the ability to solve design problems without having to fall back on basic gate truth tables. We are not ignoring our need for the truth table, but we are expanding our insight into logic device operation. We now investigate the notion of gate **enable** and **disable** modes of operation.

To enable a logic device means that the device output is allowed to change level. To disable a logic device means that the device output is prevented from changing level. In both cases at least one input line determines the course of action. The terms **inhibit** and **degate** are also used synonymously with "disable."

The word "control" is the essence of the enable and disable concept. We often designate a logic gate input signal to be a controlling input, which then determines when the device is enabled or disabled. This also implies that any other input signals, whether or not their values affect the output values, are controlled by the enable/disable line. Before examining the specifics, let's consider why this concept is useful.

A logic designer does not sit down at a desk and randomly stack gates together. Rather, he or she solves a design problem or designs a circuit to perform a specific task that is generally understood and somewhat well defined, such as "End Data Transfer When Data Timer Equals Zero." A word definition such as this identifies the operational characteristics of a specific piece of logic; in this case a control line terminates the transfer of data when a timer, called the data timer, equals zero. The designer can create a truth table outlining this function, but more likely, the designer will treat the desired operation as an enable/disable application. The data signal is transferred or not transferred as directed by a control line designated for that purpose. It is apparent from the word definition that data transfer continues until a data timer circuit provides a zero level. The data timer signal is the likely candidate as the necessary control signal. As long as the timer output is a one (we must assume binary levels, of course), then data transfer continues. Therefore, when the data timer is high, it enables data transfers and when it is low, it inhibits or disables data transfers. As we will soon see, enable and disable operations are merely extensions of basic gate theory.

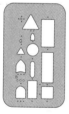

DESIGN EXAMPLE 2.20

The ignition system of a car will not operate unless a 4-digit code is correctly pushed on a key pad. Under what condition is the ignition system enabled? When is it disabled?

Solution   The ignition system is enabled when the correct code is entered. The ignition system is disabled when the code is entered incorrectly.

How AND/NAND Gates Are Enabled and Disabled

The fact that AND and NAND gates each has a unique input combination and several nonunique input combinations is the key idea behind enable and disable operations. For AND and NAND gates all inputs at the high logic level constitute the unique combination. Consider the AND gate in Figure 2.36. You can designate input A as your signal or data line and input B as your control

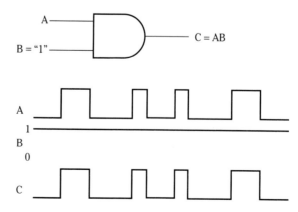

Figure 2.36 AND Gate Enable Operation

line. Notice that when input B is held high, the output follows the input A signal. In this case the B input enables the gate, allowing the data signal through to the output. Relate each timing diagram time segment to the AND gate truth table and you can verify the output. Simply stated, the enable level for an AND gate is one.

Figure 2.37 shows an AND gate disable operation. Input B is now low. As you know, the AND output is zero when any input is also zero (nonunique conditions). Therefore, control line B now dictates gate operation and inhibits any output activity from taking place, even though input A's signal is changing. Notice that a disabled gate still has a valid logic level output. Disabling does not render the device inoperable, but it does restrict output operation.

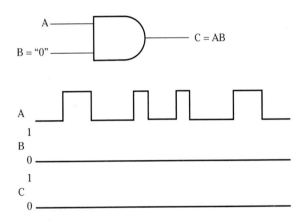

Figure 2.37
AND Gate Disable
Operation

In both enable and disable operations a basic AND truth table confirms output conditions. As previously stated, however, it is usually more effective and expedient to apply logic design ideas by utilizing the insight gained from an understanding of fundamental concepts. Enable and disable concepts are merely alternate ways to consider basic gate operation. In fact, the word gate accurately describes logic gate operation in a manner analogous to the operation of a common backyard fence gate. That is, an input signal is prevented from passing through to a logic device output by appropriate control, just as a closed gate prevents you from entering your neighbor's yard.

Compare AND gate enable and disable operations to the NAND gate. Figures 2.38 and 2.39 demonstrate enable and disable properties of the NAND gate by comparing them to the AND gate. Can you see the similarities and differences? Both AND and NAND gates are enabled with one level inputs and disabled with zero level inputs. When enabled, both gate outputs follow the input A waveform, with the difference that the NAND gate and AND gate outputs are complements of each other. When disabled, both gates rest at a constant output level—a zero level for the disabled AND and at a one level for the disabled NAND. Therefore, the only difference between the AND and NAND gate enable/disable operation is simply a level of inversion.

Now we can return to the initial design example mentioned earlier. We want to enable or disable the transfer of data according to the state of a data

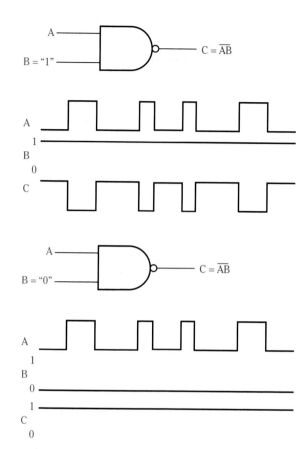

Figure 2.38
NAND Gate Enable
Operation

Figure 2.39
NAND Gate Disable
Operation

timer. Figure 2.40 shows how an AND gate is used to create the desired function. Since an AND gate is enabled with a high input level, the data timer control line is utilized to enable the gate; data transfers continue when this control line is high. The associated timing diagram indicates when data transfers are enabled.

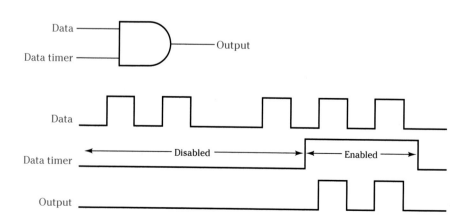

Figure 2.40
AND Gate Enabling
Function

How OR/NOR Gates Are Enabled and Disabled

Enable and disable capabilities also apply to NOR and OR gates. As you may already surmise, the only difference between enable and disable operations for these two functions is a level of inversion. Figure 2.41 shows an OR and a NOR gate, both possessing a data input designated as "A" and a control line identified as "B." The high level input at B forces the OR output to a high level and the NOR output to a low level. The A input cannot affect the output under this condition; thus, input A as well as the gate are considered disabled. For both the OR and NOR gates the nonunique input conditions determine when the gates are disabled. Notice as well, how high level inputs disable OR/NOR logic but enable AND/NAND logic. This is a helpful comparison to keep in mind.

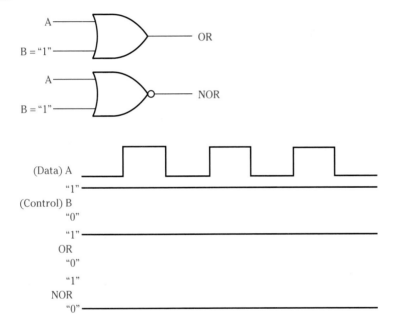

Figure 2.41
OR/NOR Disable
Timing

Enable operation for the OR and NOR gates is illustrated in Figure 2.42. In this case a zero level input on control line B enables both the OR and NOR functions (unique input condition). The OR gate, when enabled, allows the data present at input A to appear at the output. The NOR gate, when enabled, inverts the input data as they pass through to the output. Compared to AND/NAND gates, only the input level differs for the OR/NOR gates' enable operation. Zero input levels enable the OR/NOR function, whereas AND/NAND functions are enabled with one levels.

The examples demonstrating enable and disable operations have all used 2-input gates, although the enable/disable operation is certainly not limited to 2-input devices. In fact, you can appreciate the power of this concept by realizing that it is possible for a single input line to control the operation of the remaining seven inputs of a 8-input gate. This kind of capability is at the heart of most digital systems.

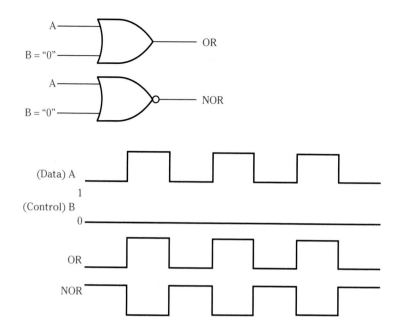

(Data) A

1
(Control) B
0

OR

NOR

Figure 2.42
OR/NOR Enable Timing

DESIGN
EXAMPLE 2.21

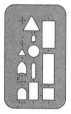

Draw the timing diagram for the following circuit; label on the diagram the points in time where the OR gate is enabled and disabled by signal a:

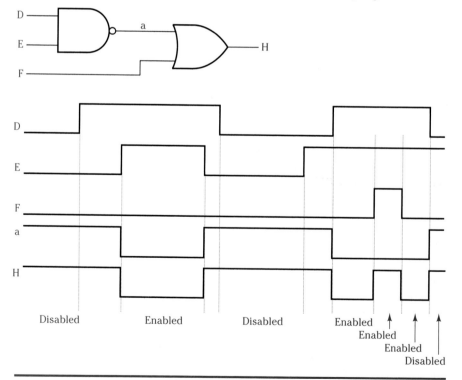

FOR YOUR
INFORMATION

Gate	Input condition to:		Output when:	
	Enable	Disable	Enabled	Disabled
Enable/$\overline{\text{Disable}}$ / Clock (AND)	1	0	CLOCK	0
Enable/$\overline{\text{Disable}}$ / Clock (NAND)	1	0	$\overline{\text{CLOCK}}$	1
$\overline{\text{Enable}}$/Disable / Clock (OR)	0	1	CLOCK	1
$\overline{\text{Enable}}$/Disable / Clock (NOR)	0	1	$\overline{\text{CLOCK}}$	0

- AND and NAND gates are enabled/disabled by the same input conditions.
- OR and NOR gates are enabled/disabled by the same input conditions.

The Active Level Concept

The enabling and disabling concept naturally leads to the idea of an **active level**. As we have seen, an enable level, for instance, can be high or low, depending on the logic gate used. **An active level is the binary level (one or zero) that activates a circuit.** The previously discussed design task, "End Data Transfer When Data Timer Equals Zero," for example, was created using a 2-input AND gate. A data input line was enabled or disabled by a second control line. Associated with this control line was an active level determining when the specific enabling or disabling action took place. If our main intent for this circuitry is the continuation of data transfer, then a high active level may be appropriate for the control line. If, on the other hand, the circuitry is designed to inhibit data transfer, we consider the line to be active level low, since a low degates this particular circuit. As we can see, there are two points of view for this control signal operation. The circuit designer knows which level is considered the active level (and, obviously, which level is inactive).

Often the signal name assigned by the designer gives the reader of the circuit schematic useful information about the active level. Figure 2.43a shows

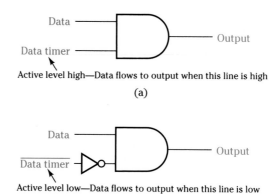

Figure 2.43
(a) Active Level High
Usage (b) Active Level
Low Usage

this circuit functioning with an active level high input, whereas Figure 2.43b shows the same circuit using an active level low. Both circuits allow data to move from input to output, but the overbar across the words $\overline{\text{data timer}}$ (read as: "not data timer") indicates that the circuit designer expects a low level on the input for the designated circuit operation. This is a clue that a low level input allows data transfer for that particular circuit.

An alternate method indicating an active level is the inverting bubble. The presence of a bubble, such as that in Figure 2.44a, suggests that an active level low signal is expected. The lack of a bubble (Figure 2.44b) implies that an active level high input condition is expected. Both bubbles and overbars are used frequently.

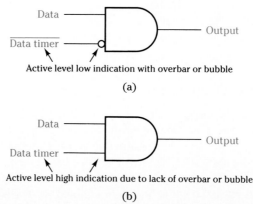

Figure 2.44
(a) Active Level Low
Bubble/Overbar
Notation (b) Active
High Notation

DESIGN
EXAMPLE 2.22

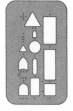

Identify the active level of each signal on the following gate:

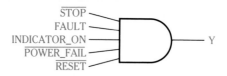

Solution   Active level low signals are identified by an overbar placed over the signal name. These include $\overline{\text{STOP}}$, $\overline{\text{POWER\_FAIL}}$, and $\overline{\text{RESET}}$.

Active level high signals do not have the overbar placed over the signal name. The active high level is implied. These include FAULT, INDICATOR_ON, and Y.

2.12 ■ THE EXCLUSIVE-OR FUNCTION

Consider the circuit and accompanying truth table shown in Figure 2.45. The C output comprises an extremely useful function having applications in mathematics, binary coding, and error correction. You may also notice the resemblance between this truth table and the truth table for a 2-input OR gate. The only difference occurs for the input combination A,B = 11. Since this circuit is similar to the OR function, except for the condition noted, it is called an Exclusive-OR (EX-OR) function. The circuit acts like an OR gate for the exclusive case where only one input is high.

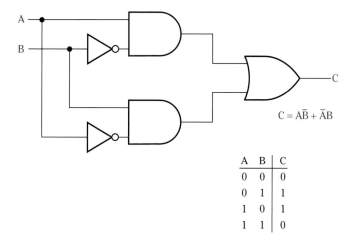

$$C = A\bar{B} + \bar{A}B$$

A	B	C
0	0	0
0	1	1
1	0	1
1	1	0

Figure 2.45
Exclusive-OR Circuit

Formally defining the Exclusive-OR gate function, we state: A 2-input Exclusive-OR gate output is high when either input A or B is high, but not both. The Exclusive-OR gate output is low when inputs A and B are both high or both low. Because this circuit function is so useful, it is identified with a unique logic symbol and treated as a basic gate rather than as a circuit. The symbol for a 2-input Exclusive-OR gate is shown in Figure 2.46.

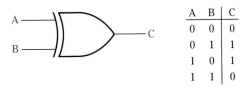

A	B	C
0	0	0
0	1	1
1	0	1
1	1	0

Figure 2.46
EX-OR Symbol and
Truth Table

Notice that the EX-OR definition limits the function to a 2-input device. This is not a strict limitation. The EX-OR function may have any number of inputs, but the multi-input EX-OR operation is treated differently. Refer to the 2-input EX-OR truth table and note that the device output is high for an odd number of input ones (01 or 10). Conversely, when the EX-OR is producing a zero output, the input combinations contain an even number of ones (00 is considered an even combination). Therefore, the Exclusive-OR function is also an **even/odd-checking** circuit. (We are checking to see if the input combination contains an even or odd number of ones, not if the decimal equivalent of the combination is even or odd.) Increase the number of EX-OR inputs and we can expect a high output whenever the input combinations possess an odd number of ones. Figure 2.47 shows how a 3-input Exclusive-OR function is created with 2-input Exclusive-OR gates. A 3-input EX-OR symbol is shown as well, although it is more common in practice to use 2-input symbols. (Practical EX-OR gates are typically purchased as 2-input devices.) The truth table accompanying the figure clearly shows the even and odd checking capabilities of the device. Any number of Exclusive-OR gates can be connected together to expand the Exclusive-OR function. We will see how this is effectively utilized in later chapters.

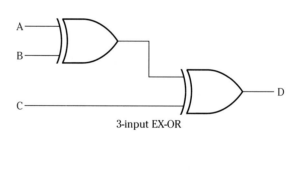

3-input EX-OR

A	B	C	D
0	0	0	0
0	0	1	1
0	1	0	1
0	1	1	0
1	0	0	1
1	0	1	0
1	1	0	0
1	1	1	1

Odd number of ones in these combinations

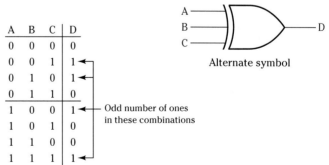

Alternate symbol

Figure 2.47
Multiple Input
Exclusive-OR Function

The Exclusive-OR Boolean expression was created by analyzing the Exclusive-OR circuit in Figure 2.45. The expression, $C = A\overline{B} + \overline{A}B$, is often expressed in a shortened form as $C = A \oplus B$, where the $\oplus$ sign indicates the EX-OR operation.

Complete the timing diagram for the following Exclusive-OR gate:

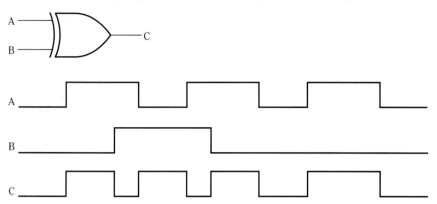

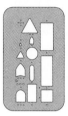

Create the truth table for the following circuit. Indicate the output level that denotes input combinations containing an odd number of ones; an even number of ones:

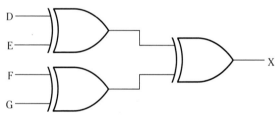

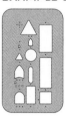

Solution   Create a truth table on a gate-by-gate basis to determine the output response.

D E F G	D $\oplus$ E	F $\oplus$ G	X = D $\oplus$ E $\oplus$ F $\oplus$ G	
0 0 0 0	0	0	0	
0 0 0 1	0	1	1	
0 0 1 0	0	1	1 ←	X = 1 indicates input
0 0 1 1	0	0	0	combinations with an
0 1 0 0	1	0	1	odd number of ones.
0 1 0 1	1	1	0	
0 1 1 0	1	1	0	
0 1 1 1	1	0	1 ←	
1 0 0 0	1	0	1	
1 0 0 1	1	1	0 ←	X = 0 indicates input
1 0 1 0	1	1	0	combinations with an
1 0 1 1	1	0	1	even number of ones.
1 1 0 0	0	0	0	
1 1 0 1	0	1	1	
1 1 1 0	0	1	1	
1 1 1 1	0	0	0 ←	

Exclusive-NOR

A logic operation similar to the Exclusive-OR function is the **Exclusive-NOR (EX-NOR)** operation. The EX-NOR property is the complement function of an EX-OR. That is, **the output of a 2-input EX-NOR gate is low only when one input is high; when both inputs are both high or both low, the output will be high.** Symbolically, an Exclusive-NOR gate is drawn the same as an Exclusive-OR with a bubble (indicating inversion) added to the output pin. Thus, the Boolean expression for the Exclusive-NOR function would read: $C = \overline{A \oplus B}$.

Controlled Inversion

One useful application of the EX-OR or EX-NOR gates is **controlled inversion.** Figure 2.48 illustrates the timing diagram for a controlled inverter. A continuous stream of pulses is applied to input X while input Y is designated as a control line. When input Y is zero, the EX-OR output follows the input X signal. When input Y is placed at the one level, then the output signal is the complement of input X. The inversion of the input X signal is controlled by the logic level at Y. This example also illustrates the fact that enable and disable conditions do not exist for the Exclusive-OR function. Regardless of the controlling input level, an output signal related to input X always exists in this example. The output cannot be forced to a steady logic level as occurs when other logical functions are disabled. This certainly should not be construed as a limitation, since the disable/enable capability is easily accomplished with other gates. The ability to control inversion increases our design flexibility.

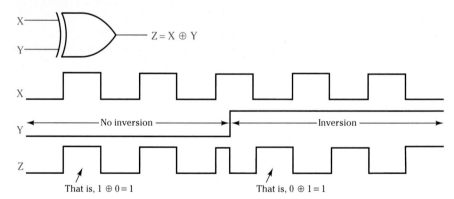

Figure 2.48
EX-OR Timing
Illustrating Controlled
Inversion

Magnitude Comparator

Both Exclusive-OR and Exclusive-NOR gates are frequently utilized to compare the magnitude of two binary numbers. Consider a 2-input Exclusive-OR function. If one input represents a binary number while the second input represents another binary number, then the Exclusive-OR output level will indicate if the two numbers are equal or not equal in value. If both inputs are either high or low (equal), the Exclusive-OR output will be low. If both inputs differ (unequal),

Gate Summary

	A	B	C	TTL Part Number	CMOS Part Number
AND $C = AB$	0	0	0	7408	74HC08
	0	1	0		
	1	0	0		
	1	1	1		
NAND $C = \overline{AB}$	0	0	1	7400	74HC00
	0	1	1		
	1	0	1		
	1	1	0		
OR $C = A + B$	0	0	0	7432	74HC32
	0	1	1		
	1	0	1		
	1	1	1		
NOR $C = \overline{A + B}$	0	0	1	7402	74HC02
	0	1	0		
	1	0	0		
	1	1	0		
EX-OR $C = A \oplus B$	0	0	0	7486	74HC86
	0	1	1		
	1	0	1		
	1	1	0		

	A	$\overline{A}$	TTL Part Number	CMOS Part Number
$\overline{A}$	0	1	7404	74HC04
	1	0		

TYPICAL PART NUMBERS

Some Common Part Numbers

FUNCTION	PART NUMBER	TECHNOLOGY
2-input AND	7408	TTL
	74C08	CMOS
	4081	CMOS
	10104	ECL
3-input AND	7411	TTL
	74C11	CMOS
	4073	CMOS
2-input OR	7432	TTL
	74C32	CMOS
	4071	CMOS
	10101	ECL
2-input NAND	7400	TTL
	74C00	CMOS
	4011	CMOS
2-input NOR	7402	TTL
	74C02	CMOS
	4001	CMOS
	10102	ECL
2-input Exclusive-OR	7486	TTL
	74C86	CMOS
	10107	ECL
2-input Exclusive-NOR	4077	CMOS
	10107	ECL

then the Exclusive-OR output will be high, indicating the inequality. In practical **magnitude comparators,** multiple inputs are common. For example, a 74HC688 can compare two 8-bit binary numbers. Other comparators determine equality as well as less than or greater than comparisons of the two input numbers.

DESIGN
EXAMPLE 2.25

Design a 3-bit magnitude comparator.

Solution   By extending the comparing capability of the Exclusive-OR function, we can design the comparator. The numbers to be compared are identified as A2 A1 A0 and B2 B1 B0. Each pair of bits may be compared with an Exclusive-OR gate to determine if the bits of the same significance are equal (i.e., compare A2 to B2). Since there are 3-bits, three Exclusive-OR gates are necessary for these comparisons. The outputs of all three EX-OR gates can then be combined to determine the equality of both numbers. Since EX-OR gates produce a low level output when the two inputs are equal, an OR gate is used to combine all EX-OR gate outputs—the OR output can only be low when all inputs are low. Therefore, a low level output from the circuit implies equality. (If Exclusive-

NOR gates are used for the comparisons, then an AND gate is used to combine the results of the individual bit comparisons.)

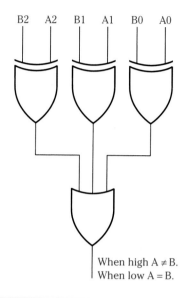

When high A ≠ B.
When low A = B.

SUMMARY

■ Several simple logic functions form the basis for complex digital circuit design. These include the AND, OR, INVERT, NAND, and NOR functions.

■ Logic functions may be described in several ways, including truth tables, logic symbols, and Boolean equations.

■ Truth tables represent the logical operation of a logic function or logic circuit and contain the output response for every input combination applied to the function or circuit.

■ Timing diagrams convey information about a logic function or circuit's response to specific input signals over a period of time.

■ A higher level of design intuition is possible by an understanding of the basic logic gates' enable and disable modes of operation.

■ Devices that are enabled allow their output signals to change.

■ Devices that are disabled prevent their output signals from changing.

■ The Exclusive-OR operation has the logical capability to distinguish between an even or odd number of ones in an input combination.

■ Exclusive-OR gates may be combined to extend the Exclusive-OR operation.

PROBLEMS

Sections 2.3–2.5, 2.10

*1. Write the equation and draw the symbol for the following AND truth table. Identify all unique and nonunique input combinations:

X1	W5	T9	Z12
0	0	0	0
0	0	1	0
0	1	0	0
0	1	1	0
1	0	0	0
1	0	1	0
1	1	0	0
1	1	1	1

2. Write the equation and draw the symbol for the following OR truth table. Identify all unique and nonunique input combinations:

N9	E7	A3	PC
0	0	0	0
0	0	1	1
0	1	0	1
0	1	1	1
1	0	0	1
1	0	1	1
1	1	0	1
1	1	1	1

*3. Create the truth table for a 5-input OR gate. Draw the symbol and write the Boolean equation.

4. Create a truth table for a 5-input NOR gate. Draw the symbol and write the Boolean equation.

5. Create the truth table for a 5-input AND gate. Draw the symbol and write the Boolean equation.

*6. Create the truth table for a 5-input NAND gate. Draw the symbol and write the Boolean equation.

7. How many input lines are required to produce an OR gate capable of handling 512 input combinations? How many of the input combinations produce a high level output? How many produce a low level output? Repeat for a NOR gate.

---

* See Appendix F: Answers to Selected Problems.

*8. How many input combinations are possible with a 10-input AND gate? How many combinations produce a high level output? How many combinations produce a low level output?

9. Repeat problem 8 using a 10-input NAND gate.

Section 2.7

10. Develop a complete circuit truth table for the following circuit:

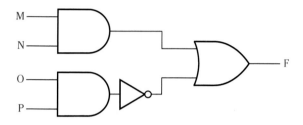

*11. For the circuit shown below, determine the output level for input combinations W,X,Y = 001; W,X,Y = 101; W,X,Y = 010. Draw the complete circuit truth table to verify:

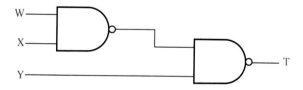

12. For the following circuit:
   (a) Write the circuit equation.
   (b) Determine the output level when ABC = 101; when ABC = 011.
   (c) Create the complete circuit truth table.

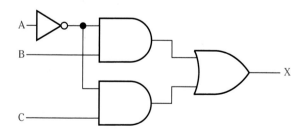

---

* See Appendix F: Answers to Selected Problems.

*13. For the following circuit:
    (a) Write the circuit equation.
    (b) Determine the output level when WXYZ = 1010; when WXYZ = 0010.
    (c) Create the complete circuit truth table.

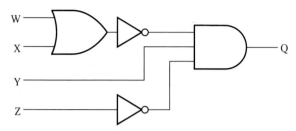

14. Write the expression for the following network. Develop a complete circuit truth table:

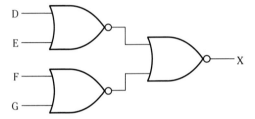

15. Write the circuit equation for the following network:

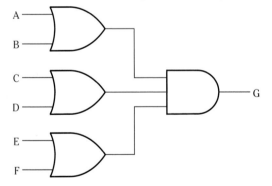

*16. Determine the circuit equation for the following network:

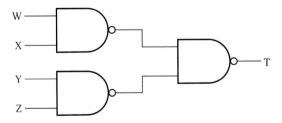

* See Appendix F: Answers to Selected Problems.

17. Develop complete truth tables for the circuits given in problems 15 and 16.

18. Write the equations for the following network:

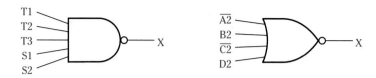

*19. Using the following circuit, write the output equation for T:

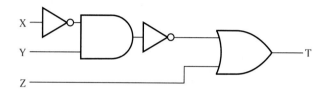

20. A computer system uses the following circuits to enable communications to printers, disk drives, and so on. Which binary values of A7–A0 are required to produce the high level enabling signal? Write the equation representing these values:

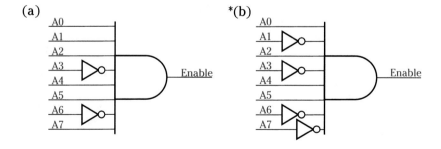

Sections 2.8–2.9

*21. Determine the output for Y using the following circuit and input waveform:

---

* See Appendix F: Answers to Selected Problems.

*22. Draw the waveforms present at all inverter outputs assuming no propagation delay:

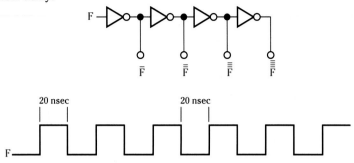

23. Repeat the previous problem assuming that each inverter has a 6 nsec propagation delay time.

*24. Using the input waveforms shown below, determine the output waveform for SELECT:

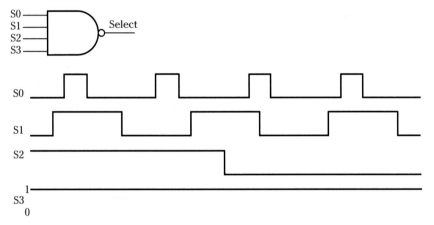

25. Repeat problem 24 with the input waveforms feeding a 4-input NOR gate.

*26. Draw the timing diagram for the following circuit given the inputs for W, X, and Y as shown:

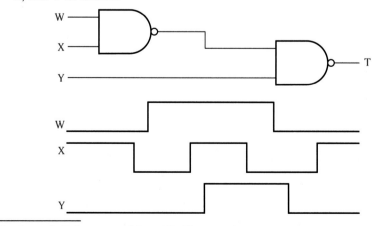

* See Appendix F: Answers to Selected Problems.

27. Complete the timing diagram for the following circuit:

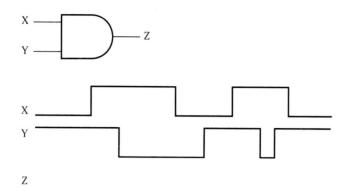

28. Redraw the preceding timing diagram when inputs X and Y feed a NAND gate.

*29. The following waveforms are the inputs to a 4-input OR gate. Complete the diagram:

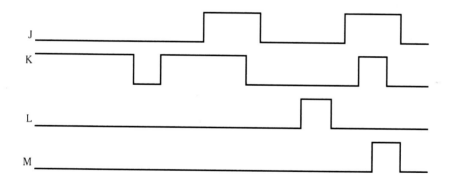

30. Use the waveforms in problem 29 as the inputs to a 4-input NAND gate. Redraw the timing diagram.

*31. Draw the output waveform C, using the following inputs, for a 2-input Exclusive-OR gate:

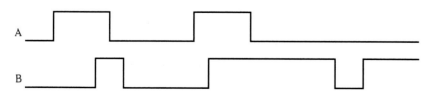

---

* See Appendix F: Answers to Selected Problems.

32. Using the input waveforms shown here, draw the output waveform:

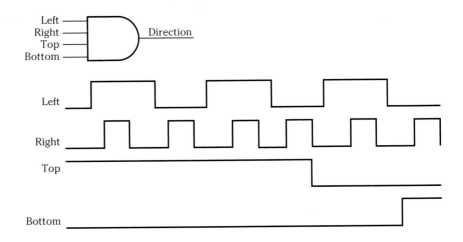

33. Draw the output waveforms for the following gates if the input signals for A, B, and C are as shown:

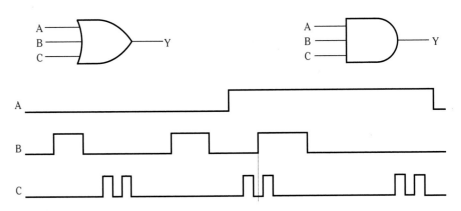

34. Draw the output timing waveform for:

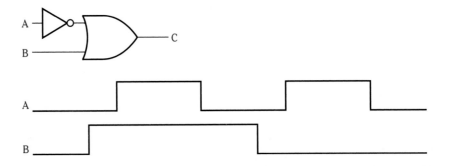

*35. Complete the timing diagram for:

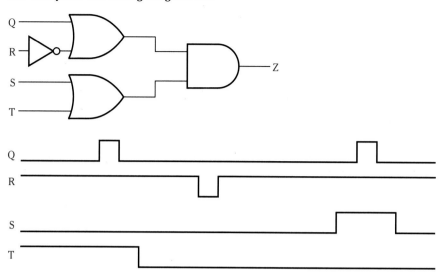

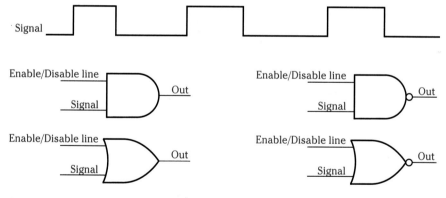

Section 2.11

36. What is the output for the following gates when the gates are enabled? What is the output when the gates are disabled? What input levels are required to enable and disable each gate?

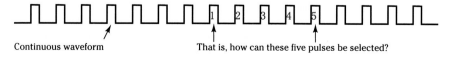

*37. A high logic level is required to activate an electronic circuit. The activating signal consists of five pulses that are selected from a continuous running waveform as shown. Using the enable/disable concept, design a simple circuit to show how a control line can select five pulses when desired. Draw timing diagrams to illustrate the circuit operation:

Continuous waveform   That is, how can these five pulses be selected?

---

* See Appendix F: Answers to Selected Problems.

Section 2.12

38. Show how an Exclusive-OR gate can be used to convert waveform A into waveform B:

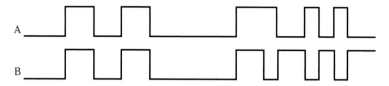

39. For the circuit shown below, what would happen to the output if:
    (a) the circuit acted normally?
    (b) input A became disconnected and shorted to input B?
    (c) input A became disconnected and shorted to input C?
    (d) the inverter was left out of the signal A path?
    (e) If the output was low, which OR gate input should be tested as the most likely point of error? Why?

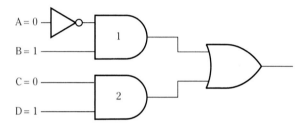

CHAPTER 3

BOOLEAN ALGEBRA THEOREMS AND CIRCUIT DESIGN

OBJECTIVES

To explain how basic Boolean algebra principles relate to digital logic design and to explain how these principles differ from and compare to those of standard algebra.

To discuss the application of Boolean algebra as a useful digital design tool.

To demonstrate how Boolean algebra reduces circuit design complexity.

To illustrate how Boolean algebra techniques apply to circuit schematics.

To explain DeMorgan's Theorem and to demonstrate how logic circuits can be converted from one form to another through its application.

To demonstrate how to extract a circuit equation from a truth table using several design techniques.

PREVIEW

In the last chapter Boolean algebra was introduced as we studied both logic gate and logic circuit equations. In this chapter we dramatically enhance our ability to simplify digital circuits using the formal properties of Boolean algebra. We also investigate how to convert truth tables into functional circuit equations using various circuit design techniques.

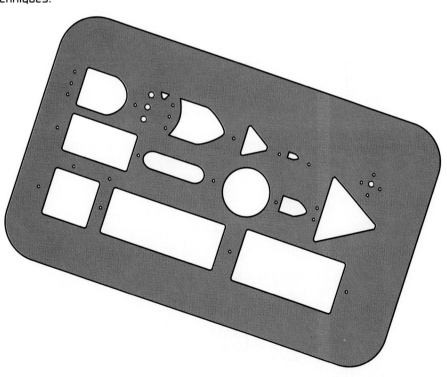

3.1 ■ THE FUNDAMENTAL RULES OF BOOLEAN ALGEBRA

Boolean algebra is a particularly useful tool for the digital designer since its use allows us to describe logic circuits mathematically. Once so described, mathematical techniques applied to equations will alter the circuit, leading to design improvements.

Several mathematical rules govern and guide not only the use of familiar standard algebra, but proper Boolean algebra usage as well. These rules are the commutative, associative, and distributive laws.

The Commutative Law

Algebraically, the commutative law states that two numbers can be added together without regard to their order. For example, $7 + 2 = 9$ just as $2 + 7 = 9$. The same property holds true for the logical OR operation when expressed as a Boolean algebra equation. Thus, $X + Y = Z$ can be restated as $Y + X = Z$, without any loss in logical meaning. The commutative law, extended to the AND operation, illustrates the equivalence of $UV = W$, or $VU = W$. This algebraic property proves useful when manipulating Boolean equations for circuit simplification. As you will see, rearranging a logic equation is a proven way to obtain a simplified circuit.

DESIGN EXAMPLE 3.1

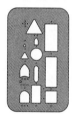

Write the logic expression for the following gate. Rearrange the equation according to the commutative law and redraw the new circuit:

Solution The commutative law states that the order of two variables ORed or ANDed together is unimportant. Therefore, the gate expression $Y = A + B$ may be rearranged to read $Y = B + A$. The gate symbol is:

$$Y = B + A$$

The Associative Law

The associative law directs the grouping of three or more Boolean variables. The expressions $3 + 5 + 1$ and $5 + 3 + 1$ both equal 9 because the order in which the numbers are summed is inconsequential. This associative property can be stated logically as $A + B + C = G$, $B + C + A = G$, $C + A + B = G$, and so on. In addition, the equation $A + B + C = G$ can be viewed as

$(A + B) + C = G$, or $A + (B + C) = G$. Notice how the parentheses differentiate between the logical operations. Operations within parentheses take precedence over operations outside the parentheses. In this example, however, since all operations are OR operations, the logical outcome is the same. Figure 3.1 shows how the associative law alters the connections of variables in a circuit.

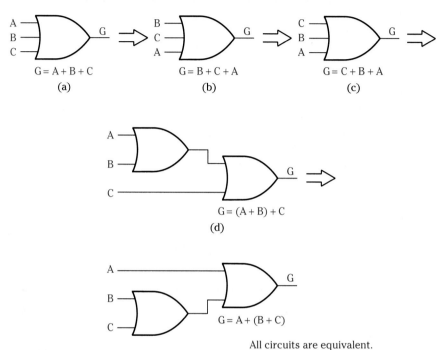

$G = A + B + C$ $G = B + C + A$ $G = C + B + A$

     (a)   (b)   (c)

$G = (A + B) + C$

(d)

$G = A + (B + C)$

All circuits are equivalent.

(e)

Figure 3.1
Associative Law

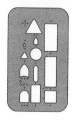

(a) Using the associative law, the expression $A_1 + A_2 + A_3 = X$ can be rewritten as $(A_1 + A_2) + A_3 = X$.

Draw the circuit for the second equation to show that both equations are equivalent:

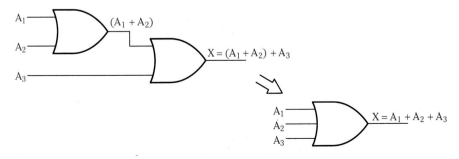

$(A_1 + A_2)$

$X = (A_1 + A_2) + A_3$

$X = A_1 + A_2 + A_3$

For both circuits, $X = 1$ whenever A_1 OR A_2 OR $A_3 = 1$.

(b) Repeat step (a) for the expression:

$$Y = ABC \quad \text{is equivalent to} \quad Y = (CB)A$$

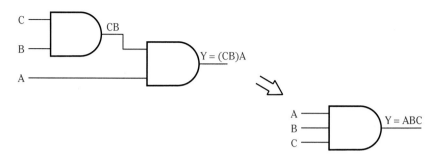

For both circuits, $Y = 1$ when $A = B = C = 1$.

The Distributive Law

The distributive law is widely used to simplify Boolean equations. In both Boolean algebra and standard algebra this law details how equations are expanded and factored. For instance, $A(B + C) = Z$ may be rewritten as $AB + AC = Z$ using an operation similar to algebraic multiplication. In this instance we say that the expression $A(B + C) = Z$ is "**expanded**" into the equivalent expression, $AB + AC = Z$. In a similar fashion, also invoking the distributive law, we can "**factor**" an expression such as $AB\overline{C} + BD + BE = X$ into $B(A\overline{C} + D + E) = X$ by pulling the common variable B out of the terms comprising the Boolean expression. No doubt, you readily appreciate the application of these algebraic laws from your previous mathematical studies.

FOR YOUR
INFORMATION   **Summary of Fundamental Boolean Laws**

Commutative	Associative	Distributive
Variables can be ANDed or ORed in any logical order.	Variables can be ANDed or ORed in any logical grouping.	Guides factoring and expansion of equations.
Examples:	Examples:	Examples:
$AB = BA$	$A(BC) = AB(C) = (A)(B)(C)$	$X = AB + BC = B(A + C)$
$A + B = B + A$	$A + (B + C) = (A + B) + C$	

Verify the distributive law by proving that:

$$Y = A(B + C) \quad \text{is equivalent to} \quad Y = AB + AC$$

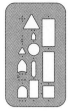

Solution Equivalence can be shown using a truth table:

A	B	C	B + C	A(B + C)	AB	AC	AB + AC
0	0	0	0	0	0	0	0
0	0	1	1	0	0	0	0
0	1	0	1	0	0	0	0
0	1	1	1	0	0	0	0
1	0	0	0	0	0	0	0
1	0	1	1	1	0	1	1
1	1	0	1	1	1	0	1
1	1	1	1	1	1	1	1

Both A(B + C) and AB + AC have the same output and are equivalent
equations.

Useful Reduction Theorems

The creative power the designer acquires through Boolean algebra usage orig-
inates from several simple theorems. These theorems reduce complex, unwieldy circuit designs into sleek, efficient logic structures. The theorems are
easy to comprehend since they relate to basic logic gate operation.

Nine Basic Boolean Theorems Promoting Logic Reduction

Many of the nine simplification theorems that follow deal with OR gate and
AND gate operations. Since AND and OR functions are fundamental logic
operations, even in NAND and NOR gates, we will see that the following theorems constitute the majority of Boolean reduction techniques:

Theorem 1: $A \cdot 0 = 0$
Theorem 2: $A \cdot 1 = A$
Theorem 3: $A \cdot A = A$ AND theorems
Theorem 4: $A \cdot \overline{A} = 0$

Theorem 5: $A + 0 = A$
Theorem 6: $A + 1 = 1$
Theorem 7: $A + A = A$ OR theorems
Theorem 8: $A + \overline{A} = 1$

Theorem 9: $\overline{\overline{A}} = A$ Inverter theorem

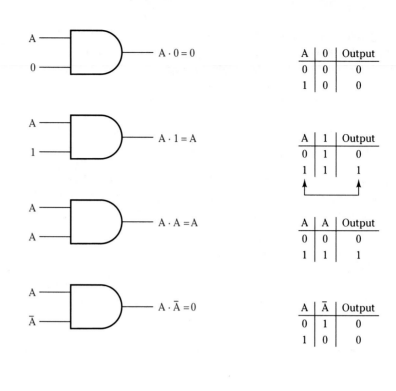

Figure 3.2
AND Gate Reduction
Theorems

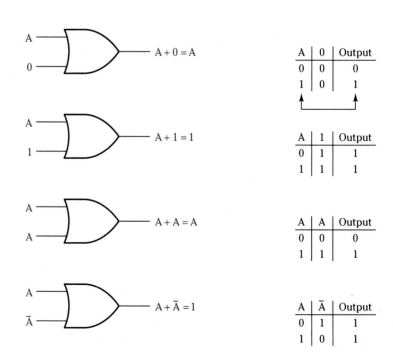

Figure 3.3
OR Gate Reduction
Theorems

Figure 3.2 schematically explains the AND gate theorems; Figure 3.3 explains the OR gate theorems; Figure 3.4 summarizes the inverter theorem. As you can see, the logic symbols, coupled with their associated truth tables, illustrate how to apply these basic theorems. The AND gate, for instance, always has a zero level output for any zero input. This is true whether the input at zero is a variable that momentarily happens to be low or an input at a permanent low level, as illustrated in Theorem 1 ($A \cdot 0 = 0$) and Theorem 4 ($A \cdot \overline{A} = 0$). In both cases at least one input in the equations is always at the low logic level, forcing the output low as well. For Theorem 4, if input A is high, its complement $\overline{A}$ must be low, resulting in a low output. If A is low, $\overline{A}$ is high, but the output remains in the low state. These formal explanations simply illustrate the AND gate disable operation.

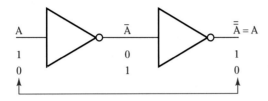

Figure 3.4
Inversion Theorem

Determine the output for the following gates:

DESIGN
EXAMPLE 3.4

$X \cdot 1 \cdot \overline{X} = 0$
$X \cdot \overline{X} = 0$

(inputs: X, 1, $\overline{X}$)

Theorems 2 and 4   If $X = 1$, then $\overline{X} = 0$. Therefore output $= 0$ since any AND input at zero forces the output to zero.

$1 \cdot \overline{X} \cdot \overline{X} = \overline{X}$
$1 \cdot \overline{X} = \overline{X}$

(inputs: 1, $\overline{X}$, $\overline{X}$)

Theorems 2 and 3   A variable ANDed with one equals the variable.

$X \cdot 1 \cdot X = X$
$X \cdot X = X$

(inputs: X, 1, X)

Theorems 2 and 3   A variable ANDed with one equals the variable.

The OR theorems $A + 1$ and $A + \overline{A}$ simply indicate the various cases where at least one high input forces the output high. The theorems $A + A$ and $A + 0$ show the input possibilities where the output follows the value of input variable A. Theorem 5, for instance, illustrates the OR gate enable case where a zero level on one input allows the value on the other input to control the gate output.

DESIGN
EXAMPLE 3.5

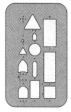

Determine the output for the following gates:

$$0 + X + \overline{X} = 1$$
$$X + \overline{X} = 1$$

Theorems 5 and 8   If X = 0, then $\overline{X}$ = 1.
Any OR input at 1 forces the output to 1.

$$X + X + 1 = 1$$
$$X + 1 = 1$$

Theorems 6 and 7   Any OR input at 1 forces the output to 1.

$$X + 0 + X = X$$
$$X + X = X$$

Theorems 5 and 7   If X = 0, then X + 0 + X = 0.
If X = 1, then X + 0 + X = 1.
Therefore, the output equals X.

Next, we examine the important points showing how these theorems can be applied to logic simplification. When we begin to extract logic equations from truth tables, we will see that the first step in this process produces a correct logical expression, but not necessarily a simple logic expression. Constructing circuits represented by nonsimplified expressions results in costly, redundant logic structures. By applying the theorems mentioned, as well as the associative, commutative, and distributive laws, we can usually reduce equations to a simpler and more economical form. The following example exemplifies these ideas.

DESIGN
EXAMPLE 3.6

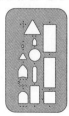

Reduce Y = $\overline{X}$X + Z to its simplest form.

$$Y = \overline{X}X + Z \qquad [\overline{X}X \text{ reduces to } 0, \text{ Theorem 4}]$$
$$Y = 0 + Z \qquad [0 + Z \text{ reduces to } Z, \text{ Theorem 5}]$$
$$Y = Z$$

Reduce T = XY + U(Z + $\overline{Z}$) to its simplest form.

$$T = XY + U(Z + \overline{Z}) \qquad [Z + \overline{Z} \text{ reduces to } 1, \text{ Theorem 6}]$$
$$T = XY + U(1) \qquad [U(1) \text{ reduces to } U, \text{ Theorem 2}]$$
$$T = XY + U$$

Several other theorems assist in simplifying variable arrangements that often occur in logic equations. These theorems are equations derived from the eight AND and OR theorems originally mentioned, but when treated and recognized as separate entities, they help to speed up the reduction process:

Theorem 10: X + XY = X

Figure 3.5a shows the logic diagram for the equation given as Theorem 10 to illustrate why this reduction makes sense. Notice that input variable X connects to an AND gate and also feeds ahead to an OR gate. When you observe such a connection, it is a good sign that circuit simplification is in order. Assume for a moment that input X is high (Figure 3.5b). Then the OR gate output is high since X disables the gate. If, on the other hand, X is low (Figure 3.5c), then the AND gate is disabled, resulting in a low AND output. The low from the AND gate and the low from X both feed the OR gate. With all OR inputs low, you can expect a corresponding low output. Whether X is high or low, the OR gate output is always equal to X. Input Y has no effect on the logical output and is therefore unnecessary. The output equation can be reduced using Boolean theorems and laws as well:

$$
\begin{aligned}
X + XY &= X \\
X(1 + Y) &= X \qquad \text{[Factoring X]} \\
X(1) &= X \qquad \text{[Apply Theorem 6]} \\
X &= X \qquad \text{[Apply Theorem 2]}
\end{aligned}
$$

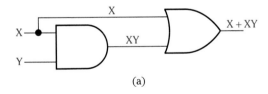

(a)

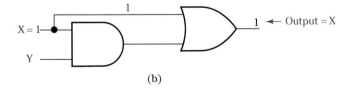

(b)

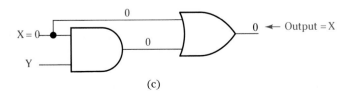

(c)

Figure 3.5

Circuit Reduction

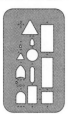

Apply Theorem 10 to reduce Z = B + AC + BC.

Solution   Theorem 10 states that X + XY = X. In the problem equation the B variable is substituted for variable X in Theorem 10 while variable C is substituted for Y. With this in mind, the equation is rearranged to read:

$$Z = (B + BC) + AC$$

- Y
- X
- X

$$Z = B + AC$$

Theorem 11: X(X + Y) = X

Theorem 11 is given schematically in Figures 3.6a, 3.6b, and 3.6c. Figure 3.6d is the logical simplification for expression X(X + Y) = X. This is similar to

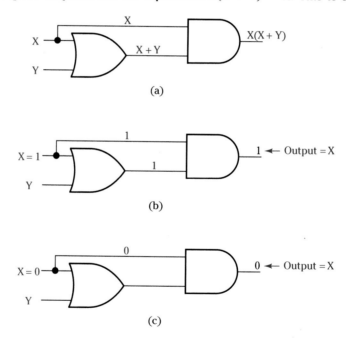

Figure 3.6
Circuit Reduction via
Theorem 11

the example discussed previously for Theorem 10. That is, a single input variable, X, controls the circuit output. If X is high, then the AND gate receives two high level inputs and the circuit output is high as well. If X is low, the AND gate is degated and the circuit output is low. The circuit output follows X under both circumstances. Variable Y in this circuit is "redundant," meaning that it serves no useful purpose. Generally, it is best to eliminate redundant logic. (In some designs, however, redundant logic enhances reliability by providing backup circuits or by correcting timing problems, a situation discussed in a later chapter.) We will attempt to reduce all equations to their simplest form unless otherwise stated.

Theorems 12 and 13 are not obvious reductions. Interesting algebraic proofs verify these reductions, but we will avoid perplexing algebraic proofs and rely on truth table proofs instead. Also note that these equations do not reduce to single variables, but rather, reduce to simpler equations.

Theorem 12: $X + \bar{X}Y = X + Y$
Theorem 13: $\bar{X} + X\bar{Y} = \bar{X} + \bar{Y}$

Figure 3.7 is a truth table proving the equality of Theorem 12. The truth table shows that the output for column $X + \bar{X}Y$ is the same as the output column for $X + Y$ for all input combinations, verifying the theorem. Figure 3.8 provides a similar proof for Theorem 13. In both theorems every possible input combination of X and Y is recorded and a complete circuit truth table is constructed. Comparing the output columns verifies equivalence.

X	Y	$\bar{X}$	$\bar{X}Y$	$X + \bar{X}Y$	$X + Y$
0	0	1	0	0	0
0	1	1	1	1	1
1	0	0	0	1	1
1	1	0	0	1	1

$X + \bar{X}Y = X + Y$ Proof

Figure 3.7
Proof of Equality—
Theorem 12

X	Y	$\bar{X}$	$\bar{Y}$	$X\bar{Y}$	$\bar{X} + X\bar{Y}$	$\bar{X} + \bar{Y}$
0	0	1	1	0	1	1
0	1	1	0	0	1	1
1	0	0	1	1	1	1
1	1	0	0	0	0	0

$\bar{X} + X\bar{Y} = \bar{X} + \bar{Y}$ Proof

Figure 3.8
Proof of Equality—
Theorem 13

Theorem 14: $XY + X\bar{Y} = X$

Theorem 14 is the heart of many of the reduction processes we will discuss throughout this text. The theorem shows how a variable may be eliminated when two AND terms differ by only a single variable. The following algebraic process confirms the proof of this theorem:

$$XY + X\bar{Y} = X$$
$$X(Y + \bar{Y}) = X \quad \text{[Factoring X]}$$
$$X(1) \quad\quad = X \quad \text{[Eliminate Y using Theorem 8]}$$
$$X \quad\quad\quad = X \quad \text{[X(1) = X from Theorem 2]}$$

DESIGN EXAMPLE 3.8

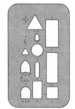

Reduce the following equation through the application of Theorem 14:

$$Y = A\bar{B}CD + ABCD$$

Solution   Only one variable differs from one AND term to the other—variable B. It may be eliminated. Thus, $Y = ACD$.

This can be confirmed by factoring:

$$Y = A\bar{B}CD + ABCD$$
$$Y = ACD(\bar{B} + B)$$
$$Y = ACD(1)$$
$$Y = ACD$$

Theorem 14 can be extended to simplify OR functions differing by a single variable:

$$(X + Y)(X + \bar{Y}) = X$$

DESIGN EXAMPLE 3.9

Apply Theorem 14 to the following expression to obtain a reduction. Prove equality:

$$Y = (A + \bar{B} + C)(A + B + C)$$

Solution   The theorem states that the variable that differs between the two OR functions may be eliminated. Thus, $Y = A + C$.

A truth table will prove equality.

A	B	C	$\bar{B}$	C	$A + \bar{B} + C$	$A + B + C$	Y	$A + C$
0	0	0	1	1	1	1	1	1
0	0	1	1	0	0	1	0	0
0	1	0	0	1	1	1	1	1
0	1	1	0	0	1	0	0	0
1	0	0	1	1	1	1	1	1
1	0	1	1	0	1	1	1	1
1	1	0	0	1	1	1	1	1
1	1	1	0	0	1	1	1	1

Equal

3.2 ■ APPLYING THE BOOLEAN THEOREMS

Now that we are armed with substantial Boolean reduction power, work can begin on equation simplification. Keep in mind that applying the correct theorem is only part of the reduction process. As with standard algebra, correctly manipulating the equation according to the distributive, associative, and commutative laws is equally important.

Factoring or expanding a Boolean equation leads to an equation that is equivalent in function but rearranged in mathematical order. Rearranging equations frequently contributes to the elimination of variables or terms by promoting the subsequent use of Boolean theorems. As long as we do not change the logical meaning of an equation, it becomes desirable to eliminate variables. Determining whether the equation needs expansion or factoring is generally the first goal. This, of course, means that the difference between an expanded or factored equation is clear. A factored equation is shown in Figure 3.9. The factored equation has many variations, but a noticeable characteristic is the presence of parentheses indicating that a variable or term outside the parentheses is ANDed with the variables or terms within the parentheses. The portion of Figure 3.9 having this characteristic is $A(BC + DE)$. In this example variable A is a factor that is common to both terms within the parentheses; A has been factored out from these two terms.

$$Y = A(BC + DE) + EFG$$

Figure 3.9
Factored Equation
Form

The original equation, in expanded form, is given in Figure 3.10. Here the various terms are shown in a fully expanded form where variable A is present in terms ABC and ADE. Therefore, factoring an equation merely removes common variables (or groups of variables) from two or more terms. This tends to condense equations, making other reductions more obvious. On the other hand, expanding an equation often has the benefit of logically grouping variables with their complements, allowing for a reduction.

$$Y = ABC + ADE + EFG$$

Figure 3.10
Expanded Equation
Form

Expand $Y = (A + B)(A + C)$.

Solution   Use the distributive law to expand the equation:

$$Y = (A + B)(A + C) \qquad [(A + B)(A + C) \text{ expanded}]$$
$$Y = AA + AC + BA + BC$$
$$Y = A + AC + BA + BC \qquad [\text{Factor A}]$$
$$Y = A(1 + C + B) + BC$$
$$Y = A(1) + BC$$
$$Y = A + BC$$

DESIGN
EXAMPLE 3.10

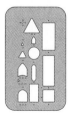

As you may surmise, both expansion and factoring are useful techniques. These methods are merely applications of the associative, commutative, and distributive laws used frequently with standard algebra. What this all boils down to is simply: **If an equation is expanded and no obvious reduction is evident, factor it; if an equation is factored and no obvious reduction is evident, expand it.** The factoring/expansion process, combined with Boolean theorems, leads to simplified equations.

Figure 3.11 illustrates the reduction process steps utilizing the techniques just discussed. Four terms comprise the expanded equation shown. First, look for obvious reductions; for example, are there any variables and their complements ANDed or ORed? If so, they can be reduced. This is not the case in Figure 3.11, however. Even though variables and their complements are present within the equation, they are not logically placed for reduction. Therefore, the next step is factoring, and, of course, we must determine the most advantageous variables to factor. Generally, factoring as many variables as possible from a term proves most useful. Here the variables A and B comprise a portion of two terms. Since this appears to be our best bet, we proceed to factor so that the equation $Y = ABC + AB\bar{C} + ACD + \bar{C}F$ becomes $Y = AB(C + \bar{C}) + ACD + \bar{C}F$.

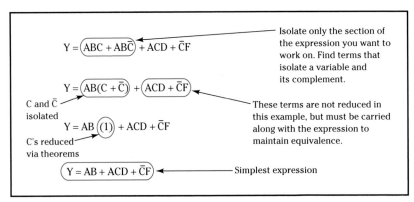

Figure 3.11
Reduction Process

Two points are worth mentioning here. First, the factoring has rearranged the equation and isolated variable C with its complement. Since we know that a variable ORed with its complement is reducible, we are on our way to a simpler equation. Second, the remaining two terms, ACD and $\bar{C}F$ are not modified in this stage of the process. However, they still must be carried along with the equation throughout the reduction process. Always keep track of variables and terms or an incorrect expression will result.

Referring back to the factored equation (specifically, variables C and $\bar{C}$), we proceed with the reduction. ORing a variable with its complement always produces an output of one. The equation now reads: $Y = AB(1) + ACD + \bar{C}F$. Another reduction is possible. A AND B AND 1 reduces to A AND B (AB). The final equation is $Y = AB + ACD + \bar{C}F$. This equation cannot be reduced further, although that may not be immediately obvious. Let's try factoring again (since we still have an expanded expression) just to be sure. The only variable left to factor is A. Therefore, $Y = A(B + CD) + \bar{C}F$. Within the parentheses the

AND-OR expression is not reducible because all the variables are unique, and thus are essential to the equation and not removable. In case you are looking at C and $\overline{C}$, notice that these do not factor because they are not the same variable—they are complements. Furthermore, they cannot be eliminated by any theorem since they are part of separate terms. It should be evident that none of the theorems discussed can simplify the equation any further, so that the reduction is complete.

Figure 3.12 illustrates the circuitry differences between the original equation and reduced equations. The circuit saving is obvious. The number of gates is not only reduced, but the size of the gates (number of inputs) is also cut down.

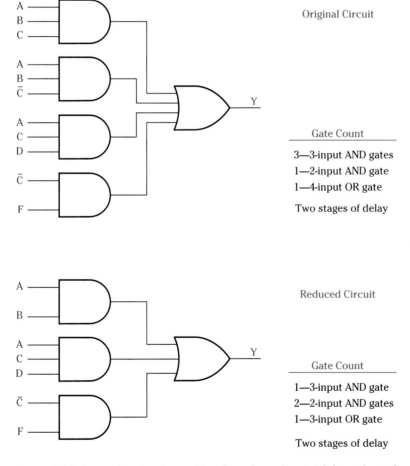

Original Circuit

Gate Count

3—3-input AND gates
1—2-input AND gate
1—4-input OR gate

Two stages of delay

Reduced Circuit

Gate Count

1—3-input AND gate
2—2-input AND gates
1—3-input OR gate

Two stages of delay

Figure 3.12
Circuit Reduction
Comparison

Figure 3.13 shows the circuit resulting from factoring out A from the reduced equation. (By doing this, we attempted to reduce the equation further.) Although equivalent, the circuit complexity is noticeable. The most undesirable circuit characteristic resulting from this reduced equation is the creation of four stages of delay facing the signals C and D from input to output. Since the first reduced equation only has two stages of delay, it is generally the preferred design.

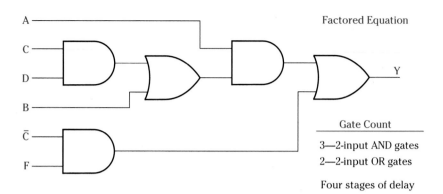

Factored Equation

Y

Gate Count

3—2-input AND gates
2—2-input OR gates

Four stages of delay

Figure 3.13
Factored Circuit

Therefore, from a practical standpoint, design considerations other than just reduced gate count often determine the final circuitry configuration chosen. We will investigate many other practical design constraints and considerations throughout this text.

**DESIGN
EXAMPLE 3.11**

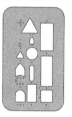

(a) Find the minimum expression for $R = \overline{S}P\overline{Q} + S\overline{P}Q + SP\overline{Q}$.

$$R = \overline{S}P\overline{Q} + S\overline{P}Q + SP\overline{Q} \qquad \text{[Factor } P\overline{Q}\text{]}$$
$$R = P\overline{Q}(\overline{S} + S) + S\overline{P}Q$$
$$R = P\overline{Q}(1) + S\overline{P}Q$$
$$R = P\overline{Q} + S\overline{P}Q$$

(b) Find the minimum expression for $A = (X + Y)(\overline{X} + \overline{Z})(Y + Z)$.

$$A = (X + Y)(\overline{X} + \overline{Z})(Y + Z) \qquad\qquad \text{[Expand expression]}$$
$$A = (X\overline{X} + X\overline{Z} + Y\overline{X} + Y\overline{Z})(Y + Z) \qquad \text{[Simplify with theorems]}$$
$$A = (0 + X\overline{Z} + Y\overline{X} + Y\overline{Z})(Y + Z)$$
$$A = (X\overline{Z} + Y\overline{X} + Y\overline{Z})(Y + Z) \qquad\qquad \text{[Expand again]}$$
$$A = X\overline{Z}Y + Y\overline{X}Y + Y\overline{Z}Y + X\overline{Z}Z + Y\overline{X}Z + Y\overline{Z}Z$$

$$A = XY\overline{Z} + \overline{X}Y + Y\overline{Z} + 0 + \overline{X}YZ + 0 \qquad \text{[Simplify with theorems]}$$
$$A = XY\overline{Z} + \overline{X}Y + Y\overline{Z} + \overline{X}YZ \qquad\qquad \text{[Factor YZ and XY]}$$

$$A = Y\overline{Z}(X + 1) + \overline{X}Y(1 + Z) \qquad\qquad \text{[Simplify with theorems]}$$
$$A = Y\overline{Z}(1) + \overline{X}Y(1)$$
$$A = Y\overline{Z} + \overline{X}Y$$

3.3 ■ DeMORGAN'S THEOREM

Now we explore a theorem that provides substantial circuit reduction and design power. DeMorgan's Theorem enables us to change AND-type logic to OR-type logic, or vice versa, yet maintain circuit equality. Invoking this theorem often leads to simpler circuitry or circuitry in a more usable form, teaches us to view logical devices from different perspectives, and extends our under-standing of gate operation.

The typical DeMorgan's Theorem explanation goes something like $\overline{AB} = \overline{A} + \overline{B}$, which means that the complement of a product (AND) is equal to the sum (OR) of the complements. This shows how AND logic is converted to OR logic, but this alone probably is not enough to regard the theorem highly. Let's dispense with the classical theorem definition and concentrate on putting it to good use. We start by defining a procedure that allows us to use the theorem on anything from a basic gate to a lengthy equation.

To DeMorganize:

1. Complement all input variables.
2. Change AND gates to OR gates, or change OR gates to AND gates.
3. Complement the output.

Simple. Now we apply this procedure to the NAND gate expression $Y = \overline{AB}$. The expression inputs are A and B. Since A and B are represented in true form, they are both complemented following procedure step 1. Thus, A and B become $\overline{A}$ and $\overline{B}$. Following step 2, the AND function inherent in the NAND operation converts to an OR function. We complement the output in step 3. Since the NAND output is considered complemented (as defined by the presence of an overbar), we recomplement it according to the procedure by removing the overbar. The new expression is now $Y = \overline{A} + \overline{B}$. Figure 3.14 details this process step by step and proves equality with a truth table proof for the two expressions. In a nutshell, this simple three-step process transforms logic from one form to another.

Expression to be DeMorganized

$Y = \overline{A\,B}$ ← Output—complement

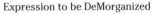
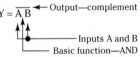

Inputs A and B
Basic function—AND

DeMorganization

$Y = \overline{AB}$

$Y = \overline{\overline{A}\,\overline{B}}$ Step 1:   Complement inputs

$Y = \overline{\overline{A} + \overline{B}}$ Step 2:   Change AND to OR

$Y = \overline{\overline{\overline{A} + \overline{B}}}$ Step 3:   Complement output

$Y = \overline{A} + \overline{B}$ Cleanup

Proof of equality

A	B	$\overline{AB}$	$\overline{A}$	$\overline{B}$	$\overline{A} + \overline{B}$
0	0	1	1	1	1
0	1	1	1	0	1
1	0	1	0	1	1
1	1	0	0	0	0

Equal

Figure 3.14
Proof of DeMorgan's
Theorem

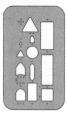

DESIGN EXAMPLE 3.12

DeMorganize $Y = \overline{A + B}$.

Solution

Complement input variables	$Y = \overline{\overline{A} + \overline{B}}$
Change OR to AND	$Y = \overline{\overline{\overline{A}\overline{B}}}$
Complement output	$Y = \overline{\overline{\overline{\overline{A}\overline{B}}}}$
Clean up	$Y = \overline{A}\overline{B}$

DeMorganizing Expressions

The DeMorganization process can also be utilized to alter an entire expression or any part of an expression, as outlined in Figure 3.15. Notice that the OR function is chosen for DeMorganization. This means that the NAND gate outputs are the inputs to the OR gate and are treated as a single variable during the process. Therefore, when the inputs are complemented, an overbar is drawn over $\overline{XU} \rightarrow \overline{\overline{XU}}$ and $\overline{WYT} \rightarrow \overline{\overline{WYT}}$. Any portion of a circuit can be DeMorganized as long as care is taken to identify clearly the inputs and outputs of the function to be DeMorganized. As is evident from the expression and accompanying logic diagram, DeMorgan's Theorem produces two uniquely different but equal circuits. Gate count is reduced in this example from three gates to one, and the signal delay is also reduced from two stages to one. DeMorgan's Theorem assists us with circuit modification as well as with circuit reduction.

$$A = \overline{\overline{XU} + \overline{WYT}}$$
$$A = \overline{\overline{\overline{XU}} + \overline{\overline{WYT}}} \qquad \text{Step 1: Complement inputs}$$
$$A = \overline{\overline{\overline{XU}} \; \overline{\overline{WYT}}} \qquad \text{Step 2: Change OR to AND}$$
$$A = \overline{\overline{\overline{XU}} \; \overline{\overline{WYT}}} \qquad \text{Step 3: Complement output}$$
$$A = \overline{XU} \; \overline{WYT} \qquad \text{Cleanup}$$

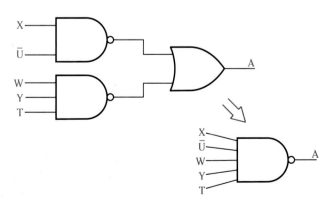

Figure 3.15 DeMorganization Process

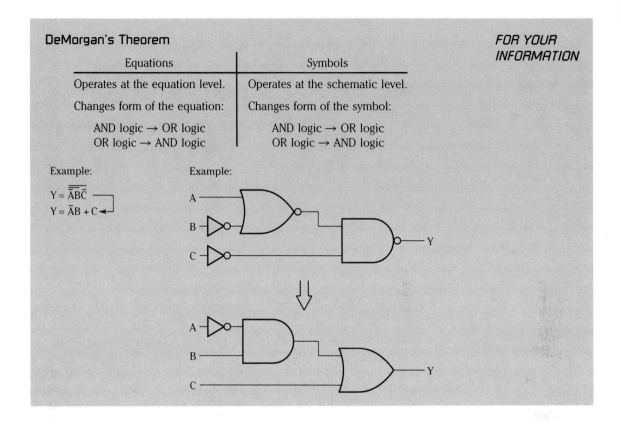

FOR YOUR INFORMATION

DeMorgan's Theorem

Equations	Symbols
Operates at the equation level.	Operates at the schematic level.
Changes form of the equation:	Changes form of the symbol:
AND logic → OR logic OR logic → AND logic	AND logic → OR logic OR logic → AND logic

Example:

$$Y = \overline{\overline{A}B\overline{C}}$$
$$Y = \overline{A}B + C$$

Example:

Rearrange the following equation using DeMorgan's Theorem:

$$Z = \overline{(\overline{A} + C)(B + \overline{D})}$$

$$Z = \overline{\overline{(\overline{A} + C)}\,\overline{(B + \overline{D})}} \qquad \text{[Complement inputs]}$$

$$Z = \overline{\overline{(\overline{A} + C)} + \overline{(B + \overline{D})}} \qquad \text{[Change AND to OR]}$$

$$Z = \overline{\overline{\overline{(\overline{A} + C)} + \overline{(B + \overline{D})}}} \qquad \text{[Complement output]}$$

$$Z = \overline{(\overline{A} + C)} + \overline{(B + \overline{D})} \qquad \text{[Clean up]}$$

DESIGN EXAMPLE 3.13

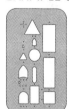

We have just begun to see DeMorgan's power at work. A real appreciation of the theorem's usefulness comes with learning how to apply the theorem symbolically. Figure 3.16 shows how to manipulate a logic symbol with the theorem. The process steps are the same as those used to DeMorganize an expression. First, the inputs are complemented. Symbolically, this means adding bubbles if there are none on the inputs or removing bubbles if they already exist on the inputs. Of course, this is equivalent to complementing the input

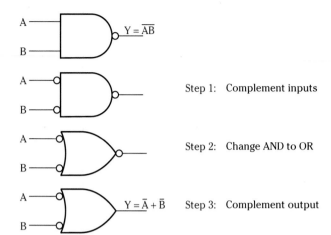

Step 1:   Complement inputs

Step 2:   Change AND to OR

Step 3:   Complement output

Figure 3.16
DeMorganizing a Logic
Symbol

variables. Then the logic function is changed from AND to OR or OR to AND. This is accomplished simply by redrawing the symbol. Last, the output is complemented by adding or removing a bubble.

One interesting point that is clear in this figure is the unique symbol resulting from DeMorgan's process. This outcome is so important, we will discuss it separately in a moment.

Basic Gates—A Different Perspective

The odd-looking gate resulting from DeMorganizing the NAND gate in Figure 3.16 perfectly illustrates that logical functions can be analyzed in various ways. Figure 3.17 shows the DeMorgan representation for all the basic gates we have studied. The three-step process described earlier was used to DeMorganize the

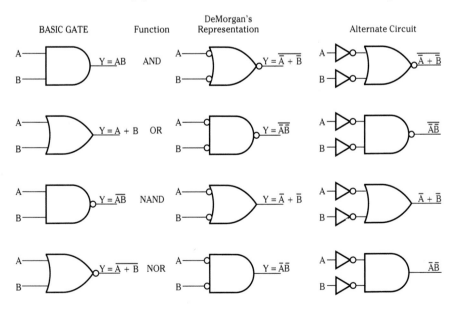

Figure 3.17
Gate Comparison

gates and equations shown; all are equivalent to their original corresponding gates and equations. Any one of these representations is acceptable and a useful way to indicate basic logic functions.

Now we really put the theorem to work by learning how to **read logic symbols.** Starting from input to output and associating bubbles with logic zeros, or the lack of bubbles with logic ones, we begin a symbolic DeMorganization process. For instance, the NAND gate inputs have no bubbles, which associates these inputs with the high logic level. The output, on the other hand, has a bubble that associates the NAND output with a low logic level (refer to Figure 3.18). Since the fundamental logic operation for the NAND gate is the AND operation, we read the NAND symbol by noting the following: When inputs A AND B are high, the output is low. You should recognize this as the unique condition from the NAND gate truth table.

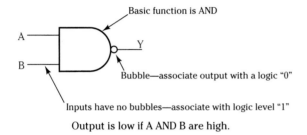

Output is low if A AND B are high.

Figure 3.18
Reading a NAND
Symbol

Figure 3.19 shows the DeMorganized NAND gate symbol. Also noted are the logic levels associated with the inputs and output as well as the basic logic function. This symbol still represents a NAND function, even though the basic symbol is now an OR and reads as: If input A OR B is low, the output is high. This version of the NAND gate graphically shows us the nonunique NAND gate conditions. This is summarized in Figure 3.20. Now we have learned not only

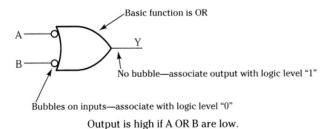

Output is high if A OR B are low.

Figure 3.19
DeMorganized NAND
Symbol

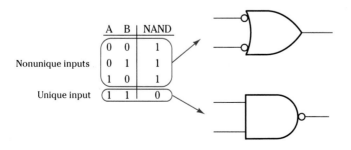

Figure 3.20
NAND Truth Table—
Symbol Comparison

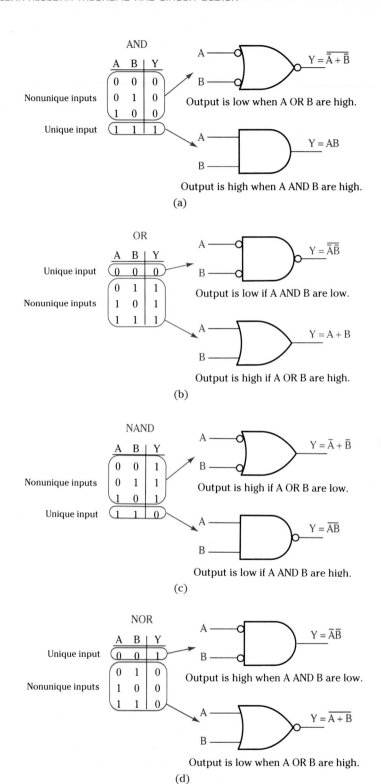

Figure 3.21
DeMorganized Symbols—Truth Table Comparisons

how to read logic symbols, but also how DeMorgan's Theorem helps us to
recall a gate's complete logical operation.

Figures 3.21a, b, c, and d summarize the truth table/symbol relationship
for all basic logic gates. The NAND gate is included again for the sake of
completeness. In each case the standard gate logic symbol and the De-
Morganized symbol is given. Also, in each case one symbol corresponds to
the unique truth table conditions, whereas the others correspond to the nonunique conditions. Determining any gate input–output relationship becomes
very easy when you learn to read the logic symbols. This technique also assists
in analyzing circuit diagrams.

DeMorganize the following logic symbols:

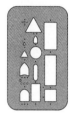

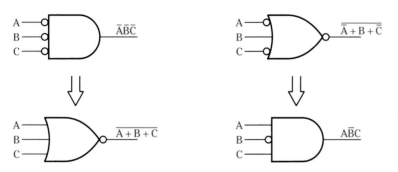

Solution DeMorganize by complementing inputs, changing function, and
complementing output.

Taking all of this one step further, we apply what we have studied at the
gate level using the circuit in Figure 3.22a. The circuit output equation in this
figure looks rather complex because of the many overbars. We may wish to
DeMorganize such equations to clarify and simplify analysis. As we see from
both the equation and the schematic, the basic logic function governing the
output is the OR function (from the NOR gate). This function is the target for
DeMorganization (although any function within the equation can be De-
Morganized). The inputs to the NOR gate are outputs from other circuits and
are not single variables, but rather, are logical groups of variables. For the
DeMorganization process these inputs are treated as single variables. For instance, the input to the NOR from the AND-inverter portion of the circuit is
$\overline{ABC}$. Since this term represents the input to the NOR, when DeMorganizing,
the entire term is complemented, not the individual variables comprising the
term. The other NOR input is $\overline{D} + E$, which is treated as a single variable. Both
the algebraic and circuit results are shown in Figure 3.22a. By comparing the
expression and schematic, we can see which portion of the logic structure is
modified by the DeMorganization process.

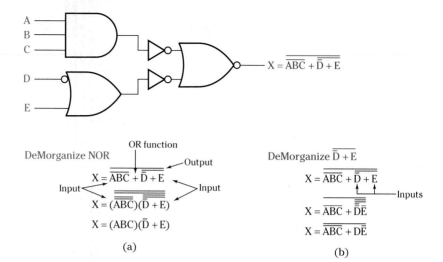

Figure 3.22
Equation
DeMorganization

As mentioned before, other portions of the circuit can be DeMorganized if desired. For instance, the gate combining inputs $\overline{D}$ and E is DeMorganized without affecting any other circuitry, as shown in the expression in Figure 3.22b. In this case the inputs are variables $\overline{D}$ and E; the basic logic function is OR. Notice how input $\overline{D}$ and the output have their overbars removed during the complementing process. Input E, on the other hand, acquires an overbar since it is uncomplemented to begin with.

DeMorganizing Circuit Schematics

We can continue to apply DeMorgan's Theorem to portions of the circuit equation, modifying it as much as we choose, but we will be up to our necks in algebra before long. An alternative approach is to DeMorganize the circuit schematic. The DeMorganization example just described is shown in Figure 3.23, but this time circuit diagrams are used to carry out the process using the symbolic approach applied to individual gates. For example, the output NOR gate is DeMorganized by assigning bubbles to the two inputs (complement inputs), changing the OR symbol to an AND (change function), and removing the bubble from the output (complement output). Then by recognizing that an inverter and bubble in series is double inversion, we can clean up the circuit to yield a new but equivalent circuit.

We can apply the theorem to any gate on the schematic with the same end results, with the benefit that the process is considerably easier using only the symbols. For instance, it is generally simpler to spot instances of double inversion and other obvious reductions in the schematic rather than in the equation. Working at the circuit level brings our design skills and insight to bear right where they accomplish the most and gives us the opportunity to modify any portion of a circuit to meet the design objectives at hand. This method also facilitates substituting components if required parts are not available.

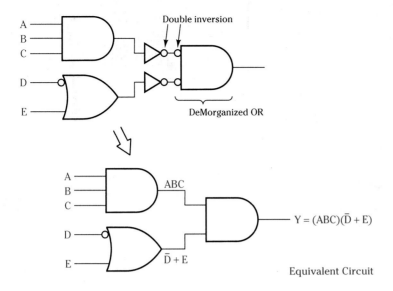

Double inversion

DeMorganized OR

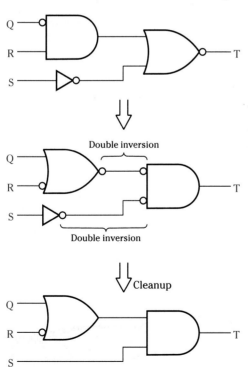

ABC

$Y = (ABC)(\bar{D} + E)$

$\bar{D} + E$

Equivalent Circuit

Figure 3.23
Circuit
DeMorganization
Process

DeMorganize the AND and OR functions on the following schematic:

DESIGN
EXAMPLE 3.15

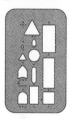

Double inversion

Double inversion

Cleanup

Solution   DeMorganize the AND symbol and the OR symbol (NOR gate). Look for subsequent reductions.

3.4 ■ DESIGNING YOUR OWN CIRCUITS

Thus far we have focused on acquiring the requisite knowledge needed to understand logic design, comprehending the complexities involved in logic reduction, and appreciating the benefits gained. However, we have depended on circuits and equations designed by others. How were these circuits and equations designed in the first place? Naturally, any design begins with an idea. In the case of combinatorial logic (defined below), basic design ideas can usually be expressed in truth table form. That is, by using a truth table, the designer states the desired circuit outputs for each input combination. Once the truth table is complete, a circuit can always be obtained. We will study several techniques that allow us to transform truth tables into equations and, consequently, into practical working circuits.

Combinatorial logic is a term used to describe a logic network having no storage capability. This means that input variables can change level, causing outputs to also change level, but that when the inputs are restored to their original levels, the outputs also revert to their original levels. If outputs are monitored after all input changes occur, you will find no evidence that the inputs did in fact change, since the outputs only reflect the current state of the inputs. A combinatorial network cannot store a logical value. Therefore, the state of the output(s) at any point in time depends on the input levels existing at the same point in time. That is, a truth table adequately describes combinatorial circuit operation since it relates specific input values to specific output values. **Sequential circuits,** which do possess storage capability, will be discussed in a later chapter.

3.5 ■ FUNDAMENTAL PRODUCTS AND THE SUM OF PRODUCTS METHOD

The first and most widely used design technique we will study is the **sum of products (SOP) method.** The key ingredient used in this logic equation-producing technique is the **fundamental product.** A fundamental product is nothing more than an AND function (often referred to as a product term) combining a unique combination of input variables to produce a one output level. Of course, an AND gate output is high only when all inputs are high. Therefore, defining the fundamental product for any input combination means identifying the input levels that cause a one output to occur. Figure 3.24 illustrates several fundamental products. The first gate in the figure is a 3-input AND gate with inputs A, B, and C all at the one level. It is apparent that this input combination places the output at the one level. The fundamental product is the logic equation ABC. The significance of this becomes clear if we consider the second AND gate. The inputs are A = 0, B = 1, and C = 1. In order to obtain a one output for this particular input combination, the complement of A must be ANDed with the true levels of B and C. The fundamental product is $\overline{A}BC$. It should be clear that only $\overline{A}BC$ can produce the one level output for

A = 1
B = 1
C = 1

For Y = 1 ⟶ ABC = 1 1 1
Fundamental product = ABC

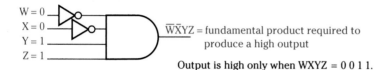

A = 0
B = 1
C = 1

$Y = \bar{A}BC$ = fundamental product
Y is 1 only when ABC = 0 1 1.

A = 1
B = 0
C = 0
D = 1

$Y = A\bar{B}\bar{C}D$ = fundamental product
Y is 1 only when ABCD = 1 0 0 1.

W = 0
X = 0
Y = 1
Z = 1

$\overline{W}\overline{X}YZ$ = fundamental product required to produce a high output
Output is high only when WXYZ = 0 0 1 1.

Figure 3.24
Fundamental Products

this circuit. Therefore, the fundamental product is a unique AND gate circuit whose logical expression identifies a specific input condition. Presenting any other input combination to a fundamental product circuit produces a low output level. This property will be utilized as an integral part of the sum of products method. (Several other circuits are illustrated in Figure 3.24 for additional clarification.)

What is the fundamental product required to produce a high level output for the following variables:
(a) W,X,Y,Z = 0101
(b) W,X,Y,Z = 1011
(c) W,X,Y,Z = 0000
(d) M,N,P,Q,T,Y,Z = 1000110

DESIGN EXAMPLE 3.16

Solution   The fundamental product is formed by ANDing the complement of all variables at the low state with the true of all variables at the high state. Thus:
(a) $\overline{W}X\overline{Y}Z$
(b) $W\overline{X}YZ$
(c) $\overline{W}\overline{X}\overline{Y}\overline{Z}$
(d) $M\overline{N}\overline{P}\overline{Q}TY\overline{Z}$

DESIGN EXAMPLE 3.17

Verify that the fundamental product $\overline{A}BC$ produces a high level output only for input combination 011.

Solution   Create a truth table to verify the fundamental product:

A	B	C	$\overline{A}$	$\overline{A}BC$	
0	0	0	1	0	
0	0	1	1	0	
0	1	0	1	0	
0	1	1	1	1	←—$\overline{A}BC$
1	0	0	0	0	
1	0	1	0	0	
1	1	0	0	0	
1	1	1	0	0	

Now that we understand the idea behind fundamental products, we will examine how they apply to truth tables. Refer to Figure 3.25. The circuit described by this truth table has three input variables forming eight input combinations. Since all eight combinations have an assigned associated output condition, Figure 3.25 shows a complete truth table. Let's consider what the table is telling us. The table describes a circuit with an output that is high when A, B, and C are equal to 001, or 011, or 100. How do these three particular combinations produce the one level output? The fundamental product should come to mind. Investigating the combination A, B, C = 001, we see that this combination yields a one level output from an AND circuit having the fundamental product $\overline{A}\overline{B}C$. The two other combinations produce one level outputs for fundamental products $\overline{A}BC$ and $A\overline{B}\overline{C}$. Since any one of these combinations can provide the desired one level output, ORing (summing) the fundamental products together produces the complete logic expression representing the truth table. This is given in Figure 3.26, along with a complete circuit truth table verifying the expression.

Since the term "sum" refers to OR gates and the term "product" refers to AND gates, we can expect to see these functions in the actual circuit. In fact, every SOP expression results in a circuit similar in form to that in Figure 3.27. Circuitry with AND gates feeding an OR gate are called **AND-OR networks**.

A	B	C	Y	
0	0	0	0	
0	0	1	1	→ Fundamental product = $\overline{A}\overline{B}C$
0	1	0	0	
0	1	1	1	→ Fundamental product = $\overline{A}BC$
1	0	0	1	→ Fundamental product = $A\overline{B}\overline{C}$
1	0	1	0	
1	1	0	0	
1	1	1	0	

Figure 3.25
Obtaining
Fundamental Products
from a Truth Table

A	B	C	$\bar{A}$	$\bar{B}$	$\bar{C}$	$\bar{A}\bar{B}C$	$\bar{A}BC$	$A\bar{B}\bar{C}$	$Y = \bar{A}\bar{B}C + \bar{A}BC + A\bar{B}\bar{C}$
0	0	0	1	1	1	0	0	0	0
0	0	1	1	1	0	1	0	0	1
0	1	0	1	0	1	0	0	0	0
0	1	1	1	0	0	0	1	0	1
1	0	0	0	1	1	0	0	1	1
1	0	1	0	1	0	0	0	0	0
1	1	0	0	0	1	0	0	0	0
1	1	1	0	0	0	0	0	0	0

$$Y = \bar{A}\bar{B}C + \bar{A}BC + A\bar{B}\bar{C}$$

— Same as truth table in Figure 3.25—proof of equality

Figure 3.26
ORing Fundamental Products to Produce a SOP Expression

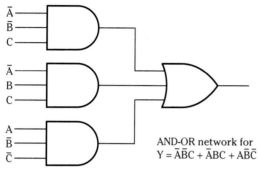

AND-OR network for
$Y = \bar{A}\bar{B}C + \bar{A}BC + A\bar{B}\bar{C}$

Figure 3.27
AND-OR Network

The SOP method always produces a logical expression, but usually it is not the simplest or "minimum" expression. Boolean reduction is often required as illustrated in Figure 3.28. The original SOP expression has been simplified, by Boolean reduction techniques, to the reduced equation and circuit shown. Notice that the reduced expression is still an SOP expression, and consequently the circuit is also an AND-OR network.

$$Y = \bar{A}\bar{B}C + \bar{A}BC + A\bar{B}\bar{C}$$
$$Y = \bar{A}C(\bar{B} + B) + A\bar{B}\bar{C}$$
$$Y = \bar{A}C(1) + A\bar{B}\bar{C}$$
$$Y = \bar{A}C + A\bar{B}\bar{C}$$

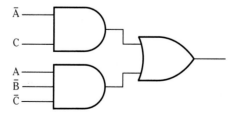

Figure 3.28
Reduced AND-OR Network

To summarize, the following steps convert a truth table into a sum of products expression:

- Obtain the fundamental product for each truth table one level output.
- OR the fundamental products together.
- Simplify the resulting SOP expression.

DESIGN
EXAMPLE 3.18

Determine the minimum SOP expression for the following truth table:

W	X	Y	Z	T	
0	0	0	0	0	
0	0	0	1	0	
0	0	1	0	0	
0	0	1	1	0	
0	1	0	0	0	
0	1	0	1	1	$\overline{W}X\overline{Y}Z$
0	1	1	0	0	
0	1	1	1	1	$\overline{W}XYZ$
1	0	0	0	0	
1	0	0	1	0	
1	0	1	0	0	
1	0	1	1	0	
1	1	0	0	0	
1	1	0	1	1	$WX\overline{Y}Z$
1	1	1	0	0	
1	1	1	1	1	$WXYZ$

Solution   Determine each fundamental product from the truth table, OR the fundamental products together, and simplify.

$T = \overline{W}X\overline{Y}Z + \overline{W}XYZ + WX\overline{Y}Z + WXYZ$
$T = \overline{W}XZ(\overline{Y} + Y) + WXZ(\overline{Y} + Y)$
$T = \overline{W}XZ + WXZ$
$T = XZ(\overline{W} + W)$
$T = XZ$

DESIGN
EXAMPLE 3.19

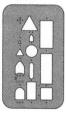

Memory circuits in a computer system are activated by a high-level signal when the following condition occurs: input signals READ or WRITE are low and ADDRESS13 is low. Create the truth table for this circuitry and determine the sum of products expression. Draw the circuit.

READ	WRITE	ADDRESS13	OUTPUT	
0	0	0	1	$FP = \overline{READ}*\overline{WRITE}*\overline{ADDRESS13}$
0	0	1	0	
0	1	0	1	$FP = \overline{READ}*WRITE*\overline{ADDRESS13}$
0	1	1	0	
1	0	0	1	$FP = READ*\overline{WRITE}*\overline{ADDRESS13}$
1	0	1	0	
1	1	0	0	
1	1	1	0	

OUTPUT = $\overline{READ}*\overline{WRITE}*\overline{ADDRESS13} + \overline{READ}*WRITE*\overline{ADDRESS13} +$
         $READ*\overline{WRITE}*\overline{ADDRESS13}$
OUTPUT = $\overline{READ}*\overline{ADDRESS13}(\overline{WRITE} + WRITE) +$
         $READ*\overline{WRITE}*\overline{ADDRESS13}$
OUTPUT = $\overline{READ}*\overline{ADDRESS13}$ (1) $+ READ*\overline{WRITE}*\overline{ADDRESS13}$
OUTPUT = $\overline{READ}*\overline{ADDRESS13} + READ*\overline{WRITE}*\overline{ADDRESS13}$
OUTPUT = $\overline{ADDRESS13}$ ($\overline{READ} + READ*\overline{WRITE}$) [Variation of Theorem 12]
OUTPUT = $\overline{ADDRESS13}$ ($\overline{READ} + \overline{WRITE}$)
OUTPUT = $\overline{ADDRESS13}*\overline{READ} + \overline{ADDRESS13}*\overline{WRITE}$

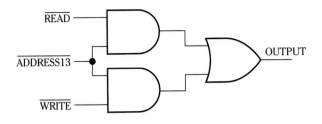

3.6 ■ FUNDAMENTAL SUMS AND THE PRODUCT OF SUMS METHOD

An alternate circuit design technique is the **product of sums (POS) method.** The POS technique creates Boolean equations consisting of OR terms, which are ANDed together; POS equations result in **OR-AND networks.**

The **fundamental sum,** the key ingredient for this method, is an OR gate equation describing the unique input combination required to produce a zero output level. Take, for example, input combination A,B,C,D = 1100. What must be done with this combination to produce the unique low OR output? Since the OR output is low only when all inputs are low, any input bits in the high state must be conditioned to appear low at the OR gate input. Therefore, any input bit that is one must be inverted before it is presented to the OR gate. In this example variables A and B are high, whereas C and D are low. Thus, A and B are inverted, yielding an OR output of $\overline{A} + \overline{B} + C + D$. Figure 3.29 graphically depicts this process. The output is low for this input combination when inputs A and B are inverted. Of course, any other input combinations presented to the circuit cause the output to go high because at least one OR gate input will always be high. Therefore, the fundamental sum produces a low output level for only one specific input condition. We will utilize this property with the POS technique.

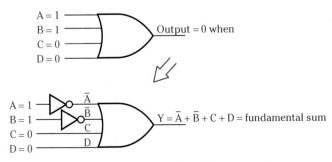

OR output is zero only when all inputs equal zero.

Figure 3.29
Fundamental Sum

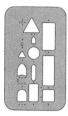

DESIGN
EXAMPLE 3.20

What is the fundamental sum required to produce a low level output for the following variables:
(a) W,X,Y,Z = 0101
(b) W,X,Y,Z = 1011
(c) W,X,Y,Z = 0000
(d) M,N,P,Q,T,Y,Z = 1000110

Solution   The fundamental sum is formed by ORing the complement of all variables in the high state with the true of all variables in the low state. Thus:
(a) $W + \bar{X} + Y + \bar{Z}$
(b) $\bar{W} + X + \bar{Y} + \bar{Z}$
(c) $W + X + Y + Z$
(d) $\bar{M} + N + P + Q + \bar{T} + \bar{Y} + Z$

For the sake of comparison, we will design a POS circuit using the truth table from the SOP example. This table is given in Figure 3.30, along with the fundamental sum for each zero in the truth table. Please keep in mind that fundamental products are obtained from high levels, whereas fundamental sums are obtained from low levels.

A	B	C	Y	
0	0	0	0	← $A + B + C$
0	0	1	1	
0	1	0	0	← $A + \bar{B} + C$
0	1	1	1	
1	0	0	1	
1	0	1	0	← $\bar{A} + B + \bar{C}$
1	1	0	0	← $\bar{A} + \bar{B} + C$
1	1	1	0	← $\bar{A} + \bar{B} + \bar{C}$

Fundamental sums

POS expression

Fundamental sums

$$Y = (A + B + C)(A + \bar{B} + C)(\bar{A} + B + \bar{C})(\bar{A} + \bar{B} + C)(\bar{A} + \bar{B} + \bar{C})$$

AND functions

Figure 3.30
Obtaining a POS
Expression from a
Truth Table

Since five zero outputs are in the truth table of Figure 3.30, there are also five fundamental sums. These sums are then ANDed together to form a product of sums expression, also listed in Figure 3.30. This expression identifies the groupings of variables that provide a low level output on the truth table. The OR terms are ANDed together because the circuit output is low when any OR term output is low. In addition, the circuit configuration resulting from a POS circuit is an OR-AND network, that is, a group of OR gates feeding into an AND gate. This is true for all product of sums expressions.

You have noticed, no doubt, that the POS expression is substantially more complicated than the equivalent SOP expression. The reason should be clear: More zero outputs are in the truth table than one outputs. Therefore, this POS equation has more terms than the equivalent SOP equation. This can be a good reason to select one method over the other. Fewer terms (product or sum) imply fewer gates in the actual circuit. But since Boolean and other reduction techniques can be applied to the equations, this may not be an overriding consideration—just one to keep in mind. The actual implementation of a POS or SOP circuit is often dictated by the logic family used. Some technologies simply lend themselves to an SOP implementation rather than the equivalent POS representation, or vice versa. Also notice that an SOP expression is the expanded expression discussed previously in relation to Boolean reduction, whereas the POS expression is the factored expression. Since we can mathematically change the form of an expression, yet still maintain equivalence, the relationship between POS and SOP is clear.

To summarize, the following steps convert a truth table into a product of sums expression:

- Obtain the fundamental sum for each truth table zero level output.
- AND the fundamental products together.
- Simplify the resulting POS expression.

Determine the minimum POS expression for the following truth table:

DESIGN EXAMPLE 3.21

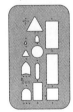

F	G	H	I	T	
0	0	0	0	0	$F + G + H + I$
0	0	0	1	0	$F + G + H + \bar{I}$
0	0	1	0	1	
0	0	1	1	1	
0	1	0	0	1	
0	1	0	1	1	
0	1	1	0	1	
0	1	1	1	1	
1	0	0	0	0	$\bar{F} + G + H + I$
1	0	0	1	0	$\bar{F} + G + H + \bar{I}$
1	0	1	0	1	
1	0	1	1	1	
1	1	0	0	1	
1	1	0	1	1	
1	1	1	0	1	
1	1	1	1	1	

Solution Find the fundamental sum for each low output, AND these together, and simplify.

$$T = (F + G + H + I)(F + G + H + \bar{I})(\bar{F} + G + H + I)(\bar{F} + G + H + \bar{I})$$
$$T = (F + G + H)(\bar{F} + G + H) \qquad \text{Theorem 14}$$
$$T = G + H \qquad \text{Theorem 14}$$

3.7 ■ THE COMPLEMENT METHOD

Another technique used to extract a circuit expression from a truth table is the complement method. Using this procedure, we can complement the normal truth table output column so that all zero output levels on the original truth table become ones and are used to derive a normal sum of products expression. The resulting SOP expression represents the complement of the desired truth table output and must be recomplemented to yield the correct expression. This may sound a bit complex, but, in actual practice, it is easy to implement. If a truth table has fewer zero outputs than one outputs, this technique can be advantageous, as was mentioned in the POS section. However, the complement method uses the generally easier SOP process, combining the best of both. Figure 3.31 outlines the process.

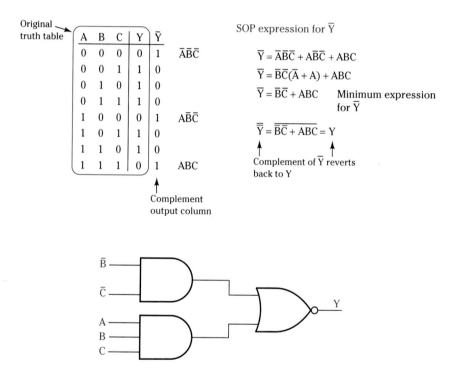

Original truth table

A	B	C	Y	$\bar{Y}$	
0	0	0	0	1	$\bar{A}\bar{B}\bar{C}$
0	0	1	1	0	
0	1	0	1	0	
0	1	1	1	0	
1	0	0	0	1	$A\bar{B}\bar{C}$
1	0	1	1	0	
1	1	0	1	0	
1	1	1	0	1	ABC

Complement output column

SOP expression for $\bar{Y}$

$$\bar{Y} = \bar{A}\bar{B}\bar{C} + A\bar{B}\bar{C} + ABC$$

$$\bar{Y} = \bar{B}\bar{C}(\bar{A} + A) + ABC$$

$$\bar{Y} = \bar{B}\bar{C} + ABC \qquad \text{Minimum expression for } \bar{Y}$$

$$\overline{\bar{Y}} = \overline{\bar{B}\bar{C} + ABC} = Y$$

Complement of $\bar{Y}$ reverts back to Y

Figure 3.31
Complement Method

Figure 3.31 gives a truth table; output Y is the desired output column. To extract an equation from the truth table utilizing the complement method, we simply complement output column Y to $\bar{Y}$. This changes the zero outputs to ones, as illustrated. Fundamental products are obtained for all one outputs in complement column $\bar{Y}$ and are used to produce an SOP expression. This expression, which is for $\bar{Y}$, is also given in Figure 3.31. We follow standard Boolean reduction practices to minimize the expression, of course. Keep in mind that our first priority is not to obtain an expression for $\bar{Y}$, but rather, to use $\bar{Y}$ as a shortcut to the expression for Y. What is the difference between $\bar{Y}$ and Y? Only a level of inversion. Therefore, if we complement the $\bar{Y}$ equation we will obtain Y, which means we are back to an equation representing the original Y column output. Equation simplification is easier, since the complement method produces fewer terms to manipulate.

To summarize, the following steps convert a truth table into a logical expression using the complement method:

- Complement the truth table output column.
- Obtain the fundamental product for each one level output on the newly created complement column.
- OR the fundamental products together.
- Simplify the resulting SOP expression.
- Recomplement the expression.

A burglar alarm is designed to protect a factory. Two doors are equipped with sensors that produce a high level output when open. A timer produces a high level output whenever the factory is closed for the day. Design a circuit to sound an alarm if either of the two doors is opened when the factory is closed. The alarm requires an active low signal.

DESIGN EXAMPLE 3.22

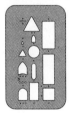

Use the complement method. For simplicity, the timer is A; the doors are B and C; and the alarm output is X.

A	B	C	X	$\overline{X}$	
0	0	0	1	0	
0	0	1	1	0	
0	1	0	1	0	
0	1	1	1	0	
1	0	0	1	0	
1	0	1	0	1	FP = $A\overline{B}C$
1	1	0	0	1	FP = $AB\overline{C}$
1	1	1	0	1	FP = ABC

$$\overline{X} = A\overline{B}C + AB\overline{C} + ABC \qquad \text{Expression for } \overline{X}$$
$$\overline{X} = AC(\overline{B} + B) + AB\overline{C}$$
$$\overline{X} = AC(1) + AB\overline{C}$$
$$\overline{X} = AC + AB\overline{C}$$
$$\overline{X} = A(C + B\overline{C})$$
$$\overline{X} = A(C + B)$$
$$\overline{\overline{X}} = \overline{A(C + B)} \qquad \text{Complement } \overline{X}$$
$$X = \overline{A(C + B)} \qquad \text{X output equation}$$

(Timer) A

(Door 1) B

(Door 2) C

X (Alarm)

The three methods discussed all provide Boolean expressions that produce functional, working logic circuits. Great design flexibility is possible having multiple design methods at your disposal. In the next chapter we will examine advanced techniques to produce equations without some of the Boolean reduction headaches.

3.8 ■ NAND-NAND AND NOR-NOR NETWORKS

Now that we have learned how to extract equations from truth tables, we will invoke DeMorgan's Theorem to provide additional circuit flexibility. Since the three design methods covered earlier create circuits that are either AND-OR or

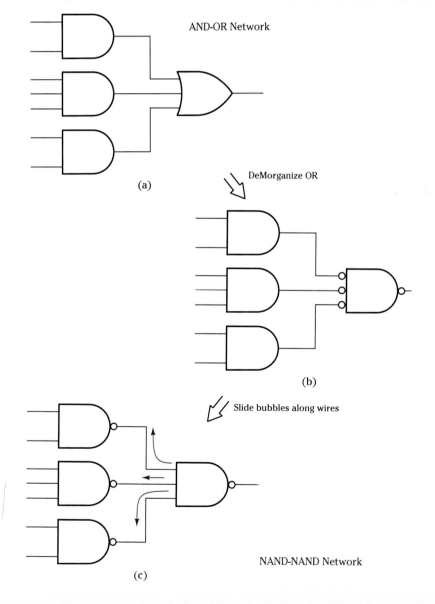

Figure 3.32
Producing a NAND-
NAND Network

OR-AND networks, it is beneficial to investigate how DeMorganization affects these circuit forms. Refer to Figure 3.32.

Figure 3.32a shows an AND-OR network. Using this network, we symbolically DeMorganize the OR gate (Figure 3.32b) and then clean up the diagram by logically "sliding the bubbles" along the OR input wires back to the AND outputs as in Figure 3.32c. A NAND-NAND network results. This is a significant concept because it shows that every AND-OR network can be replaced, gate for gate, with NANDs to produce a NAND-NAND network. Therefore, every SOP expression can be implemented in NAND-NAND circuitry, an especially useful technique. The NAND-NAND implementation may not alter our gate count but it could reduce our design's **package count** since all the required NAND gates could be on a single chip. Reducing the number of physical packages in a design has significant cost and reliability advantages. In addition, some technologies, such as a logic masterslice, are often built exclusively around the NAND gate, with no other logical function used—all logic circuits are NAND based. The NAND gate, as well as the NOR gate, is sometimes referred to as a **universal gate** because any other logical function can be derived from it.

Draw the AND-OR and NAND-NAND circuits for the following equation:

$$Y = A\overline{B}C + \overline{A}BC + \overline{A}\overline{B}C$$

Solution   Draw the AND-OR network directly from the equation. Substitute each AND and OR gate with an equivalent size NAND to create the NAND-NAND network.

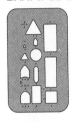

DESIGN EXAMPLE 3.23

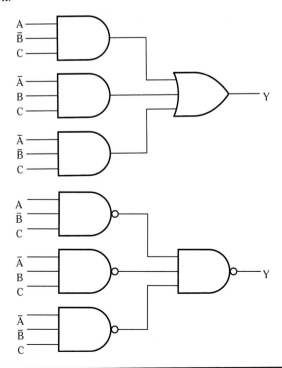

The advantages discussed also hold true for NOR-NOR networks obtained from OR-AND networks. Figure 3.33 shows the conversion process from an OR-AND circuit to a NOR-NOR circuit. Again, as in the NAND-NAND case, each gate in the OR-AND network is replaced, gate by gate, with a NOR.

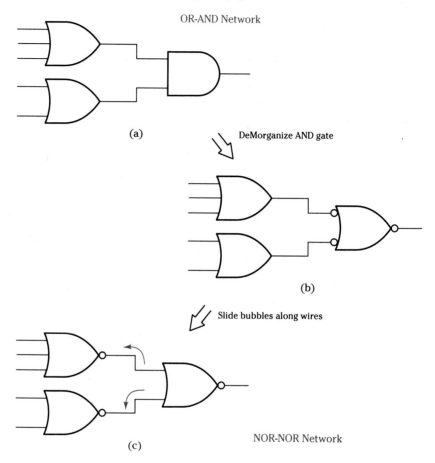

Figure 3.33
Producing a NOR-NOR
Network

DESIGN
EXAMPLE 3.24

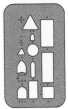

Draw the OR-AND and NOR-NOR networks for the following equation:

$$Y = (\overline{A} + B + \overline{C})(A + \overline{B} + C)(\overline{A} + \overline{B} + \overline{C})$$

Solution   Draw the OR-AND network directly from the equation. Substitute each OR and AND gate with an equivalent size NOR to create the NOR-NOR network.

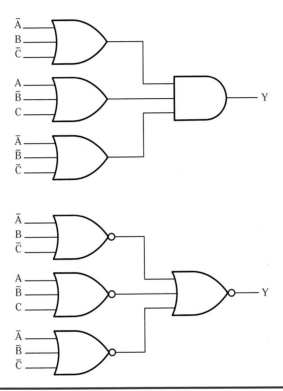

SUMMARY

- Logic circuits can be represented as Boolean algebra expressions. Applying Boolean reduction techniques to the circuit equations minimizes the complexity of digital circuit design.

- Boolean equations are simplified using a small number of basic simplification principles. These principles are often applied through repetitive expanding and factoring of circuit equations.

- DeMorgan's Theorem alters the form of digital logic. Through its use, AND logic is converted into OR logic or OR logic is converted into AND logic.

- DeMorgan's Theorem can be applied at the equation level or at the schematic level.

■ Once a design idea is expressed in truth table form, a logic circuit is assured.

■ Several design techniques, including the sum of products, the product of sums, and the complement methods, are used to convert a truth table into a digital circuit.

■ Using fundamental products, a sum of products expression may be formed. This form of logic produces AND-OR networks; AND-OR networks may be directly replaced with NAND-NAND networks.

■ Using fundamental sums, a product of sums expression may be formed. This form of logic produces OR-AND networks; OR-AND networks may be directly replaced with NOR-NOR networks.

PROBLEMS

Sections 3.1–3.2

1. Expand and reduce the following expressions:
   *(a) $T = (X + Y)(X + Z)$
   (b) $Y = (A + B + C)(A + \bar{B} + C)(\bar{A} + B + \bar{C})$
   (c) $X = (D + E + F + G)(D + F)$
2. Factor and reduce the following expressions:
   *(a) $X = ABC + A\bar{B}C + AB\bar{C} + \bar{A}B\bar{C}$
   (b) $Y = A + AC + AB + BC$
   (c) $T = M\bar{N}O\bar{P} + MNO\bar{P} + \bar{M}NOP + M\bar{N}O\bar{P}$
3. Simplify the following Boolean expressions. Draw the original and reduced circuits:
   *(a) $Q = XY + \bar{X}Y + \bar{X}\bar{Y}$
   (b) $T = (AB + C)(\bar{A} + B)$
   (c) $Y = \bar{J}\bar{K}\bar{L} + \bar{J}\bar{K}L + J\bar{K}\bar{L} + JK\bar{L}$
   (d) $A = WXY + Y(W + XY) + WY(WXY + \bar{W}\bar{X})$
   (e) $U = X + XZ + XY + YZ$

Section 3.3

4. Simplify the following equations using DeMorgan's Theorem when appropriate:
   (a) $X = \bar{A}B\bar{C}D + \bar{A}BC\bar{D} + \overline{(\bar{A} + \bar{B} + \bar{C} + \bar{D})}$
   *(b) $X = \overline{\bar{A}\bar{B}CD}$
   (c) $Y = \overline{(\overline{MN} + O)(N + \bar{O})}$
   (d) $Z = \bar{S}\bar{T}U + \bar{S}T\bar{U} + \bar{S}TU$

---

* See Appendix F: Answers to Selected Problems.

5. Using the following equation:

$$T = \overline{(\overline{UV} + V\overline{X})(\overline{UV} + U\overline{X})}$$

*(a) DeMorganize the NAND function.
(b) DeMorganize the NOR function.
(c) Develop a truth table to prove that all three forms of the expression
   are equivalent.
6. DeMorganize the following circuits:

*(a)

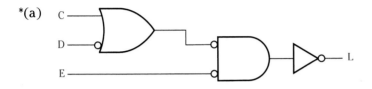

(b)

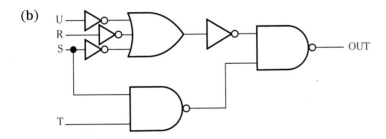

7. For the original and DeMorganized circuits in problem 6(a, b), determine:
   (a) The standard TTL part numbers required.
   (b) The standard CMOS part numbers required.
   (c) Does a gate savings result between the original circuit and De-
       Morganized circuit? For TTL? For CMOS?
   (d) Does the chip package count improve between the original and
       DeMorganized circuits? For TTL? For CMOS?
   (e) Using the chip propagation delay times cited in the manufacturers'
       data books, what differences in the worst case circuit delay exist be-
       tween the original and DeMorganized circuits?

---

* See Appendix F: Answers to Selected Problems.

8. "Read" the following logic symbol/circuits:

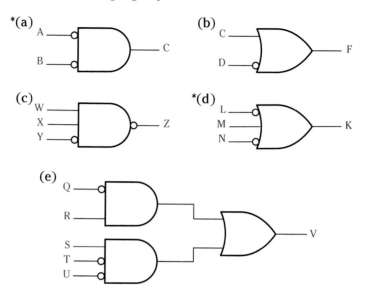

Section 3.4

9. (a) The control circuitry for a computer memory chip performs a "write" or storage operation when two control signals—Write_Enable and Chip_Enable—are both low. A "read" or retrieval operation takes place when Chip_Enable is low, Write_Enable is high, and a third control line—Output_Enable—is low. Create the truth tables, simplified equations, and circuits for active high read and write signals.
   (b) Re-create truth tables, equations, and circuits for active low read and write signals.

Sections 3.5–3.8

10. Draw the following circuits:
    *(a) Output is low when SW1 is high, SW2 is low, SW3 is high.
    (b) Output is high when SW1 is low, SW2 is low, SW3 is high.
    (c) Output is high when SW1 is low, SW2 is low, SW3 is low.
    (d) Output is low if SW1 is low or SW2 is high or SW3 is low.
    (e) Output is high if SW1 is high or SW2 is low or SW3 is low.
    (f) Output is low if SW1 is high or SW2 is high or SW3 is high.

---

* See Appendix F: Answers to Selected Problems.

11. Find the equation and draw the circuit for the following truth tables using the product of sums method:

*(a)

A	B	C	F
0	0	0	0
0	0	1	0
0	1	0	1
0	1	1	1
1	0	0	1
1	0	1	0
1	1	0	0
1	1	1	1

(b)

R	S	T	Q
0	0	0	0
0	0	1	1
0	1	0	1
0	1	1	0
1	0	0	1
1	0	1	0
1	1	0	1
1	1	1	1

(c)

J	K	L	T1
0	0	0	1
0	0	1	1
0	1	0	0
0	1	1	0
1	0	0	1
1	0	1	1
1	1	0	0
1	1	1	0

12. Convert the POS circuits in problem 3.11 to NOR-NOR networks.

13. Use the sum of products method to create the expression and circuit for the following truth tables:

*(a)

C	B	A	X
0	0	0	0
0	0	1	1
0	1	0	1
0	1	1	1
1	0	0	0
1	0	1	1
1	1	0	1
1	1	1	0

(b)

N	O	P	T
0	0	0	1
0	0	1	1
0	1	0	0
0	1	1	0
1	0	0	0
1	0	1	0
1	1	0	1
1	1	1	1

(c)

A	B	C	Y
0	0	0	0
0	0	1	0
0	1	0	0
0	1	1	1
1	0	0	1
1	0	1	0
1	1	0	0
1	1	1	0

(d)

W	X	Y	Z
0	0	0	1
0	0	1	0
0	1	0	1
0	1	1	0
1	0	0	1
1	0	1	0
1	1	0	1
1	1	1	0

(e)

D	C	B	A	S1
0	0	0	0	0
0	0	0	1	0
0	0	1	0	1
0	0	1	1	0
0	1	0	0	0
0	1	0	1	0
0	1	1	0	1
0	1	1	1	1
1	0	0	0	1
1	0	0	1	0
1	0	1	0	0
1	0	1	1	1
1	1	0	0	0
1	1	0	1	0
1	1	1	0	1
1	1	1	1	1

(f)

W	X	Y	Z	Q
0	0	0	0	1
0	0	0	1	1
0	0	1	0	1
0	0	1	1	1
0	1	0	0	0
0	1	0	1	0
0	1	1	0	0
0	1	1	1	0
1	0	0	0	1
1	0	0	1	1
1	0	1	0	1
1	0	1	1	1
1	1	0	0	0
1	1	0	1	0
1	1	1	0	0
1	1	1	1	0

* See Appendix F: Answers to Selected Problems.

14. Repeat the problems in 13 above using the complement method.
15. Convert the SOP circuits from problem 13 to NAND-NAND networks.
*16. A circuit is required that detects the output of four temperature sensors. Each sensor produces a high output when the temperature exceeds a preset limit; otherwise the output is low. Design a circuit that indicates whenever any three sensors have exceeded their temperature limits.
17. A microwave oven is controlled by a logic circuit to:
    (a) Turn on the magnetron when the door is closed, when a time has been selected, and when the "GO" button is pushed.
    (b) Turn off the magnetron when the time has expired or the door is opened.

    Design the circuits to produce two separate signals—"Magnetron On" and "Magnetron Off." Assume the following active high signals are present: door closed, time selected, GO pushed. Therefore, a low level on any of the three signal lines indicates the door is opened, time has expired, or GO is not active.
18. The warm air circulation fan for two solar heated rooms (room B and room C) runs under the following conditions:
    (a) If the sun is out, the inside temperature of room B > 75°, the inside temperature of room C < 75°, and the furnace is off.
    (b) If the sun is out, the inside temperature of room B < 75°, the inside temperature of room C > 75°, and the furnace is off.
    (c) If the sun is not out, both rooms B and C are < 75°, and the furnace is on.

    Devise an equation and circuit for the Fan_on signal. Assume sensor signals are active high when the sun is out, individual room temperatures are above 75°, and the furnace is on.

---

* See Appendix F: Answers to Selected Problems.

CHAPTER 4

CIRCUIT SIMPLIFICATION TECHNIQUES

OBJECTIVES

To explore circuit simplification techniques that are alternatives to Boolean reduction.

To understand and utilize the Karnaugh map for circuit simplification.

To introduce the notion of the "don't care state."

To learn and understand the Quine–McCluskey reduction method and its advantages and disadvantages as compared to the Karnaugh map method.

To investigate the advantages of computer reduction and computer simulation for the digital designer.

PREVIEW

This chapter discusses several techniques for simplifying circuit equations. These methods rely on graphics, charts, and computer software to carry out the reduction task and are often much easier to apply to equations than Boolean algebra reduction.

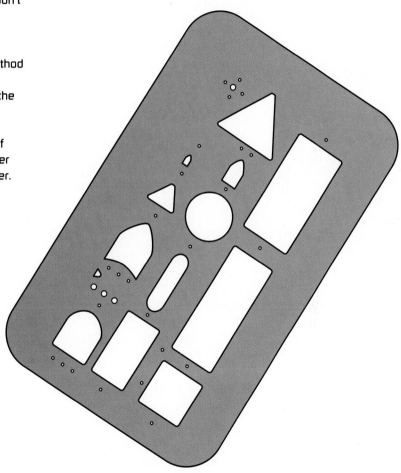

4.1 ■ THE KARNAUGH MAP

We discovered in the previous chapter that although equations developed from a truth table are functionally correct, they are not simplified. Hence, logic reduction is necessary. Our reduction tools have been Boolean theorems and mathematical rigor. Unfortunately, many people are not fluent mathematicians, so Boolean algebra is sometimes a hard road to follow. In addition, Boolean reduction is an error-prone process because of the many variables requiring mathematical manipulation. Even when the algebraic reduction process is followed correctly, a minimum equation may not be obtained. Consider the equations in Figure 4.1. Although both equations are equivalent, one is simpler than the other, and it is not immediately apparent how the lengthy equation can be reduced.

$$T1 = C\bar{D} + CE + DE \qquad\qquad T2 = C\bar{D} + DE$$

C	D	E	$\bar{D}$	$C\bar{D}$	CE	DE	T1	T2
0	0	0	1	0	0	0	0	0
0	0	1	1	0	0	0	0	0
0	1	0	0	0	0	0	0	0
0	1	1	0	0	0	1	1	1
1	0	0	1	1	0	0	1	1
1	0	1	1	1	1	0	1	1
1	1	0	0	0	0	0	0	0
1	1	1	0	0	1	1	1	1

Equal

Figure 4.1
Difficult Boolean
Reduction

$T1 = C\bar{D} + CE + DE$	Original equation
$T1 = C\bar{D} + CE[D + \bar{D}] + DE$	AND second term with $1[D + \bar{D} = 1]$
$T1 = C\bar{D} + CED + CE\bar{D} + DE$	
$T1 = C\bar{D}[1 + E] + DE[C + 1]$	Reduction process
$T1 = C\bar{D}[1] + DE[1]$	
$T1 = C\bar{D} + DE$	Equation T1 = Equation T2

The Boolean algebra illustrated in Figure 4.1 shows a mathematical technique that can help the reduction process. The term CE is ANDed with $(D + \bar{D})$. This is equivalent to ANDing the term with one. Mathematically, the meaning of the expression remains the same, but the reduction sequence continues toward a simpler equation with the assistance of the additional term. Obviously, some mathematical insight is required to achieve this reduction, and for those who find the mathematical approach readily apparent, situations like this are no barrier to a simplified equation. In lieu of mathematical approaches, an alternate technique can be utilized to speed up the reduction process, minimize the chance of errors, and guarantee a simplified expression. The **Karnaugh map** is one such alternate technique and is often preferred to an algebraic approach. The Karnaugh map (K-map) is a powerful graphical method used for straightforward reduction and as a design aid in many other facets of advanced logic design.

A K-map is drawn as the first step in the process. The contents of a truth table are placed on the K-map in a way that promotes logic simplification. Since truth tables vary in length, it should not be surprising that K-maps also vary in size. Two-, three-, and four-variable maps are used in practical design. The mapping technique becomes rather complex with more than four variables and requires three-dimensional mapping. In fact, a paper* written by M. Karnaugh on the subject describes a three-dimensional plastic frame used with roulette chips, which assists with large map reduction. Even Mr. Karnaugh concludes that "mental gymnastics" are required to reduce sizable maps. Our discussions will be confined to maps of four variables or less. For larger reductions we will use other techniques.

Illustrated in Figure 4.2 are the layouts for two-, three-, and four-variable K-maps. The maps consist of an arrangement of squares. Each square will have placed into it one of the logic levels from the truth table output column. The number of squares on the map relates to the number of input variables. For instance, a three-variable map has $2^3 = 8$ squares, just as a three-variable truth table has eight input combinations. Since the K-map is simply an alternative representation of a truth table, a one-to-one correspondence between K-map squares and truth table input combinations always exists.

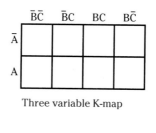

Three variable K-map

Two variable K-map

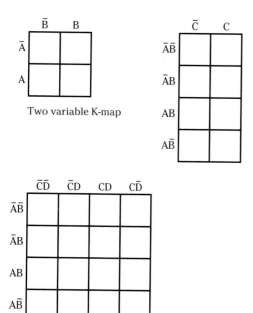

Four variable K-map

Figure 4.2
K-Map Layouts

* M. Karnaugh, "The Map Method for Synthesis of Combinatorial Logic Circuits," Transactions AIEE Communications and Electronics Vol. 72, Pt. 1, Nov. 1953.

Variables listed along the sides of the K-map also represent the input variable combinations. The map identifies and eliminates unnecessary variables, so it is crucial to represent the input variables in a particular order. **Gray code** order, the order used to list the variables, means that from one adjacent variable combination to another only a single variable changes level. For example, $\overline{A}\overline{B}$ occupies the top left side of the map. The variable grouping immediately below is $\overline{A}B$. Only the B variable changes from complement notation to true notation. Checking further, you will notice that this characteristic is maintained throughout the map. This order is essential for correct reduction.

Figure 4.3 indicates how K-map squares and truth tables relate. As mentioned earlier, each square corresponds to a specific truth table combination. This figure shows that every map square in a four-variable map is identified by reading two variables along the left and two along the top to determine the input combination that corresponds to a particular square. On a typical map, given in Figure 4.4, each map square contains the truth table output binary

	$\overline{C}\overline{D}$	$\overline{C}D$	CD	$C\overline{D}$
$\overline{A}\overline{B}$	$\overline{A}\overline{B}\overline{C}\overline{D}$	$\overline{A}\overline{B}\overline{C}D$	$\overline{A}\overline{B}CD$	$\overline{A}\overline{B}C\overline{D}$
$\overline{A}B$	$\overline{A}B\overline{C}\overline{D}$	$\overline{A}B\overline{C}D$	$\overline{A}BCD$	$\overline{A}BC\overline{D}$
AB	$AB\overline{C}\overline{D}$	$AB\overline{C}D$	$ABCD$	$ABC\overline{D}$
$A\overline{B}$	$A\overline{B}\overline{C}\overline{D}$	$A\overline{B}\overline{C}D$	$A\overline{B}CD$	$A\overline{B}C\overline{D}$

Figure 4.3
K-Map Variable
Assignments

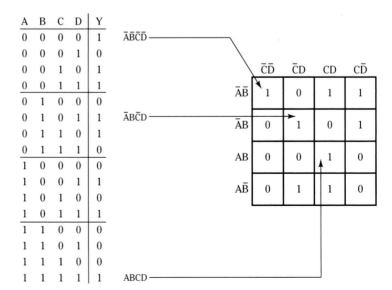

A	B	C	D	Y
0	0	0	0	1
0	0	0	1	0
0	0	1	0	1
0	0	1	1	1
0	1	0	0	0
0	1	0	1	1
0	1	1	0	1
0	1	1	1	0
1	0	0	0	0
1	0	0	1	1
1	0	1	0	0
1	0	1	1	1
1	1	0	0	0
1	1	0	1	0
1	1	1	0	0
1	1	1	1	1

Figure 4.4
Assigning Truth Table
Values to the K-Map

value assigned according to the input combination producing it. In actual practice, all the one level outputs are first placed on the map, with zero outputs filling the remaining vacant squares. Regardless of map size, the following steps are always taken:

1. Draw the appropriate size map based on truth table size.
2. Assign variables to the sides of the map following Gray code order.
3. Assign truth table output levels to the squares.

Draw the K-map used to reduce a truth table containing variables X_1, X_2, X_3, and X_4. Identify the K-map squares representing:
(a) $\overline{X}_1 X_2 \overline{X}_3 X_4$
(b) $\overline{X}_1 \overline{X}_2 \overline{X}_3 \overline{X}_4$
(c) $X_1 \overline{X}_2 \overline{X}_3 X_4$

DESIGN EXAMPLE 4.1

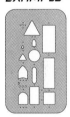

Solution Assign variables to the sides of the map following Gray code order. Since the truth table contains four variables, the K-map will be composed of 16 squares.

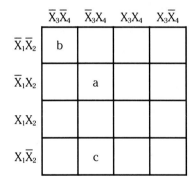

	$\overline{X}_3\overline{X}_4$	$\overline{X}_3 X_4$	$X_3 X_4$	$X_3\overline{X}_4$
$\overline{X}_1\overline{X}_2$	b			
$\overline{X}_1 X_2$		a		
$X_1 X_2$				
$X_1\overline{X}_2$		c		

Fill in a K-map with the following truth table data:

DESIGN EXAMPLE 4.2

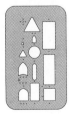

R	S	T	P
0	0	0	1
0	0	1	0
0	1	0	0
0	1	1	1
1	0	0	0
1	0	1	0
1	1	0	1
1	1	1	1

Solution   Draw a three-variable K-map. Fill the K-map squares with the truth output column ones and zeroes.

	$\overline{T}$	T
$\overline{R}\,\overline{S}$	1	0
$\overline{R}S$	0	1
RS	1	1
$R\overline{S}$	0	0

After filling the map, circuit simplification can begin. All ones on the map in groups of two, four, or eight and horizontally or vertically aligned are identified. A group of two ones, as shown in Figure 4.5, is known as a **pair**. Notice how the pairs are encircled for identification. A pair leads to the elimination of one variable from the logic equation because a pair identifies two product terms where only a single variable changes level from one term to the next. In essence, the K-map graphically puts reduction Theorem 14 ($X\overline{Y} + XY = X$) to work for us. Figure 4.6 illustrates how to obtain the reduced equation.

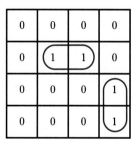

Figure 4.5
Encircling Pairs on a K-Map

Variable C changes from $\overline{C}$ to C

	$\overline{C}\overline{D}$	$\overline{C}D$	CD	$C\overline{D}$
$\overline{A}\overline{B}$	0	0	0	0
$\overline{A}B$	0	0	0	0
AB	0	1	1	0
$A\overline{B}$	0	0	0	0

Variables → AB
AB remain constant

$$Y = AB\overline{C}D + ABCD$$
$$Y = ABD(\overline{C} + C)$$
$$Y = ABD(1)$$
$$Y = ABD$$

Boolean reduction

Figure 4.6
Reducing a K-Map Pair

From K-map:
    $Y = ABD$ Variable C is eliminated since it changes from $\overline{C}$ to C.

The ones in the example shown in Figure 4.6 represent $AB\overline{C}D$ and ABCD. Utilizing a standard sum of products approach directly from the truth table, these two fundamental products would be ORed together and reduced, as shown, yielding Y = ABD and eliminating variable C in the process. The K-map provides the same result. Using the K-map, we isolate the variable that changes. Variables A and B along the vertical side of the map, for example, are associated with both ones in the pair. This means that variables A and B are a required part of the logical expression and cannot be eliminated. Along the top of the map, the alignment of ones shows that the first one in the pair is associated with $\overline{C}D$, whereas the second is associated with CD. It is apparent that variable C changes from complement form to true form when moving from input combination $AB\overline{C}D$ to ABCD. This variable is unnecessary in the resulting logic equation. The variables retained, A, B, and D, are ANDed together, forming the expression Y = ABD. When several pairs are present on the same map, the product terms they represent are ORed to form a sum of products (SOP) expression. Incidentally, when it is impossible to enclose a lone one on the map with any other one, no simplification exists for that term, as illustrated in Figure 4.7.

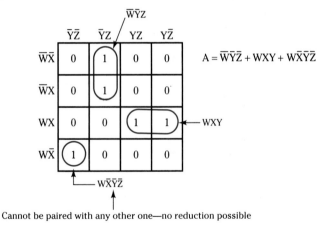

$A = \overline{W}\overline{Y}\overline{Z} + WXY + W\overline{X}\overline{Y}\overline{Z}$

Cannot be paired with any other one—no reduction possible

Figure 4.7
Creating an Equation from the K-Map

Reduce the following K-map by identifying any reductions that exist:

	$\overline{Y}\overline{Z}$	$\overline{Y}Z$	YZ	$Y\overline{Z}$
$\overline{W}\overline{X}$	0	1	0	0
$\overline{W}X$	0	0	1	1
WX	1	1	0	0
$W\overline{X}$	0	0	0	0

Equation $= \overline{W}XY + WX\overline{Y} + \overline{W}\overline{X}\overline{Y}Z$

DESIGN EXAMPLE 4.3

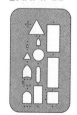

Solution
(a) Circle pairs of ones that are aligned horizontally or vertically.
(b) Determine which variable can be eliminated from the pair—the variable that changes from true to complement form or from complement to true form when moving from one square to the next.
(c) OR all reduced terms together.

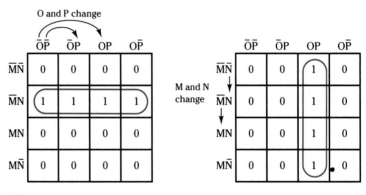

$Y = \overline{M}N$ Variables O and P eliminated $Y = OP$ Variables M and N eliminated

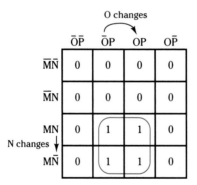

$Y = MP$ Variables N and O eliminated

Figure 4.8
K-Map Quads

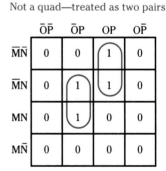

Not a quad—treated as two pairs

$Y = N\overline{O}P + \overline{M}OP$

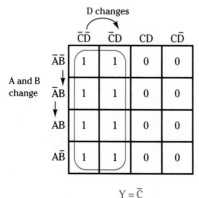

$Y = \overline{C}$

Figure 4.9
K-Map Octets

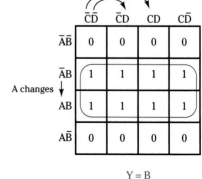

$Y = B$

Clearly, grouping ones together is beneficial. The advantages grow with larger groupings. Grouping four ones, called a **quad**, eliminates two variables, whereas a grouping of eight, called an **octet**, eliminates three variables. Figure 4.8 shows several quads and how to acquire a reduced logic expression. As with the pair, quads require strict horizontal and vertical alignment of the ones. Figure 4.9 shows the reduction process for octets. In order to ensure a minimum solution, always enclose the maximum number of ones.

Identify the quads and their corresponding reduced terms in the following K-map:

DESIGN EXAMPLE 4.4

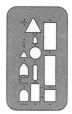

	$\overline{Y}\overline{Z}$	$\overline{Y}Z$	YZ	$Y\overline{Z}$
$\overline{W}\overline{X}$	0	1	0	1
$\overline{W}X$	0	1	0	1
WX	0	1	0	1
$W\overline{X}$	0	1	0	1

Two quads exist on this map.

$$\text{Equation} = \overline{Y}Z + Y\overline{Z}$$

Solution Quads are groups of four ones aligned in a horizontal or vertical position. Circle these and determine the two variables that change within the grouping.

Reduce the octet shown on the following K-map. The K-map represents equation Y:

DESIGN EXAMPLE 4.5

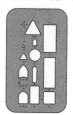

	$\overline{E}\overline{F}$	$\overline{E}F$	EF	$E\overline{F}$
$\overline{C}\overline{D}$	1	1	1	1
$\overline{C}D$	1	1	1	1
CD	0	0	0	0
$C\overline{D}$	0	0	0	0

$$Y = \overline{C}$$

Solution   Encircle the octet. Determine the reduced equation by eliminating the three variables that change within the grouping.

This technique seems very straightforward and easy. However, some variations to basic grouping must be recognized to obtain the simplest logic expression. **Overlapping** is the first variation, meaning that ones on the map already encircled may be shared with other ones. Consider Figure 4.10, for instance. In this example creating two pairs from three ones occurs simply by including a one in both pairs. Figure 4.11 shows another overlapping example. Here a quad and a pair are formed by sharing a common one. Notice that sharing ones still requires the proper horizontal and vertical alignment mentioned earlier. Using overlapping to enclose "stray" ones within pairs, quads, or octets is essential to obtain a minimum reduction.

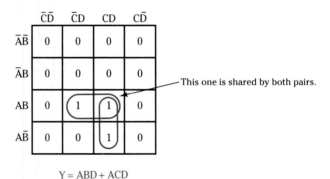

Figure 4.10
Overlapping

$$Y = ABD + ACD$$

This one is shared by both pairs.

Figure 4.11
Overlapping a Pair and a Quad

$$Z = EG + \bar{D}\bar{F}G$$

This one is shared by the pair and quad.

Consider Figure 4.12 as an example of what overlapping is not. Of the three quads enclosed, one is marked as unnecessary. The marked quad is unnecessary because the ones within this quad are already completely enclosed by the others—none of the ones is a stray. The useless quad adds nothing to the logic equation even though a term can be written for it; the quad represents a redundant term. Enclosing ones that are already encircled is not helpful unless a nonenclosed one is included as well. In other words, overlapping is permissible when at least one previously noncircled K-map one is included in the overlapping.

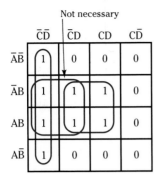

Not necessary

	$\bar{C}\bar{D}$	$\bar{C}D$	CD	$C\bar{D}$
$\bar{A}\bar{B}$	1	0	0	0
$\bar{A}B$	1	1	1	0
AB	1	1	1	0
$A\bar{B}$	1	0	0	0

$Y_1 = \bar{C}\bar{D} + BD + B\bar{C}$
$Y_2 = \bar{C}\bar{D} + BD$ } Equivalent

A	B	C	D	$\bar{C}$	$\bar{D}$	$\bar{C}\bar{D}$	BD	$B\bar{C}$	Y_1	Y_2
0	0	0	0	1	1	1	0	0	1	1
0	0	0	1	1	0	0	0	0	0	0
0	0	1	0	0	1	0	0	0	0	0
0	0	1	1	0	0	0	0	0	0	0
0	1	0	0	1	1	1	0	1	1	1
0	1	0	1	1	0	0	1	1	1	1
0	1	1	0	0	1	0	0	0	0	0
0	1	1	1	0	0	0	1	0	1	1
1	0	0	0	1	1	1	0	0	1	1
1	0	0	1	1	0	0	0	0	0	0
1	0	1	0	0	1	0	0	0	0	0
1	0	1	1	0	0	0	0	0	0	0
1	1	0	0	1	1	1	0	1	1	1
1	1	0	1	1	0	0	1	1	1	1
1	1	1	0	0	1	0	0	0	0	0
1	1	1	1	0	0	0	1	0	1	1

Figure 4.12
Proof of Unnecessary
Overlapping

Determine the minimum equation for Y as represented by the following K-map:

DESIGN
EXAMPLE 4.6

	$\bar{O}\bar{P}$	$\bar{O}P$	OP	$O\bar{P}$
$\bar{M}\bar{N}$	0	1	0	0
$\bar{M}N$	1	1	1	1
MN	0	0	1	1
$M\bar{N}$	0	0	0	0

$$Y = \bar{M}N + NO + \bar{M}\bar{O}P$$

Solution Use overlap to create two quads and a pair.

K-Map Technical Details

Formal definitions governing Karnaugh mapping mathematically explain the process and ensure correctness throughout the procedure. We will explore the theory a bit to clarify how to carry out the K-map process in an error-free manner.

Ones on a K-map are called **implicants**—the values involved in the logic equation. The primary goal of mapping is enclosing implicants in the largest enclosures possible, called **prime implicants.** If all ones are enclosed as prime implicants in the proper fashion, a minimum logic expression is assured.

As we have already seen, enclosing ones is not a problem, but enclosing ones so that prime implicants are **essential prime implicants** requires insight. An essential prime implicant is a grouping of ones enclosed by a single prime implicant. An essential prime implicant is a prime implicant, but the term

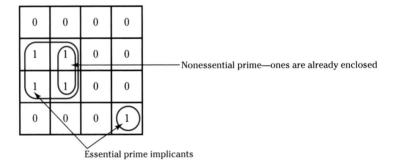

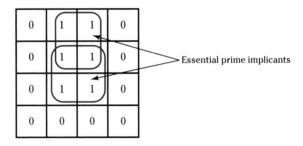

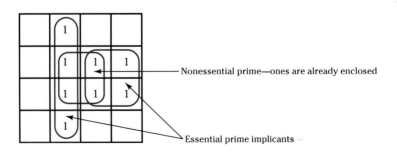

Figure 4.13
Prime Implicant
Details

means that the ones enclosed within the essential prime implicant are not **all** included within other enclosures. This doesn't prevent overlapping; it simply indicates that enclosing previously enclosed ones is not beneficial. When the map is made up entirely of essential prime implicants, the minimum solution is found. Figure 4.13 diagrams these terms.

4.2 ■ ROLLING K-MAP EDGES

One final K-map form that must be recognized is **rolling** the K-map edges together. The two vertical edges of a K-map are the same edge. This is also true for the map's horizontal edges—the top and bottom of the map are the same edge. In both cases Gray code order is maintained from one side of the map to the other. (Along the top the variables change from $\overline{C}\overline{D}$ to $C\overline{D}$—only C changes—and along the sides $\overline{A}B$ changes to $A\overline{B}$.) Therefore, moving from one side of the map to the other is the same as moving from one square to another. For example, a one along the left edge of the map (e.g., $AB\overline{C}\overline{D}$) is logically placed next to the one in the same horizontal row along the right side of the map ($AB\overline{C}D$), as shown in Figure 4.14.

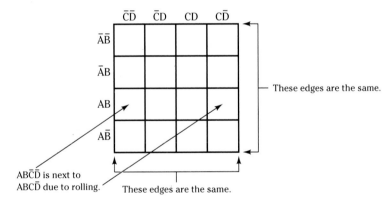

$AB\overline{C}\overline{D}$ is next to
$AB\overline{C}D$ due to rolling.   These edges are the same.

These edges are the same.

Figure 4.14
K-Map Edges

Since this arrangement confirms that any ones residing along opposite edges are logically next to each other, ones so aligned may be enclosed by rolling the edges of the map together. The rolling technique, like overlapping, is used to extract a minimum logic expression from a K-map. An example is given in Figure 4.15. The two pairs along the left and right map edges are rolled

	$\overline{J}\overline{K}$	$\overline{J}K$	JK	$J\overline{K}$
$\overline{H}\overline{I}$	0	0	0	0
$\overline{H}I$	1	0	0	1
HI	1	0	0	1
$H\overline{I}$	0	0	0	0

$Y = I\overline{K}$ With rolling a minimum solution is obtained.

$Y = IJ\overline{K} + I\overline{J}\overline{K}$ Treated as two pairs without rolling—reduction is required.

$Y = I\overline{K}(\overline{J} + J)$
$Y = I\overline{K}(1)$
$Y = I\overline{K}$

Figure 4.15
Rolling K-Map Edges

together to form a quad. If treated only as individual pairs, further Boolean reduction would be evident, indicating that a minimum solution was not found. Rolling encloses the four ones to form an essential prime implicant. Therefore, complete K-map reduction involves the following steps:

1. Encircle ones in octets, quads, or pairs, if possible.
2. Use overlap and rolling whenever possible to attain the maximum size enclosure.
3. Make sure to enclose all ones within essential prime implicants.

DESIGN EXAMPLE 4.7

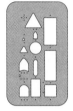

Determine the minimum equation for Z as represented by the following K-map:

	$\overline{F}\overline{G}$	$\overline{F}G$	FG	$F\overline{G}$
$\overline{D}\overline{E}$	0	1	1	0
$\overline{D}E$	1	0	0	1
DE	0	0	0	0
$D\overline{E}$	0	1	1	0

$$Z = \overline{E}G + \overline{D}E\overline{G}$$

Solution   Use rolling to create a quad and a pair.

FOR YOUR INFORMATION

Here's an alternate way to set up a K-map:

		B		$\overline{B}$		
	$ABCD$	$ABC\overline{D}$	$A\overline{B}C\overline{D}$	$A\overline{B}CD$	C
A					
	$AB\overline{C}D$	$AB\overline{C}\overline{D}$	$A\overline{B}\overline{C}\overline{D}$	$A\overline{B}\overline{C}D$	
					$\overline{C}$
	$\overline{A}B\overline{C}D$	$\overline{A}B\overline{C}\overline{D}$	$\overline{A}\overline{B}\overline{C}\overline{D}$	$\overline{A}\overline{B}\overline{C}D$	
$\overline{A}$					
	$\overline{A}BCD$	$\overline{A}BC\overline{D}$	$\overline{A}\overline{B}C\overline{D}$	$\overline{A}\overline{B}CD$	C
	D	$\overline{D}$	D		

Determine the minimum expression for the following truth table using the Karnaugh map method.

A	B	C	D	Y
0	0	0	0	1
0	0	0	1	0
0	0	1	0	1
0	0	1	1	1
0	1	0	0	1
0	1	0	1	0
0	1	1	0	1
0	1	1	1	1
1	0	0	0	1
1	0	0	1	0
1	0	1	0	1
1	0	1	1	1
1	1	0	0	0
1	1	0	1	0
1	1	1	0	1
1	1	1	1	1

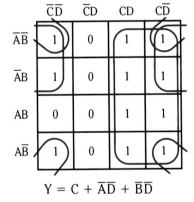

1 Octet = C
1 quad with roll
and overlap = $\overline{A}\overline{D}$
1 quad—all four corners
rolled together = $\overline{B}\overline{D}$

$$Y = C + \overline{A}\overline{D} + \overline{B}\overline{D}$$

4.3 ■ HOW DON'T CARE STATES LEAD TO ADDITIONAL REDUCTION

An interesting facet in some logic designs is that certain input combinations may be unnecessary. That is, proper circuit operation is not enhanced or hindered by these combinations. Refer to the example given in Figure 4.16. In this circuit only one switch may be on (high level) at a time. Any truth table input combination containing two or three ones is meaningless for this particular circuit design, since these combinations imply that several switches are on simultaneously. Such situations occur often in logic design. For many technical reasons, some input conditions can be ignored. These states are identified

S_1 S_2 S_3	Light (with don't cares)	Light (without don't cares)
0 0 0	0	0
0 0 1	1	1
0 1 0	1	1
0 1 1	d	0
1 0 0	1	1
1 0 1	d	0
1 1 0	d	0
1 1 1	d	0

Figure 4.16
Don't Care State
Reduction Comparison

K-map without don't cares
Light = $\bar{S}_1 \bar{S}_2 S_3 + \bar{S}_1 S_2 \bar{S}_3 + S_1 \bar{S}_2 \bar{S}_3$

K-map with don't cares
Light = $S_1 + S_2 + S_3$

as **don't care states** or **don't care conditions.** Circuit reduction can occur by taking advantage of don't care states. The logic designer treats these conditions as designer wildcards, since they are of no concern and cannot affect circuit performance or reliability. However, since the don't cares represent unused input combinations, they can often be grouped with valid input combinations to generate a logic reduction. Utilizing don't care conditions is easy with a K-map.

According to the truth table in Figure 4.16, a zero output indicates that the light is off, whereas a one indicates that the light is on. The circuit under design will turn on the light when one of the switches (and only one) is on. The input combinations treated as don't care states have the letter "d" assigned in the output column, since we don't care if these combinations produce a one or zero. These input combinations, 011, 101, 110, and 111, represent conditions when more than one switch is on—the don't cares. Once the designer transfers the truth table ones, zeros, and d's to a K-map, he or she treats the d's as ones or zeros—whatever is more advantageous. Naturally, combining don't care conditions with ones is most beneficial since it can lead to a maximum-sized enclosure. In this example all d's are considered ones to produce the largest groupings possible (quads). Don't care states treated as zeros are ignored during the mapping process. In the switch example the don't care states contribute to a simple circuit design—a 3-input OR gate. The OR gate produces a one output for any switch that is on. If two or three switches were on simultaneously, an output would also occur. But this cannot happen based on our initial design assumption that only one switch is on at a time.

Compare the OR gate equation and associated K-map to the truth table, K-map, and equation (also in Figure 4.16) obtained without using don't care states. This equation is more complicated, and consequently requires additional parts. However, in using the more complicated circuit, the light remains off when two or more switches are on simultaneously. We did not want to design the circuit to operate this way, but these circuit differences indicate the importance of understanding when a don't care condition is appropriate. If don't care states can be assured, then simpler circuitry results.

Use a K-map to determine the reduced logic equation for the following truth table:

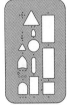

DESIGN EXAMPLE 4.9

A	B	C	D	Y
0	0	0	0	0
0	0	0	1	0
0	0	1	0	0
0	0	1	1	0
0	1	0	0	0
0	1	0	1	1
0	1	1	0	0
0	1	1	1	0
1	0	0	0	0
1	0	0	1	1
1	0	1	0	d
1	0	1	1	d
1	1	0	0	d
1	1	0	1	d
1	1	1	0	d
1	1	1	1	d

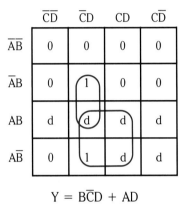

$$Y = B\overline{C}D + AD$$

Solution Assign truth table output values, including don't care states, to the K-map. Find the reduced equation using don't care states as ones wherever possible.

4.4 ■ THE QUINE–McCLUSKEY SIMPLIFICATION METHOD

Our K-map discussion pointed out the reduction difficulties occurring when designing with more than four input variables—multiple K-maps are required. Obtaining a minimim circuit solution from such complex maps rests solely with the designer's ability to recognize patterns not only on a single map, but also on several maps at a time. Fortunately, since the technique is not limited by the number of variables, the Quine–McCluskey method is a practical alternative. However, pitfalls exist.

The Quine–McCluskey procedure is time-consuming and repetitious, and care must be taken to prevent errors. Generally, designers prefer K-maps for designs with four variables or less; the Quine–McCluskey method is used for larger designs.

Using the Quine–McCluskey method generally implies that we will be facing truth tables of significant size. Rather than fill pages of paper with a single truth table, we will use a short-hand notation to represent input combinations yielding an output of one. This notation is shown in Figure 4.17. The equation $Y = \Sigma(1, 2, 6, 10, 13, 19, 24, 29, 30)$ indicates, with decimal numbers, truth table outputs equal to one. The number 30, for example, represents the five binary bits 11110. Thus, our notation in this example refers to a five-variable truth table, with each output stemming from a five-variable input combination. Thus, the decimal one in the short-hand notation represents that the binary input combination 00001 produces a high level output.

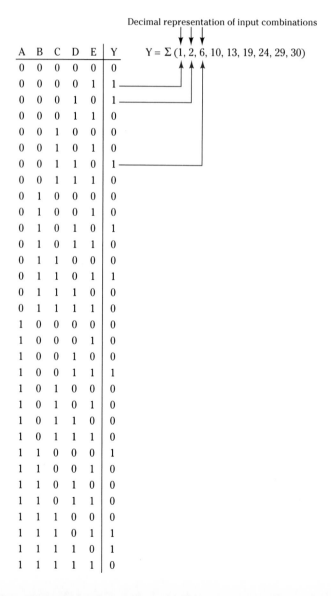

A	B	C	D	E	Y
0	0	0	0	0	0
0	0	0	0	1	1
0	0	0	1	0	1
0	0	0	1	1	0
0	0	1	0	0	0
0	0	1	0	1	0
0	0	1	1	0	1
0	0	1	1	1	0
0	1	0	0	0	0
0	1	0	0	1	0
0	1	0	1	0	1
0	1	0	1	1	0
0	1	1	0	0	0
0	1	1	0	1	1
0	1	1	1	0	0
0	1	1	1	1	0
1	0	0	0	0	0
1	0	0	0	1	0
1	0	0	1	0	0
1	0	0	1	1	1
1	0	1	0	0	0
1	0	1	0	1	0
1	0	1	1	0	0
1	0	1	1	1	0
1	1	0	0	0	1
1	1	0	0	1	0
1	1	0	1	0	0
1	1	0	1	1	0
1	1	1	0	0	0
1	1	1	0	1	1
1	1	1	1	0	1
1	1	1	1	1	0

Decimal representation of input combinations

$Y = \Sigma\,(1, 2, 6, 10, 13, 19, 24, 29, 30)$

Figure 4.17
Short-Hand Truth
Table Notation

Draw the truth table representing:

$$T = \Sigma(3, 5, 9, 12)$$

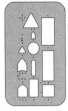

Solution Since 12 is the largest number in the list and is 1100 in binary, a four-variable truth table is required. Assign ones on the table for input combinations 0011 (3), 0101 (5), 1001 (9), and 1100 (12).

A	B	C	D	Y
0	0	0	0	0
0	0	0	1	0
0	0	1	0	0
0	0	1	1	1
0	1	0	0	0
0	1	0	1	1
0	1	1	0	0
0	1	1	1	0
1	0	0	0	0
1	0	0	1	1
1	0	1	0	0
1	0	1	1	0
1	1	0	0	1
1	1	0	1	0
1	1	1	0	0
1	1	1	1	0

The Quine–McCluskey method is carried out using various tables, which are drawn to eliminate unnecessary circuit variables methodically. The number of tables used (we call these tables **reduction tables**) depends on the complexity of the design and leads the designer to the prime implicants for the circuit. As with the K-map process, the designer searches for a minimum number of prime implicants—the essential prime implicants. A final table (called the prime implicant table) clearly points these out.

The reduction tables lead to logic simplification by implementing some of our favorite logic theorems. Specifically, $X\overline{Y} + XY = X(\overline{Y} + Y) = X(1) = X$ (Theorem 14) is carried out through visual inspection of input variable combinations. This is how it is done. Take, for example, the input combinations A,B,C,D = 0100, and A,B,C,D = 0101. Within these two binary combinations, only the least significant bit (LSB) differs. The two groups combine into one on a reduction table by writing them as 010_. The dash identifies the bit position differing in value. This notes that the variable associated with the dash (D in this case) is unnecessary. We can verify the accuracy of this method by reducing the SOP expression associated with the combinations 0100 and 0101— $\overline{A}B\overline{C}\overline{D} + \overline{A}B\overline{C}D$. Thus, $\overline{A}B\overline{C}\overline{D} + \overline{A}B\overline{C}D = \overline{A}B\overline{C}(\overline{D} + D) = \overline{A}B\overline{C}(1) = \overline{A}B\overline{C}$, or 010_. This process is carried out many times in the Quine–McCluskey reduction and is the same process utilized on a K-map.

The Initial Listing Table

Now we will formalize the procedure using the truth table and accompanying tables in Figure 4.18. We convert the truth table or short-hand notation for the desired circuit into an **initial listing table**, which lists only the input combinations producing a one level output. However, we do not indiscriminately list the input combinations. Instead, we order the combinations according to the number of ones contained within the individual combinations. Thus, all combinations with a single one occur first. Those combinations with two ones are second, and so on. The input combination's equivalent decimal number is also noted on the list for convenience. Lastly, we write down the input variable letters along the top of the list aligned with the data on the table.

A	B	C	D	Y
0	0	0	0	1
0	0	0	1	0
0	0	1	0	0
0	0	1	1	0
0	1	0	0	1
0	1	0	1	1
0	1	1	0	0
0	1	1	1	1
1	0	0	0	1
1	0	0	1	0
1	0	1	0	0
1	0	1	1	0
1	1	0	0	1
1	1	0	1	1
1	1	1	0	0
1	1	1	1	1

$Y = \Sigma\,(0, 4, 5, 7, 8, 12, 13, 15)$

Initial Listing Table

	A	B	C	D	← Input variables
0	0	0	0	0	✓
4	0	1	0	0	✓
8	1	0	0	0	✓
5	0	1	0	1	✓
12	1	1	0	0	✓
7	0	1	1	1	✓
13	1	1	0	1	✓
15	1	1	1	1	✓

Grouping of single ones { 4, 8

Grouping of three ones { 7, 13, 15

Decimal numbers ⌐

Checkmarks indicate a comparison when creating the first reduction table (Figure 4.19).

Figure 4.18
Forming the Initial
Listing Table

For the example given the first entry is all zeros. Separate this combination from the other with a horizontal line. Next, list any combinations containing single ones. This includes input combination 4 (0100) and 8 (1000) in this example. Another horizontal line is drawn to separate these combinations from the next group of combinations, those with two ones. This procedure continues until the entire list of truth table entries is exhausted. The initial listing table now contains each input combination contributing a one output level to the circuit listed in an ordered sequence.

Create an initial listing table for variables W, X, Y, and Z representing the logic function:

$$T = \Sigma(1, 3, 5, 9, 11, 12)$$

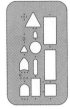

Solution The initial listing table orders the one level outputs variable combinations according to the number of ones within the input combinations. Thus:

1 has one one	0001
3 has two ones	0011
5 has two ones	0101
9 has two ones	1001
12 has two ones	1100
11 has three ones	1011

Initial listing table for $T = \Sigma(1, 3, 5, 9, 11, 12)$:

	W	X	Y	Z
1	0	0	0	1
3	0	0	1	1
5	0	1	0	1
9	1	0	0	1
12	1	1	0	0
11	1	0	1	1

Reduction Tables

Reduction tables are created from the initial listing table following the reduction procedure outlined earlier. Figure 4.19 shows the reduction tables that result from the Quine–McCluskey process using the initial listing table data just discussed. The following section explains how the reduction tables are formed.

Starting with the initial listing table, we can make comparisons between input combinations differing by only a single bit. For instance, the 0000 combination is compared with each combination in the next lower group (0100 and 1000 in this case) to search for any single bit variation. Since we are looking only for single bit variations, we will compare only those groups that are adjacent to each other. (These groupings were ordered according to the number of ones in each input combination when we created the initial listing table—the express purpose of the initial listing table.) When we compare 0000 to 0100 (decimal 0 and 4), a single bit difference is entered on the first reduction table as 0_00. That is, the third bit is low in one combination and high in the other. The comparison between 0000 and 1000 (decimal 0 and 8) yields _000, which is also entered appropriately on the first reduction table. In this comparison the differing bits are in the most significant bit position.

Next to the first reduction table entries, we indicate, in decimal form, the two combinations that lead to each entry. Thus, 0,4 is entered next to 0_00

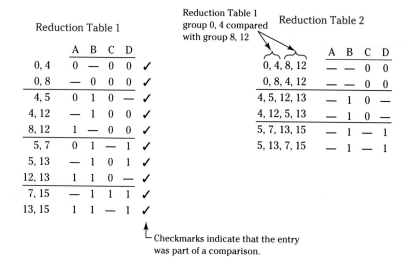

Figure 4.19
Forming the Reduction
Tables

and 0,8 is entered next to __000. For clarity, a horizontal line is drawn on the first reduction table after these entries, since these comparisons are all that are possible from the initial listing table's first two groups.

Comparisons continue, as shown, for all combinations within adjacent groups on the intitial listing table. For instance, comparisons between the group containing combinations 4 and 8 and the group containing combinations 5 and 12 yield single bit differences for 4,5, 4,12, and 8,12. Dashes, of course, indicate the bit position containing the single bit difference within these comparisons. Any valid comparisons are entered onto the first reduction table and every combination involved in a comparison is checkmarked on the initial listing table (checkmarks are shown in Figure 4.18). This is crucial, since at the end of the process the circuit prime implicants are all combinations lacking a checkmark.

After the first reduction table is complete, a second reduction table is created. Now comparisons are made between the adjacent combinations on the first reduction table and comparisons with a single bit difference are entered onto the second reduction table. For example, compare the 0,4 (0__00) combination with each combination in the next lower group. Two characteristics will determine if a comparison yields a reduction. The dashes in the compared combinations must match up in the same bit position and there must be only a single bit difference among the remaining bits. Combination 0,4 (0__00) from the higher group meets this criterion when compared with combination 8,12 (1__00) in the lower group. This successful comparison is noted on the second reduction table as 0,4,8,12, and the combination is listed as __ __00. Checkmarks are placed on the first reduction table next to 0,4 and 8,12 to show that the combinations have been used in a valid comparison.

The comparisons continue through all remaining combinations on the table. Notice that combination 0,8 when compared to combination 4,12 yields _ _00, the same result as the comparison between 0,4 and 8,12. The result is identical because both comparisons involve combinations 0, 4, 8, and 12. Since both comparisons yield the same result, only one requires consideration. Typically, the combination listing the decimal numbers in ascending order is the combination utilized in the final analysis. In this instance the combination used is 0,4,8,12.

After the second reduction table is complete, a third table is created if comparisons are possible between the combinations on the second reduction table. Comparisons are made between combinations containing two dashes in the same bit positions and a single bit difference. In this example no comparisons can meet this criteria so that there is no need for a third table and the comparison process ends. In general, reduction tables are created as needed until comparisons become impossible.

After all comparisons are complete and all tables are created, the prime implicants, those entries that lack checkmarks, are easily recognized. Remember that the prime implicant is a product term required in the final logic equation. Each prime implicant's Boolean term is derived from the reduction table in which it resides. For example, all prime implicants in Figure 4.19 are in reduction table 2. Combination 0,4,8,12 is the first. Any dashes in the combination imply that the variables associated with those bits are unnecessary in the product term. The binary bits remaining in the combination determine the prime implicant. If the bit is one, then the variable is used; if the bit is zero, the complement of the variable is used. Combination 0,4,8,12 has variables A,B,C,D = _ _00. Therefore the prime implicant is $\overline{C}\overline{D}$, and the remaining prime implicants are _10_ = $B\overline{C}$ and _1_1 = BD.

The prime implicants can be ORed to obtain a valid SOP logic equation. The equation has been simplified since some variables are eliminated, but the result is not necessarily the simplest equation possible. A **prime implicant table** (our last table, honest) is used to eliminate any nonessential prime implicants.

Using the initial listing table shown, develop reduction tables as necessary while evaluating comparisons between the input combinations. Identify the prime implicants.

Initial listing table for $T = \Sigma(1, 3, 5, 9, 11, 12)$:

	W	X	Y	Z	
1	0	0	0	1	✓
3	0	0	1	1	✓
5	0	1	0	1	✓
9	1	0	0	1	✓
12	1	1	0	0	
11	1	0	1	1	✓

DESIGN EXAMPLE 4.12

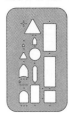

Solution   Compare adjacent groups on the initial listing table for single bit differences. Enter these onto the first reduction table using a dash to indicate the differing bit position. Place a checkmark next to each combination used in a valid comparison. Create a second reduction table to complete the comparison process. Prime implicants are the combinations lacking a checkmark. The prime implicant is obtained by comparing the combination to the variables listed along the top of the reduction tables. If the combination bit position has a dash, then the variable listed above it is not used. If the combination bit position is a one, then use the variable above it. If the combination bit position is a zero, then use the complement of the variable listed above it.

First Reduction Table

	W	X	Y	Z	
1,3	0	0	—	1	
1,5	0	—	0	1	
1,9	—	0	0	1	✓
3,11	—	0	1	1	✓
9,11	1	0	—	1	

Second Reduction Table

	W	X	Y	Z
1,9,3,11	—	0	—	1

Prime implicants are:

$$12 = WX\overline{Y}\,\overline{Z}$$
$$1,3 = \overline{W}\,\overline{X}Z$$
$$1,5 = \overline{W}\,\overline{Y}Z$$
$$9,11 = W\overline{X}Z$$
$$1,9,3,11 = \overline{X}Z$$

The Prime Implicant Table

The prime implicant table for the problem we are discussing (Figures 4.18, 4.19) is given in Figure 4.20. Listed along the top of the table are decimal numbers taken from the initial listing table. As you recall, these decimal num-

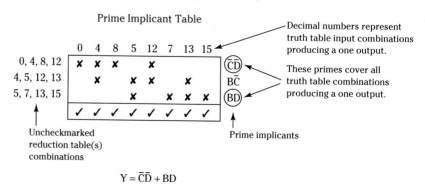

Prime Implicant Table

Figure 4.20
Forming the Prime Implicant Table

$$Y = \overline{C}\,\overline{D} + BD$$

bers represent the input binary combinations producing a one output for the desired circuit.

Along the left side of the prime implicant table are the uncheckmarked decimal combinations obtained from the initial listing and reduction tables. These, of course, represent the prime implicants for the circuit we are designing. The Boolean terms for these prime implicants (also obtained from the reduction tables) are placed along the right side of the prime implicant table. Naturally, the prime implicants should align with their corresponding decimal combination on the left.

The table is filled (on a row-by-row basis) by placing crosses under each decimal number (listed across the top), which correspond to the decimal prime implicant combinations (listed on the left). For example, crosses are placed in the top row under numbers 0, 4, 8, 12 to correspond with the prime implicant combination 0,4,8,12. This process is repeated for every prime implicant row. When complete, at least one cross is present in every column. The table now identifies every input combination contributing to the prime implicants. Incidentally, in this example every decimal combination listed on the left is comprised of four numbers. This is not always the case. Depending on the prime implicants involved, there might be more or fewer numbers. Decimal combinations should be listed in order of size, that is, listing the largest first and working toward the smallest. This not only organizes the prime implicant table, but it is also helpful during analysis.

After all crosses are placed on the table, the essential prime implicants are determined. Initially, you will inspect the table for columns containing a single cross. In Figure 4.20 this occurs in columns 0, 8, 7, and 15, so a checkmark is placed along the bottom of the table under these columns. This indicates that the prime implicant terms associated with these columns (listed on the right) are absolutely required in the final logic equation. Your objective is to make sure that every column is represented in the final logic equation. Only one prime implicant can satisfy columns with a single cross, and for clarity these prime implicants are circled. Circles identify the circuit equation essential prime implicants and also assist you in the next step.

After all single cross columns are identified, any nonchecked columns remaining must be associated with a prime implicant. If any crosses in the nonchecked columns align with prime implicants already circled, their columns are also checked. This means that these columns and associated input combinations are associated with previously selected prime implicants. In our example columns 4 and 12 can be associated with previously circled prime implicant $\overline{CD}$, whereas columns 5 and 13 can be associated with prime implicant BD. Finding these matches is important since our objective is to associate each column with a prime implicant. When all columns are checkmarked, all circled prime implicants are ORed together to form the final, simplified logic equation. Thus, $Y = \overline{CD} + BD$.

Alas, analysis is not always this straightforward. There are situations in which the association between crosses and prime implicants is not obvious. Consider the prime implicant table in Figure 4.21. Column 3 is immediately selected since it contains only a single cross. A checkmark is placed under

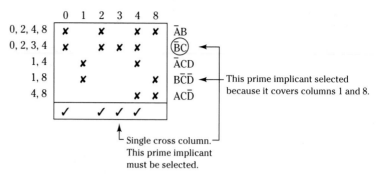

Figure 4.21
Selecting Prime
Implicants

$$Y = \bar{B}C + B\bar{C}\bar{D}$$

this column and the prime implicant associated with the cross in column 3, $\bar{B}C$, is circled. Columns 0, 2, and 4 are also checked off because the crosses in these columns line up with prime implicant $\bar{B}C$. This leaves columns 1 and 8 outstanding. Since the selected prime implicant does not align with any cross within these columns, we need additional prime implicants to complete the equation. Keep in mind that the fewer prime implicants we select, the simpler the resulting logic equation will be. Therefore, we do not take all prime implicants associated with the unselected columns; we attempt to find the fewest prime implicants required.

Prime implicant $B\bar{C}\bar{D}$ aligns with crosses in both columns 1 and 8 and is chosen because it is the only prime implicant that covers both columns. Hence, every column on the table is associated with at least one prime implicant, resulting in a complete equation. When a choice between two or more prime implicants is possible, the prime implicant with the fewest variables is chosen.

DESIGN EXAMPLE 4.13

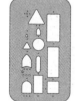

Draw the prime implicant table for the prime implicants found in Design Example 4.12. Determine the simplified output equation.

Solution Along the top of the prime implicant table list the decimal input combinations from the initial listing table. Along the left side list all unchecked decimal combinations (prime implicants) from the initial listing and reduction tables. Along the right side list the actual prime implicants associated with each decimal combination. Find the final equation by associating each column with a prime implicant.

	1	3	5	9	12	11	
1,9,3,11	×	×		×		×	$\bar{X}Z$
1,5	×		×				$\bar{W}\bar{Y}Z$
1,3	×	×					$\bar{W}\bar{X}Z$
9,11				×		×	$W\bar{X}Z$
12					×		$WX\bar{Y}\bar{Z}$
		✓		✓			

Columns 5 and 12 are checked because they contain only one cross. Prime implicants $\overline{W}\overline{Y}Z$ and $WX\overline{Y}\overline{Z}$ associated with these two columns are essential to the final equation. Columns 1, 3, 9, and 11 remain and are all associated with prime implicant $\overline{X}Z$. $\overline{X}Z$ is selected to cover these columns in the final equation. The final equation reads:

$$T = \overline{X}Z + \overline{W}\overline{Y}Z + WX\overline{Y}\overline{Z}$$

More on Finding Essential Prime Implicants

Unfortunately, with very large prime implicant tables the problem of selecting the essential prime implicants increases in complexity. Some trial-and-error processes can identify prime implicants for use when the choice is not obvious, but most of them are too complicated and time-consuming to present here. However, we will discuss one method that uses some Boolean algebra to assist in determining the required primes. This particular method does not guarantee a minimum equation, but provides a reasonable approximation to one.

Figure 4.22 shows how to convert the prime implicant table in Figure 4.21 into a Boolean equation and how to reduce it. First, the prime implicant table is redrawn, eliminating the decimal number combinations. In their place we substitute letters—U, V, W, X, and Y (chosen randomly for convenience)—to identify the rows in the prime implicant table. Since selecting prime implicants is equivalent to selecting rows from the table, a minimum solution equation is obtained from a minimum number of rows selected. The objective of this procedure is to select the minimum number of rows when they cannot be determined directly from the prime implicant table.

Any equation derived from a prime implicant table is associated with at least one cross from each column. Using this line of reasoning, notice that column 0 requires that either row U or row V be represented in the equation. Column 1 requires row W or row X. These are stated as simple OR statements, such as Column 0 = (U + V) and Column 1 = (W + X). Therefore, the rows representing each column in a prime implicant table can be expressed as an

---

Summary of the Quine–McCluskey Procedure

FOR YOUR INFORMATION

1. Create a truth table or truth table equation for the design problem.
2. Translate the truth table into an initial listing table.
3. Reduce the initial listing table into the first reduction table.
4. Continue reductions into second, third, . . . reduction tables if possible.
5. Create a prime implicant table using information from the initial listing and reduction tables (uncheckmarked combinations).
6. Obtain the essential prime implicants from the prime implicant table.
7. Use the essential prime implicants to form the final equation.

	0	1	2	3	4	8	
U	✗		✗		✗	✗	$\overline{A}B$
V	✗		✗	✗	✗		$\overline{B}C$
W		✗			✗		$\overline{A}CD$
X		✗				✗	$B\overline{C}\overline{D}$
Y					✗	✗	$A\overline{C}D$

$T = (U + V)(W + X)(U + V)(V)(U + V + W + Y)(U + X + Y)$ ◄— This equation represents every

$T = (U + V)(U + V)(W + X)(VU + VV + VW + VY)(U + X + Y)$ column on the prime implicant table.

$T = (U + V)(W + X)(VU + V + VW + VY)(U + X + Y)$

$T = (U + V)(W + X)\ V(U + 1 + W + Y)(U + X + Y)$

$T = (U + V)(W + X)\ V(U + X + Y)$

$T = (VU + VV)(W + X)(U + X + Y)$

$T = (VU + V)(W + X)(U + X + Y)$

$T = V(U + 1)(W + X)(U + X + Y)$

$T = V(W + X)(U + X + Y)$

$T = V(WU + WX + WY + XU + XX + XY)$

$T = V(WU + WX + WY + XU + X + XY)$

$T = V(WU + WY + X(W + U + 1 + Y))$

$T = V(WU + WY + X)$

$T = VWU + VWY + \boxed{VX}$

Figure 4.22
Selecting a Minimum Row Set

└─ Use—represents the fewest rows
required to satisfy the expression.

OR function. Furthermore, since all columns are represented in the final equation, the OR terms we are describing must be ANDed together since each term represents one column. Thus, the equation that describes the prime implicant table in Figure 4.22 is:

$$(U + V)(W + X)(U + V)(V)(U + V + W + Y)(U + X + Y)$$

Note, this is not the final equation used to build a circuit. This is only an interim equation used to select prime implicants when the selection process is not obvious. The equation is simplified following standard Boolean reduction procedures to identify the minimum number of prime implicants possible.

After the equation has been simplified, we must try to determine a minimum **row set**, which means selecting the fewest rows possible. (The arbitrary letters assigned to the table represent rows.) In this example the simplified interim equation is VWU + VWY + VX. This reads: The required row sets for a minimum logic equation are rows V, W, and U, or rows V, W, and Y, or rows V and X. Since V and X represent the fewest rows, they are selected. V represents prime implicant $\overline{B}C$, whereas X represents prime implicant $B\overline{C}\overline{D}$, providing the final logic equation, $\overline{B}C + B\overline{C}\overline{D}$.

Confirm that the prime implicants found from the prime implicant table in Design Example 4.13 form a minimum row set.

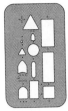

Solution Redraw the prime implicant table using letters (i.e., A, B, C, D, E) in place of the decimal combinations. OR-AND the columns to create an equation representing the table. Reduce to determine the minimum row set and final equation.

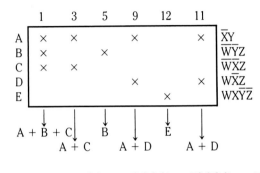

$$T = (A + B + C)(A + C)(B)(A + D)(E)(A + D)$$

$$T = (A + C)(B)(E)(A + D)$$

$$T = (A + CD)(B)(E)$$

$$T = ABE + CDBE$$

↑ Minimum row set because this term has the fewest variables.

Since A represents row $\overline{X}Y$ and B represents row $\overline{W}\overline{Y}Z$ and E represents row $WX\overline{Y}Z$, then

$$T = \overline{X}Y + \overline{W}\overline{Y}Z + WX\overline{Y}Z$$

Using Don't Care States with the Quine–McCluskey Method

As you recall from the discussion on K-maps, don't care states can simplify logic equations and are employed when the circuit output is not affected by some input combinations. Since the Quine–McCluskey reduction is generally used when the number of circuit variables exceeds four, some of the resulting input combinations will possibly be considered as don't care states. Therefore, it is advantageous to utilize don't care's with the Quine–McCluskey reduction process for additional circuit simplification.

The Quine–McCluskey process changes very little with don't care terms. First, all truth table one outputs and all don't care outputs are entered onto the initial listing table. Comparisons are made as usual and reduction tables are formed as needed. Naturally, the don't care combinations help the procedure by providing matches with other combinations. The matches provided by utilizing don't care states eliminate variables and lead to circuit simplification.

When all reduction tables are complete, the prime implicant table is produced. It is formed **only** with the truth table input combinations that produce one outputs. Don't care states are omitted from the prime implicant table. From this point on prime implicant selection continues as normal. If the don't care states were able to simplify the equation, the resulting prime implicants would contain fewer input variables.

DESIGN EXAMPLE 4.15

Use the Quine–McCluskey method to find the reduced equation for the following truth table:

A	B	C	D	Y
0	0	0	0	0
0	0	0	1	0
0	0	1	0	0
0	0	1	1	0
0	1	0	0	0
0	1	0	1	1
0	1	1	0	0
0	1	1	1	0
1	0	0	0	0
1	0	0	1	1
1	0	1	0	d
1	0	1	1	d
1	1	0	0	d
1	1	0	1	d
1	1	1	0	d
1	1	1	1	d

Initial Listing Table

	A	B	C	D	
5	0	1	0	1	✓
9	1	0	0	1	✓
d 10	1	0	1	0	✓
d 12	1	1	0	0	✓
d 11	1	0	1	1	✓
d 13	1	1	0	1	✓
d 14	1	1	1	0	✓
d 15	1	1	1	1	✓

— Indicates a don't care state.

First Reduction Table

	A	B	C	D	
5,13	—	1	0	1	
9,11	1	0	—	1	✓
10,11	1	0	1	—	✓
10,14	1	—	1	0	✓
12,13	1	1	0	—	✓
12,14	1	1	—	0	✓
11,15	1	—	1	1	✓
13,15	1	1	—	1	✓
14,15	1	1	1	—	✓

Second Reduction Table

	A	B	C	D	
9,11,13,15	1	—	—	1	
10,11,14,15	1	—	1	—	Use 10,11,14,15
10,14,11,15	1	—	1	—	
12,13,14,15	1	1	—	—	Use 12,13,14,15
12,14,13,15	1	1	—	—	

Prime Implicant Table

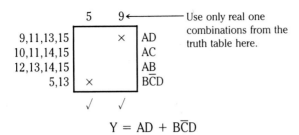

$$Y = AD + B\overline{C}D$$

Solution   Carry out the Quine–McCluskey process as usual until the prime implicant table is formed. Only include the real one level outputs from the truth table on the prime implicant table. The don't care states are used only during the comparison steps with the initial listing and reduction tables.

An automated assembly line is monitored by the following sensors:

Pressure	P
Temperature	T
Motion	M
Humidity	H
Speed	S

The assembly line starts up for the following sensor conditions (order of variables is P, T, M, H, S):

$$\text{Startup} = \Sigma(1, 2, 6, 8, 9, 14, 15, 16, 19, 20, 25, 26, 30)$$

Determine the circuit equation for the system.

Solution   Use the Quine–McCluskey method to reduce this five-variable circuit.

Initial Listing Table

	P	T	M	H	S	
1	0	0	0	0	1	✓
2	0	0	0	1	0	✓
8	0	1	0	0	0	✓
16	1	0	0	0	0	✓
6	0	0	1	1	0	✓
9	0	1	0	0	1	✓
20	1	0	1	0	0	✓
14	0	1	1	1	0	✓
19	1	0	0	1	1	
25	1	1	0	0	1	✓
26	1	1	0	1	0	✓
15	0	1	1	1	1	✓
30	1	1	1	1	0	✓

First Reduction Table

	P	T	M	H	S	
1,9	0	—	0	0	1	
2,6	0	0	—	1	0	No second
8,9	0	1	0	0	—	reduction table
16,20	1	0	—	0	0	
6,14	0	—	1	1	0	
9,25	—	1	0	0	1	
14,15	0	1	1	1	—	
14,30	—	1	1	1	0	
26,30	1	1	—	1	0	

Prime Implicant Table

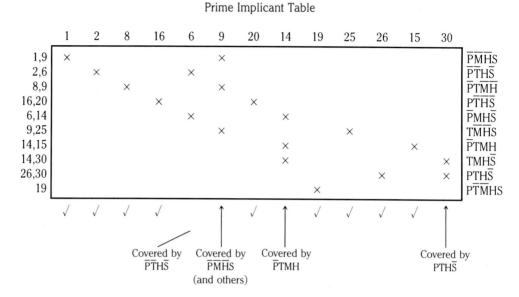

Covered by $\overline{PTHS}$

Covered by $\overline{PMHS}$ (and others)

Covered by $\overline{PTMH}$

Covered by $PTH\overline{S}$

$\sqrt{}$ = single × columns. Therefore, associated prime implicants are required.

$$\text{Startup} = \overline{PMHS} + \overline{PTHS} + \overline{PTMH} + P\overline{THS} + T\overline{MHS}$$
$$+ \overline{P}TMH + PTH\overline{S} + P\overline{TM}HS$$

4.5 ■ COMPUTER SIMPLIFICATION

The simplification methods covered are lifesavers when it comes to reducing circuitry, but they are time-consuming processes. Computer simplification programs greatly accelerate not only logic simplification, but the overall design process as well. In fact, any major logic design effort carried out by industry today is most certainly centered around an automated design process. Automated design is a necessity for companies competing in a cost-sensitive and rapidly changing market. This important consideration drives high technology business and should be understood by the new design technician or engineer. The following sections provide an introduction into the methods and requirements necessary for computer assisted design.

Computer simplification programs generally are not used as stand-alone programs. Simplification software is usually integrated within a full **computer aided design** (CAD) system or is found as a part of other specialty software packages. This is reasonable since any major design effort involves considerably more design steps than just logic simplification. Computer aided or computer assisted design systems range from simple and inexpensive personal computer based packages to powerful and sophisticated mainframe, minicomputer, or workstation based systems. Figure 4.23 outlines the functional aspects of a logic design CAD system.

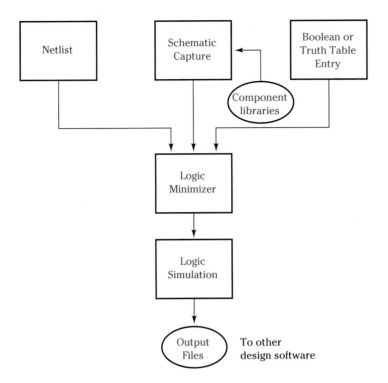

Figure 4.23
Typical CAD System
Organization for Logic
Circuit Development

The input or front end for a logic circuit CAD system is usually flexible. That is, Boolean equations or truth tables are suitable forms of input information. An alternate entry mechanism is **schematic capture**. Schematic capture lets the designer draw the schematic of the circuit under development. Systems with this capability include "libraries" of common logic symbols easily accessed and positioned on screen by the designer. **Netlist** input is also possible. A netlist describes all the connections in a logic circuit. Schematic capture software also generates a netlist, making it possible to use a schematic capture program from one system as the input into another CAD system. In any case, we can describe a logic circuit using commonly understood symbols and employ this as the input to computer reduction software.

More advanced systems include symbols and pin outs for common logic family part numbers. If, for instance, a designer is designing a circuit using a CMOS logic family, entering a CMOS part number automatically places the correct logic symbol onto the screen. Since pin-out information is also known, component interconnections can be checked by the system. In some cases the part number libraries also contain component behavioral characteristics, allowing for circuit simulation and testing. This means that factors such as propagation delay time can be included in a circuit simulation run to determine the impact of delay time on circuit performance.

As far as circuit simplification goes, once the CAD system receives the circuit description, the logic minimizer software can act on it. The software interprets the design equations for the circuit and eliminates any unnecessary variables or functions. The methodical checking process of the Quine–McCluskey method lends itself to computer programming. Many minimizing programs are based on the Quine–McCluskey method. Other reduction packages are relatively simple, using only the most basic Boolean reduction theorems to achieve their goal. Still other reduction software packages are proprietary programs written by the manufacturers of the CAD software. In this case the "**software algorithm**" is created and protected by the manufacturer. In any event, reduction software speeds up circuit minimization, although it is unlikely that you will purchase only a circuit simplification package.

Some Computer Minimization Examples

Figure 4.24 illustrates a four-variable truth table, the full sum of products equation, and corresponding reduced logic equation. The reduced equation can be verified through the K-map or Quine–McCluskey process. Computer simplification also accomplishes this reduction as explained in the following sections.

A	B	C	D	Y
0	0	0	0	1
0	0	0	1	0
0	0	1	0	1
0	0	1	1	1
0	1	0	0	1
0	1	0	1	0
0	1	1	0	0
0	1	1	1	1
1	0	0	0	1
1	0	0	1	0
1	0	1	0	1
1	0	1	1	0
1	1	0	0	1
1	1	0	1	1
1	1	1	0	0
1	1	1	1	0

Full circuit equation:

$$Y = \bar{A}\bar{B}\bar{C}\bar{D} + \bar{A}\bar{B}C\bar{D} + \bar{A}\bar{B}CD + \bar{A}B\bar{C}\bar{D} + \bar{A}BCD + A\bar{B}\bar{C}\bar{D} + A\bar{B}C\bar{D} + AB\bar{C}\bar{D} + AB\bar{C}D$$

Figure 4.24
Reduced Four-Variable
Equation

Reduced circuit equation:

$$Y = \bar{C}\bar{D} + \bar{B}\bar{D} + AB\bar{C} + \bar{A}CD$$

The quality of computer reduction software determines the effectiveness of the reduction process. Furthermore, some reduction packages are written for specific packaging technologies and may not be appropriate or completely adequate for general simplification. Thus, it is important to understand the limitations and intended application of the software package you are using. With these words of caution, we will examine Figure 4.25.

Figure 4.25 shows how to enter a logic equation into a hardware design program. This program, CUPL, is used to design programmable logic devices

```
/**  Inputs  **/
Pin 2          = a      ;      /*                                    */
Pin 3          = b      ;      /*                                    */
Pin 4          = c      ;      /*                                    */
Pin 5          = d      ;      /*                                    */


/**  Outputs  **/
Pin 19         = Y      ;      /*                                    */


/**  Logic Equations  **/
x = a&!b&!c&!d # !a&!b&c&!d # !a&b&!c&!d # a&!b&!c&!d # !a&!b&c&d #
    a&!b&c&!d # a&b&!c&!d # !a&b&c&d # a&b&!c&d;
```

Note:
 ! = inversion
 & = AND
 # = OR

Figure 4.25
CUPL Equation Entry

(discussed extensively in later chapters) and includes many of the necessary design process steps for programmable device technology. We are interested in the CUPL program section labeled "Logic Equations." The CUPL program expects the equation for a circuit to be expressed in this section. The logic expression shown in Figure 4.25 is an SOP expression taken directly from the truth table without reduction in Figure 4.24. The computer software will reduce the equation and relieve us from manual reduction work.

No doubt, you have noticed the odd appearance of the equation. Since standard computer keyboards are used for equation entry, many logic symbols, such as the overbar for negation, are not available. Alternate symbols are used instead. CUPL uses the exclamation point (!) to indicate negation. Therefore, !A is read as "not A" and is equivalent to $\overline{A}$. The ampersand symbol (&) indicates the AND function, whereas the pound symbol (#) indicates OR. Therefore, Y = !A&!B&!C&!D is equivalent to $Y = \overline{ABCD}$. The fundamental logic symbols may vary from program to program, but are easy to use and understand.

Figure 4.26 shows the output of the CUPL logic minimizer software modeled around the Quine–McCluskey process. This reduced equation provided by

Expanded Product Terms

```
x =>
      a & b & !c & d
    # b & !c & !d
    # !b & !d
    # !a & c & d
```

```
Y =>
    !a & c & d
    # a & b & !c      ← Minimized output equation    or    Y = ACD + ABC + BD + CD
    # !b & !d
    # !c & !d
```
$$Y = \overline{A}CD + AB\overline{C} + \overline{B}\,\overline{D} + \overline{C}D$$

Y.oe$_1$ =>

Figure 4.26
Computer Minimized Equation

computer reduction is identical to that obtained by the K-map or the Quine–McCluskey process. The program took the full SOP logic equation, and in only a few seconds of computer time, produced the simplified result.

As mentioned before, not all reduction software works equally well. The output from a minimizer designed for quick reduction (an option in the CUPL system) is shown in Figure 4.27. It is obvious that the reduced equation is simpler than the full-blown SOP expression, but not so simple as the run shown in Figure 4.26. Clearly, the quality of the software determines the quality of the reduction. The logic designer must possess a strong sense of design to use computerized design systems effectively. Automated design does not mean that anyone can do logic design. It is incorrect to assume that automated design systems will solve every design problem; the burden is on the designer to evaluate the results from automated design.

Expanded Product Terms

x =>
    a & b & !c & d
  # b & !c & !d
  # !b & !d
  # !a & c & d

Y =>
    a & b & !c & d
  # b & !c & !d   ←Minimized output equation   or $Y = AB\overline{C}D + B\overline{C}\overline{D} + \overline{B}\overline{D} + \overline{A}CD$
  # !b & !d
  # !a & c & d

Figure 4.27
Partially Minimized
Equation

DESIGN
EXAMPLE 4.17

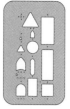

Write the full SOP expression, in a form compatible with CUPL software, for the following truth table:

W	X	Y	Z	A
0	0	0	0	0
0	0	0	1	1
0	0	1	0	0
0	0	1	1	1
0	1	0	0	0
0	1	0	1	1
0	1	1	0	0
0	1	1	1	0
1	0	0	0	0
1	0	0	1	1
1	0	1	0	0
1	0	1	1	0
1	1	0	0	1
1	1	0	1	0
1	1	1	0	0
1	1	1	1	1

Solution Write each fundamental product using the CUPL keyboard symbols for negation (!) and AND (&). OR the fundamental products using the CUPL symbol for OR (#).

A = !W&!X&!Y&Z # !W&!X&Y&Z # !W&X&!Y&Z # W&!X&!Y&Z
   # W&X&!Y&!Z # W&X&Y&Z

Simulation

Once logic simplification is satisfactory, the designer tests his or her design using a logic simulator. This step is increasingly more important as industry moves toward programmable logic devices and highly integrated logic packages. The integrated circuit fabrication process can place a significant amount of complex circuitry in a single integrated circuit package. It is impractical to prototype designs using common breadboarding (handwiring) techniques when several thousand logic gates are involved because breadboarding is time-consuming and error-prone. Furthermore, a breadboarded circuit cannot be expected to perform as well as an integrated circuit since the physical characteristics of the devices and interconnections in the two technologies are far different. Integrated circuit performance is superior because the logic devices are closer together, fabricated at the same time, and connected on silicon rather than by a jumble of wires. The differences in electrical performance are enormous.

Simulators take into account the exact logic and electrical characteristics of the circuit technology in use. Depending on the complexity of the simulator, anything from simple functional testing to an exact timing analysis can be performed. Working from only a logic description of a circuit, a designer can thoroughly evaluate a design's operation with the aid of a simulator. Simulator output is also often used as part of a **test generation** system. Test vectors created by the simulator are used to test the actual logic circuits when they are built.

Once these steps are complete, other design software may be used to create printed circuit boards (PCB), develop programmable logic hardware, or create design documentation.

Computer aided design is cost-effective and a necessary part of a typical design cycle. The new technician or engineer should be familiar with and expect to use these tools of the trade. CAD systems do not minimize the designer's responsibility to understand logic design. Rather, the rapid changes in technology mandate that the designer know his or her business completely in order to keep up wtih changing technology and effectively utilize it.

SUMMARY

- Karnaugh map (K-map) reduction provides a graphical alternative to the algebraic circuit reduction process.
- K-maps represent truth table data in a form such that the reduction process $X\bar{Y} + XY = X$ can be carried out by inspection.
- K-map reduction takes place by systematically placing ones on the map in groups of eight, four, two, or one. Rolling map edges and overlapping

groups of ones are techniques also used to assist in the minimization process.

■ Identifying don't care states in a system design can lead to additional circuit reduction.

■ When more than four variables are required in a design, the simplification process is made easier by employing the Quine–McCluskey process or computer simplification programs.

■ Quine–McCluskey is a tabular reduction method often used as the basis for computer reduction software.

■ Computer simplification is generally used as part of an overall computer aided logic design process.

PROBLEMS

Sections 4.1–4.3

1. Use a Karnaugh-map (K-map) to extract the Boolean expression from the following truth tables:

*(a)

A	B	C	Y
0	0	0	0
0	0	1	0
0	1	0	0
0	1	1	0
1	0	0	1
1	0	1	0
1	1	0	1
1	1	1	1

(b)

A	B	C	Y
0	0	0	1
0	0	1	0
0	1	0	1
0	1	1	1
1	0	0	1
1	0	1	0
1	1	0	0
1	1	1	0

(c)

A	B	C	Y
0	0	0	0
0	0	1	1
0	1	0	1
0	1	1	0
1	0	0	1
1	0	1	1
1	1	0	1
1	1	1	0

2. Reduce the following K-maps:

*(a)

	$\bar{C}\bar{D}$	$\bar{C}D$	CD	$C\bar{D}$
$\bar{A}\bar{B}$	1	1	0	0
$\bar{A}B$	1	1	0	0
AB	d	0	0	0
$A\bar{B}$	0	0	0	d

(b)

	$\bar{C}\bar{D}$	$\bar{C}D$	CD	$C\bar{D}$
$\bar{A}\bar{B}$	0	0	0	d
$\bar{A}B$	0	0	1	0
AB	d	0	0	1
$A\bar{B}$	1	0	0	1

---

* See Appendix F: Answers to Selected Problems.

3. Determine the simplified logic equation represented by each of the following K-maps:

*(a)

	C̄D̄	C̄D	CD	CD̄
ĀB̄	0	0	0	1
ĀB	0	0	1	0
AB	0	1	0	0
AB̄	1	0	0	0

(b)

	C̄D̄	C̄D	CD	CD̄
ĀB̄	1	0	1	0
ĀB	0	1	0	1
AB	0	0	0	0
AB̄	0	0	0	0

(c)

	C̄D̄	C̄D	CD	CD̄
ĀB̄	0	0	0	1
ĀB	0	0	1	1
AB	1	0	1	1
AB̄	0	0	0	0

(d)

	C̄D̄	C̄D	CD	CD̄
ĀB̄	0	0	0	0
ĀB	0	1	1	1
AB	1	1	1	0
AB̄	0	0	1	0

(e)

	C̄D̄	C̄D	CD	CD̄
ĀB̄	0	1	1	0
ĀB	1	0	1	1
AB	0	0	0	0
AB̄	1	1	1	1

(f)

	C̄D̄	C̄D	CD	CD̄
ĀB̄	1	1	1	1
ĀB	0	1	1	1
AB	0	0	0	0
AB̄	1	0	0	0

(g)

	C̄D̄	C̄D	CD	CD̄
ĀB̄	0	0	1	0
ĀB	0	0	1	1
AB	1	1	0	0
AB̄	0	1	0	0

(h)

	C̄D̄	C̄D	CD	CD̄
ĀB̄	1	0	0	1
ĀB	1	0	1	1
AB	0	0	1	0
AB̄	1	0	0	1

*(i)

	C̄D̄	C̄D	CD	CD̄
ĀB̄	1	0	0	1
ĀB	1	1	1	1
AB	1	1	1	1
AB̄	1	0	0	1

4. Four temperature sensors are used to determine when the temperature of a vat of liquid exceeds a preset limit. The sensors are set to provide a high level output when the temperature is exceeded, otherwise a low level output is given. Using a K-map, design a circuit that will indicate when any two of the sensors have exceeded their temperature limits.

---

* See Appendix F: Answers to Selected Problems.

5. Simplify the following truth tables with K-maps:

*(a)

W	X	Y	Z	G
0	0	0	0	1
0	0	0	1	1
0	0	1	0	0
0	0	1	1	1
0	1	0	0	0
0	1	0	1	0
0	1	1	0	0
0	1	1	1	0
1	0	0	0	0
1	0	0	1	0
1	0	1	0	1
1	0	1	1	1
1	1	0	0	0
1	1	0	1	1
1	1	1	0	0
1	1	1	1	0

(b)

W	X	Y	Z	G
0	0	0	0	1
0	0	0	1	1
0	0	1	0	0
0	0	1	1	0
0	1	0	0	1
0	1	0	1	1
0	1	1	0	0
0	1	1	1	1
1	0	0	0	1
1	0	0	1	1
1	0	1	0	0
1	0	1	1	0
1	1	0	0	1
1	1	0	1	1
1	1	1	0	0
1	1	1	1	1

(c)

W	X	Y	Z	G
0	0	0	0	1
0	0	0	1	0
0	0	1	0	1
0	0	1	1	0
0	1	0	0	0
0	1	0	1	0
0	1	1	0	0
0	1	1	1	0
1	0	0	0	1
1	0	0	1	0
1	0	1	0	1
1	0	1	1	0
1	1	0	0	0
1	1	0	1	0
1	1	1	0	0
1	1	1	1	0

6. Use K-maps to reduce the following sum of products expressions:
   (a) $\overline{A}\overline{B}\overline{C}D + \overline{A}\overline{B}C\overline{D} + \overline{A}BC\overline{D} + AB\overline{C}\overline{D} + A\overline{B}\overline{C}D$
   (b) $\overline{W}\overline{X}Y\overline{Z} + \overline{W}X\overline{Y}Z + \overline{W}XYZ + WX\overline{Y}Z + WXYZ$
   (c) $\overline{Q}\overline{R}S\overline{T} + \overline{Q}RST + \overline{Q}RS\overline{T} + \overline{Q}RST + QR\overline{S}\overline{T} + QR\overline{S}T$
7. Reduce the following truth table using the Quine–McCluskey process:

A	B	C	D	T
0	0	0	0	1
0	0	0	1	0
0	0	1	0	1
0	0	1	1	1
0	1	0	0	1
0	1	0	1	0
0	1	1	0	0
0	1	1	1	1
1	0	0	0	1
1	0	0	1	0
1	0	1	0	1
1	0	1	1	0
1	1	0	0	1
1	1	0	1	1
1	1	1	0	0
1	1	1	1	0

---

* See Appendix F: Answers to Selected Problems.

8. (a) Verify the results of problem 7 using a K-map.
    (b) Verify the results of problem 7 using a computer simplification program, if available.

Section 4.4

9. Use the Quine–McCluskey reduction process to minimize:
  *(a) $Y = \Sigma(3, 7, 8, 9, 12, 13)$
  (b) $Y = \Sigma(2, 6, 8, 10, 12)$
  (c) $Y = \Sigma(1, 2, 9, 10, 16, 17, 22, 25)$
  (d) $Y = \Sigma(0, 1, 2, 3, 4, 5, 6, 7, 16, 17, 18, 19, 20, 21, 22, 23)$

Section 4.5

10. If computer software is available to do logic equation minimization, then reduce the following equation:

$$Z = ABC + A\bar{B}C + AB\bar{C} + \bar{A}B\bar{C}$$

(a) Determine how to write the equation for software use.
(b) Determine how to execute the program.
(c) Determine how to obtain the simplified result.

---

* See Appendix F: Answers to Selected Problems.

CHAPTER 5

BINARY ARITHMETIC AND MATHEMATICAL CIRCUITS

OBJECTIVES

To explain how the mathematical operations addition, subtraction, multiplication, and division are carried out using the binary number system.

To explore the important aspects of representing binary numbers in 1's and 2's complement notation.

To explain how complement arithmetic is used in arithmetic computations.

To explore the notion of signed binary numbers and how they are used in digital systems.

To explore how logic circuitry is designed and utilized to implement mathematical operations.

To investigate carry look ahead designs and how they enhance the performance capabilities of mathematical circuitry.

To discuss the uses of an arithmetic and logic unit (ALU).

PREVIEW

People are fascinated by computers because of their ability to perform arithmetic at amazing speeds. Digital circuitry is specifically developed to implement mathematical functions that are carried out at electronic speeds. Since the digital gates switch rapidly, mathematical computing is also rapidly accomplished. With this in mind, our two primary chapter objectives can be explored—how common mathematical functions are carried out using binary numbers and how mathematical digital circuits are designed.

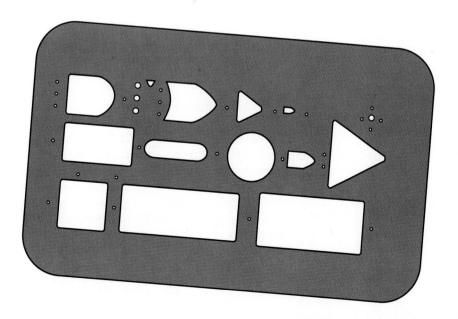

5.1 ■ BINARY ADDITION

Binary addition and decimal addition are similar processes, making the binary techniques easy to master. Four possible outcomes (shown in Figure 5.1) occur when two binary bits are added together. Two-bit addition always produces a **sum** and a **carry**. The sum results from adding 2-bits together and, on paper, the sum is recorded directly below the bits being added. A carry is obtained when the addition of 2-bits results in a number equal to 2 (10). The carry is recorded as the next most significant bit. (Sometimes the lack of a carry is referred to as a carry of zero, whereas a carry is often referred to as a carry of one.)

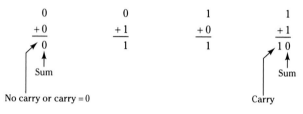

a) Single-bit addition

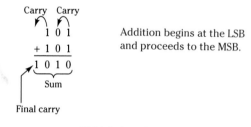

Figure 5.1
Binary Addition
(a) Single Bit
(b) Multibit

b) Multi-bit addition

When two multibit binary numbers are added, sums are produced and recorded for each bit. Carries generated during the process are added to the next most significant bits (MSBs). This process begins with the least significant bits (LSBs) and continues through to the most significant. The final carry becomes the MSB of the answer. This process is also detailed in Figure 5.1. Naturally, the digital addition circuitry we will design follows these principles.

The idea of carries in decimal math is very natural for us since we have applied the method throughout our lives. When faced with carries in the binary system, sometimes the concept needs to be refreshed. Design Example 5.1 compares addition, both decimal and binary, using very simple number combinations to illustrate further carries in binary addition.

Add the following binary numbers. Check answers by also adding the decimal equivalent numbers:

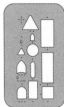

(a) 10 + 10   (b) 100 + 101   (c) 101 + 011
(d) 111 + 011   (e) 110 + 101   (f) 110 + 111

Solution   Add the two numbers bit by bit starting with the LSBs. When a carry is generated, add it to the next more significant bits in the number. Carries occur when 10 (2) or greater is obtained by any 2-bit addition. A carry out of the MSBs is recorded as the MSB of the answer.

(a) | 10 | 2 | (b) | 100 | 4 | (c) | 101 | 5 |
| + 10 | +2 | | + 101 | +5 | | + 011 | +3 |
| 100 | 4 | | 1001 | 9 | | 1000 | 8 |

(d) | 111 | 7 | (e) | 110 | 6 | (f) | 110 | 6 |
| + 011 | + 3 | | + 101 | + 5 | | + 111 | + 7 |
| 1010 | 10 | | 1011 | 11 | | 1101 | 13 |

Figure 5.2 contains several binary addition examples. Mathematically speaking, two binary numbers are aligned along the binary point—similar to decimal math where numbers are aligned along the decimal point. The binary point is usually omitted for whole numbers. The top number, the augend, is added to the bottom number, the addend, starting with the LSBs. The final result may contain the same number of bits or 1-bit more than the largest of the original numbers added. This fact determines the amount of circuit hardware used. That is, if an addition circuit is designed to add two 8-bit numbers, then the capability to display or record a 9-bit answer must be included in the design.

```
                              Binary point
                                   ↓
            Augend →    1 0 1 0.
            Addend → + 1 0 0 1.
              Sum →   1 0 0 1 1.


                       1 1 1 0 0 1 1
                     + 1 0 1 1 0 1
                     1 0 1 0 0 0 0 0


                       1 0 1 1.1 0 1
                     + 1 1 0 1 0.1
                     1 0 0 1 1 0.0 0 1
```

Figure 5.2
Adding Multibit Binary Numbers

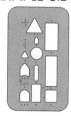

DESIGN
EXAMPLE 5.2

Add the following numbers:
(a) 111110 + 10111
(b) 0011 + 0011
(c) 10111.101 + 1001.011

Solution   Add the following normal binary addition procedures. Align the augend and addend with the binary point.

(a)   111110   (b)   0011   (c)   10111.101
     +   10111   +0011   +   1001.011
       ‾‾‾‾‾‾‾‾‾   ‾‾‾‾‾‾   ‾‾‾‾‾‾‾‾‾‾‾‾‾‾
       1010101   0110   100001.000

Digital addition circuits are generally constructed to add only two multibit binary numbers simultaneously. In other words, it is unlikely that circuits will be developed to add columns of binary numbers together. Additions such as this still use normal adder circuits but require circuitry to store partial sums created during a repetitive addition process. The length or **word size** (the number of bits comprising the binary number) of the numbers being added is limited by hardware. For instance, an 8-bit adder cannot add two 10-bit numbers together. Computing machines are designed with a specific word size in mind; all operations within the machine must then conform to the specified word size since it is set by the hardware.

5.2 ■ SUBTRACTION

Binary subtraction is similar to decimal subtraction. The four possible outcomes occurring from subtracting two single bits are given in Figure 5.3. Subtracting 1-bit from another produces the **difference** between the two. The difference between the 2-bits is recorded in the same bit position as the 2-bits being subtracted. Similar to decimal subtraction, **borrows** are often required

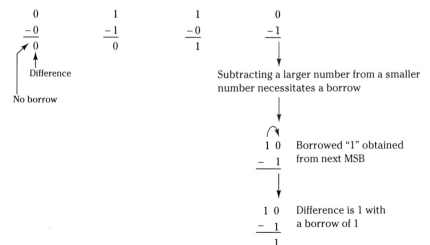

Figure 5.3
Binary Subtraction

during calculations and are necessary when one (in the subtrahend) is subtracted from zero (in the minuend). The borrow is achieved by adding 10 to the minuend bit and subtracting 1 from the minuend's next most significant place. If the borrow cannot be obtained because the next most significant place is zero, then keep checking to the left toward the more significant places until a borrow can be obtained. All intermediate zeros, between the original bit being subtracted and the place providing the borrow, are changed to one. Since borrowing presents some initial difficulty, Design Example 5.3 compares binary and decimal subtraction using recognizable binary numbers.

Subtract the following binary numbers. Check answers by subtracting the decimal equivalent numbers:

(a) 100 − 011   (b) 101 − 010   (c) 1100 − 0110
(d) 1001 − 0111   (e) 1010 − 0101   (f) 1000 − 0111

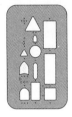

Solution Subtract the numbers bit by bit beginning at the LSB. A borrow is required when subtracting one from zero. When borrowing, the more significant bit position(s) supply the borrow.

(a)	100	4	(b)	101	5	(c)	1100	12
	−011	−3		−010	−2		−0110	− 6
	001	1		011	3		0110	6

(d)	1001	9	(e)	1010	10	(f)	1000	8
	−0111	−7		−0101	− 5		−0111	−7
	0010	2		0101	5		0001	1

Figures 5.4a, b, and c show several additional multibit subtraction problems requiring borrows. The numbers subtracted, the subtrahend from the minuend,

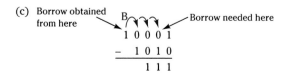

(a)

Binary point
1 0 1 1. ← Minuend
−1 0 0 1. ← Subtrahend
0 0 1 0. ← Difference

Borrows required here

(b)

B   B
1 0 1 1 0 1
−0 1 0 0 1 0
0 1 1 0 1 1

(c) Borrow obtained from here   Borrow needed here
B
1 0 0 0 1
− 1 0 1 0
1 1 1

Figure 5.4
Multibit Binary Subtraction

are aligned along the binary point and subtraction takes place from right to left, 1-bit at a time. Numbers subtracted without borrowing have their difference recorded below the bits. When borrowing is required, the next more significant bit position equal to one provides the borrow. Intermediate bit positions equal to zero are changed to ones during the borrowing process, similar to the way zeros are changed to nines in decimal subtraction.

DESIGN EXAMPLE 5.4

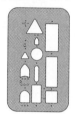

Subtract the following:
(a) 11011.01 − 10110.101
(b) 100000 − 1
(c) 1110011 − 1100101

Solution   Subtract following the normal procedure. Align the minuend and subtrahend along the binary point.

(a)	(b)	(c)
11011.010	100000	1110011
−10110.101	−000001	−1100101
00100.101	011111	0001110

```
                                                           1 0 1
                       1 0 1                          1 1 )1 1 1 1
                     ×   1 1                              − 1 1
                       1 0 1                                1 1
                     + 1 0 1                              − 1 1
                       1 1 1 1                              0 0


                                                           1 1 0 1
                     1 1 0 1                      1 0 1 )1 0 0 0 0 0 1
                   ×   1 0 1                            − 1 0 1
                     1 1 0 1                              1 1 0
                   + 1 1 0 1 0                          − 1 0 1
                     1 0 0 0 0 0 1                        1 0 1
                                                       − 1 0 1


                                                           1 0 1 0 1
                   1 0 1 0.1              1 0 0.0 1 )1 0 1 1 0 0.1 0 1
                 ×   1 0 0.0 1                       − 1 0 0 0 1
                   1 0 1 0 1                            1 0 1 0 1
               + 1 0 1 0 1 0 0 0                      − 1 0 0 0 1
                 1 0 1 1 0 0.1 0 1                        1 0 0 0 1
                                                       − 1 0 0 0 1
```

Figure 5.5
Binary Multiplication and Division

5.3 ■ BINARY MULTIPLICATION AND DIVISION

Binary numbers are also multiplied and divided using the same processes for decimal numbers. Figure 5.5 shows multiplication and division examples. Multiplication is achieved by multiplying each bit in the multiplicand with each bit in the multiplier. Begin by multiplying the least significant multiplier bit with each multiplicand bit. Continue the process by successively multiplying every multiplier bit with each bit in the multiplicand. This creates partial products that equal the multiplicand when the multiplier bit is one or that equal zero when the multiplier bit is zero. The first partial product obtained is recorded. Each succeeding partial product is recorded below the first. However, before recording the partial product, we must first shift it to the left by 1-bit. The partial products are then added to obtain the result.

Multiply the following:
(a) 101101 × 110
(b) 110101.1 × 101.01
(c) 11111 × 11

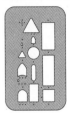

Solution   Multiply following the normal procedure. Record each partial product by left-shifting. Account for the decimal point.

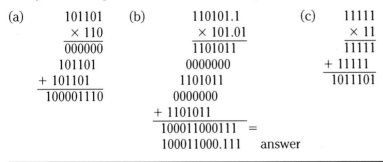

```
(a)       101101      (b)          110101.1     (c)        11111
          × 110                    × 101.01                × 11
          000000                   1101011                 11111
          101101                  0000000               + 11111
        + 101101                   1101011                1011101
         100001110                0000000
                                + 1101011
                                 100011000111   =
                                 100011000.111    answer
```

Binary division occurs by subtracting the divisor from the dividend, beginning with the dividend's MSBs. For instance, if the divisor is 2-bits long, then the subtraction takes place 2-bits at a time. When a subtraction is possible, a one is recorded above the LSB of the subtraction to become part of the result (quotient). If the subtraction cannot take place (without producing a negative difference), a zero is entered in the result. This process continues in the typical "long-hand method" until all bits in the dividend have been "compared" with the divisor.

DESIGN
EXAMPLE 5.6

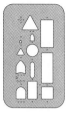

Divide the following:
(a) $100001110 - 110$
(b) $1001.11 - 11$

Solution   Divide using the long-hand method, comparing divisor to dividend.

$$
\begin{array}{r}
101101 \\
\text{(a)} \quad 110\overline{)100001110} \\
-110 \\
\hline
1001 \\
-110 \\
\hline
111 \\
-110 \\
\hline
110 \\
-110 \\
\hline
0
\end{array}
\qquad
\begin{array}{r}
11.01 \\
\text{(b)} \quad 11\overline{)1001.11} \\
-11 \\
\hline
11 \\
-11 \\
\hline
011 \\
-11 \\
\hline
0
\end{array}
$$

Multiplication and division hardware designs vary in complexity and speed. Multiplication, for instance, can be carried out simply by combining addition and shifting. This approach provides basic multiplication capability, but not with the speeds or flexibility of a specialized multiplier chip. Dedicating hardware for these functions, as well as higher level math functions, is an expensive design consideration that is not taken lightly. Computers often combine hardware and software to carry out advanced mathematical processes.

5.4 ■ REDUCING CIRCUITRY WITH COMPLEMENT MATH TECHNIQUES

Complement math techniques save circuitry because they allow binary subtraction to be accomplished with addition circuits. This means that addition circuits, fundamental to most mathematical operations, can also perform subtraction and thereby reduce the hardware expense and space requirements of mathematical circuitry. Complement numbers are the standard way to represent negative numbers in most computing equipment. Two methods are commonly used—1's complement and 2's complement math.

One's Complement Subtraction

The 1's complement representation of a binary number is just that—an alternate way to represent a binary number. The 1's complement of a binary number is obtained in one simple step:

■ Complement each binary bit in the number to produce the 1's complement representation of the number.

Thus, the 1's complement representation of 110011101 is:

001100010

Convert the following numbers into 1's complement form:
(a) 011011010
(b) 1111111
(c) 00000
(d) a 4-bit representation of binary 3

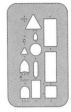

*DESIGN
EXAMPLE 5.7*

Solution Complement each bit in the number to obtain the 1's complement. The 4-bit representation of 3 requires that leading zeros be added to the number to make 3 (11) into a 4-bit number (0011). All bits are then complemented.

(a) 011011010 ⟶ 100100101
(b) 1111111 ⟶ 0000000
(c) 00000   ⟶ 11111
(d) 0011   ⟶ 1100

The simple 1's complement process makes it possible to perform 1's complement subtraction using the following procedure:

- Convert the subtrahend to a 1's complement number.
- Leave the minuend alone.   #5 in 1st Row
- Add the minuend to the 1's complement subtrahend.
- Perform an end around carry.
- Positive answers are indicated by a carry of one. Negative answers are indicated by a carry of zero.
- If the answer is negative, 1's complementing the answer will provide the true magnitude of the difference.

These steps are illustrated in Figure 5.6. The 1's complement subtrahend is added with the original minuend. The final carry from this addition (the MSB of the result) has two purposes. First, it indicates if the result of the subtraction is a positive or negative difference. When the final carry is one, the difference is positive. A zero indicates a negative difference. That is, a larger number was subtracted from a smaller number (more about this below). Second, the final carry is re-added to the sum from the same addition process, a procedure known as **end around carry**. The sum from the end around carry addition is the final answer—the difference between the original minuend and subtrahend.

For the situation when a larger number is subtracted from a smaller number, a negative difference results. We already realize that the final carry being equal to zero (or the lack of a carry, if you prefer) identifies a negative result. In addition, the negative answer is given in 1's complement form. Therefore, if the true magnitude of the answer is desired, obtain the 1's complement of the answer. (The 1's complement of the 1's complement is the original number.)

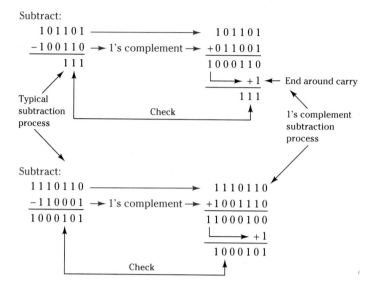

Subtract:

```
  101101  ──────────────────────────→    101101
 -100110  → 1's complement →            +011001
 ───────                                 ───────
     111                                1000110
                                         └────→ +1  ←── End around carry
                                          ──────
Typical                                     111
subtraction              Check
process
                                                          1's complement
                                                          subtraction
                                                          process

Subtract:

 1110110  ──────────────────────────→   1110110
 -110001  → 1's complement →            +1001110
 ───────                                 ────────
 1000101                                11000100
                                         └────→ +1
                                          ───────
                                          1000101
                    Check
```

Figure 5.6
1's Complement
Subtraction

As you can see, 1's complement subtraction produces results identical to conventional subtraction.

The process, inversion and two additions, may seem more complicated than the typical subtraction process, but these are common procedures in basic computer designs. The circuit savings resulting from the complement method makes 1's or 2's complement circuits "standard equipment" in computing systems.

DESIGN EXAMPLE 5.8

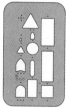

Subtract the following using the 1's complement subtraction process. Subtract the numbers conventionally to verify your answer:
(a) 1011 − 1001
(b) 1110101 − 101101

Solution   Add the 1's complement of the subtrahend to the original minuend. Complete the process with an end around carry.

(a)
```
   1011   ──────→       1011
  -1001   —1's→    +   0110
  ─────              ───────
   0010              10001 —final carry = 1
                       └→+1 —end around carry
                      ─────
                      0010    answer
```

(b)
```
   1110101  ──────→       1110101
 -  101101  —1's→     +   1010010
 ────────             ─────────
  1001000             11000111
                        └──→+1
                       ─────────
                       1001000
```

Subtract the following using the 1's complement subtraction process:
(a) $1001 - 1011$
(b) $100110 - 111010$

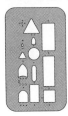

Solution Add the 1's complement of the subtrahend to the original minuend. A negative result is indicated by the final carry equal to zero. Take the 1's complement of the answer to find the true magnitude.

(a) $\begin{array}{r} 1001 \\ -1011 \\ \hline 0010 \end{array}$ ⟶ —1's→ $\begin{array}{r} 1001 \\ +\ 0100 \\ \hline 01101 \end{array}$ —final carry = 0, neg
$\quad\quad\quad\quad\quad\quad\quad\quad\quad$ ↳+0—end around carry is not necessary
$\quad\quad\quad\quad\quad\quad\quad\quad\quad$ $\overline{1101}$ answer in 1's complement form

True magnitude obtained by 1's complementing 1101. Thus, $1101 \rightarrow -0010$.

(b) $\begin{array}{r} 100110 \\ -111010 \\ \hline 10100 \end{array}$ ⟶ —1's→ $\begin{array}{r} 100110 \\ +\ 000101 \\ \hline 0101011 \end{array}$
$\quad\quad\quad\quad\quad\quad\quad\quad\quad$ ↳→+0
$\quad\quad\quad\quad\quad\quad\quad\quad\quad$ $\overline{101011}$
$\quad\quad\quad\quad\quad\quad\quad\quad\quad$ -010100 answer

Two's Complement Subtraction

The most frequently used subtraction method in digital systems is 2's complement subtraction. A 2's complement representation for a binary number is required in the process and is obtained in two steps:

■ Create the 1's complement of the binary number.
■ Add 1 to the 1's complement.

Thus, the 2's complement representation of 110011101 is:

$$001100010 + 1 = 001100011$$

Convert the following numbers into 2's complement numbers:
(a) 110110
(b) 00001
(c) 00000
(d) a 5-bit representation of binary 4.

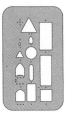

Solution Complement each bit and add 1 to the result. A 5-bit representation of 4 is 00100. Take the 2's complement of this number.

(a) $110110 \longrightarrow 001001 + 1 = 001010$
(b) $00001 \longrightarrow 11110 + 1 = 11111$
(c) $00000 \longrightarrow 11111 + 1 = 100000$
(d) $00100 \longrightarrow 11011 + 1 = 11100$

Using the 2's complement number, 2's complement subtraction is possible with the following procedure:

■ Convert the subtrahend into its 2's complement equivalent number.
■ Leave the minuend alone.
■ Add the minuend to the 2's complement subtrahend.
■ A carry equal to one indicates a positive answer. A carry equal to zero indicates a negative answer. The carry is not used as part of the answer.
■ If the answer is negative, it will be in 2's complement form. Take the 2's complement of this number to determine the true magnitude of the difference.

Figure 5.7 shows the details. The 2's complement version of the subtrahend is added to the minuend, providing the difference between the original minuend and subtrahend. No end around carry is needed, so the answer is obtained immediately after the addition. As was the case with 1's complement subtraction, the final carry generated from the addition of the minuend and 2's complement subtrahend indicates when a negative or positive difference is obtained. When the final carry is one, a positive difference is indicated and the answer is the output of the addition process. (The final carry is ignored as part of the answer.)

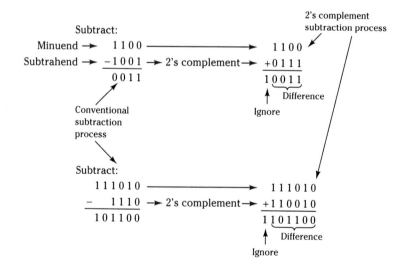

Figure 5.7
2's Complement
Subtraction

The final carry is zero (referred to as no carry) when the difference between the original minuend and subtrahend is negative. If this is the case, then one additional step is required because the negative difference is given in 2's complement form. Thus, 2's complementing this answer will produce the magnitude of the difference.

Subtract the following using the 2's complement subtraction process:
(a) 1101 − 1011
(b) 1110101 − 101101

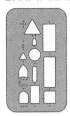

Solution   Add the 2's complement subtrahend to the original minuend.

(a)   1101 ⟶ 1101   (b)   1110101 ⟶ 1110101
    −1011 —2's→ + 0101   − 101101 —2's→ + 1010011 —2's→
    ‾‾‾‾   ‾‾‾‾‾‾‾   ‾‾‾‾‾‾‾‾
     0010   10010   1001000   11001000
                  ↑   ↑
          positive answer   positive answer
          (ignore carry as   (ignore carry as
          part of the answer)   part of the answer)

Subtract using the 2's complement process.
(a) 1001 − 1011
(b) 100110 − 111010

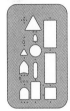

Solution   Add the 2's complement subtrahend to the original minuend. Negative answers require 2's complementing to be displayed in true magnitude form.

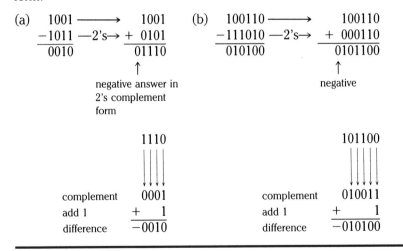

(a)   1001 ⟶ 1001   (b)   100110 ⟶ 100110
    −1011 —2's→ + 0101   −111010 —2's→ + 000110
    ‾‾‾‾   ‾‾‾‾‾‾   ‾‾‾‾‾‾
     0010   01110   010100   0101100
                  ↑   ↑
          negative answer in   negative
          2's complement
          form

                    1110   101100
                    ↓↓↓↓   ↓↓↓↓↓↓
    complement   0001   complement   010011
    add 1   +   1   add 1   +   1
    difference   ‾‾‾‾‾‾   difference   ‾‾‾‾‾‾‾
                  −0010   −010100

5.5 ■ SIGNED NUMBERS

The numbers used for illustration purposes so far have been **unsigned numbers;** we just assumed they are positive, except for a few specific cases when a negative result was obtained. Negative numbers, of course, are found in many calculations and the various addition and subtraction methods used in digital systems must handle them. In the decimal world a negation sign is assigned

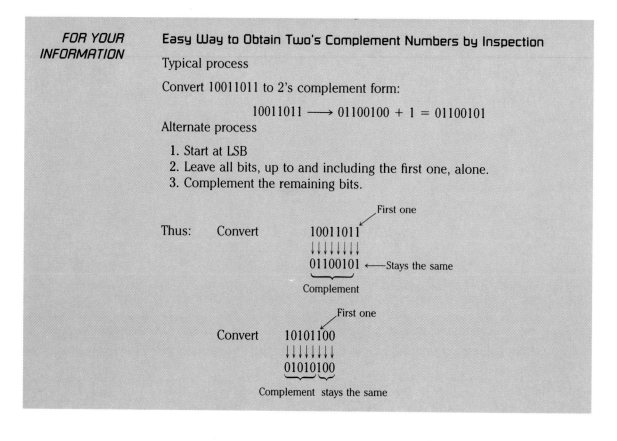

Easy Way to Obtain Two's Complement Numbers by Inspection

Typical process

Convert 10011011 to 2's complement form:

$$10011011 \longrightarrow 01100100 + 1 = 01100101$$

Alternate process

1. Start at LSB
2. Leave all bits, up to and including the first one, alone.
3. Complement the remaining bits.

Thus: Convert   First one
                       10011011
                       ↓↓↓↓↓↓↓↓
                       01100101 ←—Stays the same
                       ‿‿‿‿‿‿
                       Complement

          Convert   First one
                       10101100
                       ↓↓↓↓↓↓↓↓
                       01010100
                       ‿‿‿‿ ‿‿
              Complement stays the same

to any number considered negative. In the binary world an additional bit is used to represent the sign of the number. Typically, a zero indicates a positive number and a one indicates a negative number. Numbers containing a **sign bit** are **signed numbers**.

We must now face the fact that digital circuitry limits the magnitude of numbers. The number 3, for instance, can be written in binary in many ways—11, 011, 00011, 00000000011, and so on. The number of bits used in the representation is constrained by the fact that each bit requires physical digital hardware to store, add, and otherwise manipulate the number. Economics and practicality limit the maximum "word size" or length of a number. For example, many computers deal with 8-bit (one byte) numbers simply because the computer circuitry is designed for 8-bit quantities. The maximum length of an unsigned number in these machines will be 8-bits, which allows for the representation of the decimal numbers 0 through 255 (00000000 through 11111111). When signed numbers are represented in the same system, 1-bit is dedicated to the sign, leaving only 7-bits for the magnitude.

Figure 5.8 shows several examples of signed positive and negative numbers. For positive numbers the sign bit is zero and is placed before the magnitude. Negative numbers can be represented in three ways—signed magni-

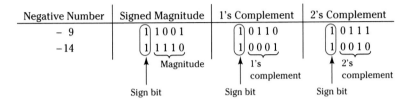

Positive Number	5-bit Representation	8-bit Representation
+12	Sign → ⎡0⎤ 1 1 0 0	⎡0⎤ 0 0 0 1 1 0 0
+ 5	bits ⎢0⎥ 0 1 0 1	⎢0⎥ 0 0 0 0 1 0 1
+ 9	⎣0⎦ 1 0 0 1	⎢0⎥ 0 0 0 1 0 0 1
+25	Not possible	⎣0⎦ 0 0 1 1 0 0 1

Magnitude

Sign bits

Negative Number	Signed Magnitude	1's Complement	2's Complement
− 9	⎡1⎤ 1 0 0 1	⎡1⎤ 0 1 1 0	⎡1⎤ 0 1 1 1
−14	⎣1⎦ 1 1 1 0	⎣1⎦ 0 0 0 1	⎣1⎦ 0 0 1 0
	Magnitude	1's complement	2's complement
	Sign bit	Sign bit	Sign bit

Converting −9 to 2's complement form:

① Start with +9 ———————————→   1 0 0 1
② Assign a positive sign bit ———→ 0 1 0 0 1
③ Convert to 2's complement form   1 0 1 1 0
                                   + 1
                              ──────────
                                 1 0 1 1 1 ——→ −9

Figure 5.8
Signed Number
Representation

tude, 1's complement, and 2's complement forms. Signed magnitude notation simply tacks a sign bit onto the number. This is easy to do, but this notation doesn't take advantage of the complement subtracters common to most machines. Therefore, 1's and 2's complement notations are the preferred representation for negative numbers. Converting a negative number to complement representation is accomplished using previously discussed procedures, with the addition that a sign bit is included. The easiest way to convert a negative number into complement form is to start with the positive representation of the number, then convert the number and sign bit into the 1's or 2's complement form desired. Figure 5.8 also shows this procedure.

Express the following positive numbers as 8-bit signed magnitude binary numbers:
(a) +12   (b) +8   (c) +25

DESIGN
EXAMPLE 5.13

Solution   Use 1-bit for the sign and 7-bits to represent the magnitude.

(a) 00001100   (b) 00001000   (c) 00011001
    ↑
   sign

DESIGN
EXAMPLE 5.14

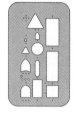

Express the following negative numbers as 8-bit signed numbers in signed magnitude, 1's complement, and 2's complement notation:
(a) -15 (b) -38 (c) -106

Solution Seven-bits are used for the magnitude in each notation. Start with the basic positive magnitude of the number and then convert into the appropriate form.

signed magnitude:

(a) 10001111   (b) 10100110   (c) 11101010
    ↑
   sign

1's complement:

(a) 00001111 $\longrightarrow$ 11110000
(b) 00100110 $\longrightarrow$ 11011001
(c) 01101010 $\longrightarrow$ 10010101

2's complement:

(a) 00001111 $\longrightarrow$ 11110001
(b) 00100110 $\longrightarrow$ 11011010
(c) 01101010 $\longrightarrow$ 10010110

As mentioned earlier, the digital hardware used in a system limits the possible range of numbers. In a signed magnitude system the range of numbers permissible is 2^n, where n is the number of bits (excluding the sign bit). Therefore, if 7-bits are used to represent the magnitude of the number, the range is from $+127$ (01111111) through -127 (11111111). If only three magnitude bits were used, the range would be from $+7$ (0111) to -7 (1111).

The range differs slightly with complement notation. Using 7-bits for the magnitude, the permissible range for numbers in a 1's complement system is $+127$ (01111111) through -127 (10000000). In this system negative numbers are represented in complement form; positive numbers are listed normally.

A 7-bit representation in a 2's complement system yields a range from $+127$ (01111111) through -128 (10000000). Notice that one additional number is possible with 2's complement notation.

List the complete sequence of 4-bit signed numbers allowed using signed magnitude, 1's complement, and 2's complement notation.

Solution   One bit is required for the sign, leaving 3-bits for the magnitude. List decimal numbers first to establish the range. Then convert to the appropriate form:

Decimal	Signed	1's Complement	2's Complement
+7	0111	0111	0111
+6	0110	0110	0110
+5	0101	0101	0101
+4	0100	0100	0100
+3	0011	0011	0011
+2	0010	0010	0010
+1	0001	0001	0001
0	0000	0000	0000
−1	1001	1110	1111
−2	1010	1101	1110
−3	1011	1100	1101
−4	1100	1011	1100
−5	1101	1010	1011
−6	1110	1001	1010
−7	1111	1000	1001
−8		not possible	1000

DESIGN EXAMPLE 5.15

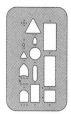

Addition and Subtraction with Signed Numbers

Figures 5.9a and b illustrate several examples of how 2's complement addition and subtraction are carried out. When adding or subtracting signed numbers, all negative numbers must be expressed in 2's complement form prior to the addition. Positive numbers are used in their normal positive representations. The results of the addition or subtraction process, when positive, include a zero sign bit and the magnitude answer. When the results are negative, the sign bit is one and the magnitude is a 2's complement number. Taking the 2's complement of the magnitude produces the absolute value of the answer. Any carries generated beyond the sign bit during the process are ignored. Although not specifically shown, the process using 1's complement math is similar.

Adding

There are several sign possibilities for addition. That is, numbers may be added together that have the same sign (both positive or both negative) or opposite

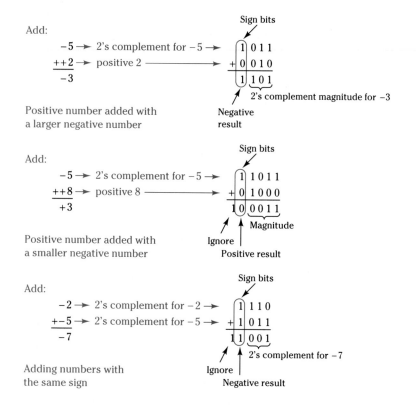

Add:

$$
\begin{array}{rcl}
-5 & \rightarrow \text{2's complement for } -5 \rightarrow & \boxed{1}\,0\,1\,1 \\
++2 & \rightarrow \text{positive 2} \longrightarrow & +\boxed{0}\,0\,1\,0 \\
\hline
-3 & & \boxed{1}\,1\,0\,1 \\
\end{array}
$$

Sign bits

2's complement magnitude for −3

Positive number added with
a larger negative number

Negative
result

Add:

$$
\begin{array}{rcl}
-5 & \rightarrow \text{2's complement for } -5 \rightarrow & \boxed{1}\,1\,0\,1\,1 \\
++8 & \rightarrow \text{positive 8} \longrightarrow & +\boxed{0}\,1\,0\,0\,0 \\
\hline
+3 & & 1\boxed{0}\,0\,0\,1\,1 \\
\end{array}
$$

Sign bits

Magnitude

Positive number added with
a smaller negative number

Ignore

Positive result

Add:

$$
\begin{array}{rcl}
-2 & \rightarrow \text{2's complement for } -2 \rightarrow & \boxed{1}\,1\,1\,0 \\
+-5 & \rightarrow \text{2's complement for } -5 \rightarrow & +\boxed{1}\,0\,1\,1 \\
\hline
-7 & & 1\boxed{1}\,0\,0\,1 \\
\end{array}
$$

Sign bits

2's complement for −7

Adding numbers with
the same sign

Ignore

Negative result

Subtract:

$$
\begin{array}{rl}
-9 & \text{Convert } -9 \text{ to} \\
-+5 & \text{2's complement} \\
\hline
-14 & \text{form}
\end{array}
\qquad
\begin{array}{l}
1\,0\,1\,1\,1 \\
-0\,0\,1\,0\,1 \rightarrow \text{2's complement} \rightarrow
\end{array}
\qquad
\begin{array}{l}
1\,0\,1\,1\,1 \\
+1\,1\,0\,1\,1 \\
\hline
1\,1\,0\,0\,1\,0
\end{array}
$$

2's complement form

Ignore

Negative

Thus:

$0\,0\,1\,0 = -1\,1\,1\,0$

$$
\begin{array}{rl}
+8 & \text{Convert } -3 \text{ to} \\
--3 & \text{2's complement} \rightarrow \\
\hline
\times 11 & \text{form}
\end{array}
\qquad
\begin{array}{l}
0\,1\,0\,0\,0 \\
-1\,1\,1\,0\,1 \rightarrow \text{2's complement} \rightarrow
\end{array}
\qquad
\begin{array}{l}
0\,1\,0\,0\,0 \\
+0\,0\,0\,1\,1 \\
\hline
0\,1\,0\,1\,1 = +1\,1
\end{array}
$$

Positive

Figure 5.9
(a) Adding Signed
Numbers
(b) Subtracting
Signed Numbers

$$
\begin{array}{rl}
-5 & \text{Convert } -5 \\
--7 & \text{AND } -7 \text{ to} \\
\hline
+2 & \text{2's complement form}
\end{array}
\qquad
\begin{array}{l}
1\,0\,1\,1 \\
-1\,0\,0\,1 \rightarrow \text{2's complement} \rightarrow
\end{array}
\qquad
\begin{array}{l}
1\,0\,1\,1 \\
+0\,1\,1\,1 \\
\hline
1\,0\,0\,1\,0 = +2
\end{array}
$$

Ignore

Positive

signs (one positive and one negative). Two positive numbers are added together in the typical fashion. Both the magnitudes and sign bits are included in the addition. The result will be a positive number as indicated by the zero sign bit. (Note that the sign bit meaning is the opposite of that discussed earlier with unsigned numbers.)

Adding two negative numbers necessitates that both numbers be expressed in 2's complement form. Then addition takes place in the normal fashion and a negative answer is obtained. Since negative sums are in 2's complement form, 2's complementing the sum restores the number to actual magnitude form.

Adding numbers with unlike signs will produce either a positive or a negative number, depending on the relative magnitudes of the two numbers. When a negative number is added to a larger positive number, a positive result occurs. The negative number, expressed in 2's complement form, is added to the larger positive number. The resulting sign bit will indicate a positive sum.

A negative result is obtained when adding a negative number to a smaller positive number. As always, the negative number is obtained from the addition process in 2's complement form and must be restored by 2's complementing if the actual magnitude is desired.

Add the following 8-bit (sign bit and 7-bit magnitude) signed numbers using 2's complement arithmetic.
(a) $+12 + -25$ (b) $-38 + -57$ (c) $+58 + -3$

Solution   Express negative numbers in 2's complement form prior to adding. Negative sums will be obtained in 2's complement form as well.

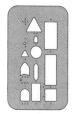

(a)
$$
\begin{array}{r}
+12 \longrightarrow \quad 00001100 \\
+ -25 \text{ —2's→ } +11100111 \\
\hline
-13 \qquad \underline{11110011} = 0001101 \text{ magnitude} \\
\uparrow \\
\text{sign}
\end{array}
$$

(b)
$$
\begin{array}{r}
-38 \text{ —2's→ } \quad 11011010 \\
+ -57 \text{ —2's→ } + \ 11000111 \\
\hline
-95 \qquad \underline{110100001} = 01011111 \text{ magnitude} \\
\uparrow\uparrow \\
\text{ignore} \quad\quad \text{sign}
\end{array}
$$

(c)
$$
\begin{array}{r}
+58 \longrightarrow \quad 00111010 \\
+ - 3 \text{ —2's→ } + \ 11111101 \\
\hline
+55 \qquad \underline{100110111} = \text{ magnitude (55)} \\
\uparrow\uparrow \\
\text{ignore} \quad\quad \text{sign}
\end{array}
$$

Subtraction

The subtraction examples illustrated in Figure 5.9b are carried out using 2's complement subtraction. The method used is the same one described earlier for unsigned numbers—2's complement the subtrahend and add it to the minuend—except that we now must use the sign bit and express any negative numbers in 2's complement form. This 'means that a 2's complement subtrahend will be used for the basic subtraction process and any other negative numbers will also require 2's complementing.

For instance, when subtracting a positive number from a negative number (such as $-9 - 5$), the negative minuend (-9) is converted into 2's complement form and added to the 2's complement subtrahend (5). A negative number will result in this example as indicated by the resulting sign bit. As with unsigned numbers, any negative result shows up as a 2's complement number and must be converted (2's complementing) to obtain the true magnitude.

The process gets interesting when subtracting a negative number in the subtrahend from a minuend. The negative number will undergo the 2's complement process twice. Once because negative numbers are represented in their 2's complement form; second because the subtrahend is always 2's complemented in 2's complement subtraction. Two's complementing a 2's complement number reverts the number back to its original form.

In addition to the examples in Figure 5.9, Design Example 5.17 shows several combinations of positive and negative numbers undergoing 2's complement subtraction.

DESIGN EXAMPLE 5.17

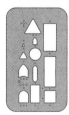

Using 2's complement methods, subtract the following 8-bit (sign bit and 7-bit magnitude) signed numbers:
(a) $-28 - -16$
(b) $-43 - -93$
(c) $-18 - +6$
(d) $+45 - -33$

Solution   Use the 2's complement subtraction process including the sign bits. All negative numbers are processed in 2's complement form. Negative results are also in 2's complement form. Any carries generated beyond the sign bit are ignored.

(a) $+28 \longrightarrow 00011100$ $-28 = -2\text{'s} \rightarrow 11100100$
       $+16 \longrightarrow 00010000$ $-16 = -2\text{'s} \rightarrow 11110000$

$$
\begin{array}{lll}
-28 \longrightarrow & 11100100 \longrightarrow & 11100100 \\
- -16 \longrightarrow & -11110000 \;-2\text{'s} \rightarrow & +00010000 \\
\hline
-12 & & 11110100
\end{array}
$$

sign $(-)$ $\rfloor 0001100 = -12$

(b) $+43 \longrightarrow 00101011 \qquad -43 = -2\text{'s} \rightarrow 11010101$
     $+93 \longrightarrow 01011101 \qquad -93 = -2\text{'s} \rightarrow 10100011$

$$
\begin{array}{r}
-43 \longrightarrow 11010101 \longrightarrow \\
- \; -93 \longrightarrow -10100011 -2\text{'s} \rightarrow + \; 01011101 \\
\hline
+50 \qquad\qquad\qquad 11010101 \\
\end{array}
$$

$$100110010 = +50$$

ignore ⌐ ¬ sign (+)

(c) $+18 \longrightarrow 00010010 \qquad -18 = -2\text{'s} \rightarrow 11101110$
     $+6 \longrightarrow 00000110$

$$
\begin{array}{r}
-18 \longrightarrow 11101110 \longrightarrow 11101110 \\
- \; + \; 6 \longrightarrow -00000110 -2\text{'s} \rightarrow + \; 11111010 \\
\hline
-24 \qquad\qquad\qquad 111101000 \\
\end{array}
$$

ignore ⌐ $0011000 = -24$
       ¬ sign (−)

(d) $+45 \longrightarrow 00101101$
     $+33 \longrightarrow 00100001 \qquad -33 = -2\text{'s} \rightarrow 11011111$

$$
\begin{array}{r}
+45 \longrightarrow 00101101 \longrightarrow 00101101 \\
- \; -33 \longrightarrow -11011111 -2\text{'s} \rightarrow +00100001 \\
\hline
+78 \qquad\qquad\qquad 01001110 = +78 \\
\end{array}
$$

sign (+) ⌐

Overflow

One problem that can occur when adding or subtracting signed numbers is overflow, shown by example in Figure 5.10. Overflow occurs when the calculated result contains more bits than the circuitry can handle. In the figure 9 added to 8—both 4-bit numbers plus a sign bit—generates an answer of 17. A binary 17 (10001) is 5-bits long (6-bits including the sign bit). Since the suggested circuitry processing this example can only handle four magnitude bits and a sign bit, a 6-bit result is impossible, accounting for the incorrect level of the sign bit. An overflow condition is possible when adding two positive numbers or two negative numbers together since these combinations produce

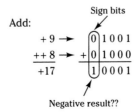

Add:

Sign bits

$$
\begin{array}{r}
+9 \rightarrow \quad 0\,1\,0\,0\,1 \\
++8 \rightarrow + 0\,1\,0\,0\,0 \\
\hline
+17 \qquad 1\,0\,0\,0\,1 \\
\end{array}
$$

Negative result??

Figure 5.10
Overflow

a result larger in magnitude than the individual numbers. Using circuitry that compares the sign of the result to the sign of the input numbers, we can detect overflow conditions.

DESIGN EXAMPLE 5.18

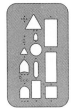

Develop the equation for a circuit that detects an adder overflow condition.

Solution   Overflow occurs when:
(a) adding two positive numbers produces a negative result.
(b) adding two negative numbers produces a positive result.
Therefore, comparing the sign bits of the addend and the augend to the sign bit of the result will determine when overflow occurs. The following truth table lists all possible sign bit combinations—six are normal, two are overflow:

Sign Bit Addend = A	Sign Bit Augend = B	Result Sign Bit = C	Overflow Bit
0	0	0	0
0	0	1	1 $\overline{A}\overline{B}C$
0	1	0	0
0	1	1	0
1	0	0	0
1	0	1	0
1	1	0	1 $AB\overline{C}$
1	1	1	0

$$\text{Overflow} = \overline{A}\overline{B}C + AB\overline{C}$$

5.6 ■ ADDITION CIRCUITS

This chapter has clearly established the need for addition circuitry since we have demonstrated that both binary addition and subtraction, signed and unsigned, may be carried out using the addition process. This important section develops the addition circuitry required for these tasks and also provides a major milestone in our study of digital design and computer circuitry. Now we will see how simple gates can actually carry out mathematical operations. This transition takes us beyond the study of simple logic gates to the development of the computing circuitry that makes high speed computers possible.

The Half Adder

The **half adder** is the fundamental building block of addition circuitry. We design this simple, yet powerful building block by examining the truth table in Figure 5.11. Described by this table are all possible outcomes occurring through the addition of two binary bits. As we can see, half adders have two outputs, since the addition of 2-bits yields a sum and the potential for a carry. For instance, zero added to one produces a sum equal to one and a carry

Adding two binary bits—X with Y

X	Y	Carry	Sum	Equivalent Math Operation
0	0	0	0	$0 + 0 = 0\,0$
0	1	0	1	$0 + 1 = 0\,1$
1	0	0	1	$1 + 0 = 0\,1$
1	1	1	0	$1 + 1 = 1\,0$

AND   Exclusive-OR function
function

Figure 5.11
Half Adder Truth Table

equal to zero. The column labeled "equivalent math operation" shows standard addition equations for convenience. Exercise care, since both OR and addition are represented by the plus (+) sign. Naturally, since we are indicating addition, $1 + 1 = 10$.

Close inspection of the table in Figure 5.11 shows that the sum column is an Exclusive-OR function. Thus, the sum of 2-bits is obtained by Exclusive-ORing the bits together. The carry is created by ANDing the two input bits since the only time a carry is generated when adding two single bits is when both bits are ones. The combination of an Exclusive-OR and an AND function creates a half adder, as shown in Figure 5.12. Although elementary in function, the half adder is utilized as a building block for larger addition circuits. In other words, we can design adders with greater computational capacity using half adders instead of resorting to individual gate design for each possible combination of input bits. In this way we see how logic circuits are improved and designed using previous concepts and proven ideas.

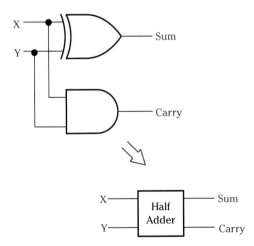

Figure 5.12
Half Adder Circuit

The Full Adder

The half adder clearly demonstrates that binary addition is possible with the simple logic gates we have studied. Since many mathematical functions are derived from addition (multiplication, for instance, is repetitive addition), it makes sense that advanced adder capability will contribute to advanced mathematical capability. The half adder is limited in scope since most arithmetic operations involve more than two binary bits. Therefore, we use half adders to construct a **full adder**—a circuit capable of adding three binary bits. The full adder is still limited in terms of overall computational power, but it represents another stepping stone toward a full-scale adder design.

First, we define all possible outcomes occurring when three binary bits are added so that we may note the functional capabilities of the full adder. The truth table in Figure 5.13 lists these capabilities—the sums and carries resulting from 3-bit addition. (Notice that 3-bit addition generates a single sum and carry; the full adder has two outputs.) We design a full adder to match these outcomes.

Addition of three binary bits—$X + Y + Z$

X	Y	Z	Carry	Sum	Equivalent Math Operation
0	0	0	0	0	$0 + 0 + 0 = 0\ 0$
0	0	1	0	1	$0 + 0 + 1 = 0\ 1$
0	1	0	0	1	$0 + 1 + 0 = 0\ 1$
0	1	1	1	0	$0 + 1 + 1 = 1\ 0$
1	0	0	0	1	$1 + 0 + 0 = 0\ 1$
1	0	1	1	0	$1 + 0 + 1 = 1\ 0$
1	1	0	1	0	$1 + 1 + 0 = 1\ 0$
1	1	1	1	1	$1 + 1 + 1 = 1\ 1$

Figure 5.13
Full Adder Truth Table

We could begin the design by using truth tables and Karnaugh maps to produce a working circuit—an approach we will use later. But initially, let's apply our understanding of binary addition and new knowledge of half adder circuitry to change our approach from rote design techniques to intuitive design. You will find that an intuitive design approach based on the fundamental understanding of a process is very helpful as you tackle more advanced projects.

Figure 5.14 shows one way to view 3-bit addition that is consistent with the information on the full adder truth table. Here, the addition of $1 + 1 + 1$ is carried out on paper in a conventional manner, 2-bits at a time. This implies that 3-bit addition can be accomplished using 2-bit addition, so we adapt half adder building blocks to the process and see what happens. The first step toward this design is shown in Figure 5.15.

Adding $X + Y + Z$ when   X = 1
                                                            Y = 1
                                                            Z = 1

$$
\begin{array}{rl}
X \rightarrow & 1 \\
Y \rightarrow & +\ 1
\end{array} \Big\} \text{2-bit add}
$$

Intermediate carry $\rightarrow$ 1 0 $\longleftarrow$ Intermediate sum

$$
\begin{array}{rl}
 & 1\ 0 \\
Z \rightarrow & +\ 1
\end{array} \Big\} \text{2-bit add}
$$

                        1 1

Final carry ———↑   ↑—— Final sum

Figure 5.14
Adding Three Single Bits

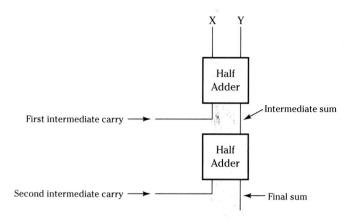

Figure 5.15
Cascading Half Adders

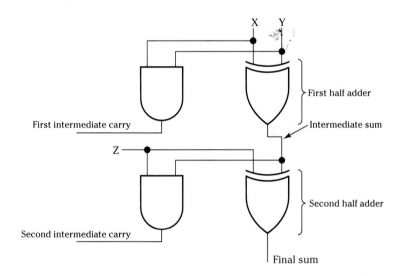

X	Y	Z	Carry	Sum	Intermediate Sum	Final Sum	First Carry	Second Carry	First & Second Carries ORed
0	0	0	0	0	0	0	0	0	0
0	0	1	0	1	0	1	0	0	0
0	1	0	0	1	1	1	0	0	0
0	1	1	1	0	1	0	0	1	1
1	0	0	0	1	1	1	0	0	0
1	0	1	1	0	1	0	0	1	1
1	1	0	1	0	0	0	1	0	1
1	1	1	1	1	0	1	1	1	1

Match

Match

Figure 5.16
Determining the Carry
Function

Inputs X and Y are added together with a half adder to produce an intermediate sum and carry. From the Figure 5.14 problem description it is clear that the intermediate results are not the final results since only 2 of the 3-bits have been added. Notice that on paper the intermediate sum is added to input Z. That is, the sum of the first 2-bits is added to the third bit. This requires 2-bit addition on paper and is accomplished with circuitry using a second half adder, as indicated in Figure 5.15. It appears that the final sum of the 3-bit addition is the sum output of the second half adder, but the status of the final carry is unclear since our design still leaves us with two intermediate carries. We can be highly confident that the final sum is correctly obtained because our circuit follows the paper addition process quite closely. However, the same cannot be said for the final carry. What can we do?

We get down to the circuit level to determine the nature of the final carry. Our circuit design to this point is redrawn and shown in Figure 5.16. The full adder truth table used earlier to predict the final sums and carries is expanded to include all the partial circuit outputs shown (for both sums and intermediate carries). The truth table data is acquired by testing the circuit for every possible input combination, then recording results. For instance, input combination XYZ = 000 produces an intermediate sum of zero and a first intermediate carry value of zero. The final sum and second intermediate carry are also both zero. These values are entered onto the truth table in appropriate columns, and every input combination is tested in turn. We examine the partial outputs, particularly the intermediate carries, and combine them logically to match the desired outcomes to complete our full adder design.

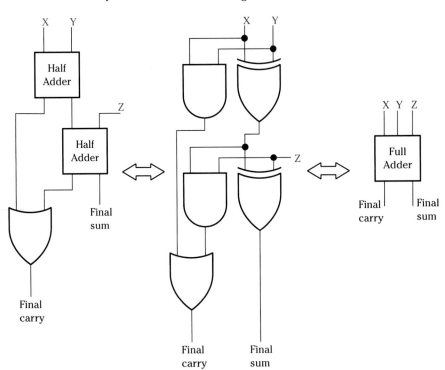

Figure 5.17
Full Adder Circuitry

As you can see, the final sum column matches the projected final sum output column, completing the final sum output design. The first and second individual intermediate carries are not the final carry since their columns do not match that of the projected final carry. However, close inspection shows that ORing the intermediate carries together produces the proper final carry response, completing the full adder design shown in Figure 5.17. Thus, using two half adders and an OR gate provides the 3-bit addition capability that will be necessary for more advanced addition circuits.

Show that the full adder will correctly carry out the addition of $1 + 0 + 1 = 10$.

Solution   Assign the numbers to be added to the full adder inputs—A = 1, B = 0, C = 1—on the logic schematic. Determine each gate's output level to confirm the sum and carry output levels.

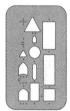

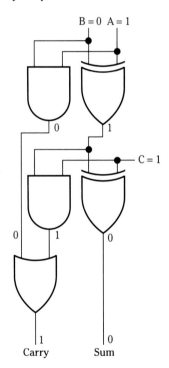

An Alternate Full Adder Design

Figure 5.18 shows a full adder that was designed using Karnaugh maps (K-maps). Using two maps, one for the sum output and the other for the carry output, we can design directly from the full adder truth table. Notice that the sum K-map shows an alternating pattern of ones and zeros, an indication that

X	Y	Z	Carry	Sum
0	0	0	0	0
0	0	1	0	1
0	1	0	0	1
0	1	1	1	0
1	0	0	0	1
1	0	1	1	0
1	1	0	1	0
1	1	1	1	1

Carry:

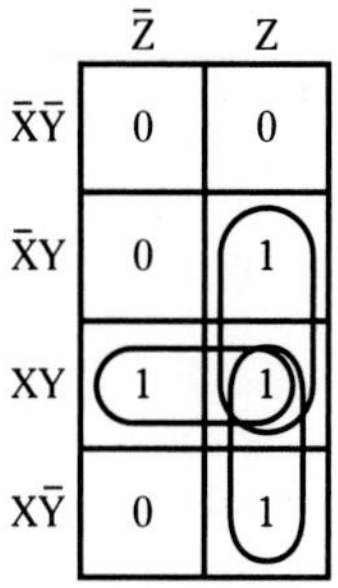
$\bar{Z}$ Z
$\bar{X}\bar{Y}$ 0 0
$\bar{X}Y$ 0 1
XY 1 1
$X\bar{Y}$ 0 1

$\text{Carry} = YZ + XY + XZ$

Sum:

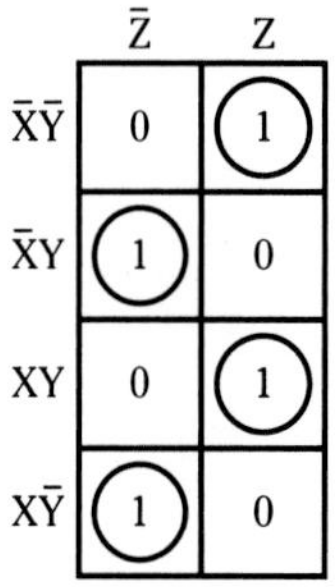
$\bar{Z}$ Z
$\bar{X}\bar{Y}$ 0 1
$\bar{X}Y$ 1 0
XY 0 1
$X\bar{Y}$ 1 0

$\text{Sum} = \bar{X}\bar{Y}Z + \bar{X}Y\bar{Z} + XYZ + X\bar{Y}\bar{Z}$

$\text{Sum} = \bar{X}(\bar{Y}Z + Y\bar{Z}) + X(YZ + \bar{Y}\bar{Z})$

Define $\bar{Y}Z + Y\bar{Z}$ as $Y \oplus Z = A$

Define $YZ + \bar{Y}\bar{Z}$ as $\overline{Y \oplus Z} = \bar{A}$

Then: $\text{Sum} = \bar{X}(A) + X(\bar{A})$

$\text{Sum} = \bar{X}A + X\bar{A} = X \oplus A$

Therefore: $\text{Sum} = X \oplus Y \oplus Z$

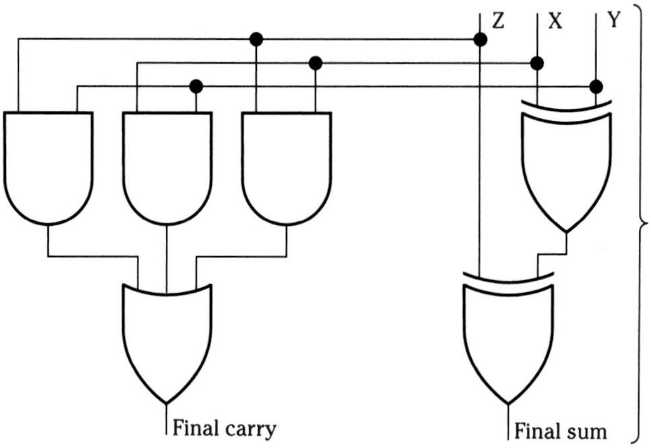
Z
X
Y
Full adder
Final carry
Final sum

Figure 5.18
Alternate Full Adder Design

the Exclusive-OR (or Exclusive-NOR—complement of the EX-OR) function is required. The accompanying Boolean algebra equations show how the Exclusive-ORing of the three input bits leads to the sum circuitry, the same sum circuitry we designed earlier.

The carry circuitry is slightly more complex than our earlier design, but it is easy to obtain using standard sum of products design techniques. Combine both the sum and carry circuits as shown in the circuit diagram and another full adder design is realized.

Confirm that the K-map-derived full adder will correctly add $1 + 1 + 1 = 11$.

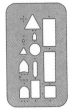

Solution   Apply the input levels to A, B, and C on the full adder schematic. Determine the individual gates' output levels in order to find the final sum and carry outputs.

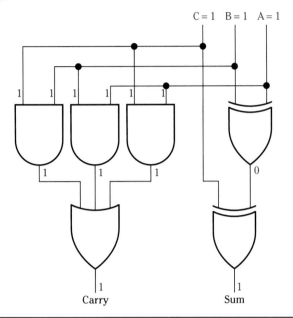

The Ripple Carry Adder

You may be wondering, "Now what? Do I design a circuit to add 4-bits?" No, fortunately. Consider for a moment how two multibit binary numbers are added (e.g., $1101 + 0110$). First, the two LSBs are added together. Carries generated by this addition are added with the 2-bits in the next most significant column. The carry from this addition is added in with the 2-bits of the next most significant column in an identical process, a process that is repeated for as many columns as there are numbers. Viewed as a step-by-step process, other than the least significant column, multibit addition is a series of 3-bit adds. Full adders add 3-bits, so by cascading full adders together, you can create multibit addition circuits. Figure 5.19 illustrates the key concepts behind this

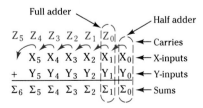

Figure 5.19
Addition of Two 6-Bit Binary Numbers

process. (The LSBs in this example are added using a half adder since only 2-bits are involved.)

Figure 5.20 illustrates how half and full adders are cascaded to create a **ripple carry adder**. The 2-bits from each place in the binary numbers are added with the carry from a previous stage in a full adder. The full adder sum output is retained as part of the complete answer. The full adder carry bit is passed along for further addition to the next most significant stage, except for the last full adder carry. This carry is used as the MSB of the answer. The adder's length is extended by attaching additional full adders onto the circuit so that any length adder is possible. The term ripple carry indicates that a final answer, specifically the MSB of the answer, is not produced by the circuit until all

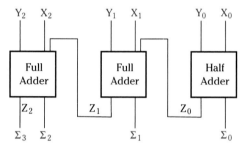

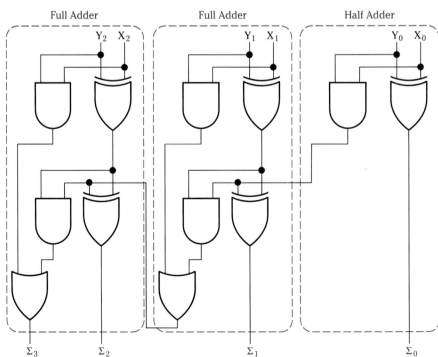

Figure 5.20
3-Bit Ripple Carry Adder

Figure 5.21
Ripple Carry Adder Logic

intermediate carries propagate to the output. Since electrical signals are delayed by a finite amount of time as they pass through each logic gate, a certain amount of time will elapse between the moment numbers are presented to the adder circuit and the moment a stable answer appears at all output pins. We will discuss this aspect of circuit operation in a moment.

Figure 5.21 shows the logical circuitry for the ripple carry adder depicted in Figure 5.20. Each half adder and full adder building block has been replaced by the necessary half adder and full adder logic gates. You can see that a lengthy adder requires a considerable amount of logic, and clearly, we are now dealing with complex and sophisticated logical functions. Nonetheless, it is interesting that all this complexity and sophistication is still derived from basic logic gates.

Design a 5-bit ripple carry adder using half and full adders.

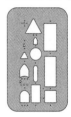

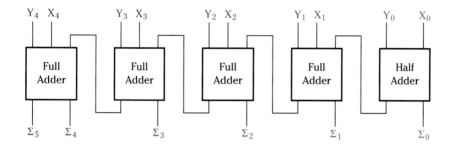

Speed Limitations of the Ripple Carry Adder

The need for arithmetic circuitry is satisfied by the ripple carry adder design. Binary addition can be extended to any length simply by cascading full adders onto the basic adder design. However, ripple carry adders are not suitable for high speed arithmetic calculations since the binary numbers presented to the adder circuit for calculation are subjected to gate propagation delays as their values move through the logic circuitry. The worst propagation delay occurs for the last carry generated in the ripple adder as it ultimately becomes the MSB of the result. The MSB's delay determines the minimum time required for an addition since it is the last bit computed by the adder.

A **critical path analysis** is used to determine the delay times most likely to affect system performance. The critical path is the electrical path from input to output (gates and connections) that determines the overall speed of a digital system. There may be one or more critical paths in a digital system contributing to the overall performance level. In the ripple carry adder the carry path is the critical one since a final answer is not stable until the individual full adder carries filter through the circuitry. In other words, the adder output is invalid until the final carry arrives. Figure 5.22 outlines the critical path for the ripple carry adder shown in Figure 5.21.

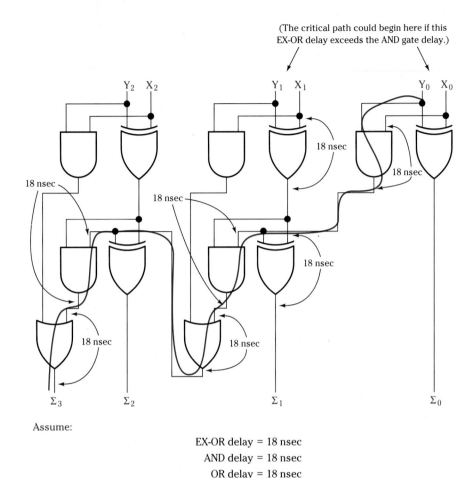

Figure 5.22
Ripple Carry Adder
Delay

Assume:

EX-OR delay = 18 nsec
AND delay = 18 nsec
OR delay = 18 nsec

Total carry delay = 18 nsec × 5 = 90 nsec
Σ_1 delay = 18 nsec × 2 = 36 nsec
Best possible addition rate = 1/90 nsec = 11.1 MHz

The carry from the half adder is the signal facing the longest delay in this example. Using the individual gate delay times given (all gates have 18 nsec delays), the total delay for the carry path is 90 nsec. Note that the delay for the Σ_1 output is only 36 nsec; the carry delay is well over twice the sum output delay. Of course, the carry delay increases as the adder length increases, so a longer adder slows the carry path even more.

The delay time restricts the number of repetitive additions the circuit is capable of performing. Using a 90 nsec carry delay as an example, the best possible repetitive addition rate is 1/90 nsec = 11.1 MHz. This means additions can be carried out at an 11.1 MHz rate if data can be provided to the circuit X and Y inputs that fast. In practical circuits the rate is considerably reduced since time is required to collect and store input and output data. Even so, the primary factor responsible for reducing addition time is the ripple carry delay.

FOR YOUR INFORMATION

Calculating Ripple Carry Delay

Total delay for n stages:

$$\text{Total delay} = n(\text{AND gate delay}) + (n - 1)(\text{OR gate delay})$$
$$\text{Maximum addition rate} = 1/\text{total delay}$$

Example:

$$\text{If AND gate delay} = 18 \text{ nsec}$$
$$\text{and OR gate delay} = 18 \text{ nsec}$$

What is the total delay for a 16-bit adder?

$$\text{Total delay} = 16(18 \text{ nsec}) + (16 - 1)(18 \text{ nsec}) = 558 \text{ nsec}$$
$$\text{Maximum addition rate} = 1/558 \text{ nsec} = 1.79 \text{ MHz}$$

And, with ripple carry design, the delay grows as the adder length is increased. For example, a 16-stage adder with similar gate delays has a carry delay time of 558 nsec. This reduces the maximum addition rate to 1.79 MHz.

Determine the carry delay time for the 5-bit adder design given in Design Example 5.21. Assume an AND gate delay time of 6 nsec and an OR gate delay time of 8 nsec.

DESIGN EXAMPLE 5.22

Half adder = 1 AND gate delay - - - - - - - - - - - - - - - - 6 nsec
Full adder = 1 AND gate delay per stage - - 6 nsec × 4 = 24 nsec
Full adder = 1 OR gate delay per stage - - - 8 nsec × 4 = 32 nsec
Total delay = 62 nsec

Or:

$$\text{Delay} = n \text{ (AND gate delay)} + (n - 1)(\text{OR gate delay})$$
$$\text{Delay} = 5(6 \text{ nsec}) + (5 - 1)(8 \text{ nsec})$$
$$\text{Delay} = 30 \text{ nsec} + 32 \text{ nsec}$$
$$\text{Delay} = 62 \text{ nsec}$$

5.7 ■ USING CARRY LOOK AHEAD CIRCUITS TO INCREASE ADDER SPEED

Any high performance circuit designs are restricted by the slowest critical paths. For high speed mathematical circuitry the carry delay problem must be resolved. Using high speed logic families when constructing the addition circuits may help because a reduction in propagation delay times reduces the carry delay. However, advanced logic circuits are costly and still have delay times. Clever circuit design is needed to put a dent in the delay time problem since the ripple effect is an inherent part of the delay. **Carry look ahead** circuits are used to reduce the delays associated with ripple carry adders.

Carry look ahead circuits take advantage of the fact that carries created in the addition process are predictable. So, rather than waiting for the carries to propagate through the adder circuitry, we use carry look ahead circuitry, which produces the carries directly. Delay times are substantially reduced, but additional hardware is needed. When speed is essential, the trade-off is worthwhile.

Figure 5.23 shows a 3-bit ripple carry adder that has been modified slightly as compared to previous versions. The least significant stage, formerly a half adder, is now a full adder. The full adder input carry, C_0, allows other similar circuits to be connected into an adder cascade, a common design approach. When the least significant chip in the cascade has no input carry, C_0 is held low; when in a cascade, connections are made to other chips via the carry lines.

Figure 5.23 also introduces new terminology. The intermediate sum and intermediate carry terms used earlier are replaced by terms more appropriate for carry look ahead design—**carry propagate** and **carry generate**. These new names only represent a change in name, not a change in function. The half adder Exclusive-OR gate in each full adder stage produces the carry propagate, whereas the half adder AND gate in each stage provides the carry generate. Since there are several carry propagates and generates—one per stage—they must be clearly identified. The least significant stage carry propagate is labeled P_0; the least significant carry generate is labeled G_0. In the next most significant stage they are labeled P_1 and G_1, respectively. In each succeeding stage carry propagates and generates are labeled accordingly.

Each full adder stage in this ripple adder design sends a **circuit carry** to the next stage of the adder. These bits are labeled C_1, C_2, and C_3 in Figure 5.23. Each circuit carry has a corresponding logic expression. For instance:

$$C_1 = C_0 P_0 + G_0$$
$$C_2 = C_1 P_1 + G_1$$
$$C_3 = C_2 P_2 + G_2$$

In general, for any number of full adder stages:

$$C_{n+1} = C_n P_n + G_n$$

Each full adder stage also produces a sum. The expressions for the sums are:

$$E_0 = P_0 \oplus C_0$$
$$E_1 = P_1 \oplus C_1$$
$$E_2 = P_2 \oplus C_2$$

In general, for any number of full adder stages:

$$E_n = P_n \oplus C_n$$

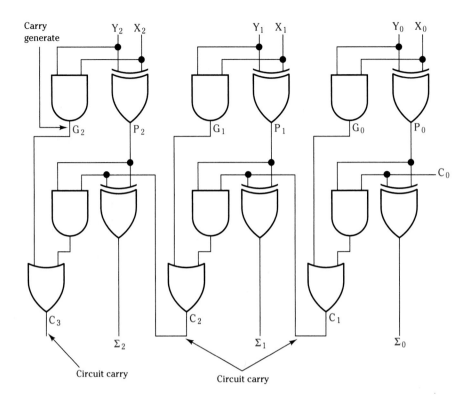

$$C_1 = C_0P_0 + G_0$$

$$C_2 = C_1P_1 + G_1 =$$
$$C_2 = (C_0P_0 + G_0)P_1 + G_1 =$$
$$C_2 = C_0P_0P_1 + G_0P_1 + G_1$$

$$C_3 = C_2P_2 + G_2 =$$
$$C_3 = (C_1P_1 + G_1)P_2 + G_2 =$$
$$C_3 = C_1P_1P_2 + G_1P_2 + G_2 =$$
$$C_3 = (C_0P_0 + G_0)P_1P_2 + G_1P_2 + G_2 =$$
$$C_3 = C_0P_0P_1P_2 + G_0P_1P_2 + G_1P_2 + G_2$$

In general:
$$C_{n+1} = C_nP_n + G_n; \text{ therefore, } C_4 = C_3P_3 + G_3$$

$$\Sigma_0 = P_0 \oplus C_0$$
$$\Sigma_1 = P_1 \oplus C_1$$
$$\Sigma_2 = P_2 \oplus C_2$$

In general:
$$\Sigma_n = P_n \oplus C_n$$

Figure 5.23
Ripple Carry Adder
Carry Equations

These equations are significant since they describe how the various sums and carries are created. Fortunately, a ripple carry adder is not the only way to create the sums and carries described by these equations. We will design a circuit using carry look ahead logic that functions as these equations describe, but with better performance.

The delays present in ripple carry adder design are attributed to the carry path. Therefore, the equations representing the circuit carries (not carry generate) must be modified if circuit performance is to improve. The equations for the circuit sums are based on simple Exclusive-OR gates and need not be changed since they do not contribute to the delay problem.

Recall that:

$$C_1 = C_0P_0 + G_0 \quad \text{and} \quad C_2 = C_1P_1 + G_1$$

Using algebraic substitution, we can change the equation for C_2 as follows:

$$C_2 = (C_0P_0 + G_0)P_1 + G_1 = C_0P_0P_1 + G_0P_1 + G_1$$

This new equation is significant, for it shows that the C_2 carry is produced from carry generates, carry propagates, and, most important, the initial input carry, C_0. The circuit carry C_1 is removed from the C_2 carry logic expression, implying that circuit carry C_1 delays do not affect C_2. Therefore, the C_2 signal is created much faster since the equation describes a circuit that is independent of the carry delay associated with the C_1 stage.

Figure 5.23 shows how circuit carry C_3 logic can also be implemented with only initial carry C_0, carry propagates, and carry generates. In general, every succeeding carry is created without any other circuit carries. This eliminates the ripple carry path and associated delays from the circuit carry logic and is the essence of carry look ahead design. Figure 5.24 shows how equations for

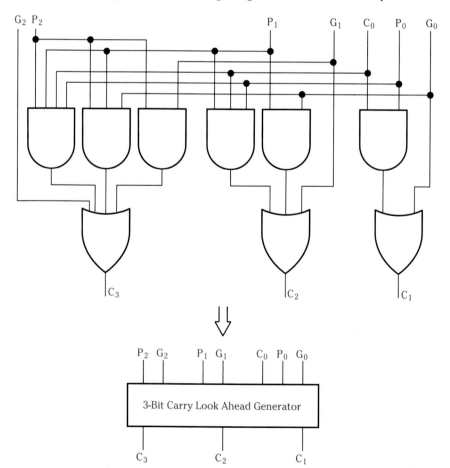

Figure 5.24
Carry Look Ahead
Generator

circuit carries C_1, C_2, and C_3 are transformed into carry look ahead logic. Now, with faster carries, we will create an adder that takes advantage of this new found speed.

Designing an adder around a carry look ahead design is relatively easy. The carry look ahead logic produces the circuit carries, completing a major portion of the adder design. The required carry propagates and carry generates are produced with half adder circuits that feed the look ahead logic. Only the sum logic needs attention. This is straightforward since we have already developed equations for the sums. Refer to Figure 5.23 for specific equations. In general, the equation for any sum output is:

$$E_n = P_n \oplus C_n$$

This general equation shows that sum outputs are obtained by Exclusive-ORing an appropriate carry propagate with an appropriate circuit carry. Therefore, the sum output for any specified adder stage is obtained by Exclusive-ORing the circuit carry for that stage with the carry propagate for that stage. Figure 5.25 shows a 3-bit adder designed with carry look ahead logic.

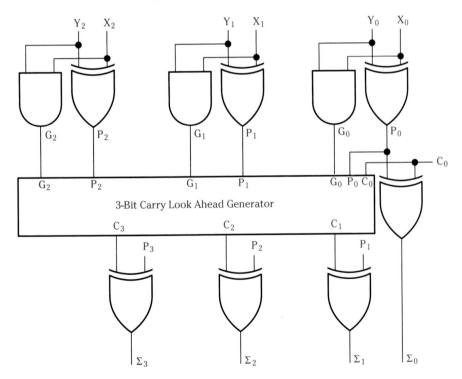

Figure 5.25
3-Bit Adder with Carry
Look Ahead

The 3-bit carry look ahead adder is faster than an equivalent ripple carry adder. The maximum delay from input to a corresponding sum output is limited to four gate delays—the half adder delay, two gate delays in the carry look ahead logic, and the sum Exclusive-OR gate delay. The speed advantage of this adder over an equivalent size ripple carry adder increases as adder size increases. As full adders are cascaded together, the ripple carry adder delay increases, while the carry look ahead adder delay remains the same. In fact, very large adders are designed around multiple carry look ahead building blocks to maximize performance.

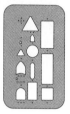

DESIGN EXAMPLE 5.23

Determine the delay for a 3-bit adder designed with carry look ahead. Compare the delay to the 3-bit ripple carry adder used in Figure 5.22.

Solution   The ripple carry delay for the circuit in Figure 5.22 was calculated to be 90 nsec. Using carry look ahead design and gates with 18 nsec delay times, the carry look ahead design delay is:

$$4 \text{ stages of delay} \times 18 \text{ nsec} = 72 \text{ nsec}$$

5.8 ■ USING ADDERS

We have examined adder circuits from the most basic to those that are fairly complex, and have also suggested that more advanced mathematical operations can be carried using the adder. Let's consider for a moment how an adder would be used in a typical computing system.

Naturally, the basic addition operation is a requirement in any computing environment, and from our past work with signed and unsigned numbers it should be evident that whether a signed or unsigned number is run through its circuitry doesn't matter. The same can be said for numbers requiring subtraction. However, taking advantage of complement math requires some circuitry attached to the adder if we are to use it for subtraction. The examples used with 1's and 2's complement math clearly show a method to carry out subtraction with addition. A practical system will have an adder/subtractor circuit based on the same set of adder hardware. Control signals determine which function is carried out.

Next comes multiplication and division. Many options are available when it comes to implementing these functions. Using adders in conjunction with other circuitry to mimic the standard handwritten multiplication and division procedures is one possibility. Designing the hardware from a truth table description is another approach that may be more expensive, but faster. Computers often carry out multiplication and division using software programs. This approach can save lots of circuitry but tends to be slower than using only hardware.

Higher level math functions also rely on software programs but are also carried out with "table lookup" utilizing previously stored values. A combination of specialized hardware and programming is also used. For instance, the sine of an angle can be computed using a MacLaurin series. This allows a fairly complex calculation to be broken down into a series of multiplications, divisions, additions, and subtractions.

The major point is that the adder forms an important starting point in digital computing circuitry. Any specific digital system will have its own set of objectives that dictate which approach to take, but the adder is likely to be an important building block in that approach.

Design a 4-bit (sign bit and three magnitude bits) 2's complement subtractor.

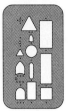

Solution The circuit must be able to 2's complement the subtrahend and add it to the original minuend. Since 4-bit numbers are used, the adder must be 4-bits in length. Rather than use two addition circuits, one for the 2's complement process and one for the numerical addition, the two steps can be combined. This is possible using an adder comprised completely of full adders. By permanently adding 1 to the least significant stage, in conjunction with complementing the subtrahend, we can obtain the 2's complement number.

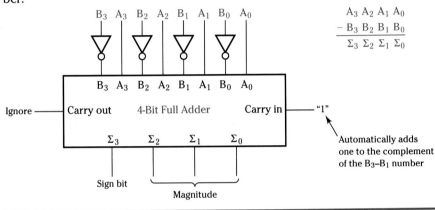

$$\begin{array}{r} A_3\ A_2\ A_1\ A_0 \\ -\ B_3\ B_2\ B_1\ B_0 \\ \hline \Sigma_3\ \Sigma_2\ \Sigma_1\ \Sigma_0 \end{array}$$

Design an 8-bit adder/subtractor circuit capable of performing 2's complement subtraction.

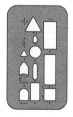

Solution The subtractor design from the previous design example will handle the subtraction portion of the design. Numbers to be added are not complemented before the adder circuitry, so that controlled inversion is required. When subtracting, the subtrahend will be complemented; when adding, it will not. Two 4-bit adders are cascaded in this example to create the required 8-bit adder.

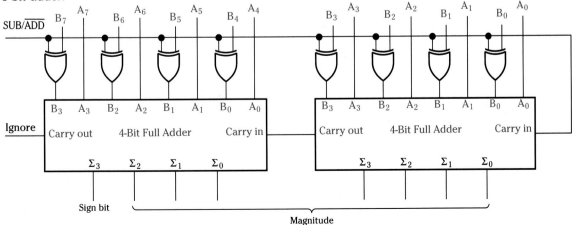

DESIGN EXAMPLE 5.26

Design a circuit capable of multiplying two 2-bit numbers. That is:

$$
\begin{array}{r}
A_1 A_0 \\
\times\ B_1 B_0 \\
\hline
A_1 A_0 B_0 \\
+\ A_1 A_0 B_1 \\
\hline
M_3 M_2 M_1 M_0
\end{array}
$$

Solution   There are many ways to implement the multiplication function. Two-bit multiplication can be accomplished with only a minimum amount of hardware. Create a truth table listing all combinations of $A_1 A_0 \times B_1 B_0$. Then use K-maps to minimize the circuits.

B_1	B_0	A_1	A_0	M_3	M_2	M_1	M_0
0	0	0	0	0	0	0	0
0	0	0	1	0	0	0	0
0	0	1	0	0	0	0	0
0	0	1	1	0	0	0	0
0	1	0	0	0	0	0	0
0	1	0	1	0	0	0	1
0	1	1	0	0	0	1	0
0	1	1	1	0	0	1	1
1	0	0	0	0	0	0	0
1	0	0	1	0	0	1	0
1	0	1	0	0	1	0	0
1	0	1	1	0	1	1	0
1	1	0	0	0	0	0	0
1	1	0	1	0	0	1	1
1	1	1	0	0	1	1	0
1	1	1	1	1	0	0	1

	$\overline{A_1}\,\overline{A_0}$	$\overline{A_1} A_0$	$A_1 A_0$	$A_1 \overline{A_0}$
$\overline{B_1}\,\overline{B_0}$	0	0	0	0
$\overline{B_1} B_0$	0	0	0	0
$B_1 B_0$	0	0	(1)	0
$B_1 \overline{B_0}$	0	0	0	0

$$M_3 = B_1 B_0 A_1 A_0$$

	$\overline{A_1}\,\overline{A_0}$	$\overline{A_1} A_0$	$A_1 A_0$	$A_1 \overline{A_0}$
$\overline{B_1}\,\overline{B_0}$	0	0	0	0
$\overline{B_1} B_0$	0	0	0	0
$B_1 B_0$	0	0	0	(1)
$B_1 \overline{B_0}$	0	0	(1	1)

$$M_2 = B_1 \overline{B_0} A_1 + B_1 A_1 \overline{A_0}$$

	$\overline{A_1}\overline{A_0}$	$\overline{A_1}A_0$	A_1A_0	$A_1\overline{A_0}$
$\overline{B_1}\overline{B_0}$	0	0	0	0
$\overline{B_1}B_0$	0	0	1	1
B_1B_0	0	1	0	1
$B_1\overline{B_0}$	0	1	1	0

	$\overline{A_1}\overline{A_0}$	$\overline{A_1}A_0$	A_1A_0	$A_1\overline{A_0}$
$\overline{B_1}\overline{B_0}$	0	0	0	0
$\overline{B_1}B_0$	0	1	1	0
B_1B_0	0	1	1	0
$B_1\overline{B_0}$	0	0	0	0

$$M_1 = B_1\overline{A_1}A_0 + B_1\overline{B_0}A_0 + \overline{B_1}B_0A_1 + B_0A_1\overline{A_0}$$
$$M_1 = A_0B_1(\overline{A_1} + \overline{B_0}) + B_0A_1(\overline{B_1} + \overline{A_0})$$
$$M_1 = A_0B_1(\overline{A_1B_0}) + B_0A_1(\overline{B_1A_0}) \qquad \text{DeMorganization}$$
$$M_1 = A_0B_1\overline{A_1B_0} + A_1B_0\overline{A_0B_1} \qquad \text{EX-OR function}$$
$$M_1 = A_0B_1 \oplus A_1B_0$$

$$M_0 = A_0B_0$$

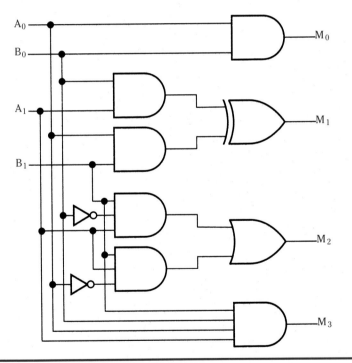

5.9 ■ ARITHMETIC LOGIC UNITS (ALU)

As mentioned earlier, once high speed adders are designed, they are frequently used to implement other mathematical functions. This requires additional circuitry, of course. A complete mathematical design includes some of the following elements: the adder, control logic to select a math function, storage elements, shifting circuits, and circuits determining if results are negative or out of range. Moving from basic adder circuitry to this kind of capability is an important design decision because there are many ways to achieve this capability, as seen in Section 5.8. Another approach that may be suitable for some computational applications is to use a predesigned chip. Most logic families already include adder chips that may be suitable for an existing application. A 7483, for instance, is a 4-bit adder chip that can be cascaded to increase the number length. Using a chip such as this, we can design adder/subtractor systems. When greater mathematical power is required, especially when reduced IC package count is a necessity, an **arithmetic logic unit (ALU)** may be preferred.

The ALU performs both arithmetic and logical functions. A simple ALU includes the provisions for addition, complement subtraction, comparison, ANDing, ORing, and Exclusive-ORing of two input numbers. More complex ALUs also include hardware to provide shifting, multiplication, division, and error checking. Control lines are used to select the operation desired. Input data is presented to the ALU and the resulting output appears after the propagation delay time of the device. Since the ALU logic is fabricated on a single chip, delay time is minimized, resulting in fast computing rates.

Many ALUs are cascadable; several identical ALUs can be connected together to increase the input bit length. Thus, three 4-bit ALUs can be cascaded to create a 12-bit ALU. Often, carry look ahead logic is used to cascade ALUs and speed up overall operation. It is important to realize that not all ALUs are purchased as ready-made chips, but that an ALU is simply a functional subsystem of any computational system. Many ALUs are designed from scratch as in the case of the ALU for a microprocessor. Microprocessor designers certainly do not buy a ready-made chip for an application such as this since the ALU must be integrated onto the same piece of silicon material as all other microprocessor subsystems. As you learn more about digital design, you will come to understand that each application is designed using a certain design philosophy. That is, decisions and trade-offs are made based on cost and performance objectives. For some applications basic logic family parts, such as those you probably use in lab, are adequate, whereas for more stringent applications custom parts and advanced design tools are required.

We now take a brief look at a typical ALU part, the 74181. Figure 5.26 shows inputs for two 4-bit numbers as well as a carry in line, four select lines, and a mode control input. Sixteen arithmetic operations and 16 logical operations are possible on two 4-bit numbers using the 74181. The operation selected is determined by the select inputs (S3–S0) and mode bit input (M = 1 for logical operations; M = 0 for arithmetic operations). The numerical results of any operation are available at the data output lines (F3–F0). In addition, the compare out line will be high when the magnitudes of both input words are equal.

Three other output lines are used for cascading purposes. The ripple carry out line may be connected to the carry in input of another 74181 to increase the word length. A carry propagate and carry generate line are also available

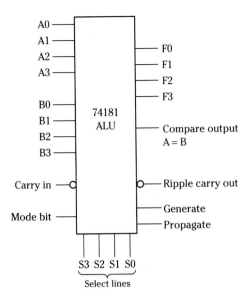

Figure 5.26
74181 ALU

for use with a carry look ahead chip. These two inputs provide for higher speed cascading of 74181s than is possible with the ripple carry expansion.

Figure 5.27 lists the functions possible with this ALU chip. As you can see, operations involving addition, subtraction, comparison, AND, NAND, OR, NOR, and Exclusive-OR can be selected with the mode bit and the select lines. (The table lists the functions available using active high inputs. It is possible to configure the chip to work with active low inputs as well.) For instance, if

Figure 5.27
74181 Functions

	Outputs F3–F0						Arithmetic Operation
Logical Operations	Arithmetic Operations Mode Bit = 0		Select Lines				Comments
Mode Bit = 1	Carry in = 1	Carry in = 0	S3	S2	S1	S0	Carry in = 1
$\overline{A}$	A	A plus 1	0	0	0	0	Output = A
$\overline{A+B}$	A + B	(A + B) plus 1	0	0	0	1	OR of A with B
$\overline{A}B$	$A + \overline{B}$	$(A + \overline{B})$ plus 1	0	0	1	0	OR of A with $\overline{B}$
0	Minus 1	0	0	0	1	1	2's complement 0(1111)
$\overline{AB}$	A plus $A\overline{B}$	A plus $A\overline{B}$ plus 1	0	1	0	0	A added with A AND $\overline{B}$
$\overline{B}$	(A + B) plus $A\overline{B}$	(A + B) plus $A\overline{B}$ plus 1	0	1	0	1	(A or B) added to A AND $\overline{B}$
$A \oplus B$	A – B – 1	A – B	0	1	1	0	A minus B (needs end around)
$A\overline{B}$	$A\overline{B} - 1$	$A\overline{B}$	0	1	1	1	A AND $\overline{B}$ minus 1
$\overline{A} + B$	A plus AB	A plus AB plus 1	1	0	0	0	A added to A AND B
$\overline{A \oplus B}$	A plus B	A plus B plus 1	1	0	0	1	A added to B
B	$(A + \overline{B})$ plus AB	$(A + \overline{B})$ plus AB plus 1	1	0	1	0	A or B added to A AND B
AB	AB – 1	AB	1	0	1	1	A AND B minus 1
1	A plus A	A plus A plus 1	1	1	0	0	Left shift bits in A
$A + \overline{B}$	(A + B) plus A	(A + B) plus A plus 1	1	1	0	1	A OR B added to A
A + B	$(A + \overline{B})$ plus A	$(A + \overline{B})$ plus A plus 1	1	1	1	0	A OR $\overline{B}$ added to A
A	A – 1	A	1	1	1	1	A minus 1

addition is desired, the mode bit is set low and the select lines are set to S3, S2, S1, S0 = 1001. Now any data presented to the A3–A0 and B3–B0 inputs appears added together at the F3–F0 outputs. The carry in line also affects the output. If there is no input carry, then the output of the addition process is simply the sum of the input data; if there is a carry, the output is the sum plus one. Note that the carry in bit is active level low so that a one on this line equals no carry.

A logical operation such as NAND takes place when the mode bit is high and the select lines are set to S3, S2, S1, S0 = 0100. Now the data output pins display the NAND of the A inputs with the B inputs. In other words, F0 = $\overline{A0B0}$, F1 = $\overline{A1B1}$, and so on.

DESIGN
EXAMPLE 5.27

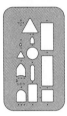

Identify the proper levels for the select and mode inputs of a 74181 to accomplish the following functions:
(a) Exclusive-NOR
(b) Subtraction

Solution   Specific mathematical functions are selected with the four select inputs and the mode bit as defined by the table in Figure 5.27.

(a) Logical operation so that mode bit = 1

$$S3, S2, S1, S0 = 1001$$

(b) Arithmetic operation so that mode bit = 0

$$S3, S2, S1, S0 = 0110$$
$$\text{Carry in} = 0$$

FOR YOUR
INFORMATION

Mathematical Capability Road Map

1. Half adder—2-bit addition
2. Full adder—3-bit addition
3. Ripple carry adder—multiple bit addition
4. Carry look ahead adder—addition with speed
5. Adders with complement circuitry—supports complement subtraction
6. Arithmetic logic units (ALU)—many math functions predesigned in hardware; fast
7. Microprocessors—mathematical operations carried out by a combination of hardware circuits and software programming
8. Math coprocessors—used with a microprocessor to provide high level math functions (i.e., trig, logs) and computational speed

Some Common Part Numbers

Function	Part Number	Technology
Two-bit adder	7482	TTL
	10180	ECL
Four-bit adder	7483	TTL
	74HC283	CMOS
Four-bit ALU	74181	TTL
	74HC181	CMOS
	4581	CMOS
	10181	ECL
Carry look ahead	74182	TTL
	74HC182	CMOS
	4582	CMOS
	10179	ECL

SUMMARY

■ Common mathematical operations, such as addition, subtraction, multiplication, and division, can be carried out in the binary number system with procedures similar to those used in the decimal number system.

■ In order to facilitate the simplification of the mathematical circuitry used in digital computers, 1's and 2's complement representations of binary numbers are used.

■ In practical digital systems it is important to represent the sign of a number as well as the magnitude of the number. Typical digital system designs will dedicate 1-bit as the sign bit.

■ Basic addition of 2-bits is accomplished with a half adder. The half adder forms the basic building block for more extensive adder designs.

■ Ripple carry adders are created using half adders and full adders. Ripple carry adders make it possible to add two multiple bit binary numbers together.

■ A ripple carry adder introduces delay time into mathematical computations that is related to the length of the adder.

■ Carry look ahead logic is used to compensate for the delay time inherent in a ripple carry adder design.

■ Arithmetic logic units (ALU) are arithmetic processing units capable of performing various math operations at high rates of speed.

PROBLEMS

Sections 5.1–5.3

1. Add the following binary numbers:
   *(a) 11011 + 10010   (b) 111111 + 11   (c) 10111.010 + 1011.101
   *(d) 1010101 + 101010   (e) 1000001 + 1001
2. Add the following decimal quantities as 5-bit binary numbers:
   *(a) 12 + 3   (b) 18 + 22   (c) 6 + 0   (d) 31 + 1
3. Subtract the following binary numbers:
   *(a) 111011 − 10010   *(b) 1101011 − 10101   (c) 10000 − 10
   (d) 11011.01 − 1100.1   (e) 111111 − 1
4. Subtract the following decimal quantities as 6-bit binary numbers:
   (a) 54 − 18   (b) 63 − 1   (c) 32 − 2   (d) 28 − 27
5. Multiply the following:
   (a) 1011 × 11.1   (b) 1111 × 1001   (c) 111111 × 1111
6. Divide the following:
   (a) 10000111 ÷ 1111   (b) 11011 ÷ 11   (c) 10101 ÷ 101

Section 5.4

7. Obtain the 1's complement of the following:
   *(a) 11111110   (b) 01010101   *(c) 0000011   (d) 0000000
8. Obtain the 5-bit 1's complement representation of binary:
   *(a) 2   (b) 13   (c) 24   (d) 31
9. Obtain the 2's complement of the following:
   *(a) 1100110   (b) 11111111   (c) 00000000   (d) 1000000
10. Obtain the 5-bit 2's complement representation of binary:
    *(a) 2   (b) 13   (c) 24   (d) 31
11. Subtract the following unsigned numbers using the 1's complement method:
    *(a) 11011 − 10010   (b) 1101011 − 10101   (c) 11111 − 1111
*12. Repeat problem 11 using the 2's complement subtraction method.

Section 5.5

13. Subtract the following unsigned numbers using the 1's complement method:
(a)	100101	(b)	111101	(c)	10011
	−110111		−111110		−100000
14. Repeat problem 13 using 2's complement subtraction.
15. Write the following numbers as 8-bit 1's complement signed numbers:
    (a) +127   *(b) +37   *(c) −24   (d) −200   (e) −1
16. Write the following numbers as 8-bit 2's complement signed numbers:
    *(a) +127   *(b) +37   (c) −24   (d) −200   (e) −1
17. Add the following 8-bit signed numbers using 2's complement math:
    *(a) 117 + −123   (b) −27 + 101   (c) −105 + 103   (d) −87 + −27

---

* See Appendix F: Answers to Selected Problems.

18. Determine if overflow occurs when adding the following 6-bit numbers using the 2's complement system:
   (a) 33 + 29   (b) 38 + 31   (c) 63 + 1

Section 5.6

*19. Draw the sum and carry output waveforms for a half adder, if the X and Y inputs add the following:

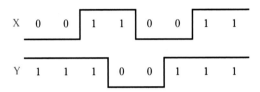

*20. Design an equivalent half adder circuit without using an Exclusive-OR gate.
21. Using either full adder circuit discussed in the text, verify the addition of:
   (a) 0 + 0 + 0   (b) 1 + 1 + 0   (c) 1 + 1 + 1
22. Determine how the sum output of a full adder can be implemented using only NAND gates.
23. Use the information in problem 22 to design a complete full adder using only NAND gates.
*24. Determine the sum and carry waveform output for a full adder design adding the following combinations X, Y, and Z:

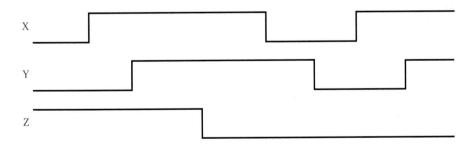

25. Redraw the waveforms for the full adder sum and carry using the waveforms in problem 24, assuming that each gate has an 8 nsec delay. Assume that the waveform times are significantly long compared to the delay time.
26. Verify that the 3-bit ripple carry adder design (Figure 5.21) will correctly add the following:
   (a) 101 + 101   (b) 100 + 010   (c) 111 + 111   (d) 011 + 000
*27. Using full adder circuits only, design a 6-bit ripple carry adder.

---

* See Appendix F: Answers to Selected Problems.

Section 5.7

28. Draw the block diagram of an 8-bit ripple carry adder. Assume each gate within the circuit has a 15 nsec delay.
    (a) What is the carry delay time for this adder?
    (b) How much does the delay increase if the adder is increased in size to a 10-bit adder?
    (c) What improvement in delay time is there if 9 nsec gates are used for both sized adders?
    (d) If the original 8-bit adder is converted to a carry look ahead design, what is the maximum delay?
    (e) What is the maximum delay for a carry look ahead circuit based on the 10-bit adder?

*29. The carry look ahead circuitry in Figure 5.23 shows the complete logic equations for circuit carries C_1, C_2, and C_3. Expand the equation given for C_4 and determine the full equation for C_5.

30. Draw the logic circuitry for carries C_4 and C_5 as developed in problem 29.

Section 5.8

*31. Design a 4-bit 1's complement subtractor using a 4-bit ripple carry adder as the basic mathematical device.

32. Modify the 1's complement circuit from problem 31 to function as a 1's complement adder-subtractor.

Section 5.9

33. Determine how the 74181 ALU should be configured to accomplish the following:
    (a) inversion
    *(b) Exclusive-OR
    (c) A inputs added to the AND of the A inputs with the B inputs

34. Using a data book, determine how a 74283/74HCT283 can be used to perform 4-bit addition; 8-bit addition.

---

* See Appendix F: Answers to Selected Problems.

CHAPTER 6

BINARY CODES, PRACTICAL COMBINATORIAL CIRCUITS, AND ERROR CHECKING SYSTEMS

OBJECTIVES

To explain why binary codes are used.

To investigate the properties of several commonly used codes.

To discuss the design and application of practical combinatorial circuits, such as decoders, encoders, and multiplexers.

To determine how errors occur in binary data.

To explore various circuits and techniques that identify or correct data errors.

PREVIEW

Many aspects of practical digital design deal with ways to represent decimal numbers in a logic environment that is only satisfied with ones and zeros. Much of the one/zero information, or data, utilized in a digital network is manipulated by commonly used circuits that move, code, and detect the data. Additionally, the circuits used are not perfect and constitute one of the many ways errors affect digital data. Often, methods are required to check for errors that may affect the integrity of information utilized by a digital system. This chapter examines some of the coding and error checking systems used in digital logic design to represent data and to detect errors in data. This chapter also examines several of the most commonly used digital circuit designs.

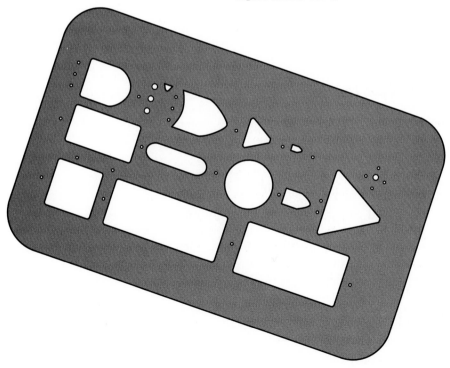

6.1 ■ THE NATURE OF BINARY CODES

In previous chapters the fact that digital systems operate on binary data has been well established. Digital systems must be able to represent, calculate, modify, display, and transfer a wide variety of information. Since the digital circuitry is confined to only two logic levels, many clever techniques are used to represent the various forms of information found in digital systems. Binary codes are employed for this purpose.

Since we as humans live in a decimal world, pure binary numbers can be construed as a code. That is, we use binary numbers to represent decimal numbers. Having used this coding technique repeatedly with truth tables, we are familiar with it. Two's complement numbers are coded representations of negative numbers. However, binary numbers are only numbers. How can non-mathematical ideas, such as the letters of the alphabet, be represented in the binary system?

This book, having been written using a computer, provides a good example. This chapter, this page, this sentence, resided at one time in computer memory circuits, were stored on a floppy disk, and were displayed on a computer monitor. In each case all information was processed using simple one and zero logic levels. The logic levels were coded in different ways by computer hardware in a manner that made sense to the computer's subsystems, since, for example, the data requirements of a floppy disk drive are far different from those of a computer monitor.

A primary use for binary codes is translating information from the digital world into an appropriate form for the analog world, and vice versa. Other uses for coded information include error checking and correcting, data compression, and mathematical operations. The actual nature and details of any code are determined by the designer; however, many standard codes have evolved over the years and are used extensively in digital system design. In all cases a particular code carries a logic circuitry requirement with it. The following sections examine several common binary codes.

Binary Coded Decimal (BCD)

As mentioned earlier, binary numbers are often used to represent decimal numbers. This common digital system application does not necessarily have to be carried out with pure binary numbers. Often, it is more expedient to code decimal numbers in alternate forms, such as **binary coded decimal** or **BCD** codes. Although there are many BCD codes, the following method is most commonly used.

The code typically called BCD is also known as the 8421 code. Each decimal digit (0 through 9) is represented as a 4-bit binary number, as shown in Figure 6.1. The 8421 refers to the binary weights for the 4-bit positions, the same weights we use to convert binary to decimal [the least significant bit (LSB) has a weight of 1, whereas the most significant bit (MSB) has a weight of 8]. The difference between BCD and pure binary is that BCD numbers are limited to

Decimal Digit	BCD Representation
0	0 0 0 0
1	0 0 0 1
2	0 0 1 0
3	0 0 1 1
4	0 1 0 0
5	0 1 0 1
6	0 1 1 0
7	0 1 1 1
8	1 0 0 0
9	1 0 0 1

1 0 1 0	
1 0 1 1	
1 1 0 0	These code groups are not used.
1 1 0 1	
1 1 1 0	
1 1 1 1	

Figure 6.1
8421 BCD Code

4-bits per digit. Figure 6.2 compares BCD and pure binary representations of the same number. Notice that each decimal digit has its own corresponding 4-bit BCD representation. As shown, the number 368 requires 12-bits using BCD as opposed to only 9-bits using a normal binary representation. Although BCD coding is not the most efficient way to represent numbers because it usually requires more bits than standard binary, we can easily convert BCD numbers into equivalent decimal numbers, which simplifies circuit design.

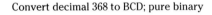

Convert decimal 368 to BCD; pure binary

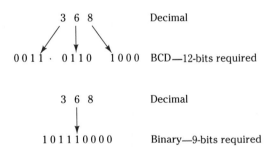

Figure 6.2
Converting Decimal
Numbers to BCD and
Binary

Since the numerical value of each BCD number is determined by the weight of the ones and zeros representing the number, BCD is an example of a **weighted code**. Other code groups that we will soon discuss have no mathematical relationship between the code and the information represented by the code. Codes bearing this property are called **unweighted codes**. As we will see, both coding schemes are useful.

DESIGN
EXAMPLE 6.1

Convert the following decimal numbers into BCD numbers:
(a) 12   (b) 15937   (c) 2007

Solution   Each decimal digital has a 4-bit BCD code equivalent.
(a) 12 = 0001 0010
(b) 15937 = 0001 0101 1001 0011 0111
(c) 2007 = 0010 0000 0000 0111

DESIGN
EXAMPLE 6.2

Convert the following BCD coded numbers into decimal numbers:
(a) 0010011001010000.00100111
(b) 000101110100
(c) 01000000000000000001

Solution   BCD numbers are 4-bit representations of decimal numbers. Starting at the binary point, convert four BCD bits at a time into the equivalent decimal digit.
(a) 0010 0110 0101 0000.0010 0111 = 2650.27
(b) 0001 0111 0100 = 174
(c) 0100 0000 0000 0000 0001 = 40001

Let's examine the unused BCD combinations. Since there are only ten decimal symbols, only ten individual BCD codes are necessary. We know that any 4-bit binary code represents 16 possible combinations. Thus, for the BCD code 6 out of the 16 combinations possible are unused. These combinations are meaningless and should never occur in a BCD system. For that matter, the presence of an unused code in a BCD system indicates that an error has occurred. Therefore, it is apparent that BCD coding provides crude error checking capability.

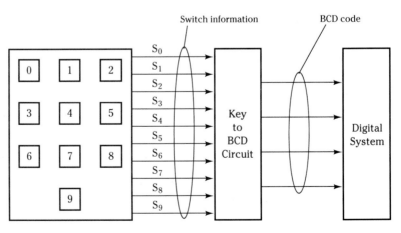

Figure 6.3
Converting Key Codes to BCD

Figure 6.3 shows how the BCD code is used with calculator keypads. As humans, we want to push buttons that make sense to us. Keys with familiar decimal numbers stamped on them do make sense to us, but they mean nothing to digital circuits. Somehow, a key strike must be converted into an appropriate binary number. In Figure 6.3 each key sends a signal to the "Key to BCD Circuit" illustrated in the diagram. This circuit converts each keystroke into a meaningful BCD number (to be elaborated upon later in this chapter). The BCD code representing the depressed key proceeds through the digital system for processing.

Design a circuit to detect the presence of any illegal BCD code.

Solution   Create a truth table with the BCD code as the input combinations. List the output as a low level for legal BCD codes and as a high level for each illegal code group.

DESIGN EXAMPLE 6.3

BCD inputs

D	C	B	A	Output
0	0	0	0	0
0	0	0	1	0
0	0	1	0	0
0	0	1	1	0
0	1	0	0	0
0	1	0	1	0
0	1	1	0	0
0	1	1	1	0
1	0	0	0	0
1	0	0	1	0
1	0	1	0	1
1	0	1	1	1
1	1	0	0	1
1	1	0	1	1
1	1	1	0	1
1	1	1	1	1

Illegal BCD codes

	$\overline{B}\,\overline{A}$	$\overline{B}A$	BA	$B\overline{A}$
$\overline{D}\,\overline{C}$	0	0	0	0
$\overline{D}C$	0	0	0	0
DC	1	1	1	1
$D\overline{C}$	0	0	1	1

Output = DC + DB

Output = 1 for any illegal BCD input

ASCII Code

In the last section we examined binary codes and how they represent numbers. Other representations are also needed in practical systems. Letters, both upper- and lowercase, require representation as do special characters such as spaces, equal signs, and dollar signs. These standard typewriter characters are called **alphanumerics**. Because of the one and zero nature of digital logic, alphanumerics require their own special code to distinguish them from binary numbers.

The **American Standard Code for Information Interchange**, or **ASCII code**, is a 7-bit unweighted code used in many microcomputing systems. Using 7-bits, 128 binary combinations are possible—certainly more than enough to represent letters, numerals, and special characters. Figure 6.4 lists a portion of the full ASCII code. Strike the letter C on an ASCII keyboard and you can expect to see an output of 1000011. This binary pattern always represents an uppercase C in ASCII. The numerical value of this number is meaningless since the 7-bit number represents a letter.

It may be worthwhile to compare briefly BCD and ASCII numerals. For example, the BCD 6 is 0110, whereas the ASCII 6 is 0110110. The seven ASCII bits representing the numeral require circuitry capable of storing and processing all 7-bits. Compared to BCD, this means additional hardware for ASCII; however, the ASCII advantage is that other characters can be represented as well. Mathematically, neither BCD nor ASCII lends itself readily to computations. Special BCD adders can be designed, for example, but the circuitry required is more complex than that for pure binary math.

---

*DESIGN
EXAMPLE 6.4*

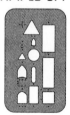

Determine the phrase represented by the following ASCII sequence:

1000001 1001100 1001100 0100000
1000110 1001111 1010010 0100000
0110001 0100000 1000001 1001110 1000100
0100000 0110001 0100000 1000110 1001111 1010010
0100000 1000001 1001100 1001100

Solution   Each 7-bit group represents one character—a letter, number, or special character. Refer to Figure 6.4 to decode the phrase.

Gray Code

We have examined several codes for numeral and alphanumeric character representation. The **Gray code** represents neither one. In fact, a single Gray code number, taken by itself, is quite meaningless. However, the Gray code has some distinct advantages in analog to digital conversion processes.

You are already familiar with the Gray code from some of the past work you have done. Variable combinations assigned along the sides of Karnaugh maps are written in a Gray code order because the Gray code is a **minimum-**

Character	ASCII Code
A	1 0 0 0 0 0 1
B	1 0 0 0 0 1 0
C	1 0 0 0 0 1 1
D	1 0 0 0 1 0 0
E	1 0 0 0 1 0 1
F	1 0 0 0 1 1 0
G	1 0 0 0 1 1 1
H	1 0 0 1 0 0 0
I	1 0 0 1 0 0 1
J	1 0 0 1 0 1 0
K	1 0 0 1 0 1 1
L	1 0 0 1 1 0 0
M	1 0 0 1 1 0 1
N	1 0 0 1 1 1 0
O	1 0 0 1 1 1 1
P	1 0 1 0 0 0 0
Q	1 0 1 0 0 0 1
R	1 0 1 0 0 1 0
S	1 0 1 0 0 1 1
T	1 0 1 0 1 0 0
U	1 0 1 0 1 0 1
V	1 0 1 0 1 1 0
W	1 0 1 0 1 1 1
X	1 0 1 1 0 0 0
Y	1 0 1 1 0 0 1
Z	1 0 1 1 0 1 0
0	0 1 1 0 0 0 0
1	0 1 1 0 0 0 1
2	0 1 1 0 0 1 0
3	0 1 1 0 0 1 1
4	0 1 1 0 1 0 0
5	0 1 1 0 1 0 1
6	0 1 1 0 1 1 0
7	0 1 1 0 1 1 1
8	0 1 1 1 0 0 0
9	0 1 1 1 0 0 1
blank	0 1 0 0 0 0 0
.	0 1 0 1 1 1 0
(	0 1 0 1 0 0 0
+	0 1 0 1 0 1 1
$	0 1 0 0 1 0 0

Figure 6.4
Partial ASCII Listing

change code. That is, from one combination to the next, only a single bit will differ. This characteristic is independent of the length of the code, an interesting feature in itself since the code length can be as long as is necessary.

Gray code alleviates the confusing results that occur when one binary number bit pattern changes to another pattern. Refer to Figure 6.5 and notice how every bit changes level when a binary 7 changes to a binary 8. If the binary numbers were indicating the reaction of some physical process, the results could be confusing since no two digital circuits change state at exactly the same time. Several false states may occur as the binary number changes from 7 to 8, including 0111, 0011, 1010, 0001, 0110, and so on. Since so many bits change, the possibility of a false indication is high.

Binary	Gray
0 0 0 0	0 0 0 0
0 0 0 1	0 0 0 1
0 0 1 0	0 0 1 1
0 0 1 1	0 0 1 0
0 1 0 0	0 1 1 0
0 1 0 1	0 1 1 1
0 1 1 0	0 1 0 1
0 1 1 1	0 1 0 0
1 0 0 0	1 1 0 0
1 0 0 1	1 1 0 1
1 0 1 0	1 1 1 1
1 0 1 1	1 1 1 0
1 1 0 0	1 0 1 0
1 1 0 1	1 0 1 1
1 1 1 0	1 0 0 1
1 1 1 1	1 0 0 0

Figure 6.5
4-Bit Binary and Gray Code Numbers

On the other hand, the Gray code combinations equivalent to binary numbers differ by only a single bit. For example, comparing the Gray code combination equivalent to a binary 7 (0100) to the Gray code equivalent for a binary 8 (1100) shows that only a single bit differs between the two. The chance of misreading the correct code is significantly reduced since only two possible code groups (7 and 8) exist for this change. Motor positioning systems are an application that can take advantage of this property.

If, for instance, a metal wheel is fabricated so that holes punched into the wheel represent Gray code numbers, and the wheel is attached to a spinning motor shaft, it becomes possible to detect the motor shaft position by placing lights and light sensors next to the wheel. Since the light sensor outputs are Gray code combinations, only 1-bit at a time changes as the wheel spins by. Thus, wheel position can be accurately detected by relating the detected Gray code to shaft position. A similar binary coded wheel is inaccurate for this application, because many bits change for even a small motion of the wheel.

Gray code numbers are easily converted to binary numbers, for computational and other processing needs, with the circuitry and process shown in Figure 6.6.

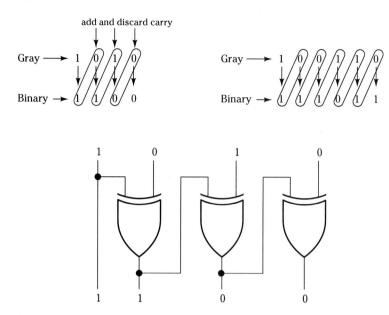

Figure 6.6
4-Bit Gray to Binary
Conversion Circuit

When the Gray code is converted to binary, regardless of the number of bits involved, the MSB remains the same. Therefore, the Gray MSB is equal to the binary MSB. Each succeeding lower order binary bit is obtained by a series of additions between the original Gray code bit and the newly created binary bit. Any carries from the addition process are ignored.

The circuit in Figure 6.6, which carries out the conversion process, should make sense. Since the conversion entails 2-bit addition, to create specifically the sum of 2-bits, an Exclusive-OR gate is used. (You will recall that half adders use Exclusive-OR gates to create the sum output, but since carries are ignored in this process, a complete half adder is unnecessary.) Cascading Exclusive-OR gates increases the conversion bit length, making it easy to convert any size number.

Convert the following Gray codes into binary numbers:
(a) 101101   (b) 011010011   (c) 101

DESIGN
EXAMPLE 6.5

Solution   Retain the MSB. Add it to the next bit of the Gray code—the sum becomes the next bit of the binary result. Repeat the addition process for the remaining bits in the code. That is:

$$\text{Gray code} \longrightarrow G_{MSB} \ldots \quad G_2 \quad G_1 \quad G_0$$

$$\text{Binary} \longrightarrow B_{MSB} \ldots \quad B_2 \quad B_1 \quad B_0$$

In general:

$$B_n + G_{n-1} = B_{n-1}$$

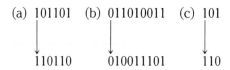

(a) 101101   (b) 011010011   (c) 101

↓   ↓   ↓

110110   010011101   110

Converting binary numbers to Gray code numbers by a similar process is also possible, although this conversion is less common than the Gray to binary process. For this conversion the MSB remains the same as well. Each binary bit, starting with the MSB, is added to the next least significant binary bit. The sum of this addition becomes the next Gray code bit; the carry is ignored. In general:

$$B_n + B_{n-1} = G_{n-1}$$

DESIGN EXAMPLE 6.6

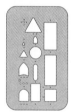

Convert the following binary numbers into Gray code numbers:
(a) 110110   (b) 10011101   (c) 10110

Solution   Retain the MSB. Add each binary bit to its least significant neighbor. The sum becomes the next Gray code bit. Apply the addition process to all binary bits.

(a) 110110   (b) 10011101   (c) 10110

↓   ↓   ↓

101101   11010011   11101

DESIGN EXAMPLE 6.7

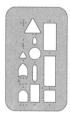

Design a circuit to convert 5-bit binary numbers into 5-bit Gray code numbers.

Solution   The MSB of both binary and Gray numbers is the same. Therefore, a wire accomplishes the conversion for this bit. The remaining bits are converted by adding two adjacent binary bits (starting with the MSB) together to obtain a sum (neglecting carries). Since an Exclusive-OR gate provides the addition capability, a network of these gates comprises the circuit.

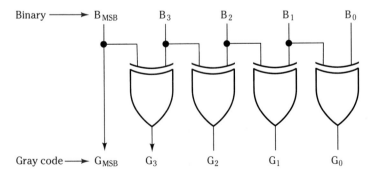

Seven Segment Code

A binary code used for display purposes is the **seven segment code**. Seven segment devices are employed extensively as displays for decimal numbers in calculators, watches, and appliances. Considering for a moment the fact that most people are unfamiliar with the intricacies of digital system, it becomes apparent that digital designers are obligated to translate digital ones and zeros into decimal numbers for the sake of the users' convenience. Seven segment devices, such as LED (light emitting diodes) and LCD (liquid crystal displays), accomplish this task. The display device is made up of seven separate segments, each capable of emitting light. Every segment responds to a binary level, which determines whether or not the segment will light. A combination of on-and-off segments outlines a decimal digit. Figure 6.7 shows the characteristics of a seven segment display.

a	b	c	d	e	f	g	Display Character
1	1	1	1	1	1	0	0
0	1	1	0	0	0	0	1
1	1	0	1	1	0	1	2
1	1	1	1	0	0	1	3
0	1	1	0	0	1	1	4
1	0	1	1	0	1	1	5
1	0	1	1	1	1	1	6
1	1	1	0	0	0	0	7
1	1	1	1	1	1	1	8
1	1	1	0	0	1	1	9

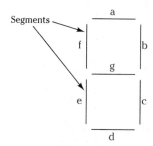

Figure 6.7
Seven Segment Display and Common Cathode Table

Schematically, each segment is identified with a lowercase letter, as shown. Since there are seven individual segments to control, a 7-bit binary pattern determines which segments light. Segments are lit using low logic levels with a **common anode device** or with high logic levels with a **common cathode device**. The binary pattern driving the device simply indicates on-and-off conditions for specific segments and is used in the digital system only to drive the display. For instance, a system designed primarily using the BCD code requires a translation from BCD to seven segment code for display purposes. The interface between the BCD system and the display device is designed with decoding circuits (discussed later in this chapter), which convert individual

BCD numbers to corresponding seven segment numbers and also supply the required electrical current. For example, a BCD 6 (0110) is converted to a seven segment (common cathode) 6 (1011111), resulting in a decimal 6 on the display. Since the BCD 6 and the seven segment code for 6 have no mathematical relationship, seven segment code is another example of an unweighted code.

DESIGN EXAMPLE 6.8

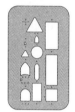

Determine the display pattern for the following common anode seven segment codes:

(a) abcdefg = 0100100

(b) abcdefg = 0000110

(c) abcdefg = 0000000

Solution Common anode display segments light when a low logic level is applied. Therefore, a zero turns on a segment while a one turns it off.

(a) 0100100 = 5

(b) 0000110 = 3

(c) 0000000 = 8

The Hexadecimal and Octal Number Systems

When working with large binary numbers, the number of bits required to represent a given number becomes excessive. It is easy to make a mistake translating or writing down a long string of ones and zeros. For this reason several other number systems related to binary, but requiring fewer digits, are employed.

Each digit in a **hexadecimal** or **hex number** is positionally weighted based on powers of 16. The 16 symbols used to represent numbers in the hexadecimal systems are shown in Figure 6.8. As you can see, because of the combination of standard numbers and letters, some unlikely looking number combinations are possible. The usefulness of this system is that a 4-bit binary number can be converted to a single hex digit by inspection and that fewer hex digits are required to represent an equivalent binary number. Remember, hex numbers are not used directly by a digital system. Hex is only a convenient way to compress large quantities of binary numbers into a manageable form for our convenience. Figure 6.8 also shows how hex and decimal numbers are related.

Figure 6.8
Hex-Decimal
Relationship

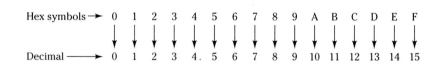

Hex symbols →	0	1	2	3	4	5	6	7	8	9	A	B	C	D	E	F
Decimal →	0	1	2	3	4.	5	6	7	8	9	10	11	12	13	14	15

The **octal** system or base 8 contains the eight symbols shown in Figure 6.9. The relationship between octal and decimal is also noted. Three-bit binary numbers can be converted to a single octal digit by inspection, and consequently offer a similar advantage as hexadecimal numbers. However, the octal system is not used as extensively as hexadecimal in current digital design. Hexadecimal is widely used in computer system design and in computer programming.

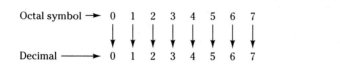

Figure 6.9
Octal-Decimal
Relationship

Figure 6.10 shows how hex and octal numbers are related to binary numbers. The ease of conversion can be noted from the figure and makes sense considering powers of 2; $2^3 = 8$ and $2^4 = 16$ are bases of the octal and hexadecimal number systems, respectively. For example, an 8-bit binary number such as 11000101 can be expressed as a 2-digit hex number. Start at the binary point and translate the first 4-bits to the left of the binary point into an equivalent hex digit. Continue to translate 4-bits at a time. Thus, 11000101 binary equals C5 hex.

Decimal	Binary	Hex	Octal
0	0 0 0 0	0	0
1	0 0 0 1	1	1
2	0 0 1 0	2	2
3	0 0 1 1	3	3
4	0 1 0 0	4	4
5	0 1 0 1	5	5
6	0 1 1 0	6	6
7	0 1 1 1	7	7
8	1 0 0 0	8	10
9	1 0 0 1	9	11
10	1 0 1 0	A	12
11	1 0 1 1	B	13
12	1 1 0 0	C	14
13	1 1 0 1	D	15
14	1 1 1 0	E	16
15	1 1 1 1	F	17
16	1 0 0 0 0	10	20

Figure 6.10
Decimal-Binary-Hex-
Octal Equivalence

The same number can be converted into octal form following the same process, but by translating 3-bits at a time. Thus, 11000101 binary = 350 octal. Notice how a ninth bit (zero) is added or used to "pad" the binary number to carry out this conversion. Zeros may be added to the binary number in hex and octal for the sake of conversion.

FOR YOUR INFORMATION

The following notation is sometimes used to eliminate confusion when using several number systems:

1001101_2 indicates a binary number
F23B38H   "H" indicates a hex number
03654_8 indicates an octal number

DESIGN EXAMPLE 6.9

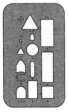

Convert the following binary numbers into hex and octal:
(a) 101011101011   (b) 11111   (c) 1011.01

Solution   Binary numbers are converted into hex 4-bits at a time, whereas the conversion into octal takes place 3-bits at a time.
(a) 1010 1110 1011 = AEB hex
    101 011 101 011 = 5353 octal
(b) 0001 1111 = 1F hex (pad with zeros)
    000 011 111 = 037 octal (pad with zeros)
(c) 1011.01 = B.4 hex (pad with zeros)
    001 011.010 = 13.2 octal (pad with zeros)

DESIGN EXAMPLE 6.10

Convert the following hex and octal numbers into binary:
(a) B78D hex   (b) 910F hex   (c) 7011 octal   (d) 3526 octal

Solution   Each hex digit is converted into a 4-bit binary number. Each octal digit is converted into a 3-bit binary number.
(a) B78D = 1011 0111 1000 1101
(b) 910F = 1001 0001 0000 1111
(c) 7011 = 111 000 001 001
(d) 3526 = 011 101 010 110

DESIGN EXAMPLE 6.11

A computer system's memory can store data in 65536 locations (addresses). Represent the range of storage locations in binary, octal, and hex notations.

Solution   The decimal range of storage locations is 0 through 65535. Convert these numbers to binary, octal, and hex.

Binary	0000000000000000 $\longrightarrow$	1111111111111111
Octal	000000 $\longrightarrow$	177777
Hex	0000 $\longrightarrow$	FFFF

Summary of Binary Codes

Code	Number of Bits	Feature	Typical Code Group
BCD	4	Binary representation of decimal numbers	Decimal 9 = BCD 1001
Gray	No limit	Minimum change code	Binary 1000 = Gray 1100
ASCII	7	Seven-bit representation of alphanumeric characters	S = ASCII 1010011
Excess-3	4	Same as BCD, except that 3 is added to each code group. Used for BCD math.	BCD 6 = 0110 Excess-3 6 = 1001
EBCDIC	8	Eight-bit representation of alphanumeric characters	S = EBCDIC 11100010

Excess-3 Code

Excess-3 code is a variation of the standard BCD code discussed earlier and is used to aid mathematical manipulation of BCD numbers. Excess-3 numbers are formed by adding 3 to each BCD number. For example, BCD 5 = 0101. The excess-3 representation is $0101 + 0011 = 1000$. A comparison of BCD and excess-3 numbers is shown in Figure 6.11.

BCD	Excess-3
0 0 0 0	0 0 1 1
0 0 0 1	0 1 0 0
0 0 1 0	0 1 0 1
0 0 1 1	0 1 1 0
0 1 0 0	0 1 1 1
0 1 0 1	1 0 0 0
0 1 1 0	1 0 0 1
0 1 1 1	1 0 1 0
1 0 0 0	1 0 1 1
1 0 0 1	1 1 0 0

**Figure 6.11
BCD—Excess-3
Equivalence**

To see the advantage of excess-3 over BCD in addition, attempt to add BCD 5 to BCD 6. The addition is $0101 + 0110 = 1011$. This result is incorrect in the BCD system since BCD eleven is 0001 0001, an 8-bit number. Standard adder circuits cannot add BCD numbers. They must be modified to detect an illegal BCD result and then compensate by adding 6 to the answer. In lieu of redesigning an adder, it may be easier to add 3 to each number being added—excess-3—and then use the standard addition circuits. Thus, BCD 0101 →

excess-3 1000 and BCD 0110 → excess-3 1001. Added together, 1000 + 1001 = 1 0001. With padding, this is read as 0001 0001 or BCD eleven. BCD answers are obtained directly by adding excess-3 numbers.

6.2 ■ PRACTICAL COMBINATORIAL CIRCUITS

We have discussed several frequently used digital circuits, such as adders and carry look aheads (CLAs). There are many other heavily utilized logical designs that crop up just about everywhere. Circuits such as decoders, encoders, multiplexers, and demultiplexers are key ingredients in many digital systems and deserve special attention.

Decoders

Of all circuits mentioned, the **decoder** is the most widely used. Decoder design complexity ranges from a simple gate to dense logic circuitry. In fact, any fundamental product is simply the decode of a specific input combination. In general, a decoder takes a number of binary inputs and separates the input combinations into individual outputs. For instance, the decoder block diagrams in Figure 6.12 each have three inputs and eight outputs. Since three input bits produce eight input combinations, a corresponding output exists for each unique input combination. This circuit is called a 3-to-8 line decoder, named after the number of inputs and outputs. If the decoder output is designed to be active high, only one output is high at any time. If the decoder output is active low, only one output is low at any time. In either case, whenever a specific input combination occurs, its corresponding output is activated.

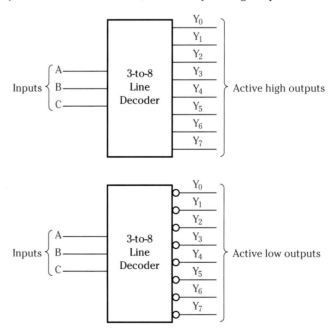

Figure 6.12
Decoders with Active High and Active Low Outputs

Figure 6.13 examines decoder operation in more detail. The 3-to-8 line decoder shown has active high outputs. The input combination presented to the circuit in this example is 011, and under these circumstances the output corresponding to the 011 input is high while all others are low. The circuitry has decoded the fact that input combination 011 is present at the three input pins. If the input combination changes to 101, the 101 output goes high while all others remain low. Again, the decoding circuitry indicates the presence of a specific input combination.

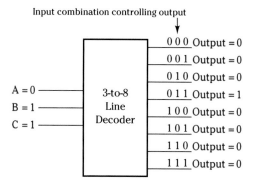

Figure 6.13
3-to-8 Line Decoder

The 3-to-8 line decoder just discussed is called an **absolute decoder** because an output line exists for each input combination. Absolute decoders decode every possible input combination, although this capability is not always required. For example, if only the presence of 101 needs to be detected, only the circuitry necessary for this combination is designed. Furthermore, some decoders produce an output if any one of several input combinations occurs. For example, a high level output may be required whenever a 010 or 111 occurs. These are examples of nonabsolute decoders. Another decoder example includes circuits converting BCD numbers to seven segment code outputs. In this case the decoder has four inputs and seven outputs; each BCD input combination may activate several outputs.

As you can see, decoders come in a wide variety of sizes and have many applications. Decoder applications vary from the simplest enable generation circuit to detailed microprocessor system memory selection circuitry. In all cases decoder logic can be designed using techniques already discussed.

If an absolute decoder were required to detect every ASCII code, what size decoder would be required? If only the capital letters were to be detected, what size decoder would be required. Would this application require an absolute decoder?

Solution Since ASCII is a 7-bit code, there are 128 possible combinations. An absolute decoder having a unique output for every ASCII input would require 128 outputs. Thus, a 7-to-128 line decoder is needed. If only capital letters are detected, a 7-to-26 line nonabsolute decoder would suffice.

DESIGN EXAMPLE 6.12

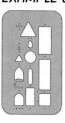

BCD Decoder

We examine the decoder operation in more detail by designing a simple BCD decoder. Assume that we wish to detect a BCD 0101. From a system perspective our circuit is required to monitor the four BCD signal lines and to provide an indication when 0101 shows up. In other words, we need a circuit to decode the presence of a BCD 0101. Figure 6.14 shows the truth table for this decoder, indicating a high level output when 0101 occurs and a low output for all other BCD inputs. Using fundamental product techniques, we can design an appropriate circuit as shown. The decoder is simply the fundamental product representing the 0101 input combination. Decoders like this are as common as snow in winter and used in most digital systems.

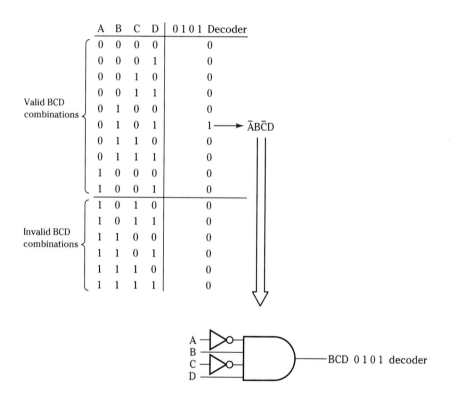

A	B	C	D	0 1 0 1 Decoder
0	0	0	0	0
0	0	0	1	0
0	0	1	0	0
0	0	1	1	0
0	1	0	0	0
0	1	0	1	1 $\longrightarrow \bar{A}B\bar{C}D$
0	1	1	0	0
0	1	1	1	0
1	0	0	0	0
1	0	0	1	0
1	0	1	0	0
1	0	1	1	0
1	1	0	0	0
1	1	0	1	0
1	1	1	0	0
1	1	1	1	0

Valid BCD combinations (rows 0000 through 1001)

Invalid BCD combinations (rows 1010 through 1111)

A
B
C
D

BCD 0 1 0 1 decoder

Figure 6.14
BCD Decoder

BCD Decoder with Don't Care States

As an aside, the decoder just designed can be simplified using don't care states, as illustrated in Figure 6.15. Recall that BCD codes 1010 through 1111 are invalid and may be treated as don't care conditions. This premise is true if the unused codes are never found in the system. Taking advantage of this fact, we can design a simpler circuit.

	A	B	C	D	0 1 0 1 Decoder
	0	0	0	0	0
	0	0	0	1	0
	0	0	1	0	0
	0	0	1	1	0
Valid BCD codes	0	1	0	0	0
	0	1	0	1	1
	0	1	1	0	0
	0	1	1	1	0
	1	0	0	0	0
	1	0	0	1	0
	1	0	1	0	d
	1	0	1	1	d
Invalid BCD codes	1	1	0	0	d
	1	1	0	1	d
	1	1	1	0	d
	1	1	1	1	d

	$\bar{C}\bar{D}$	$\bar{C}D$	CD	$C\bar{D}$
$\bar{A}\bar{B}$	0	0	0	0
$\bar{A}B$	0	1	0	0
AB	d	d	d	d
$A\bar{B}$	0	0	d	d

Decoder = $B\bar{C}D$ equation

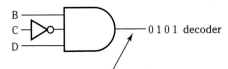

B
C
D
── 0 1 0 1 decoder

Output is actually high for combinations 0 1 0 1
and 1 1 0 1, but since 1 1 0 1 is an invalid BCD
code and should never occur, *we don't care.*

Figure 6.15
Decoder Designed
with Don't Care States

Design a circuit to decode BCD states 4, 7, and 9. Assume that only valid BCD states will occur.

Solution Treating the invalid BCD states as don't care states, the following truth table is created:

DESIGN
EXAMPLE 6.13

A	B	C	D	Output
0	0	0	0	0
0	0	0	1	0
0	0	1	0	0
0	0	1	1	0
0	1	0	0	1
0	1	0	1	0
0	1	1	0	0
0	1	1	1	1
1	0	0	0	0
1	0	0	1	1
1	0	1	0	d
1	0	1	1	d
1	1	0	0	d
1	1	0	1	d
1	1	1	0	d
1	1	1	1	d

	$\overline{CD}$	$\overline{C}D$	CD	$C\overline{D}$
$\overline{AB}$	0	0	0	0
$\overline{A}B$	1	0	1	0
AB	d	d	d	d
$A\overline{B}$	0	1	d	d

$$Y = AD + B\overline{CD} + BCD$$

Two-to-Four Line Decoder Design

A 2-to-4 line decoder has two inputs and four outputs. Figure 6.16 shows the truth table for this absolute decoder where each output is high for only one specific input combination. For instance, output X is active when inputs A,B = 0,1. The equation for this output and the remaining three (W, Y, Z) are created by determining the four fundamental products from the values listed on the truth table. Since we are really designing four separate circuits, the fundamental products are not ORed together. We are creating four distinct outputs that are related but do not interact. As seen in Figure 6.17, four AND gates comprise the decoder; only one has an active output at any time because each AND gate represents one of the four specific fundamental products. This decoder can easily be changed to an active low output decoder by replacing the AND gates with NAND gates.

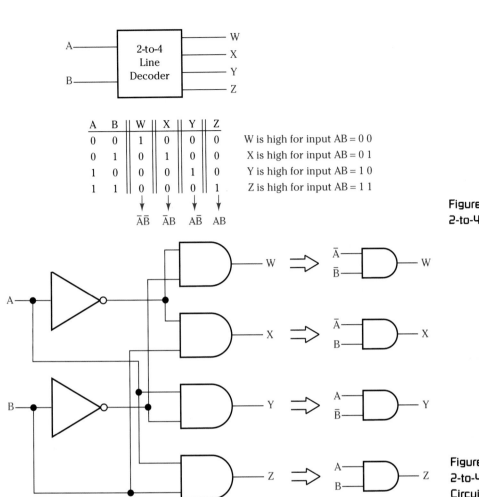

A	B	W	X	Y	Z
0	0	1	0	0	0
0	1	0	1	0	0
1	0	0	0	1	0
1	1	0	0	0	1

W is high for input AB = 0 0
X is high for input AB = 0 1
Y is high for input AB = 1 0
Z is high for input AB = 1 1

$\overline{A}\overline{B}$ $\overline{A}B$ $A\overline{B}$ AB

Figure 6.16
2-to-4 Line Decoder

Figure 6.17
2-to-4 Line Decoder
Circuitry

Design a 3-to-8 line decoder with active low outputs.

*DESIGN
EXAMPLE 6.14*

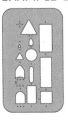

Solution Create a truth table defining the function of each output. There will
be three inputs and eight outputs.

A	B	C	O_7	O_6	O_5	O_4	O_3	O_2	O_1	O_0	$\overline{O}_7$	$\overline{O}_6$	$\overline{O}_5$	$\overline{O}_4$	$\overline{O}_3$	$\overline{O}_2$	$\overline{O}_1$	$\overline{O}_0$
0	0	0	1	1	1	1	1	1	1	0	0	0	0	0	0	0	0	1
0	0	1	1	1	1	1	1	1	0	1	0	0	0	0	0	0	1	0
0	1	0	1	1	1	1	1	0	1	1	0	0	0	0	0	1	0	0
0	1	1	1	1	1	1	0	1	1	1	0	0	0	0	1	0	0	0
1	0	0	1	1	1	0	1	1	1	1	0	0	0	1	0	0	0	0
1	0	1	1	1	0	1	1	1	1	1	0	0	1	0	0	0	0	0
1	1	0	1	0	1	1	1	1	1	1	0	1	0	0	0	0	0	0
1	1	1	0	1	1	1	1	1	1	1	1	0	0	0	0	0	0	0

Fundamental sums approach   For the complement method

That is:

$$\overline{O}_0 = \overline{A}\overline{B}\overline{C}$$

Several design approaches are possible. Using K-maps directly from the truth table will work, but it is a lengthy process for this application since any given output column has only one low output combination. If a fundamental products approach is desired, use the complement method—complement every output column, find the fundamental product, and recomplement. Thus:

$$\overline{O}_0 = \overline{A}\overline{B}\overline{C} \quad \text{therefore} \quad \overline{\overline{O}}_0 = O_0 = \overline{\overline{A}\overline{B}\overline{C}}$$

$$\overline{O}_1 = \overline{A}\overline{B}C \quad \text{therefore} \quad \overline{\overline{O}}_1 = O_1 = \overline{\overline{A}\overline{B}C}$$

$$\overline{O}_2 = \overline{A}B\overline{C} \quad \text{therefore} \quad \overline{\overline{O}}_2 = O_2 = \overline{\overline{A}B\overline{C}}$$

$$\overline{O}_3 = \overline{A}BC \quad \text{therefore} \quad \overline{\overline{O}}_3 = O_3 = \overline{\overline{A}BC}$$

$$\overline{O}_4 = A\overline{B}\overline{C} \quad \text{therefore} \quad \overline{\overline{O}}_4 = O_4 = \overline{A\overline{B}\overline{C}}$$

$$\overline{O}_5 = A\overline{B}C \quad \text{therefore} \quad \overline{\overline{O}}_5 = O_5 = \overline{A\overline{B}C}$$

$$\overline{O}_6 = AB\overline{C} \quad \text{therefore} \quad \overline{\overline{O}}_6 = O_6 = \overline{AB\overline{C}}$$

$$\overline{O}_7 = ABC \quad \text{therefore} \quad \overline{\overline{O}}_7 = O_7 = \overline{ABC}$$

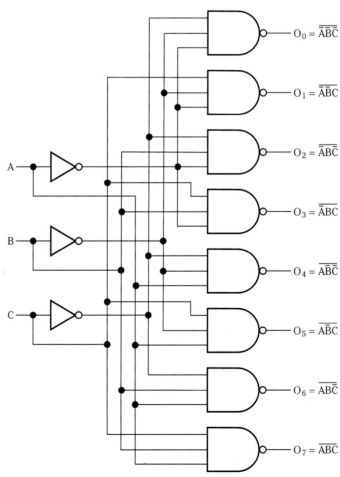

Alternatively, fundamental sums can be obtained from the original truth table outputs. Thus:

$$O_0 = A + B + C \qquad O_4 = \bar{A} + B + C$$
$$O_1 = A + B + \bar{C} \qquad O_5 + \bar{A} + B + \bar{C}$$
$$O_2 = A + \bar{B} + C \qquad O_6 = \bar{A} + \bar{B} + C$$
$$O_3 = A + \bar{B} + \bar{C} \qquad O_7 = \bar{A} + \bar{B} + \bar{C}$$

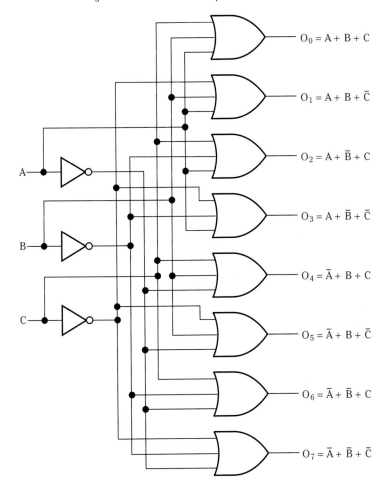

It is interesting to note that outputs in both circuit forms are equivalent by DeMorganizing. That is:

$$O_2 = \overline{\overline{A}B\overline{C}}$$
$$\downarrow$$
$$O_2 = A + \bar{B} + C$$

Encoders

Encoders are the logical opposite of decoders. Whereas a decoder breaks up an input code into multiple outputs, an encoder takes several distinct input lines and combines them into a smaller coded group. A typical encoding function might be the task of combining the various keys on a computer keyboard into representative binary codes. A practical example is an ASCII-coded keyboard. If the keyboard had 96 characters, then one wire per character—96 wires—would be an unwieldy way to transmit keyboard information to a computer. Instead, the ASCII code allows all the keyboard characters to be represented by 7-bit code groups. An encoder is one way to convert 96 different sources of information into a more manageable 7-bit code.

Figure 6.18 block diagrams an encoder, identified by its input and output lines, just as decoders are identified. The block diagram shown is an 8-to-3 line encoder, so named because there are eight inputs and three outputs. In Figure 6.18 the encoder outputs a 3-bit number representing the single input pin at a high level. We assume that only one input is active at a time, so that the remaining seven inputs would all be low. (A priority encoder circuit, on the other hand, can have more than one input active simultaneously.) If input D, the fourth input, is high while all others are low, the encoder outputs 011. The selection of input D is encoded to output a specific binary number representing the fact that input D is active. Or stated differently, if the inputs equal 00010000, the outputs equal 011.

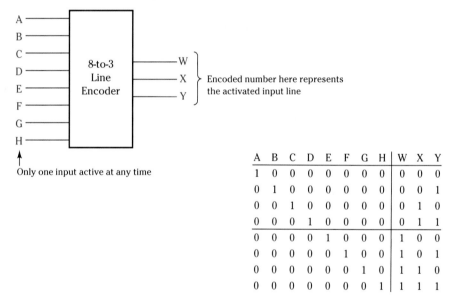

A	B	C	D	E	F	G	H	W	X	Y
1	0	0	0	0	0	0	0	0	0	0
0	1	0	0	0	0	0	0	0	0	1
0	0	1	0	0	0	0	0	0	1	0
0	0	0	1	0	0	0	0	0	1	1
0	0	0	0	1	0	0	0	1	0	0
0	0	0	0	0	1	0	0	1	0	1
0	0	0	0	0	0	1	0	1	1	0
0	0	0	0	0	0	0	1	1	1	1

Figure 6.18
8-to-3 Line Encoder

DESIGN EXAMPLE 6.15

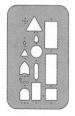

The keypad on a radio scanner contains 24 keys. If each key is wired to an encoder, how many encoded output bits are produced by the encoder circuit?

Solution   The number of encoded bits is a function of 2^n, where n is the number of bits. Six encoded bits are required in this example since $2^6 = 32$. Six is the minimum number of bits required to encode 24 combinations.

BCD Switch Encoder Design

We will design a BCD switch encoder, similar to those used for calculator keypads, to clarify our understanding of encoder principles. The encoder converts ten separate key switch closures into ten unique BCD codes to represent the decimal numerals zero through nine. This practical example also illustrates how humans and digital systems communicate. People expect familiar decimal numbers to be imprinted on their calculator keys since that is our number system of choice. Digital calculator circuitry, on the other hand, can process only binary numbers. The encoder circuit accommodates both expectations.

Figure 6.19 shows a block diagram representation for our encoder design. Our inputs will be active level low—only one input is low at a time. Each input comes from a unique switch and each switch is assigned a specific output code. The truth table in Figure 6.20 shows the assignment of the output codes.

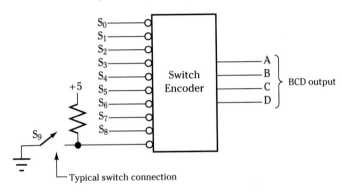

Figure 6.19
Switch Encoder

S_0	S_1	S_2	S_3	S_4	S_5	S_6	S_7	S_8	S_9	A	B	C	D
0	1	1	1	1	1	1	1	1	1	0	0	0	0
1	0	1	1	1	1	1	1	1	1	0	0	0	1
1	1	0	1	1	1	1	1	1	1	0	0	1	0
1	1	1	0	1	1	1	1	1	1	0	0	1	1
1	1	1	1	0	1	1	1	1	1	0	1	0	0
1	1	1	1	1	0	1	1	1	1	0	1	0	1
1	1	1	1	1	1	0	1	1	1	0	1	1	0
1	1	1	1	1	1	1	0	1	1	0	1	1	1
1	1	1	1	1	1	1	1	0	1	1	0	0	0
1	1	1	1	1	1	1	1	1	0	1	0	0	1

$A = 1$ when S_8 or S_9 are low
$$A = \bar{S}_8 + \bar{S}_9$$

$B = 1$ when S_4 or S_5 or S_6 or S_7 are low
$$B = \bar{S}_4 + \bar{S}_5 + \bar{S}_6 + \bar{S}_7$$

$C = 1$ when S_2 or S_3 or S_6 or S_7 are low
$$C = \bar{S}_2 + \bar{S}_3 + \bar{S}_6 + \bar{S}_7$$

$D = 1$ when S_1 or S_3 or S_5 or S_7 or S_9 are low
$$D = \bar{S}_1 + \bar{S}_3 + \bar{S}_5 + \bar{S}_7 + \bar{S}_9$$

Figure 6.20
Switch Encoder Truth Table

For example, if switch 6 (S_6) is brought low, then the encoder output equals 0110. Since the encoder has four outputs, four circuit equations are required. Also, notice that ten inputs are presented to the circuit, but (fortunately) only ten of the possible 1024 input combinations are considered, since only one switch is active at a time. Therefore, our truth table is not a complete truth table and only represents the input combinations creating the desired circuit response. We could treat the unused combinations as don't care states, but this is unwieldy since there are more than 1000 unused combinations. The K-maps or Quine–McCluskey tables that result by considering the don't care states would keep us occupied for a long time. Therefore, we will modify our design technique slightly and design from a more pragmatic point of view.

It is apparent from the truth table that output A, for instance, is high only when switch S_8 or S_9 is low. Recognizing this, an equation can easily be derived: $A = \overline{S_8} + \overline{S_9}$. Output C is high when S_2 OR S_3 OR S_6 OR S_7 are low. Thus, $C = \overline{S_2} + \overline{S_3} + \overline{S_6} + \overline{S_7}$. Equations for the remaining outputs can be defined in a similar manner and are given in Figure 6.20. Understanding how truth tables depict desired circuit response in mathematical form gives us the insight to create an equation even if a full truth table is not available or practical. We must, however, carefully evaluate the equations for simplicity since a minimized equation may not necessarily result from this intuitive approach.

$$A = \overline{S_8} + \overline{S_9} \implies \overline{S_8\ S_9}$$

$$B = \overline{S_4} + \overline{S_5} + \overline{S_6} + \overline{S_7} \implies \overline{S_4\ S_5\ S_6\ S_7}$$

$$C = \overline{S_2} + \overline{S_3} + \overline{S_6} + \overline{S_7} \implies \overline{S_2\ S_3\ S_6\ S_7}$$

$$D = \overline{S_1} + \overline{S_3} + \overline{S_5} + \overline{S_7} + \overline{S_9} \implies \overline{S_1\ S_3\ S_5\ S_7\ S_9}$$

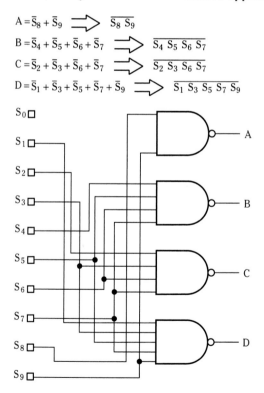

Figure 6.21
Switch Encoder Circuit

Note: S_0 is not used. Output is BCD 0 0 0 0 with or without S_0
Additional circuitry required for a unique S_0 response.

Applying DeMorgan's Theorem to the equations in Figure 6.20 yields the equations and circuitry in Figure 6.21. It makes sense to apply DeMorgan's Theorem to equations full of overbars to see if minimization is possible. In this case all equations are converted to NAND gates, saving us a considerable wiring effort in connecting nine inverters. Each switch (except S_0), when activated, creates a unique BCD output through the encoder circuit. The physical action of pressing a switch is now converted into useful BCD codes that can be utilized by a digital system.

The BCD encoder we just designed has a minor flaw. The encoder produces a BCD 0000 output when we strike the "0" key as well as when none of the keys is struck. How can the encoder circuit still be utilized to encode all keys, yet distinguish between a real "0" and a "no key struck" situation?

Solution Any circuit attached to the encoder BCD output lines senses the encoder state. Since the encoder information is not always valid, a separate signal, notifying the attached circuitry when the encoder output is good, is required. ORing all input switches together in this example produces this signal since the OR output will only be high when a key is struck.

DESIGN
EXAMPLE 6.16

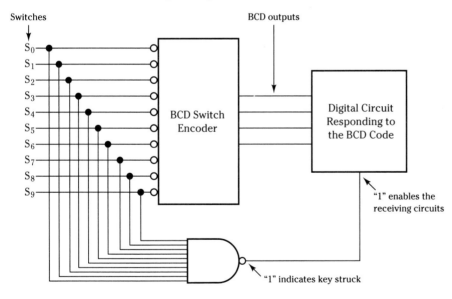

Multiplexers (MUX)

Multiplexers, also known as **data selectors**, are very useful circuits that allow a single wire to carry multiple signals. Multiplexers (MUXs) have application in other areas of digital design as well (discussed in the following sections and in other chapters).

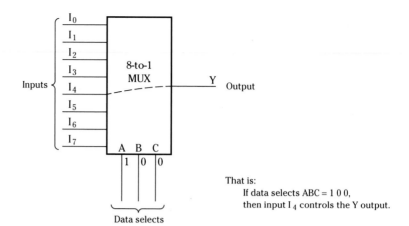

Figure 6.22
8-to-1 MUX

That is:
If data selects ABC = 1 0 0,
then input I_4 controls the Y output.

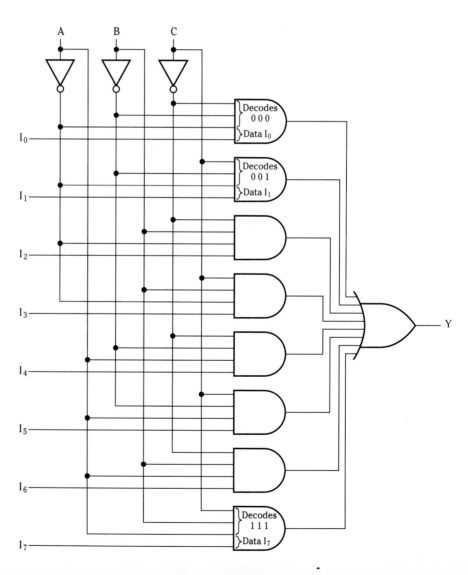

Figure 6.23
MUX Circuitry

When several signals are multiplexed, they appear on a wire on a time-sharing basis. Only one signal is present at any time, guided by data selection circuitry. In general, MUXs convert "parallel data" into "serial data." The block diagram in Figure 6.22 explains these concepts. This 8-to-1 MUX routes any input data bit selected by the data select control lines to the output. The three data select lines produce eight binary combinations; each controls a specific input line. When the data select inputs are 100, for instance, the data from input I4 appears at the output. When the data select input is changed, the newly selected input signal appears at the output.

Multiplexers are designed in a variety of sizes, and it should be evident that the number of inputs is related to the number of data select lines. MUX design is relatively straightforward; the basic circuitry is an absolute decoder of the data select inputs to which data input lines are added. Since we already understand decoders, we will design the 8-to-1 MUX using a building block approach.

The full 8-to-1 MUX is shown in Figure 6.23. Data select inputs A, B, and C are decoded by the AND gates so that each AND gate responds to a specific data select line combination. The input data lines feed individual AND gates, and since only one AND gate is enabled by the select inputs at any time, only one input line controls the MUX output. All AND gate outputs are ORed together to produce the single MUX output.

Design a 2-to-1 MUX.

Solution A MUX is easily designed by adding the necessary data lines to a suitable decoder. All decoder outputs are then ORed together.

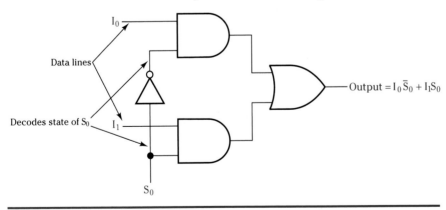

$$\text{Output} = I_0 \bar{S}_0 + I_1 S_0$$

A designer wishes to drive a single BCD-to-seven segment decoder chip with BCD information from two separate sources. Use the 2-to-1 MUX designed in Design Example 6.17 to complete this design application.

Solution Use a MUX for each BCD input line. This allows BCD data from either BCD source to be "steered" to the decoder chip.

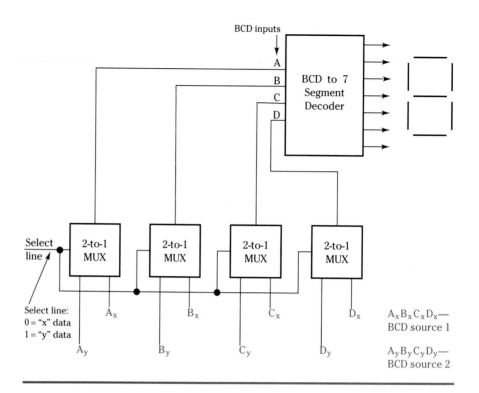

Data Selector Logic

Multiplexers are great for implementing truth table functions. Using data selector or multiplexer logic, we can translate a truth table into a functional circuit in seconds. The number of gates used with data selector logic may be greater than an equivalent K-map reduced design, but the data selector circuit is easier to design. And since MUXs are available in all technologies, this is a viable approach. Data selector design also embodies the philosophy of modern logic design. That is, reducing the chip or package count may be as important as minimizing the gate count. As chip densities increase because of technological improvements, it becomes more practical and economical to adapt a highly integrated chip for a specific application. Such is the case with data selector logic. A truth table function can be turned into working hardware using a single MUX chip. Although the chip's internal logic is often more complex than the specific application demands, the engineering benefits of reduced cost and package count accrue. This design aspect will be discussed more thoroughly in later chapters on programmable logic.

Figure 6.24 shows how an 8-to-1 MUX implements the truth table function illustrated. The truth table output column shows a pattern of ones and zeros for the various input combinations. Attaching input variables A, B, and C to the data select lines causes specific MUX input lines to be selected based on the levels of the truth table input combinations. By tying the MUX input lines high or low to correspond to the truth table input–output relationship, the truth table function can be implemented. For instance, the truth table shows input combination 110 producing a zero output. By tying the MUX input selected by

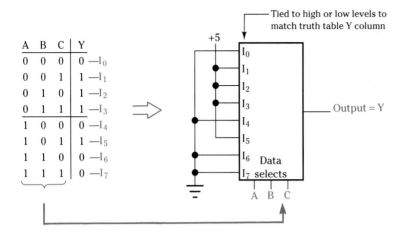

Figure 6.24
Data Selector Logic
Implementation

combination 110 to a low level, a low will appear at the MUX output when truth table variables A, B, and C = 110. This is exactly what any design for the truth table must do. All MUX inputs are simply tied to high or low levels that correspond to the truth table entries while the input variables are connected to the MUX select lines. This completes the circuit.

Design a circuit using data selector logic to indicate when any three of four pressure sensors are exceeding their pressure limits. The sensors are identified as P_0, P_1, P_2, P_3 and a low level sensor output indicates normal pressure while a high level output indicates that the pressure limit has been exceeded.

DESIGN
EXAMPLE 6.19

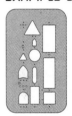

Solution   Define the function of the switches using a truth table. Since there are four inputs, a 16-to-1 MUX is required to implement the logic.

P_3	P_2	P_1	P_0	Output
0	0	0	0	0
0	0	0	1	0
0	0	1	0	0
0	0	1	1	0
0	1	0	0	0
0	1	0	1	0
0	1	1	0	0
0	1	1	1	1
1	0	0	0	0
1	0	0	1	0
1	0	1	0	0
1	0	1	1	1
1	1	0	0	0
1	1	0	1	1
1	1	1	0	1
1	1	1	1	0

DESIGN EXAMPLE 6.20

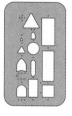

Use a 16-to-1 MUX to redesign the BCD decoder given in Design Example 6.13.

A	B	C	D	Output
0	0	0	0	0
0	0	0	1	0
0	0	1	0	0
0	0	1	1	0
0	1	0	0	1
0	1	0	1	0
0	1	1	0	0
0	1	1	1	1
1	0	0	0	0
1	0	0	1	1
1	0	1	0	d
1	0	1	1	d
1	1	0	0	d
1	1	0	1	d
1	1	1	0	d
1	1	1	1	d

Does this design reduce the number of standard TTL or CMOS integrated circuit packages compared to the design in Design Example 6.13?

Yes. One 74150 16-to-1 MUX provides the equivalent function. It requires one 7411 (74C11) and one 7432 (74C32) for the alternate design.

Demultiplexers (DEMUX)

Demultiplexers, also known as **data distributors**, perform an operation opposite that of the MUX. Demultiplexers (DEMUXs) distribute a single data input signal among many output lines. The distribution of signals is controlled by

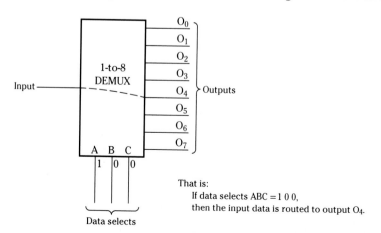

That is:
If data selects ABC = 1 0 0,
then the input data is routed to output O_4.

Figure 6.25
1-to-8 DEMUX

data select inputs similar to those in the MUX. Figure 6.25 is a block diagram of a 1-to-8 line DEMUX.

This particular DEMUX has a single input and eight outputs as indicated by the 1-to-8 line designation. One of the eight outputs receives the input data as directed by the three data select lines. Naturally, three select lines are necessary to determine which one of the eight outputs is to be activated. For example, if the data select lines ABC = 100, then the input data is routed to output O4. Change the data select code and the input data appears at another output line.

A decoder circuit is an integral part of DEMUX design, decoding the state of the select inputs. In fact, from a hardware standpoint, the DEMUX is little more than a decoder with an added data input line. For this reason an 1-to-8 line DEMUX is often called a 3-to-8 decoder. Figure 6.26 explains. As is evident,

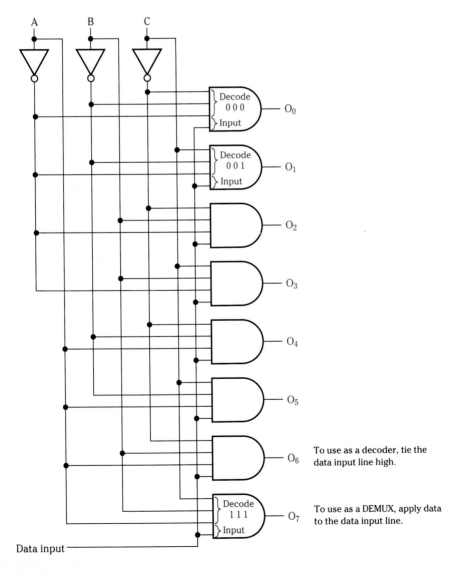

To use as a decoder, tie the data input line high.

To use as a DEMUX, apply data to the data input line.

Figure 6.26
DEMUX Circuitry

the only difference between this 1-to-8 line DEMUX and a 3-to-8 decoder is the input line common to all AND gates. The data select lines enable a single AND gate, whereas the data input line determines the output level of the selected AND gate. All other AND gates (those not selected) produce a low level output. Use the select lines to route data to a specific output line and you have a DEMUX in action; tie the data input line high and the select inputs are used to form a basic decoder.

DESIGN
EXAMPLE 6.21

Determine the output levels for the following 1-to-4 line DEMUX using a timing diagram.

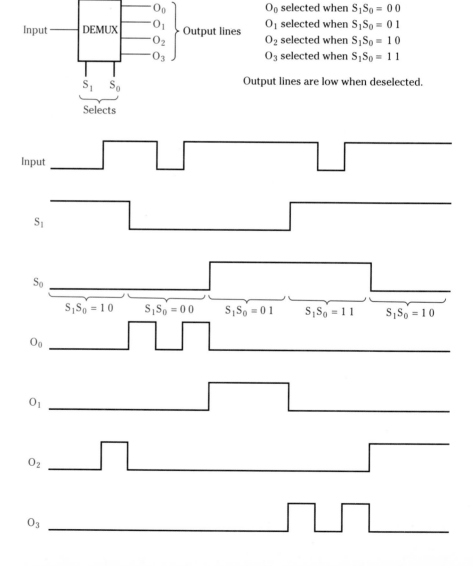

O_0 selected when $S_1S_0 = 0\,0$
O_1 selected when $S_1S_0 = 0\,1$
O_2 selected when $S_1S_0 = 1\,0$
O_3 selected when $S_1S_0 = 1\,1$

Output lines are low when deselected.

Solution   Determine when the individual outputs are selected. Complete the timing diagram. Each output follows the IN signal when selected.

Eight signals destined for eight computer terminals will be sent a considerable distance through a single fiber optic cable. Assuming that devices are present to convert digital signals to light and back, draw a block diagram of how this signal transfer could take place.

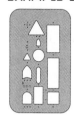

Solution   An 8-to-1 line MUX can convert the eight sources of information into a single source for presentation to the fiber optic cable. At the other end of the system an 1-to-8 line DEMUX could reseparate the signals. Some synchronization of the select signals between the two would be required in a practical system.

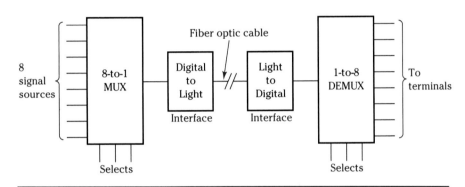

6.3 ■ ERRORS IN DIGITAL SYSTEMS

What a comfortable feeling to be 100% confident that our digital designs will respond perfectly once they are built. Unfortunately, the digital life is not like that. We can, through systematic testing and careful design, ensure that our products are reliable and working according to specification. As design technicians and engineers, we are obligated to do this as an everyday business practice. However, the electronic world facing our designs is a noisy, dirty, trying one for the individual logic gates comprising our carefully constructed circuits.

Electrical noise and disturbances are always present within and around the equipment we build. Power line voltage levels fluctuate and consequently affect logic one and zero levels. Lightning and radio frequency signals roaming the atmosphere affect digital signal quality. The gates with which we design also induce noise into circuits as they switch states. The sudden demand for current from a switching NAND gate influences the logic levels on nearby gates. On the microscopic level, alpha particles are emitted from the packaging materials housing integrated circuits. The alpha particle energy is particularly troublesome for memory devices, causing them to change logic level—a problem that increases as integrated circuit geometries shrink. Obviously, this is a very nasty environment for innocent logic gates.

In a later chapter we will look at steps to minimize some of these problems, but for now we examine what can be done if an error occurs. Let's assume that electrical noise in one form or another will eventually cause an error. Naturally, we do not ignore errors because to do so means erroneous results. In a digital computer, for example, a payroll program could produce errors if lightning caused a power line disturbance during the course of program execution. If a particular logic gate is sensitive to the voltage change from the sudden voltage transient caused by the lightning, logic levels could change. This affects the payroll program. Someone ends up with more or less money than he or she should—an undesirable situation. It is important that errors be detected so that corrective measures can be taken.

Errors are particularly damaging in **real-time processes**. Digital systems monitoring or controlling real-time events must handle them as they occur—in real time. Factory automation or the control of a rocket flight are real-time applications that suffer when errors are not corrected. We will examine several error checking systems to see how digital system reliability can be improved.

Parity

Error checking systems require hardware in addition to that of the circuitry being checked. Error checking is as much an economic decision as it is a reliability consideration since there are costs associated with the error checking hardware as well as with the support circuitry and means to handle an error. **Parity** is an error checking system frequently used to detect **single bit errors**. That is, only 1-bit is likely to be in error at any time. The parity method is cost-effective because the hardware requirements are not excessive and because the most frequent errors in digital systems are single bit errors. Parity checking provides a reasonable means of error detection, assuming that the frequency of errors is low. A high error rate implies that basic system reliability is poor and detecting unreliable operation with an error checking system will only confirm that the system is unreliable and should be redesigned.

Parity checking is implemented in two basic forms—**even parity** or **odd parity**. In both cases a **parity bit**, generated with parity generation circuitry, accompanies data throughout the system. The **parity generator** examines all data bits (the number of data bits is immaterial) and generates an appropriate even or odd parity bit. The parity bit tags along with the data, and at critical points in the system the parity bit and the data are checked with a **parity checker** circuit. The parity checker determines if an error has occurred. When there is no error, processing continues as normal. If an error occurs, a signal to that effect is provided by the parity checker. An overall system error handling strategy determines the next course of action—parity checking only detects errors. Figure 6.27 block diagrams a parity checking system.

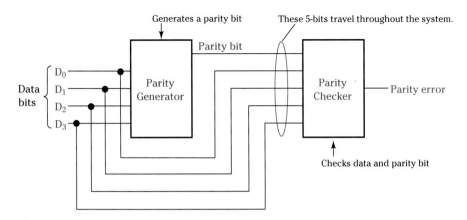

If the parity generator creates a parity bit for odd parity,
then the parity checker indicates the following:
   Odd parity detected = no error
   Even parity detected = parity error

If the parity generator creates a parity bit for even parity,
then the parity checker indicates the following:
   Odd parity detected = parity error
   Even parity detected = no error

Figure 6.27
Parity Checking
System

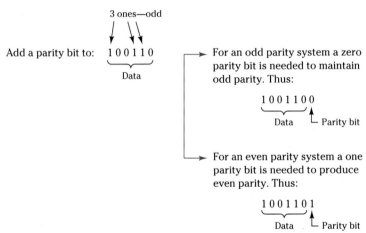

Additional examples:

| | Parity Bit | |
Data	Even	Odd
100110	1	0
101101	0	1
11110111	1	0
101	0	1
0110	0	1

Figure 6.28
Generating a Parity Bit

Even parity means that the total number of ones in a data group, including the parity bit if it exists, is an even number. Odd parity means that the total number of ones in a data group, including the parity bit if it exists, is an odd number. The actual value of the binary number is immaterial. Figure 6.28 shows several even and odd parity combinations. The parity bit is generated in such a way that an odd or even number of bits always exists, barring an error. For instance, in an odd parity system, the data group 100110 already has three ones, an odd number. The parity generator generates a zero parity bit to maintain an odd number of ones. When the parity bit is attached to the data, the data reads 1001100. With even parity, this same data group has a one parity bit generated so that the data and parity bit combination is an even number of ones—1001101. Even though these examples show the parity bit attached to the data as the LSB, there is no restriction on its placement. The parity bit could just as easily be carried throughout a digital system as the MSB.

DESIGN
EXAMPLE 6.23

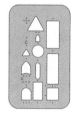

Determine the odd parity bits for the following data:
(a) 10011101   (b) 111001000   (c) 0000000

Solution Assign an odd parity bit equal to one if the number of ones in the data is an even number; otherwise make the parity bit low. All zeros are an even number.
(a) 10011101   5 ones in data; odd parity bit = 0
(b) 111001000   4 ones in data; odd parity bit = 1
(c) 0000000   0 ones in data; odd parity bit = 1

DESIGN
EXAMPLE 6.24

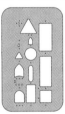

Determine the even parity bits for the following data:
(a) 10011101   (b) 111001000   (c) 0000000

Solution Assign an even parity bit equal to one if the number of ones in the data is an odd number; otherwise make the parity bit low. All zeros are an even number.
(a) 10011101   5 ones in data; even parity bit = 1
(b) 111001000   4 ones in data; even parity bit = 0
(c) 0000000   0 ones in data; even parity bit = 0

Now that we have seen how parity is assigned, we will see how this helps us detect errors. Refer to Figure 6.29. An initial design decision determines the type of parity used (even or odd), then all circuitry in the system conforms to the parity system chosen. Assume that odd parity is used to check 6-bit data groups. After parity is generated by examination of the 6-bits, 7-bits of information travel throughout the system. At all times all 7-bits should possess odd parity. Whenever a data group shows even parity, an error has occurred; 1-bit has changed level. The parity checker circuit detects this and signals an error.

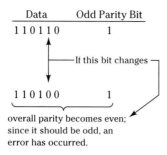

Data	Odd Parity Bit
1 1 0 1 1 0	1

—If this bit changes—

| 1 1 0 1 0 0 | 1 |

overall parity becomes even;
since it should be odd, an
error has occurred.

**Figure 6.29
Detecting a Parity
Error**

The parity checker cannot identify which bit is in error, so error correction is not possible, but the fact that an error has occurred—an important fact—is made known. Furthermore, any errors detected are single bit errors (or any odd number of errors). If 2-bits change, proper parity can be maintained even though the data is incorrect. Therefore, parity checking is useful only if the probability of multiple bit errors is low.

Assuming an even parity system, determine which data groups are in error:
(a) 01101110   (b) 1101010110   (c) 1000

**DESIGN
EXAMPLE 6.25**

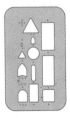

Solution Using even parity, every data group should always have an even number of ones. Therefore, an odd number of ones indicates a data error. Identifying which bit is the parity bit is unimportant since it is checked as well.
(a) 01101110   5 ones is odd—parity error
(b) 1101010110   6 ones is even—no error
(c) 1000   1 one is odd—parity error

Now that it is evident that the parity checking system can spot errors, we will design the generation and detection circuitry. Let's begin with an intuitive design.

Parity checking is accomplished by detecting an even or odd number of ones in a data group. Parity generation also follows the same premise. You may recall that the Exclusive-OR function has an even/odd combination property. That is, the EX-OR produces a low output for even input combinations (00 and 11) and a high output for odd input combinations (01 and 10). It appears that the Exclusive-OR function holds the key to the parity checking circuit. A parity checker or generator is simply a multiple input EX-OR function.

Now we examine parity generation circuitry and formalize use of the Exclusive-OR. Figure 6.30 is a four input variable truth table whose outputs are even and odd parity bits. It is apparent that the only difference between even and odd parity is a level of inversion. Commercially available parity generators often have two outputs, one for even and one for odd, since both bits are generated easily.

Figure 6.30 also includes the K-map for the odd parity bit. As you recall, the alternating pattern of ones and zeros on a K-map is indicative of the

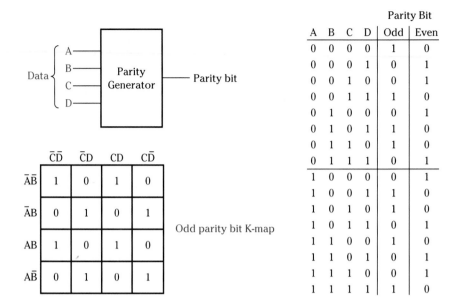

A	B	C	D	Odd	Even
0	0	0	0	1	0
0	0	0	1	0	1
0	0	1	0	0	1
0	0	1	1	1	0
0	1	0	0	0	1
0	1	0	1	1	0
0	1	1	0	1	0
0	1	1	1	0	1
1	0	0	0	0	1
1	0	0	1	1	0
1	0	1	0	1	0
1	0	1	1	0	1
1	1	0	0	1	0
1	1	0	1	0	1
1	1	1	0	0	1
1	1	1	1	1	0

Figure 6.30
Parity Generator Design

Exclusive-OR function. Since the pattern is pervasive throughout the map, the output function is the Exclusive-OR of all four variables. In other words, a parity bit is generated by Exclusive-ORing all data bits. This is true regardless of the number of data bits.

Figure 6.31 shows the circuitry required for parity generation with 4-bit data. On paper, run some test values through the circuit and confirm that the last EX-OR output produces an even parity bit. As an example, if the input data is 0111, the output is high. This would be an even parity bit since this output, combined with the original data, would now create a data group with an even number of ones. Therefore, the Exclusive-OR network produces the even parity bit. If odd parity is preferred, an inverter would be added to the output of the last EX-OR (an Exclusive-NOR function) to create an odd parity bit.

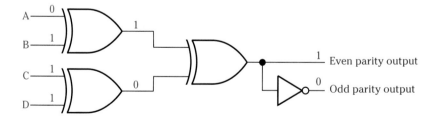

Figure 6.31
Parity Generator Circuitry

A parity checker is designed similarly to a parity generator circuit, but the circuit output is interpreted differently. That is, if even parity is produced and even parity is expected, then the output is said to be correct. Otherwise a parity error has occurred. Therefore, the parity checker output is treated as an error line. Figure 6.32 shows the parity checker. Using the previous example as a guide, we add an additional EX-OR to the checking circuitry because we are now checking 5-bits—four data bits plus a parity bit. The output of the last

EX-OR indicates even parity when high and odd parity when low. Whether or not these levels indicate a parity error depends on the specific system in use. That is, if parity opposite to that expected is indicated by the parity checker, then an error has occurred. The example shown in Figure 6.32 details this concept.

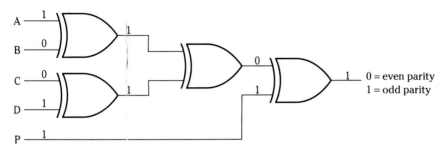

If ABCDP = 1 0 0 1 1, then odd parity exists. Assuming that this is correct, then the parity checker output = 1.

Therefore; 1 output = no error
                   0 output = parity error

Figure 6.32
Parity Checker
Circuitry

Using block diagrams, show how a parity generator would be attached to the output of an ASCII keyboard.

Solution The parity generator must sense all keyboard data to generate an appropriate parity bit. The generated parity bit would then accompany the keyboard data throughout the digital system.

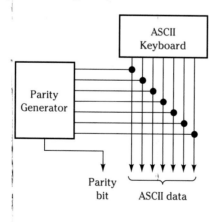

DESIGN
EXAMPLE 6.26

Checksums

Checksums are used when larger quantities of data are transferred from one digital system to another and error checking is desired. A checksum is formed by adding all the data bytes together (whether or not the data are numbers), discarding the carry, and appending what remains (typically a byte or two) to the original data. The checksum, since it is calculated from the original data, is representative of that data. A receiving digital system re-adds the data using the same process and compares the two checksums. If the checksums agree, the data has arrived without error. If the checksums do not agree, an error has occurred. Further action is required to recover from the error. Figure 6.33 diagrams this process.

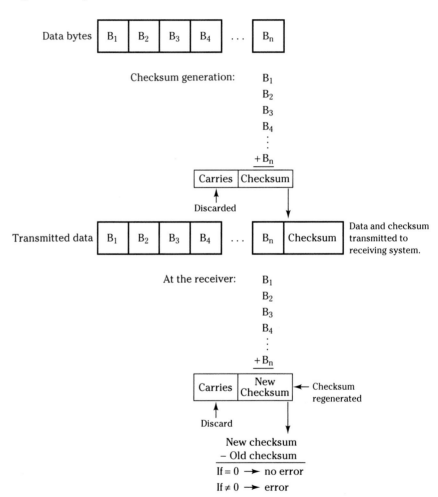

Figure 6.33
Creating a Checksum

Checksum error checking has some advantages over the parity error checking method. Parity is a good system, particularly at the circuit level, because it continually verifies the integrity of data. But when quantities of data are stored, the additional parity bits assigned to each data group require significant storage space. For instance, storing 2048 data bytes in a computer memory is equivalent to storing $2048 \times 8 = 16{,}384$-bits. If parity is assigned to each byte, an additional 2048 parity bits will also be stored. This space must be available in the memory—a costly consideration. Checksums, on the other hand, can be limited to a single byte (8-bits) in length, regardless of the number of data bytes used to generate the checksum. This requires an insignificant amount of storage. Of course, nothing is free. Checksum generation requires hardware and/or computer software with complexity beyond that required for parity generation. Addition circuits, and associated addition time delays, are factors for consideration in checksum error checking systems.

Cyclic Redundancy Checks (CRC)

Cyclic redundancy checks (CRCs) are similar to checksums, but they detect errors with greater precision. CRCs are used extensively in floppy and hard disk storage systems and in the verification of memory and programmable logic designs.

Binary division is used to generate CRCs. The process divides quantities of data by a characteristic polynomial equation. Several standard polynomials are commonly used for this procedure. The data bits destined for checking are assigned as the coefficients of a polynomial n-bits long. The data polynomial and the characteristic polynomial are then divided, producing a quotient and remainder; the remainder is sent along with the data, and reflects the contents of the data. The receiving digital system recomputes the CRC for comparison. Naturally, equality results when the data is transmitted error-free.

Since the CRC is generated on a bit basis, rather than a byte basis (as for checksums), the probability that errors can be detected is greater.

Error Correcting Codes (ECCs)

The error checking schemes discussed so far only detect errors, requiring the digital system to respond to the error in some appropriate fashion. In spite of the response, the error cannot continue through the system causing other problems. Typical system responses to errors range from simply restarting the process to elaborate "retry" schemes in which computer systems reexecute instructions over and over until the error is gone. In any case the presence of an error is far from painless since it upsets system performance.

Error checking and correcting systems go a step beyond error detection. Depending on the **error correcting code** (ECC) complexity, multiple bit errors can be detected while other errors can be identified and corrected without any appreciable time delay. Systems such as this are designed around ECCs. The most frequently used ECC is the **Hamming code**. This code describes the mathematical treatment of binary data to identify the occurrence of errors and, in many cases, the individual bits in error. The effectiveness of an ECC system is circuitry-dependent, and ECC circuitry carries a significant hardware overhead that grows as error detection and correction capability increases. For instance, a typical ECC system detects single, double, and some triple bit errors and corrects all single bit errors. Increasing capability to correct single and double bit errors represents a significant increase in circuitry. Furthermore, additional circuitry is required as the data word size is increased. That is, checking 32-bit data requires much more circuitry than that required to check 16-bit data. Since ECC is expensive, it is used only when enhanced reliability and performance warrant it, and not for general error detection.

Check bits are assigned to the data in an ECC system and are generated according to the ECC in use. Basically, check bits are multiple parity bits generated on selected portions of the data. The ECCs determine which data bits are selected. For example, the first check bit generated for a 16-bit data word is created by generating parity on data bits 1, 2, 3, 5, 8, 9, 11, and 14. The number of check bits required depends on the size of the data. In general,

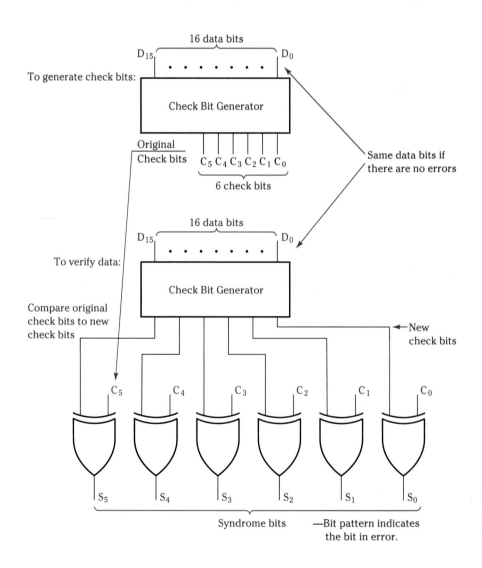

Figure 6.34
ECC System

16-bit data quantities require six check bits; 32-bit data uses seven, whereas 64-bit data uses eight. Already the check bit storage overhead is evident, not to mention all the parity generation circuitry required, but this proves worthwhile since data errors can be corrected.

The check bits always tag along with the data, and when data is eventually checked for errors, additional check bits are generated in the same manner as the first set of check bits. The new check bits are Exclusive-ORed with the original check bits producing **syndrome bits**. Syndrome bits indicate the status of the data. If the resulting syndrome pattern is all zeros, the data (and check bits) are error-free. If the syndrome bits are nonzero, the syndrome bit pattern provides information on the specific error. When the error is a multiple bit error, it is detected by the system. When the error is a single bit error, the ECC circuitry identifies the actual bit in error. The error can then be easily corrected by inverting the level of the bit as it passes through the ECC system, since any bit can only be high or low. Figure 6.34 summarizes ECC.

A Final Error Checking Wrap-Up

Despite the care used in digital design, errors still occur. More elaborate error correcting systems exist beyond those discussed. For instance, **burst errors** occur when data is transmitted at high rates of speed. If noise affects 1-bit, it is just as likely to affect several bits at once. Complex burst error processors reconstruct multiple errors in situations such as this. These systems combine CRC and syndrome capabilities to accomplish the detection and correction functions. Since errors can never be entirely eliminated, error checking systems will continue to grow in importance in many digital designs.

Some Hard/Soft Errors

Type	Problem	Cause
Soft	Circuit noise	Power line transients, switching noise, cross talk
Soft	Glitches	Design problem
Soft	Alpha particles	Energy emitted by integrated circuit packaging materials causes circuits to change state
Hard	Constant error	Component failure

Soft error—temporary errors
Hard errors—permanent errors
Glitch—occasional, change in logic level lasting for only a short duration of time

FOR YOUR INFORMATION

Error Checking Summary

Technique	Method	Capability
Parity	Even/odd parity bit appended to data; checked periodically	Detects single bit errors
Checksums	One- or two-byte sum of data appended to data; checked after transmission of data	Detects data transmission errors
CRC	One- or two-byte remainder appended to data; checked after transmission of data	Detects data transmission errors
ECC	Check bits generated according to ECCs are appended to data; comparisons produce syndrome bits, indicating validity of data	Detects and corrects errors

FOR YOUR INFORMATION

Some Common Part Numbers

Function	Part Number	Technology
4-to-16 decoder/DEMUX	74154	TTL
	4514	CMOS
3-to-8 decoder/DEMUX	74138	TTL
	10161	ECL
Encoder	74147	TTL
	4532	CMOS
	100165	ECL
MUX/data selector	74150	TTL
	10164	ECL
BCD to binary code converter	74484	TTL
BCD to seven segment decoder	7446	TTL
	4511	CMOS
Parity generator/checker	74180	TTL
	4531	CMOS
	100160	ECL
ECC	74616	TTL
	39C60	CMOS

- Binary codes are used to represent real-world information in a form suitable for digital processing.

- Weighted codes bear a mathematical relationship to binary numbers; unweighted codes are simply groups of binary bits representing information, but without any mathematical relationship to the binary number system.

- Decoders are practical digital circuits found in most digital applications. Decoders range in complexity from the simplicity of a single fundamental product (AND gate) to complete decoding of all input combinations for n input bits.

- Encoders compress digital information into as few binary bits as possible by taking advantage of the fact that n binary bits can represent 2^n combinations of the bits.

- Multiplexers and demultiplexers are useful circuits for sharing data over a single wire.

- Adapting multiplexers for data selector logic is an alternative method for implementing truth table functions with a single integrated circuit.

- Errors in digital systems occur for many reasons. Digital system performance and reliability can be improved by providing a mechanism to detect, or to detect and recover from, an error. Typical techniques are parity, checksums, cyclic redundancy checks, and error correcting codes.

Section 6.1

1. Convert the following decimal numbers into BCD code groups:
   *(a) 627   (b) 254   (c) 1278   (d) 1111   (e) 65536   (f) 7683632
   (g) 7564.6235   (h) 1001.0111
2. Change the following BCD numbers into their decimal equivalent numbers:
   *(a) 0111 1001 0101 0000   (b) 1001 1000 0100 0010
   (c) 0110 0001.1001 0101   (d) 0001.0001
3. Determine the ASCII codes required to represent the following:
   (a) HIGH SPEED CMOS LOGIC
   (b) ALPHABET SOUP
   (c) your name and age
4. Assign odd parity to the following data:
   *(a) 101010   (b) 110000110   (c) 00000100000
5. Assign even parity to the following data:
   *(a) 1000001   (b) 0001000100   (c) 11111
6. Assign odd parity to each ASCII character in the following sequence:
   RESEND DATA
7. Change the following Gray code numbers into binary numbers:
   *(a) 100101001   (b) 01101   (c) 11111111   (d) 000000

---

* See Appendix F: Answers to Selected Problems.

8. Change the following binary numbers into Gray code numbers:
*(a) 01110110   (b) 00001   (c) 1000001   (d) 1111011

Section 6.2

9. A digital system has four signals—A, B, C, and D. The following decoder circuits are present in the system as well. Determine which combinations of A, B, C, and D are decoded by these circuits:

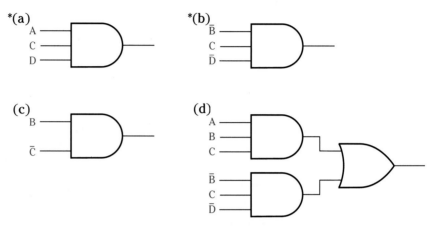

*(a)

*(b)

(c)

(d)

*10. Draw the timing diagram for the following decoder. The circuit containing the decoder has three signals—X, Y, and Z.

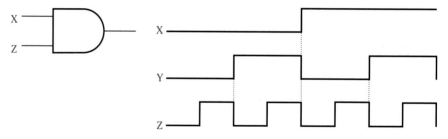

11. Design the circuitry for a 6-bit even parity generator. Show how the six data bits and the generated parity bit are checked using a parity checker circuit. Design the 7-bit parity checker. What design changes are required to modify the system to handle odd parity?

12. Design the circuitry required to convert 6-bit binary numbers into Gray code numbers.

*13. Design a decoder that will detect the presence of any legal BCD code.

14. A 7447 BCD to seven segment decoder/driver chip produces active low output signals. Determine which number will be displayed on a seven

---

* See Appendix F: Answers to Selected Problems.

segment device for the following decoder output signals. Also, determine the BCD code required to attain the display:

*(a) abcdefg = 0001111
(b) abcdefg = 1100000
(c) abcdefg = 1001100

15. Design the decoding circuitry required to light segment "d" of a seven segment display (high level output). The decoder inputs are BCD codes; the decoder output is the seven segment code. Repeat the design for the seven segment "a" light segment.

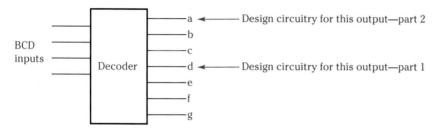

16. Design an 8-to-1 multiplexer (MUX) with both true and complementary outputs.

17. A 4-input MUX has the following signal waveforms. Determine the output waveform:

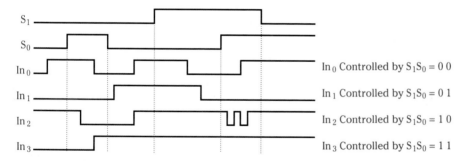

In $_0$ Controlled by $S_1 S_0 = 0\ 0$
In $_1$ Controlled by $S_1 S_0 = 0\ 1$
In $_2$ Controlled by $S_1 S_0 = 1\ 0$
In $_3$ Controlled by $S_1 S_0 = 1\ 1$

*18. Determine the output waveforms (Out$_0$,Out$_1$,Out$_2$,Out$_3$) for a demultiplexer (DEMUX) with the following input waveforms: (Out$_0$ active when S_1, S_0 = 00; Out$_1$ active when S_1, S_0 = 01; etc.)

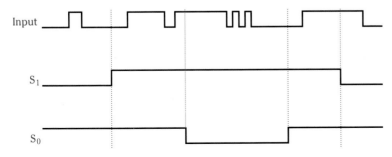

_____

* See Appendix F: Answers to Selected Problems.

19. Show how a 74150 MUX can be used to implement the following truth table using data selector logic techniques:

E	F	G	H	J
0	0	0	0	0
0	0	0	1	1
0	0	1	0	1
0	0	1	1	1
0	1	0	0	0
0	1	0	1	0
0	1	1	0	0
0	1	1	1	1
1	0	0	0	0
1	0	0	1	1
1	0	1	0	0
1	0	1	1	1
1	1	0	0	0
1	1	0	1	0
1	1	1	0	1
1	1	1	1	1

20. Design Example 6.19 used data selector logic to test when any three of four pressure switches were exceeding pressure limits. The design required the use of a 16-to-1 line MUX. Redesign the circuit using an 8-to-1 MUX. Use pressure switches P_0, P_1, and P_2 to control the select inputs. Pressure switch P_3 may be used as a data input.

*21. Design a BCD to excess-3 decoder.

22. (a) Design a 4-to-2 line switch encoder that will respond to active low inputs.

(b) Redesign the encoder to respond to active high input signals.

(c) Add the required circuitry to the basic encoder designs so that valid switch closures are detected.

23. A priority encoder is an encoder design where certain inputs have a greater priority in determining the output than other inputs. This means that if more than one input is active at a time, the input assigned the highest priority is the one that controls the output code. Design a 3-to-2 line priority encoder using inputs A, B, and C. Input A has the greatest priority and outputs 01 when active. Input B has the next highest priority and outputs 10 when active. Input C has the lowest priority and outputs 11 when active. The encoder output is 00 when none of the inputs is active. Design the priority encoder assuming active high input levels.

24. Using block diagrams (decoders, DEMUXs), show how a single set of BCD input (A, B, C, D) data can be displayed on four separate seven segment displays.

---

* See Appendix F: Answers to Selected Problems.

25. A 4-bit ALU (arithmetic and logic unit) can operate on data from several different sources:
    (a) Data is available from four signal pins DA3-DA0
    (b) Data is available from four signal pins DB3-DB0
    (c) Data is available from four "X Register" outputs X3–X0
    (d) Data is available from four "Y Register" outputs Y3–Y0
    Using block diagrams (ALU, MUXs, registers), show how X Register data or signal pin DA3–DA0 data can be routed to one set of ALU inputs. Show how Y Register data or signal pin DB3–DB0 data can be routed to the other set of ALU inputs. For example, this will allow the following ALU addition operations to take place:
    (a) X Register data + Y Register data
    (b) X Register data + DB3–DB0 data
    (c) DA3–DA0 data + Y Register data
    (d) DA3–DA0 data + DB3–DB0 data

CHAPTER 7

LATCHES, FLIP-FLOPS, AND TIMING CIRCUITS

OBJECTIVES

To introduce and explain the need for simple storage devices.

To introduce the various flip-flop configurations used in digital design.

To explain the differences between flip-flops and latches and how level sensitive inputs and edge sensitive inputs differ.

To illustrate the methods and intricacies of flip-flop timing.

To explain how hazard and race conditions adversely affect circuit operation.

To explain how one-shots can be used as timing elements in digital circuits.

PREVIEW

In previous chapters we covered many aspects of combinatorial logic design, including combinatorial circuits and associated devices. We now move to another class of logic circuitry—sequential logic. In chapters to come we will see how logic information is stored, manipulated, and controlled in a prescribed sequence of events. Flip-flops and latches are the basic logic devices used in the design of sequential circuits and are the subject of this chapter.

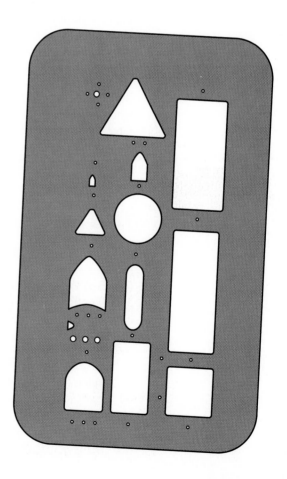

7.1 ■ AN INTRODUCTION TO FLIP-FLOPS

In addition to having an unforgettable name, the **flip-flop** is a device capable of storing a single binary bit. The flip-flop output, which may be one or zero, cannot necessarily be determined by examining the device input lines. That is, after the device is placed in a logical state the input conditions creating the particular logic state can change, but the output may remain the same. Hence, single bit storage or memory capability is available—the device remembers or retains logical information placed within it.

Storage capability is a significant leap in digital design sophistication. **Sequential** circuit design becomes possible, allowing the designer to control events in a predictable sequence. Sequential circuits are circuits where output values depend not only on input values, but also on previously stored values. Naturally, design complexity increases with sequential circuits, so a thorough understanding of basic flip-flop principles is extremely important.

A flip-flop is sometimes referred to as a **latch**. Both flip-flops and latches are electrical devices constructed from a **bistable multivibrator**. A bistable multivibrator is a transistorized circuit capable of having two distinct stable output **voltages (states)**. In the logic world we call the states one and zero. A bistable multivibrator can be controlled to be in either one state or the other by control line(s). Both flip-flops and latches may also be in either the one or zero state, as determined by control signals since they are bistable multivibrator designs. Although the distinction between the term latch and flip-flop is not always clear, the difference in terminology lies in the method used to place the devices into a logic state. Basically, latches are **level sensitive** devices, whereas flip-flops are **edge sensitive** devices. Both methods will be defined in later sections. We will use the term flip-flop to refer to 1-bit storage elements in general. However, when referring to specific circuitry, the terms latch and flip-flop will be used as appropriate.

7.2 ■ BASIC FLIP-FLOP CIRCUITRY

There are many flip-flop and latch designs to cover and they all evolve from one of the basic two gate circuits shown in Figure 7.1. These basic circuits form **Set–Reset latches**. Notice that each circuit has two inputs, one called set and the other called reset. In addition, each circuit has two outputs identified as Q and $\overline{Q}$. The only difference between the NAND-NAND Set–Reset latch and the NOR-NOR Set–Reset latch is the active level of the inputs. The NAND-NAND circuit responds to low level signals, whereas the NOR-NOR responds to high level signals. As you would expect, both are useful.

Notice the method in which gate outputs are fed back to the gate inputs on the two circuits illustrated in Figure 7.1. This is **feedback**, the mechanism by which digital storage is obtained. Using this feedback, we use the signal at the output of one gate to maintain an input signal, and consequently an output level at the other gate. This basic property allows a logic value to be held by

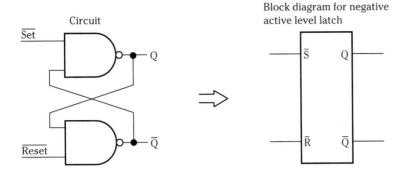

Block diagram for negative
active level latch

Circuit

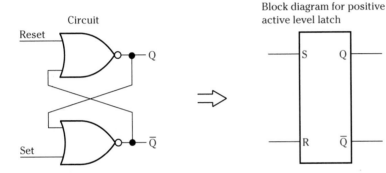

Block diagram for positive
active level latch

Circuit

Figure 7.1
NAND-NAND
Set–Reset Latch;
NOR-NOR Set–Reset
Latch

the circuit even though the signals causing the value to occur in the first place have changed. As a point worth noting, accidentally designing feedback into a system can cause inadvertent **latch-up** problems that create unstable and unreliable systems. However, in the case of a latch or flip-flop the use of feedback is intentional, so its application must be understood.

The block diagram representation shown in Figure 7.1 for the two Set–Reset latches is used frequently in digital design. Once latch operation is understood, drawing a simple rectangle to represent the circuit is far more convenient than drawing the circuitry itself. This kind of notation will be used often for all latches and flip-flops once circuit operation is understood.

Latch and flip-flop outputs are typically labeled Q and $\overline{Q}$, meaning that the true and complement output levels are available. When referring to the state of a latch or flip-flop, the Q output is the reference point. Therefore, if a **latch or flip-flop is described as being "set," we mean that the Q output is high. If the latch or flip-flop is described as being "reset," we mean that the Q output is low.**

In order for a latch or flip-flop to attain the set or reset states, proper input signals must be applied. Setting the NAND-NAND Set–Reset latch is accomplished by applying a low level signal to the set input line. When this occurs, the Q output goes high and $\overline{Q}$ goes low. This condition is maintained as long as set is low and reset is high. If the set line is then brought high, with reset still remaining high, the Q output stays in the high or set state. Even though both inputs are inactive (high levels), the flip-flop retains its previous logic state.

A similar situation occurs for the reset state when applying a low level to the reset input while maintaining a high level on the set input. When this occurs, the Q output goes low and $\overline{Q}$ goes high. This condition is held as long as set is high and reset is low. If the reset line is then brought high, with set still remaining high, the Q output stays in the low or reset state. As before, both inputs are inactive (high levels), but the flip-flop output retains its previous logic state.

We must avoid having both set and reset active, a senseless situation since it is not logical to set and reset a device simultaneously. Activating set and reset together produces Q and $\overline{Q}$ outputs that equal each other. That is, Q and $\overline{Q}$ are no longer complementary, a violation of the initial definition of Q and $\overline{Q}$. This situation may be considered "undetermined," "unstable," or "ambiguous" because the state will not be maintained when correct signals are reapplied. Furthermore, if both set and reset are deactivated simultaneously, there is no way to predict whether the latch will end up in the set or reset state—definitely ambiguous. In a case such as this the fastest logic gate in the latch circuit determines the final outcome; the designer will not be able to predict the outcome.

The condition in which both inputs are inactive is important since the output state does not change. Therefore, under this condition the state of the output depends on the last set or reset operation performed and is often characterized as a **previous state** condition. In other words, if the last operation were a set operation, Q would remain high even though the set and reset inputs are inactive; if the last operation were a reset operation, Q would remain low even though the set and reset inputs are inactive. The logic levels on set and reset do not convey any information about the latch output levels in this situation. Figure 7.2 outlines the logical operation of the NAND-NAND latch. The operation of the NOR-NOR device is similar except that the active level of the inputs are high.

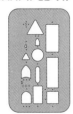

DESIGN EXAMPLE 7.1

Determine the Q and $\overline{Q}$ outputs for a NOR-NOR Set–Reset latch when:
(a) Set = 1, Reset = 0
(b) Set = 0, Reset = 0
(c) Set = 0, Reset = 1
(d) Set = 1, Reset = 1

Solution NOR-NOR Set–Reset flip-flops operate with positive active levels. Therefore, set occurs when Set = 1; reset occurs when Reset = 1.
(a) Set = 1, Reset = 0 $\longrightarrow$ Q = 1, $\overline{Q}$ = 0 set condition
(b) Set = 0, Reset = 0 $\longrightarrow$ Q and $\overline{Q}$ = previous state
(c) Set = 0, Reset = 1 $\longrightarrow$ Q = 0, $\overline{Q}$ = 1 reset condition
(d) Set = 1, Reset = 1 $\longrightarrow$ Q = $\overline{Q}$ = 0 avoid this condition

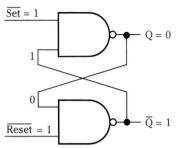

Assume Q = 0 to start. Both set and reset are inactive. Output levels make sense for the input conditions given.

(a)

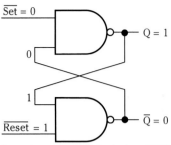

Set is low, forcing Q high (any NAND low input creates a high output). The high from Q along with the high from the reset input forces $\overline{Q}$ low. Thus, the latch output has changed state to the set condition.

(b)

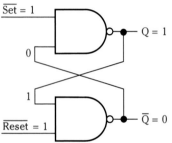

Set is made inactive (high) yet the levels on Q and $\overline{Q}$ maintain the output state of the latch. This feedback mechanism is responsible for the storage capability of the latch.

(c)

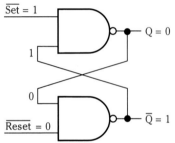

Reset is made active and the latch again changes state. In this case the low level on reset forces the lower NAND output high. The high from $\overline{Q}$ along with the high from set force the Q output to low. The device is now reset.

(d)

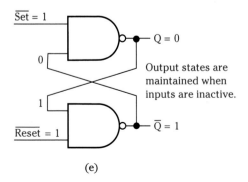

Output states are maintained when inputs are inactive.

(e)

NAND–NAND set–reset latch truth table:

Set	Reset	Q	$\overline{Q}$
0	0	unstable	
0	1	1	0
1	0	0	1
1	1	previous state	

(f)

Figure 7.2
Set–Reset Latch
Operation

Set–Reset Timing Diagrams

Timing diagrams for flip-flops are developed just as they are for combinatorial logic. Naturally, the particular flip-flop or latch operating characteristics must be considered since it is possible to see changes in input levels that do not create corresponding changes in the output level. Timing diagrams are often utilized for flip-flop and sequential circuit designs because they can convey the intricacies of flip-flop timing in a much more straightforward manner than a truth table. Figure 7.3 illustrates this fact.

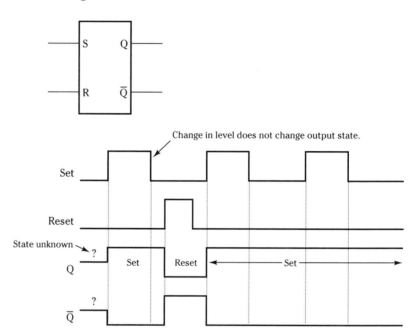

Figure 7.3
Positive Active Level
Set–Reset Latch

Initially, the output state of the latch is unknown. When power is first applied to a latch or flip-flop, the device goes to one of its two possible states. But since one gate comprising the device circuitry responds faster than the other gate, this power-on condition is not at all predictable. The uncertainty is compounded when hundreds of flip-flops or latches are used in a digital system. The designer knows that the flip-flops and latches will power-on only to a valid logic level, but will not know which level. Usually a **power-on set** or **reset** operation utilizing resistor-capacitor circuits or software programs establishes a known starting point for critical circuits. In our timing diagrams we use a question mark on a line drawn between logic levels to indicate unknown yet valid states.

As can be seen on the timing diagram in Figure 7.3, the set and reset inputs are both initially low. Since the latch used is a positive or high active level device, maintaining both set and reset low places the latch into a previous state condition. Not knowing the previous state at this point, we use a question mark to indicate our uncertainty.

Proceeding in time from left to right, we see that the first valid active signal is on the set input. As soon as this line goes active (high) (discounting propagation delay times), the Q output responds by also going high. Q remains high during the time period that set stays high. When the set input returns to the inactive low state, Q will remain high or in the set condition until it is deliberately reset. A reset does occur shortly after the set signal is deactivated in this example, and the Q output responds accordingly by falling low. Naturally, Q is low for the full duration of time in which reset is active. When reset is deactivated, Q will remain low since the latch is "storing" the last operation that took place.

The timing diagram shows two additional occurrences when set is activated. The first set returns the Q output high where it remains for the remainder of the diagram. When a set operation occurs and the latch is already set, no change in the output level takes place.

Draw the timing diagram for the following Set–Reset latch.

DESIGN EXAMPLE 7.2

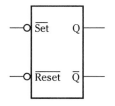

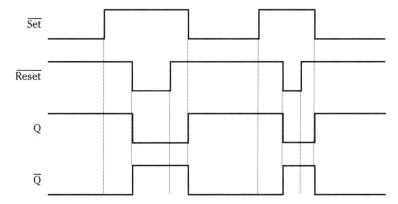

Solution   Since the latch responds to low logic levels, both set and reset operations occur when the set or reset input pins are low.

7.3 ■ GATED LATCHES

The Set–Reset latch inputs discussed are activated by high or low logic levels and are called **asynchronous inputs.** Asynchronous means that changes in the input level are not synchronized with any other activity in the digital system. The input changes are, for all practical purposes, random events. Asynchronous systems are useful in many instances because many physical and digital responses are asynchronous in nature. For instance, typing on a keyboard is an asynchronous action since the exact moment a key will be depressed is not known.

Synchronous digital systems have a specifically timed relationship. The various input and output signals in such a system are controlled by timing circuits, allowing the signals to move throughout the system in a predictable, timed manner. Usually, a **system clock** determines the timing and speed of the digital system. A clock signal is nothing more than a digital waveform occurring at a predictable rate and is often generated by a stable crystal oscillator (discussed later in this chapter). Gated latches and clocked flip-flops contain an enable or clock input line that synchronizes the setting and resetting action of the device. A gated latch is shown in Figure 7.4.

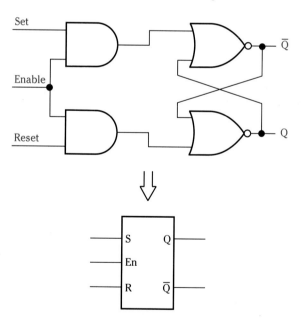

Figure 7.4
Gated Latch

In Figure 7.4 the Set–Reset circuitry has been modified to include an enable line that conditions or enables the set and reset signals. In this circuit all three inputs are active high signals, but neither a set nor reset takes effect until the enable input is high. Or, stated another way, a set or reset level can be applied to the circuit, but the specific action desired does not occur until the enable input is activated as well. This means that the enable signal determines the exact moment when the setting or resetting of the latch takes place. This is very useful when several latches (or flip-flops) are used together, as shown in Figure 7.5. The change in output for all latches occurs when the clock or enabling signal goes to the high level. This synchronizes the action of all latches simultaneously since the clock is common to all.

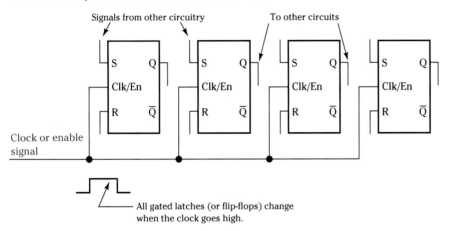

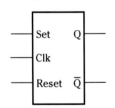

Figure 7.5
Clocking Multiple
Flip-Flops or Latches

The following gated latch controls a portion of a digital computer. Analyze the timing waveforms given:

*DESIGN
EXAMPLE 7.3*

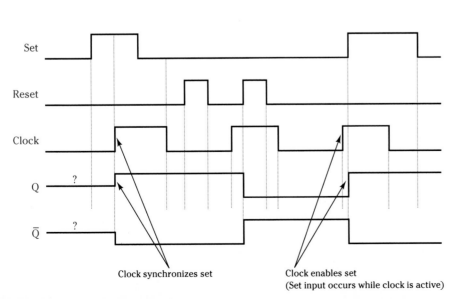

Solution   Set and reset signals are ineffective unless the enabling clock input is active as well. All signals shown are active level high.

How Clocked Flip-Flops Respond

The inputs controlling the gated latch we discussed are called **level sensitive** inputs, meaning that a solid high or low level is required for proper operation. When it comes to clocking a flip-flop, level sensitive clocks hinder performance. You may recall from previous chapters that digital waveforms contain two levels

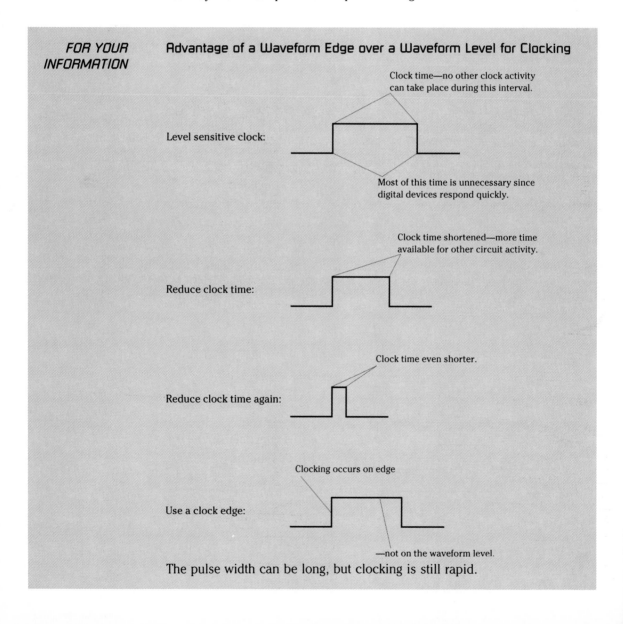

FOR YOUR INFORMATION

Advantage of a Waveform Edge over a Waveform Level for Clocking

Clock time—no other clock activity can take place during this interval.

Level sensitive clock:

Most of this time is unnecessary since digital devices respond quickly.

Clock time shortened—more time available for other circuit activity.

Reduce clock time:

Clock time even shorter.

Reduce clock time again:

Clocking occurs on edge

Use a clock edge:

—not on the waveform level.

The pulse width can be long, but clocking is still rapid.

as well as two edges. A **positive going edge** occurs as the waveform makes a transition from a zero level to a one level; a **negative going edge** occurs when the waveform makes a transition from the one level to the zero level. Edges make ideal clocking pulses because they are rapid and there is no question as to when they occur. Flip-flops and other devices responding to an edge are called **edge-triggered** devices. In fact, it is the edge triggering capability that distinguishes a flip-flop from a latch.

Edge-triggered devices are identified according to the notation used in Figure 7.6. The wedge-shaped symbol distinguishes level sensitive inputs from edge-triggered inputs; the presence of the wedge indicates an edge sensitive input. If a bubble is also present on the line, the line is negative edge sensitive; otherwise, the line is positive edge sensitive. You need to understand that an edge sensitive input responds only to the appropriate waveform transition and ignores all other edges and levels, just as a high level input ignores low level signals.

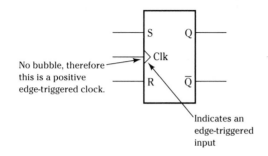

No bubble, therefore this is a positive edge-triggered clock.

Indicates an edge-triggered input

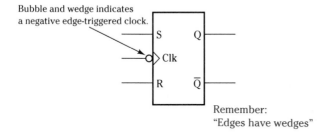

Bubble and wedge indicates a negative edge-triggered clock.

Remember: "Edges have wedges"

Figure 7.6
Edge-Triggering Notation

In order to appreciate fully edge-triggered devices, we must understand how an edge is detected by the responding device. After all, an edge-triggered flip-flop receives the same clock signal as a level sensitive latch. We know how a logic gate responds to a logic level but not to a fleeting edge. How can a logic device "see" only the edge and ignore all levels? The answer is by incorporating an **edge detection** circuit whose primary function is to allow the appropriate edge through to a flip-flop yet "filter out" all unnecessary edges and levels. Figure 7.7 illustrates a typical circuit.

A clock pulse presented to the edge detector follows two paths to an AND gate. One input to the AND is taken by the original clock signal. The other AND input also receives the clock, but it is delayed by one or more inverters. At the AND gate inputs the slight difference in phase between the original clock

Functional Circuit:

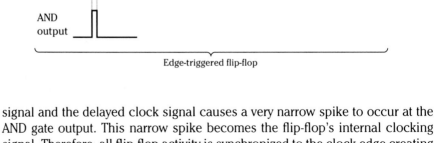

Typical Edge Detector Circuit:

Figure 7.7
Edge Detection
Circuitry

signal and the delayed clock signal causes a very narrow spike to occur at the AND gate output. This narrow spike becomes the flip-flop's internal clocking signal. Therefore, all flip-flop activity is synchronized to the clock edge creating the spike. The two most commonly used edge sensitive flip-flops, the D and J-K flip-flops, are discussed in the following sections.

DESIGN
EXAMPLE 7.4

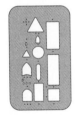

Identify the clock edge detected by the following edge detection circuit:

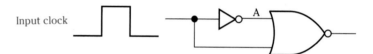

Solution   A timing diagram illustrates how delaying the clock creates a narrow spike at the NOR gate output. The delay time provided by the inverter is equal to the inverter's propagation delay time.

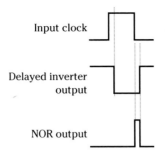

7.4 ■ THE D LATCH AND FLIP-FLOP

The D latch and D flip-flop are both common circuit arrangements designed around the basic Set–Reset latch; the major difference between the latch and the flip-flop versions is in the clocking methods. Figure 7.8 shows how a Set–Reset latch is converted into a D (delay) latch. The set and reset inputs are now replaced with a single input line called the D input. The inverter on the input line (connecting set to reset) assures that whenever the set line is activated, the reset line is deactivated. Conversely, whenever set is deactivated, the reset line is active. Therefore, the D input line controls both the set and reset operations. Latches like these are often called **transparent latches,** and a distinguishing feature of the transparent latch is that the level sensitive clock input is called an enable line. Typically, when the enable line is high, the Q output of the latch follows the D input level. This means that as data on the D input changes, so does the Q output. This is the latch's transparent mode of operation and in this mode data storage does not occur. Bringing the enable

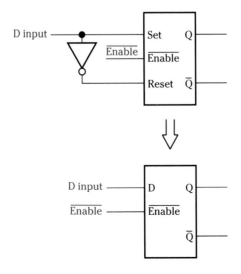

Figure 7.8
D Latch Equivalent
Circuit and D
Transparent Latch

line low causes the latch to store the binary level present on the D input. The Q output does not follow the D input under these conditions, but rather, it retains the value that was present at D when the enable line went low (the device's active level in this example since data is stored when enable is low).

DESIGN EXAMPLE 7.5

Show how D latches can be used to hold and pass information through to a 3-to-8 line decoder for subsequent decoding.

Solution   Use a D latch to store the data for each decoder input line. The latch enable input lines can be tied to a common control line that determines when transparent mode or latch mode occurs.

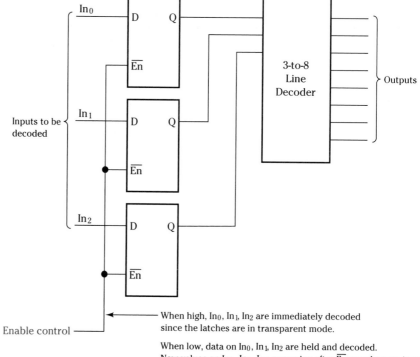

When high, In_0, In_1, In_2 are immediately decoded since the latches are in transparent mode.

When low, data on In_0, In_1, In_2 are held and decoded. New values on In_0, In_1, In_2 occurring after $\overline{En}$ goes low are ignored.

A D flip-flop is shown in Figure 7.9 along with the symbol typically used to represent the device logically. The data at the D input is stored in the flip-flop and is present at the Q output, on the positive going edge of the clock. This is the only time data at D appears at Q since there is no transparent mode of operation for the D flip-flop. Look closely at the symbol for the D flip-flop and you will notice the wedge indicating that the device will respond to a positive going clock edge. The clock and the D inputs are commonly referred to as the flip-flop's **synchronous inputs**.

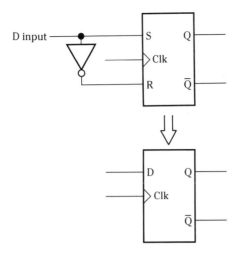

Figure 7.9
D Flip-Flop Equivalent
Circuit and D Flip-Flop

Eight-bits of data from a computer system's arithmetic logic unit (ALU) are to be stored for further computations. Show how D flip-flops can be used to store all 8-bits in a synchronized fashion.

DESIGN
EXAMPLE 7.6

Solution   Each D flip-flop receives one of the ALU output bits. A common clock synchronizes the storing action of each flip-flop. The designer will ensure that the clock signal undergoes a positive transition only when the ALU output data is valid.

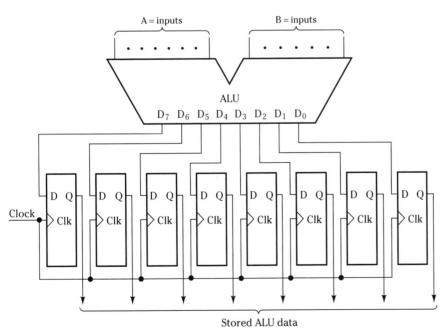

Flip-flops, when grouped for a common purpose such as this, are called "registers."

D Flip-Flop and Latch Timing

Figure 7.10 shows the timing relationship between input and output of the D flip-flop. Changes in the output level occur only when the clock line undergoes a positive going transition. Notice that the Q output resembles the D input signal except that it is delayed in time, accounting for the name of the flip-flop. Remember, the edges of the D input signal do not cause a D flip-flop to change state; only the clock edges do so. Also, notice that until the first clock transition takes place, the flip-flop output level is unknown.

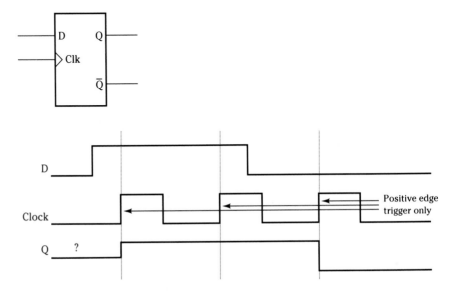

Figure 7.10
D Flip-Flop Timing
Diagram

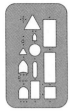

DESIGN
EXAMPLE 7.7

Determine the Q output waveform for a D flip-flop using the following waveforms for clock and D as inputs:

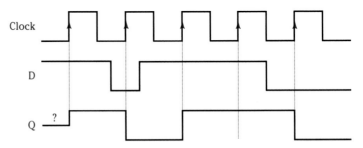

Solution Q follows the D input whenever the clock undergoes a positive transition.

Contrasting with the edge-triggered characteristic of the D flip-flop is the level sensitive timing diagram for a D latch as shown in Figure 7.11. Since the

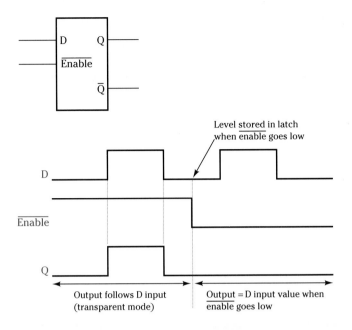

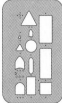

Figure 7.11
D Latch Timing
Diagram

latch is controlled by an enable input, the Q output depends on the level of enable. When the enable control line is high, the Q output follows the D input. As soon as enable is brought low, the level on D at that time is stored in the latch. Maintaining enable low prevents any other storage or transparent modes of operation from taking place.

Draw the timing diagram for a transparent D latch when presented with the following D and enable input signals:

*DESIGN
EXAMPLE 7.8*

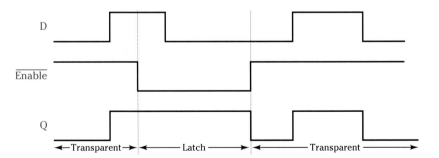

Solution Q follows D when $\overline{\text{enable}}$ is high. Q stores the value of D when $\overline{\text{enable}}$ goes low. This value remains until $\overline{\text{enable}}$ goes high again.

Figure 7.12 shows an interesting way to wire a D flip-flop and the waveform that results. The $\overline{Q}$ output is tied back to the D input line so that the logic level present on $\overline{Q}$ is also present at the input. When a clock transition occurs, the

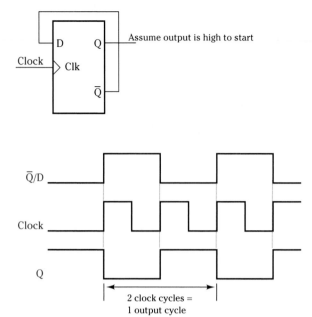

**Figure 7.12
D Flip-Flop Toggle
Operation**

logic level on the D input is transferred to the Q output. $\overline{Q}$, of course, obtains the complement level, which is fed back to the D input. The next clock transition causes the flip-flop to change state since $\overline{Q}$ controls the D input. This changing of state process continues for each clock transition and is known as **toggling**. A flip-flop's ability to toggle is crucial to the operation of counters and other sequential circuits we will discuss.

Several points are worth mentioning for the circuit arrangement in Figure 7.12. First, it may not be immediately obvious why the changing $\overline{Q}$ output signal does not affect the state of the flip-flop after a clock transition has occurred. This is not a problem because the flip-flop propagation delay time exceeds the time required for the clock to make a transition. By the time the valid logic state occurs at $\overline{Q}$ the clock edge causing the state has ended. Therefore, one clock transition will not cause multiple state changes. Note, this is not possible with level sensitive devices.

Second, the output waveform frequency from a continuously toggling flip-flop is one-half that of the clock frequency. This is referred to as a **divide by 2 capability**. **Frequency division** is an important digital application, which will be covered in later chapters.

---

*DESIGN
EXAMPLE 7.9*

If a toggling flip-flop is clocked with a 3 MHz clock signal, what is the frequency of the signal at the flip-flop Q output?

Solution Toggling divides a clock signal in half. Therefore, a 3 MHz clock produces a 3 MHz/2 = 1.5 MHz output signal.

Figure 7.13 shows the block diagram for a practical D flip-flop, the 7474. In addition to the D and clock inputs, two additional inputs are shown—**preset** and **clear**. Preset and clear are **asynchronous inputs** performing set and reset operations, respectively. The bubbles and lack of a wedge on these lines indicate that preset and clear are negative level sensitive inputs. These two inputs provide exactly the same capability as does the NAND-NAND Set–Reset latch and are frequently utilized to establish a known starting state for the flip-flop. Preset and clear always override the synchronous inputs. That is, a preset or clear operation will take precedence over a simultaneous D/clock operation. Figure 7.14 illustrates the D flip-flop timing.

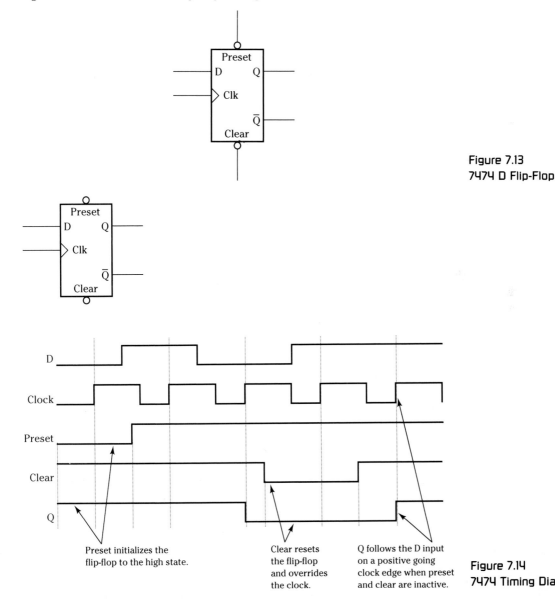

Figure 7.13
7474 D Flip-Flop

Preset initializes the
flip-flop to the high state.

Clear resets
the flip-flop
and overrides
the clock.

Q follows the D input
on a positive going
clock edge when preset
and clear are inactive.

Figure 7.14
7474 Timing Diagram

As we can see, the flip-flop responds to the clock in the normal fashion except for the occasions when either preset or clear is low. During these times the preset or clear operation dictates the state of the flip-flop and any positive clock transitions occurring during these times are ignored. Notice that preset is active on the timing diagram at the first moment, ensuring a known starting state. The active clear input immediately resets the flip-flop, a common operation in digital systems.

DESIGN
EXAMPLE 7.10

Complete the timing diagram for a 7474 D flip-flop using the following clock, D, preset, and clear input waveforms:

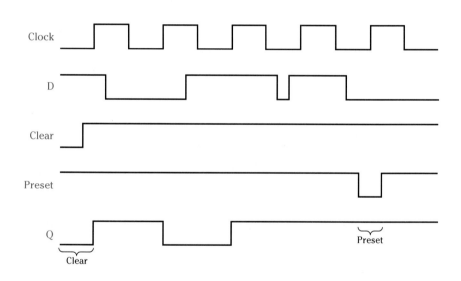

Solution Both clear and preset override normal clock activity. clear = 0 forces, Q = 0; preset = 0 forces, Q = 1.

DESIGN
EXAMPLE 7.11

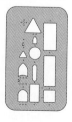

The burglar alarm circuit designed in Design Example 3.22 sounds an alarm if a door is opened but has a significant limitation—the alarm goes silent as soon as the door is reclosed. The alarm should stay on whether the door remains open or closed.

Modify the burglar alarm circuit from Design Example 3.22 to include the following capability:

Solution

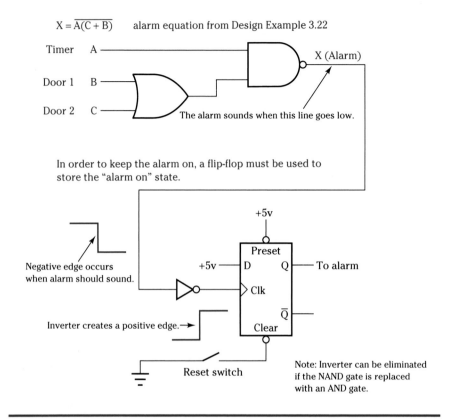

$X = \overline{A(C + B)}$ alarm equation from Design Example 3.22

Timer   A ——————————————— X (Alarm)

Door 1   B —

Door 2   C —

The alarm sounds when this line goes low.

In order to keep the alarm on, a flip-flop must be used to store the "alarm on" state.

+5v

Negative edge occurs when alarm should sound.

Preset

+5v — D   Q — To alarm

Clk

$\overline{Q}$

Inverter creates a positive edge.→   Clear

Reset switch

Note: Inverter can be eliminated if the NAND gate is replaced with an AND gate.

7.5 ■ THE J-K FLIP-FLOP

The J-K flip-flop, shown in Figure 7.15, provides flip-flop storage capability along with a considerable amount of logic flexibility. As you might expect, the basic internal storage mechanism is provided by a Set–Reset latch. Three external inputs are also available—the J input, the K input, and a clock input; typical J-K clocks are positive or negative edge sensitive.

The truth table accompanying Figure 7.15 shows that all four possible combinations of J and K produce valid logic results at the flip-flop output. Contrast this with the Set–Reset latch, which did not allow set and reset to be active simultaneously. Using the J-K device, set, reset, and previous state conditions are allowed. In addition, the input combination when J = K = 1 provides the very useful toggling feature. For a J-K input combination a change in state occurs only after a valid clock transition and corresponds to the values on the J and K inputs when the clock transition takes place.

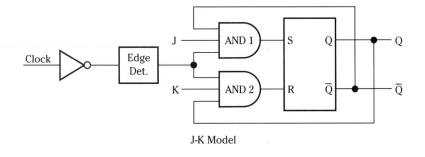

J-K Model

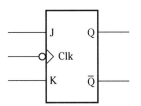

Negative edge-triggered J-K flip-flop

J	K	Q	$\overline{Q}$	
0	0	Previous state		
0	1	0	1	after a clock transition
1	0	1	0	
1	1	Toggle		

Figure 7.15
J-K Flip-Flop

Basic operation for the J-K flip-flop in Figure 7.15 is as follows:

a. *Previous State Condition (J = K = 0)* When both J and K are low, the output of both AND gates are also low. These low output levels keep both the latch set and reset inputs inactive. No change in state will take place during a clock transition.

b. *Reset Condition (J = 0, K = 1)* K equal to one enables AND2. If Q is high, the AND2 output will also go high during a clock transition. Since high is the active level for the Set–Reset latch, the latch will reset; Q will go low. If Q is already low prior to the clock pulse, AND2 is disabled and the latch reset remains inactive. However, since the latch is already reset, it will remain reset after the clock transition as well.

c. *Set Condition (J = 1, K = 0)* J equal to one enables AND1. If $\overline{Q}$ is high, the AND1 output will also go high during a clock transition. Since high is the active level for the Set–Reset latch, the latch will set; $\overline{Q}$ will go low. If $\overline{Q}$ is already low prior to the clock pulse, AND1 is disabled and the latch set remains inactive. However, since the latch is already set, it will remain set after the clock transition as well.

d. *Toggle Condition (J = K = 1)* Both AND gates are enabled by J and K during toggle mode. When a clock transition occurs, the values fed back from Q and $\overline{Q}$ determine to which state the flip-flop will toggle. If Q is high, then AND2 will create an active reset signal for the latch and Q will then go low on the following clock. If Q is low, then AND1 provides an active set signal for the latch. Q will go high on the following clock.

Practical J-K flip-flops, such as the 74LS76 shown in Figure 7.16, also have asynchronous preset and clear inputs for initialization purposes. Similar to the D flip-flop, the asynchronous inputs override the clock response, making the asynchronous capability ideal for power-on resetting and other control functions.

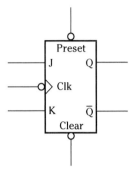

Function Table:

Preset	$\overline{\text{Clear}}$	Clock	J	K	Q	$\overline{Q}$	
0	1	X	X	X	1	0	
1	0	X	X	X	0	1	
0	0	X	X	X	1	1	← Avoid this state
1	1	↓	0	0	Q_0	$\overline{Q}_0$	
1	1	↓	1	0	1	0	
1	1	↓	0	1	0	1	
1	1	↓	1	1	Toggle		
1	1	1	X	X	Q_0	$\overline{Q}_0$	

X = don't care
↓ = negative edge
Q_0, $\overline{Q}_0$ = previous state

Figure 7.16
74LS76 Flip-Flop

Determine the output waveform at Q.

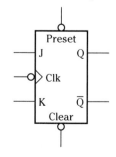

DESIGN
EXAMPLE 7.12

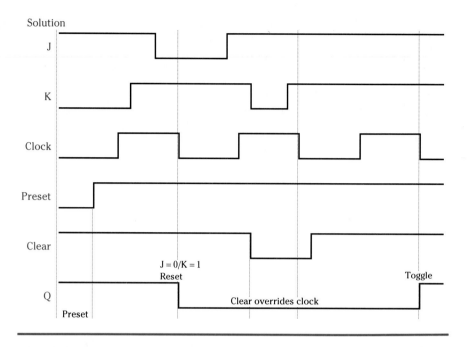

7.6 ■ FLIP-FLOP VARIATIONS

Flip-flops all have one function in common—the ability to store a single binary level. The Set–Reset latch, the D, and J-K flip-flops comprise the majority of generally used flip-flop devices, although specialized devices are available in certain logic technologies (masterslice, standard cells). Modifications made to the basic devices in the more common technologies (TTL, CMOS) create storage devices with slightly different but useful operating characteristics.

Master–Slave Flip-Flops

Figure 7.17 shows the basic logic circuitry for a **J-K Master–Slave flip-flop.** Contained within the device are two separate Set–Reset latches and associated control logic. The first latch is the master while the second is the slave. Any level present on the Q output of the slave was first stored in the master. Therefore, the slave data always reflects data previously stored in the master. A clock line is common to both master and slave except that the slave receives the complement of the master's clock. This means that the master and slave are clocked independently. The master latch receives data first. After the master latch clock is deactivated, the data is transferred to the slave since the slave clock has become active. In this manner, data presented to the master latch inputs are delayed from reaching the slave output. This effectively isolates the input from the output and is useful for solving certain timing problems (discussed in a later section).

One restriction may apply to some Master–Slave flip-flop designs—keeping the J and K inputs steady when the master clock is active. The master clock

FOR YOUR
INFORMATION

Common Flip-Flop and Latch Part Numbers

Flip-Flop/Latch	TTL	CMOS	ECL
D flip-flop	7474	74HC74	HD100131
D flip-flop	74174	74HC174	HD100151
J-K flip-flop	74107	74HC107	
J-K flip-flop	74276		

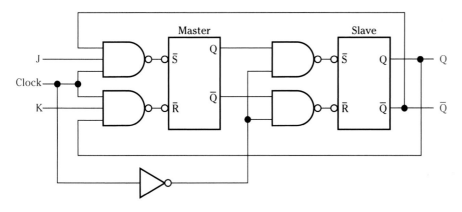

When clock is high:
   —Master flip-flop stores data as determined by the J-K inputs.
   —Slave flip-flop inputs are disabled.

When clock is low:
   —Master flip-flop data is transferred to the slave.
   —Master flip-flop inputs are disabled.

**Figure 7.17
J-K Master–Slave
Flip-Flop**

is level sensitive, so that any change in J or K levels may change the state of the master latch while the clock is active. The incorrect state resulting will eventually be transferred to the slave. In fact, clock inputs on devices such as these are more correctly referred to as **pulse sensitive.** That is, a portion of the device operation takes place while the clock is at a particular level, whereas the remainder of the device operation takes place after the clock changes level. Using a narrow clock pulse or an edge-triggered flip-flop is an alternative to this problem. In addition, **data lockout** master–slave flip-flops are edge sensitive and minimize the time period in which J and K may not change.

Draw the timing diagram for the J-K Master–Slave flip-flop's Q_{master} and Q_{slave} outputs (depicted in Figure 7.17). Use the following input waveforms and assume $Q_{master} = Q_{slave} = 0$ to start:

**DESIGN
EXAMPLE 7.13**

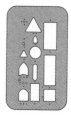

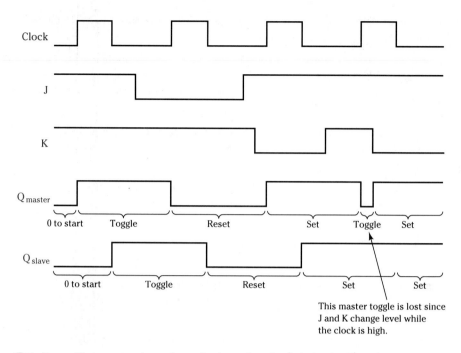

Solution The master is activated when the clock is high. The slave obtains the master data when the clock goes low.

7.7 ■ FLIP-FLOP SPECIFICATIONS

To design circuits correctly using flip-flops, you need to understand the specifications that define proper flip-flop usage. Flip-flop signals are subject to propagation delay, the same as normal gates. Naturally, this must be taken into account during detailed timing analysis. Several additional specifications are also important for proper flip-flop operation. Refer to Figure 7.18 for examples of the following:

- *Maximum Clock Frequency (f_{max})* The maximum clock frequency for a flip-flop specifies the fastest clock rate at which reliable changes in output level can occur. Clock signals faster than this rate will not produce proper flip-flop operation.
- *Setup Time (t_{su})* Setup time is the minimum time interval occurring between two signals applied to specified input terminals. For instance, the setup time for a D flip-flop may be given as 15 nsec. This means that data applied to the D input should be stable and maintained for at least 15 nsec prior to a clock transition. This allows time for signals to settle down within the flip-flop's internal circuitry.
- *Hold Time (t_h)* Hold time is similar to setup time except that it specifies

Flip-Flop Summary

Positive active level
Set–Reset latch

Negative active level
Set–Reset latch

Positive active level
Gated Set–Reset latch

D flip-flop with asynchronous
preset and clear. Positive
edge-triggered.

J-K flip-flop with asynchronous
preset and clear. Negative
edge-triggered.

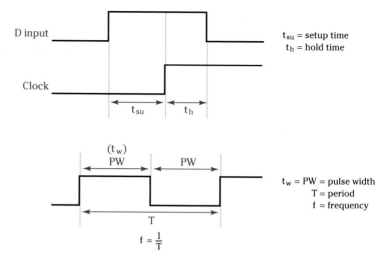

t_{su} = setup time
t_h = hold time

t_w = PW = pulse width
T = period
f = frequency

Figure 7.18
Illustration of Flip-Flop
Timing Parameters

the minimum time interval between two signals after an active signal transition takes place. For example, the hold time for a D flip-flop may be specified as 4 nsec. This means that the data signal at the D input should remain stable 4 nsec after the clock is applied.

- **Pulse Width or Pulse Duration (t_w)**   This specification guides the width of clock pulses as well as clear and preset signals. A pulse width of 16 nsec for a negative active clear signal, for instance, indicates that the clear signal must be maintained at a low level for at least 16 nsec in order for reliable operation to occur.

DESIGN
EXAMPLE 7.14

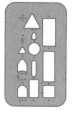

Determine if the following circuit arrangement violates the D flip-flop setup and hold time specifications:

Assume t_{pd} = 5 nsec
          t_{su} = 4.5 nsec
          t_h = 0 nsec

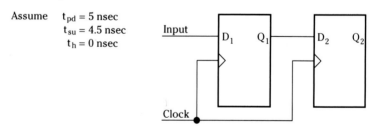

Solution   When the output of one flip-flop feeds the input of another and both are clocked by the same signal, potential timing problems arise. Draw a timing diagram to verify that all specifications are met when setting and resetting the devices. Assume $Q_1 = Q_2 = 0$ to start.

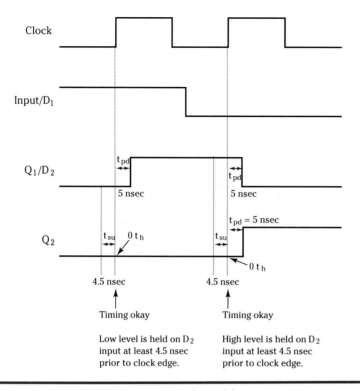

7.8 ■ PROBLEMS AFFECTING FLIP-FLOP CIRCUIT DESIGN

Both combinatorial and sequential circuit design get tricky when timing delays are considered. Recall, we stated earlier that initial circuit design is usually carried out assuming that propagation and other delays do not exist. This allows functional design to take place simply and quickly. However, we know that timing delays must be considered once the logic circuitry has reached a reasonably correct form. As a signal travels from gate to gate in a logic circuit it is delayed by the propagation delay of each individual gate and also by the rise time, fall time, and interconnection delays associated with the wiring connecting the gates. To complicate matters, the delays are not constant for similar gates nor from inputs on the same gate. This means that signals may arrive at a gate deep within the logic structure in an incorrect timing relationship to what was expected, producing inadvertent changes in logic level.

Hazards

Incorrect logic levels occurring momentarily are called **hazards** or **glitches**. The term "glitch" generally refers to any unintended change in logic level; the term "hazard" is a more formal one used to identify specific problems of this nature. For instance, a **static one hazard** occurs when a signal that is normally one momentarily goes to the zero level and then returns to one. Conversely, a

static zero hazard occurs when a signal that is normally zero momentarily goes to the one level and then returns to zero. In each case an incorrect logic level exists for a moment. This may or may not be troublesome, but it certainly has the potential to cause problems in both combinatorial and sequential circuitry. In sequential circuits the hazards can be particularly troublesome since additional edges are created by the hazards, which may cause incorrect flip-flop clocking.

Dynamic hazards occur when a line changes level several times. For instance, if a signal makes the transition from the high level to the low level and temporarily varies back and forth between these levels, then a dynamic hazard has been created.

Figure 7.19 shows how propagation delays can create hazards in a logic circuit. If the two input signals are delayed by unequal amounts of time, the output may not appear as expected. In this example the effects of the Y input change create a static zero hazard because the Y input change is delayed through the AND gate longer than the X input change. For a short moment the AND gate perceives both inputs at the high level and produces a high output level. This situation is compounded when additional gates are added to the structure and delay accumulates. Also, especially when designing with very dense logic technologies, delay problems due to both gate delays and potentially substantial interconnection delays increase the likelihood of hazards. Unlike a printed circuit layout, the layout of logic elements on silicon and the associated interconnection delays are unknown until the gate connections are routed by computer programs. Because the gates may not be placed physically next to each other on the silicon, the interconnection delays can be significant.

Since hazards will occur in logic design, the designer must determine if the hazard causes a problem (not all hazards do) or if the hazard can be corrected. If only one of the input variables controlling a circuit is changing at a time, then any **logic hazard** can be corrected. Logic hazards can be changed

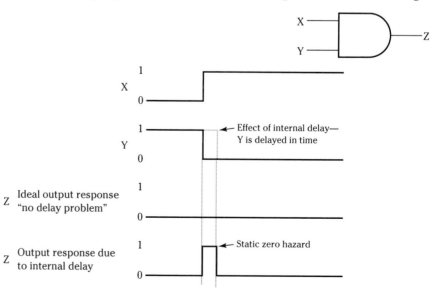

Figure 7.19
Creation of a Static Hazard

by modifying the logic circuit. This is true because it is easy to project the possible outcomes of a single variable change. If more than one variable changes at a time, or if an input variable changes before another has had time to settle down, it may be impossible to design a truly hazard-free circuit. Hazards that cannot be removed are called **function hazards** because they are a result of the logic function being designed. In other words, if you are designing a circuit and allow more than one input to change simultaneously because that is the function of your circuit, a function hazard may be created. Redesigning the circuit or reevaluating the use of the circuit outputs may be necessary to avoid the potential problems created by the hazard. In fact, analysis of any hazards that occur and their effect on circuit performance must be carried out to ensure reliable operation. Computer simulation software is an extremely useful tool in this regard.

Figure 7.20 shows how a logic hazard is created and analyzed using a simple K-map SOP (sum of products) design. The simplified circuit appears

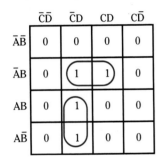

$$Y = \overline{A}BD + A\overline{C}D$$

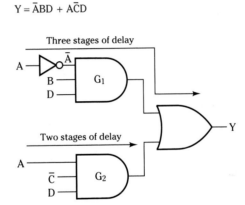

When A changes, a hazard is possible:

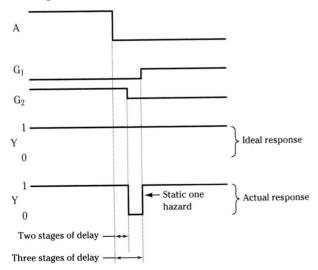

Figure 7.20
How a Logic Hazard is Created

hazard-free when we view the variables controlling the individual prime implicants. That is, when changing the input levels from within the encircled
ones of a prime implicant, only one gate—the one created by that prime
implicant—is affected. The gate output level will not change because the
variable causing the movement on the map is the variable that was eliminated
from the prime implicant during the reduction process; it is impossible for the
gate output to change. However, if an input variable changes so that we move
from the first prime implicant to the second prime implicant on the map, both
AND gates in the circuit change state and a static hazard occurs at the output.
In this example variable A changes and a static one hazard is created because
as the A input variable changes, the lower AND gate goes low before the upper
AND gate goes high due to the differing stages of delay. This forces both OR
gate inputs to be low momentarily, resulting in a brief low level output. Whether
or not this hazard is troublesome depends on how the circuit output is used,
and this requires additional analysis. Assume that the hazard is intolerable.
What can be done? Figure 7.21 shows the solution.

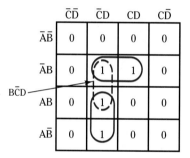

Overlapping essential prime implicants:

$$Y = \bar{A}BD + A\bar{C}D + B\bar{C}D$$

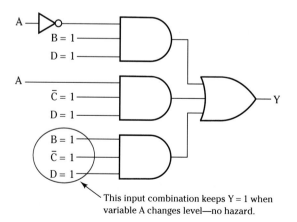

Figure 7.21
How to Circumvent a
Logic Hazard

This input combination keeps Y = 1 when
variable A changes level—no hazard.

Since the hazard occurs when the variable change causes movement from
one prime implicant to another, the transition can be smoothed out by including
another gate formed by overlapping the two necessary prime implicants. The
new circuit maintains the Y output at the one level during the change in variable

A and eliminates the hazard. Notice that the SOP expression and resulting circuit are no longer minimum solutions for the design. The addition of a redundant logic gate has increased circuit complexity but is important if the hazard must be eliminated.

Identify whether the hazards on the following K-maps are logic hazards or functional hazards:

DESIGN EXAMPLE 7.15

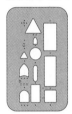

	$\overline{C}\overline{D}$	$\overline{C}D$	CD	$C\overline{D}$
$\overline{A}\overline{B}$	0	0	0	0
$\overline{A}B$	0	1	1	0
AB	0	0	0	0
$A\overline{B}$	0	1	1	0

— $\overline{A}BD$
— $\overline{A}BD$

	$\overline{C}\overline{D}$	$\overline{C}D$	CD	$C\overline{D}$
$\overline{A}\overline{B}$	0	0	0	0
$\overline{A}B$	1	0	0	0
AB	1	1	0	0
$A\overline{B}$	0	1	0	0

— $A\overline{C}D$
— $B\overline{C}\overline{D}$

Since two variables must change state when moving from one prime implicant to another, a function hazard is possible.

For example, when moving from $\overline{A}BD$, the input signals may temporarily end up in state ABD or $\overline{A}\overline{B}D$. Both these states produce a zero output level. Since the zero state is not desired, a static one functional hazard is created.

Only one variable differing in level is common between the two prime implicants (D and $\overline{D}$) so that only a logic hazard is possible. Since the circuit output could change from 1 to 0 during a change in level at D, a static one logic hazard is possible.

Solution A logic hazard may be created if only one input variable changes state. A functional hazard may be created when two or more input variables change state.

What can be done to eliminate the logic hazard found in Design Example 7.15?

DESIGN EXAMPLE 7.16

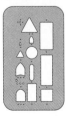

Solution Since the hazard occurs when moving from one prime implicant to another, $B\overline{C}\overline{D} \rightarrow A\overline{C}D$, adding another prime implicant that overlaps the two—$AB\overline{C}$—eliminates the hazard. The $\overline{C}$ variable maintains the output level during this transition since it is common to all three prime implicants.

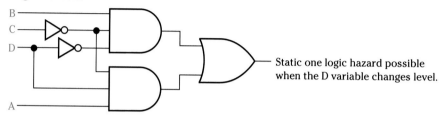

	$\overline{C}\overline{D}$	$\overline{C}D$	CD	$C\overline{D}$
$\overline{A}\overline{B}$	0	0	0	0
$\overline{A}B$	1	0	0	0
AB	1	1	0	0
$A\overline{B}$	0	1	0	0

Original Circuit:

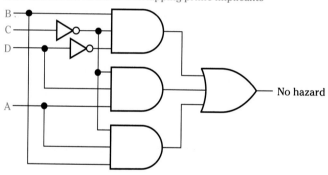

Static one logic hazard possible when the D variable changes level.

Modified circuit: includes overlapping prime implicants

No hazard

DESIGN EXAMPLE 7.17

Draw the timing diagram for the logic hazard circuit shown in Design Example 7.16 to show how a change in the D variable creates a static one logic hazard.

Solution Create the timing diagram from input to output based exclusively on a changing D variable. In other words, assume that A = B = C = constant levels to enable the AND gates; only D changes level. Account for the propagation delays on the diagram (except for the OR gate).

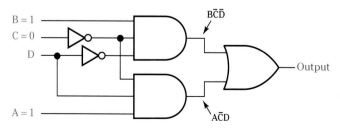

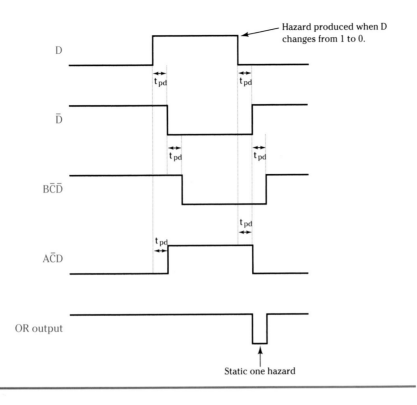

Hazard produced when D changes from 1 to 0.

Static one hazard

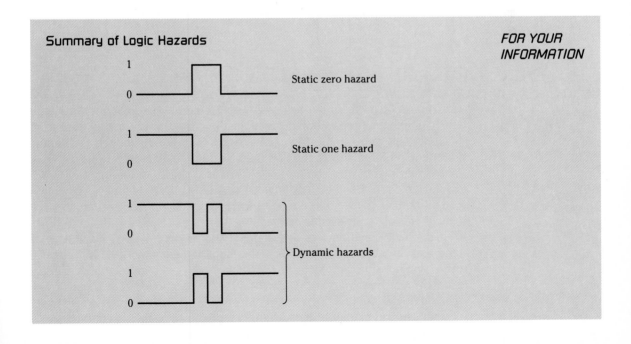

Race Conditions

A **race condition** occurs when flip-flop inputs are changed in a simultaneous or near-simultaneous manner. When a group of flip-flops is used in a sequential circuit, there are many possible states the flip-flops can attain. The circuit design defines the allowed states, but unaccounted delays may introduce some intermediate states because the flip-flops change at differing rates. The logic controlling the states reached by the flip-flops often depends on the state of the flip-flops before the change is initiated. If the initial change in all flip-flops is not simultaneous, the controlling logic may be affected by the temporary states and ultimately force the flip-flops into an incorrect state, a situation that arises as the temporary states "race" through the controlling logic network. The final state may rest with the signal that wins the race. Obviously, this is an unacceptable situation since the designer has lost control of the circuit response.

A simple example of a race is shown in Figure 7.22. Both flip-flops are clocked by the same signal with the state of Q_2 dependent on the state of Q_1. Since Q_1 is changing with Q_2 yet Q_2 depends on the previous state of Q_1, a race condition is possible. Situations such as this are often referred to as C_i problems when they are caused by clock and data signals feeding a common set of flip-flops.

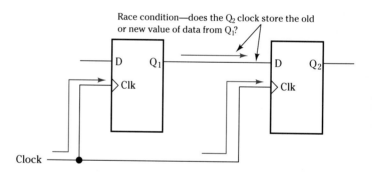

Race condition—does the Q_2 clock store the old or new value of data from Q_1?

Figure 7.22
C_i Race Condition

Clock

The problem just described can be prevented in several ways. In Figure 7.22 the inherent propagation delay of the flip-flops may prevent the race from occurring altogether. If the propagation delay of Q_1 is long enough to meet the hold time requirements of flip-flop Q_2, then the problem does not exist. This means that the previous state of Q_1 will be maintained at the input of Q_2 for a sufficient time after the clock edge passes. When the hold time specification cannot be met, Master–Slave flip-flops are frequently used since the master's change in state is delayed from reaching the output. However, the hold times of many high speed devices are so much shorter than the propagation delay time that the need for Master–Slave flip-flops is becoming less important in these situations. In fact, many flip-flops have zero hold times, so this class of problem can be eliminated. It should also be mentioned that some design methodologies prohibit C_i clocking entirely. This is more common in very large-scale integrated circuit designs where computerized testing is employed.

The solutions to hazard and race problems are not always obvious, nor is the identification of the particular problem easily carried out. Advanced design techniques are often utilized to assist the designer. We will continue to look at other problem situations as we increase our knowledge of sequential circuitry. But, as a general rule of thumb, avoid having two or more flip-flop inputs change at a time.

Identify which of the following circuits are likely to have race problems:

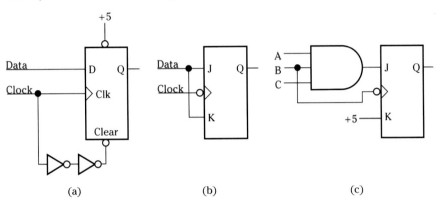

(a)   (b)   (c)

DESIGN EXAMPLE 7.18

Solution   Race conditions are likely when two or more flip-flop input signals change simultaneously and if the change can affect the state of the flip-flop.

a. A race condition exists between the clock and clear inputs. As the clock goes positive, the clear signal remains due to the inverters' propagation delay time. Removing the inverters will not eliminate the race because the flip-flop's internal delay between the two inputs may still create a race. The clock and clear signals should be separate signals.
b. No race condition exists provided that the setup and hold times are met for the clock and the J-K inputs. J and K changing simultaneously is not a race since J and K cannot initiate a change in the flip-flop state.
c. A race condition exists between the J input and the clock. The B signal controls the clock as well as the J input level.

7.9 ■ TIMING CIRCUITS

One-Shots

One-shots or **monostable multivibrators** are devices used to create pulses of specific time duration. One-shots find application as timing elements in sequential circuit design.

The one-shot shown in Figure 7.23 is a typical device. Output Q is normally in the low state—the **stable state**—and goes to the high state only when the

74121 One-Shot:

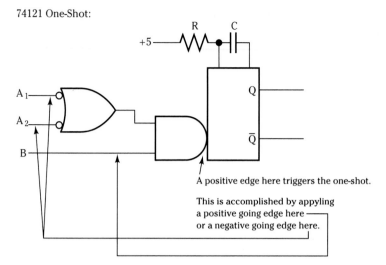

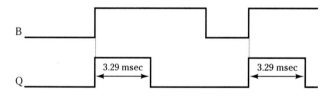

A positive edge here triggers the one-shot.

This is accomplished by appyling a positive going edge here — or a negative going edge here.

Pulse width formula for a 74121:

$$t_w = 0.7RC$$

Assume R = 4.7K and C = 1 μfd

$$t_w = (0.7)(4700)(1 \times 10^{-6})$$
$$t_w = 3.29 \text{ msec}$$

Example 1—positive edge applied to B; $A_1 = A_2$ = low to enable

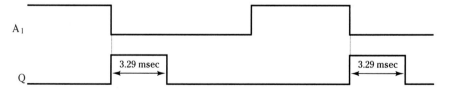

Example 2—negative edge applied to A_1; B = A_2 = high to enable

Figure 7.23
74121 One-Shot

device is **triggered**. Triggering is controlled by the various input lines and usually occurs on the negative or positive edge of a specified input. For instance, the 74121 may be triggered by a positive edge applied to the B input or to a negative edge applied to the A_1 or A_2 inputs. The edge must propagate through the input logic so that a positive going edge results at the output of the AND gate. This output is an internal point within the 74121 and cannot be monitored, but it may be controlled by the signals at either A_1, A_2, or B. The Q output remains in the high state after triggering—the **quasi-stable state**—

for a period of time determined by the designer and then falls back to the low level. The time duration for the high state depends on an externally connected R-C (resistor-capacitor) network.

Controlling One-Shot Timing

Resistor-capacitor networks are the basic timing element in many electronic circuits, making use of the property that capacitors acquire electric charge in a very predictable manner. The capacitor obtains charge through a current flow, which is limited by resistance. The rate at which the capacitor charges is determined by the resistance and capacitance in the charging circuit. This is referred to as the circuit **time constant**. Increasing either the resistance or the capacitance increases the charging time, and since the charging time can be predicted from a knowledge of the resistor and capacitor values, these elements form the basis of a timing network. The same principles apply when a capacitor discharges. That is, the rate of discharge is predictable.

Each one-shot has a formula or chart explaining how various R-C combinations affect the "on-time" for the Q output. Because of the tolerance of the timing elements, testing and verification of the output pulse width should take place. Usually, the data sheet for a one-shot will also include charts that allow the designer to pick out easily the resistor and capacitor values needed for a specific pulse width. A 74121 has the following pulse width formula: $t_w = 0.7RC$ where t_w = pulse width (one-shot on-time), R = resistor value in ohms, and C = capacitor value in farads. This formula is accurate over most of the one-shot's operating range. Data books specify the exact operating range for every one-shot, so this information should be consulted. Typically, there is a minimum and maximum resistor and capacitor value specified for the one-shots as well. Reliable operation cannot be guaranteed when exceeding these values. Many one-shots also include an internal resistor (2k for the 74121), which may be used as part of the timing network if desired.

Figure 7.23 also shows a timing diagram for the 74121 one-shot. The output of the internal AND gate must have a positive going edge to trigger the one-shot. When this occurs, the one-shot Q output goes high for the specified time interval. Using the logic provided, either a negative or positive edge signal can create the necessary trigger.

What resistor value is required with a 0.1 μfd capacitor to produce a 1 msec output pulse on a 74121 one-shot?

Solution   The component values may be determined with the 74121 formula, $t_w = 0.7RC$. Since $t_w = 0.7RC$:

$$R = \frac{t_w}{0.7C} = \frac{1 \text{ msec}}{0.7 \times 0.1 \ \mu\text{fd}} = \frac{1 \times 10^{-3}}{0.7 \times 0.1 \times 10^{-6}}$$

$$R = 14,285 \text{ ohms}$$

The nearest standard resistor value may suffice if timing requirements are not critical.

DESIGN
EXAMPLE 7.19

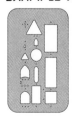

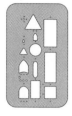

Show how the 74121 one-shot would be connected to trigger on a negative going edge. Use the resistor and capacitor values from Design Example 7.19 to create a sample output waveform.

Solution Negative edge triggering can be accomplished using the 74121 A_1 or A_2 inputs. Inputs not used must be enabled for the trigger to be effective. Numbers in parentheses are chip pin numbers.

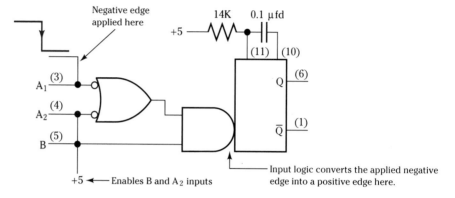

Sample Waveforms:

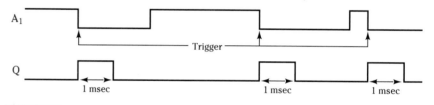

Retriggerable and Nonretriggerable One-Shots

There are two classes of one-shots—**nonretriggerable** and **retriggerable**. A nonretriggerable one-shot, such as the 74121, has limits on how frequently triggering can occur. If triggering occurs at too frequent a rate with respect to the output signal, the output signal will not maintain a constant pulse width and is said to "jitter." A **duty cycle** specification limits the triggering rate for specific combinations of resistors and capacitors. Duty cycle relates how long a waveform is high to how long it is low over the duration of one waveform cycle. Jitter is likely to occur if a one-shot is triggered shortly after the Q output has fallen back low from a previous trigger. This is an indication of too high a duty cycle. The one-shot does not have time to recover from the previous triggering cycle and produces erratic timing on subsequent pulses. Note, any triggers applied to a nonretriggerable one-shot when Q is high are ignored. Only triggers applied after Q has returned low may cause problems in the nonretriggerable class of one-shots.

Retriggerable one-shots have no duty cycle limitation. Triggering can occur as often as desired. If triggering occurs before the Q output returns to a stable state from a previous trigger, then the current timing duration is reset and another timing cycle begins anew. The Q output duration can be extended by retriggering since retriggering only resets the timing but does not force Q from the high state.

The 74122 is a retriggerable one-shot. Determine the output waveform at Q for the following input waveforms. Assume C = 0.22 μfd and R = 5.6k:

DESIGN
EXAMPLE 7.21

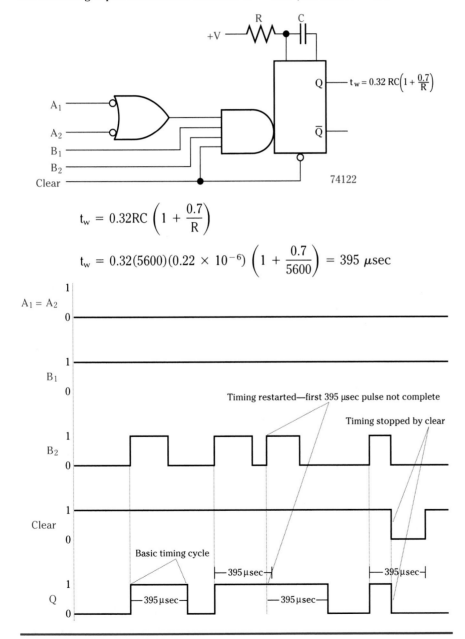

$$t_w = 0.32RC\left(1 + \frac{0.7}{R}\right)$$

$$t_w = 0.32(5600)(0.22 \times 10^{-6})\left(1 + \frac{0.7}{5600}\right) = 395\ \mu sec$$

Oscillators

Oscillators, **square wave generators**, and **astable multivibrators** all describe the various circuits used to generate continuous waveforms for both analog and digital circuits. Many sequential circuits derive their timing requirements from a system clock, which is generally obtained from an oscillator circuit, so it is logical to conclude our study of basic flip-flop devices with the study of basic oscillator circuits. A digital oscillator is a circuit used to generate precise, stable clock signals that drive sequential circuit clocking lines. Oscillator designs vary in terms of waveform accuracy and circuit complexity, but a basic understanding of oscillators can help you sort out the pros and cons of each. The design of oscillators is a complex subject involving issues of feedback theory, loop gain, and circuit stability. Fortunately, several oscillator designs have proven useful for digital system applications, and we will confine our discussion to them.

Crystal Oscillators

A **crystal oscillator** is the most widely used and the most stable oscillator utilized for digital timing. The crystal is formed from material, such as quartz, exhibiting the **piezoelectric effect.** The piezoelectric effect means that a crystal sandwiched between two metal contacts will vibrate at a resonant frequency when subjected to an applied voltage of proper frequency. In addition, the crystal vibrations create a sinusoidal output voltage at a frequency equal to the vibrating frequency. When placed in an appropriate circuit, the output waveform derived from the crystal is highly accurate and stable. If you use a 6.25 MHz crystal oscillator, for example, the output frequency will typically be 6.25 MHz $\pm$ 0.01%. In addition, the circuit's **frequency stability** will be maintained with a high degree of precision over time. This means that frequency drift of the oscillator is minimal.

Designing a crystal oscillator is a design effort not usually required by most digital designers since many common digital circuits already have provisions for oscillator requirements. For instance, most microprocessor chips have pins available for the connection of a crystal. The bulk of the oscillator circuitry is designed into the microprocessor chip and the designer merely has to specify the correct crystal for the chip in use. Many other highly integrated chips requiring oscillators provide external pins for the crystal as well.

When designing your own logic circuitry, you can use an external oscillator module. Crystal oscillator packages can be purchased from manufacturers well versed and experienced in the intricacies of oscillator design. These packages are physically small, have very good frequency stability, and provide a reasonable alternative to discrete oscillator design. For demanding applications, a discrete oscillator design may be necessary to meet the stringent performance specifications of the system. Naturally, additional study is required for proficiency in this area.

Ceramic Resonators

Ceramic resonators are formed from ceramic materials that exhibit the piezoelectric effect. Although the operational characteristics of the ceramic resonator and crystal are similar, the ceramic resonator is less accurate and less stable. However, there is a corresponding decrease in cost, so the ceramic resonator may be economically useful in less demanding timing applications.

Additional Timing Circuits

Some other timing circuits that are less accurate than the oscillators discussed, but nevertheless very useful, include timing chips, one-shots, and hybrid digital designs.

Timing chips, such as the 555 timer (Figure 7.24), are used in many applications requiring a fairly accurate, low frequency timing source. The timing elements for these chips are formed from resistor-capacitor networks. Naturally, the accuracy of the timing source is enhanced by the use of precision resistors and temperature compensated capacitors. The 555 is particularly useful since

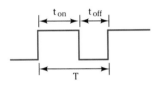

$$f = \frac{1}{T}$$
$$t_{on} = 0.693(R_A + R_B)C$$
$$t_{off} = 0.693\,R_B C$$

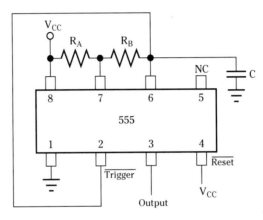

Oscillator operation:

$$\text{Frequency} = \frac{1.44}{(R_A + 2R_B)C} \quad (\text{minimum } R_A = 500\,\Omega)$$

$$\text{Duty cycle} = \frac{R_A + R_B}{R_A + 2R_B} \quad \text{for duty cycles} > 50\%$$

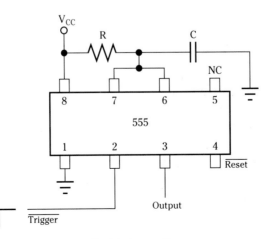

One-shot operation:
$$t_w = 1.1RC$$

Figure 7.24
555 Time
Configurations

it can function as a one-shot (monostable multivibrator) or as an oscillator (astable multivibrator). Although not strictly a digital chip, the timer's output is TTL compatible, allowing for an easy interface to most logic families. Figure 7.24 shows how the 555 can be utilized as both an oscillator and a one-shot.

Oscillator operation, or astable operation, is self-sustaining. That is, the 555 will begin oscillations at the desired frequency as soon as power is applied to the circuit. This **free running frequency** is determined by resistors R_A and R_B as well as by the selected capacitor value. The duty cycle, or the ratio of the "on-time" of the waveform to the total period of the waveform, may also be set by the resistor values.

The 555 may also function as a one-shot by applying a negative pulse to the trigger input, pin 2. The one-shot output pulse duration is set using a resistor-capacitor. The formula for the pulse width (t_w) is shown in Figure 7.24. Typically, with most integrated circuits the manufacturer's data sheets are consulted for limitations and application information.

DESIGN EXAMPLE 7.22

A computer communications circuit requires a 9600 Hz clocking signal with a 70% duty cycle. Using a 555 timer with $R_A = 1000$ ohms, determine the required values of R_B and C.

Solution   The period of a 9600 Hz signal is obtained by computing the reciprocal of the frequency, or $1/9600 = 104$ μsec. Seventy percent of the period is the on-time of the pulse, whereas 30% of the period is the pulse off-time. Thus, 104 μsec $\times 0.7 = 72.8$ μsec and 104 μsec $\times 0.3 = 31.2$ μsec. The duty cycle is computed as follows:

$$DC = \frac{R_A + R_B}{R_A + 2R_B} = 0.7 = \frac{1000 + R_B}{1000 + 2R_B}$$

$$0.7(1000 + 2R_B) = 1000 + R_B$$

$$700 + 1.4R_B = 1000 + R_B$$

$$1.4R_B - R_B = 1000 - 700$$

$$0.4R_B = 300$$

$$R_B = \frac{300}{0.4}$$

$$R_B = 750 \text{ ohms}$$

The capacitor value can be selected using:

$$t_{on} = 0.693(R_A + R_B)C$$

$$72.8 \text{ μsec} = 0.693(1000 + 750)C$$

$$C = \frac{72.8 \times 10^{-6}}{0.693 \times 1750} = 0.06 \text{ μfd}$$

Other timing circuits can be designed around some of the simple gate circuits we have discussed. For instance, two one-shots can be tied to each other so that the output of one is the controlling input of the other. When one triggers the other, an oscillator can be produced whose output frequency depends on the combined pulse width of each. Other less accurate oscillators can be designed by combining inverters and resistor-capacitor networks together.

Design an oscillator from two 74121 one-shots to produce a square wave output frequency of 150 μsec.

DESIGN EXAMPLE 7.23

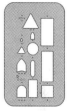

Solution Use the one-shots so that one retriggers the other when the pulse time period expires.

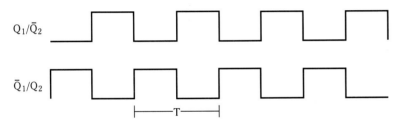

T = 150 μsec; therefore, each half of the square wave = 75 μsec:

$t_w = 0.7RC$

Using C = 0.01 μfd, we obtain:

75 μsec = 0.7R × 0.01 μfd

$R = \dfrac{75 \ \mu sec}{0.7 \times 0.01 \ \mu fd} = 10{,}714 \ ohms$

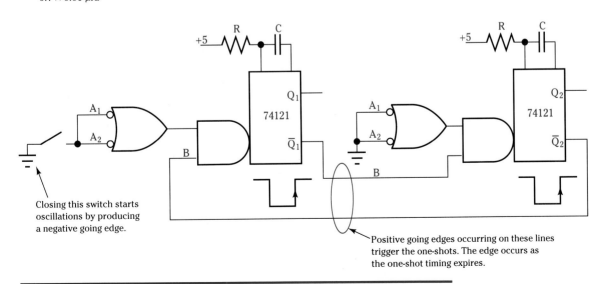

Closing this switch starts oscillations by producing a negative going edge.

Positive going edges occurring on these lines trigger the one-shots. The edge occurs as the one-shot timing expires.

SUMMARY

■ Flip-flops and latches are simple digital devices capable of storing a single binary bit.

■ In general, latches are level sensitive storage devices, whereas flip-flops are edge-triggered devices.

■ Several flip-flops and latches are commonly used in digital design including Set–Reset, D, and J-K devices.

■ The fact that flip-flops and latches can accept and retain digital information leads to the notion of sequential circuitry.

■ Sequential circuits may be classified as asynchronous or synchronous.

■ Asynchronous circuits are clocked without any signal synchronization between devices; devices in synchronous circuits are clocked in a predictable timed sequence.

■ Designing with flip-flops necessitates an understanding of basic operating principles as well as knowledge of device parameters such as setup times, hold times, and allowable clock frequencies.

■ Hazards and race conditions are potential design flaws detrimental to the performance of flip-flop based circuit design.

■ Clock pulses and other timing signals may be created using a variety of devices including one-shots, crystals, and general purpose timing chips.

PROBLEMS

Section 7.2

1. Determine the output levels for the following latches:

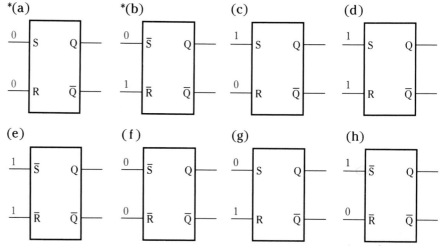

2. Complete the timing diagram for the following latch:

---

* See Appendix F: Answers to Selected Problems.

2. Complete the timing diagram for the following latch:

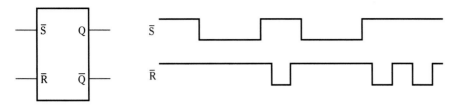

*3. Complete the timing diagram for the following latch:

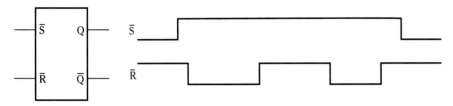

4. Complete the timing diagram for the following latch:

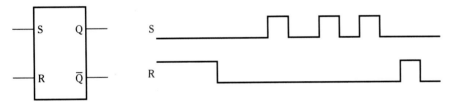

*5. Determine the output of the following latch if the waveforms shown are applied to the set and reset inputs:

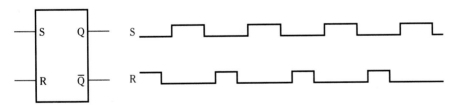

6. Determine the output waveforms for the following circuit:

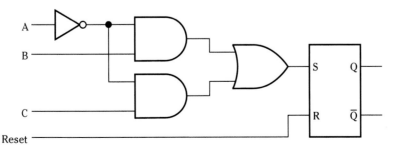

_____

* See Appendix F: Answers to Selected Problems.

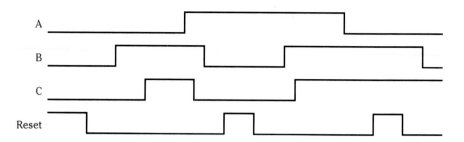

7. What is the output waveform of a NAND-NAND Set–Reset latch using the following inputs?

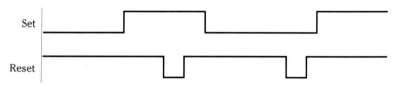

*8. Determine the Q output timing for the following inputs for a NOR-NOR Set–Reset latch:

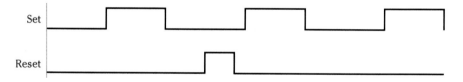

Section 7.3

9. Determine the output waveform using the following waveforms for a positive level gated NOR-NOR Set–Reset latch:

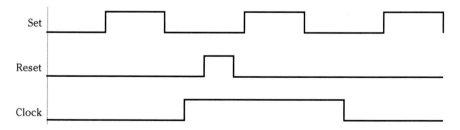

---

* See Appendix F: Answers to Selected Problems.

10. Determine the Q and $\overline{Q}$ waveforms for the following device:

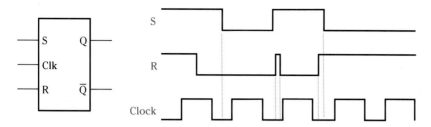

*11. Determine the Q and $\overline{Q}$ waveforms for the following device:

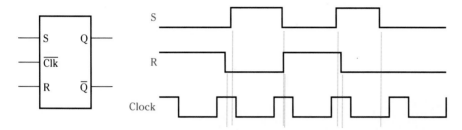

*12. In the following circuit, using the input and output conditions shown, what will be the state of each flip-flop after a clock pulse _⎍_ occurs?

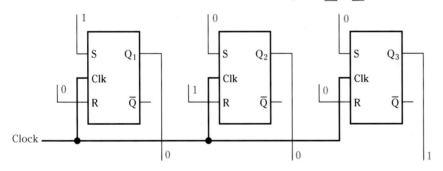

13. Draw the flip-flop waveforms for the following device:

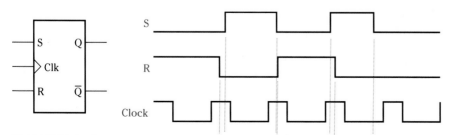

* See Appendix F: Answers to Selected Problems.

14. Determine the output waveforms for the signals and device shown:

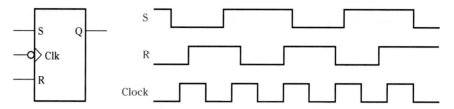

Section 7.4

*15. Show how the Q output responds to the following input waveforms:

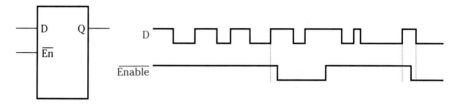

*16. How does the following D flip-flop respond to the inputs shown?

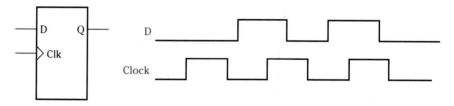

17. Draw the output waveform for the following circuit. If the clock period is 0.1 msec, what is the output period at Q? What is the frequency at Q as compared to the clock frequency?

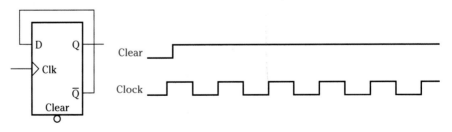

---

* See Appendix F: Answers to Selected Problems.

18. Draw the timing diagram for a positive edge-triggered D flip-flop receiving the following input waveforms:

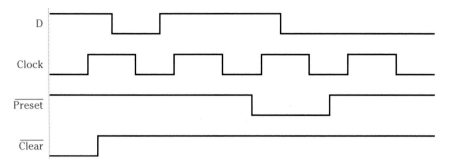

19. Determine the output for the following D flip-flops. Assume preset and clear are inactive. Assume both flip-flops are in the zero state to begin:

*(a)

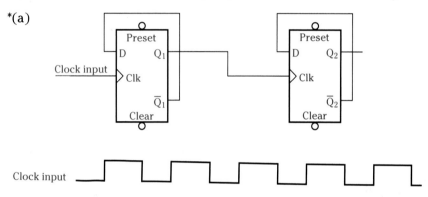

(b) What do you notice about the output pattern at Q_1 and Q_2? What do you notice about the frequency of Q_1 and Q_2 as compared to the clock input?

20. The D flip-flop shown can be used to indicate when the ideal switch is opened or closed. If the switch is attached to a door, then the fact that the door is opened or closed can be detected. Assume that three switches are wired so that a zero level indicates a closed door and a one level indicates an open door. Design a circuit to accomplish the following:

(a) Provide a high level indication if all three doors are opened at the same time.

(b) Provide a high level indication if all three doors have been opened at some time.

(c) Provide a high level indication if door 3 opens before door 2.

(d) Provide a reset switch.

---

* See Appendix F: Answers to Selected Problems.

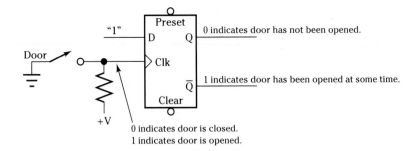

"1"

0 indicates door has not been opened.

Door

1 indicates door has been opened at some time.

+V

0 indicates door is closed.
1 indicates door is opened.

Section 7.5

*21. How does the following J-K flip-flop respond to the input signals given?

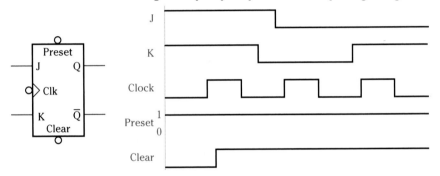

22. Show the output waveform present at the J-K flip-flop. Assume both flip-flops are in the zero state to start.

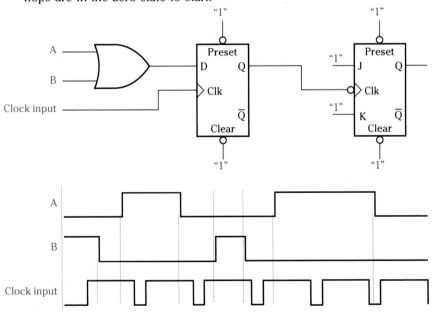

---

* See Appendix F: Answers to Selected Problems.

*23. Determine how the following J-K flip-flop will respond to the input signals also given:

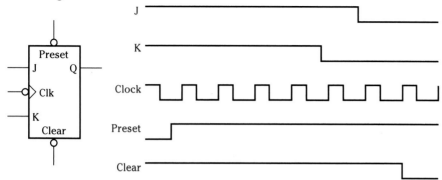

24. Determine the states of flip-flop outputs Q_1 and Q_2 after five clock pulses occur. Assume both flip-flop Q outputs are low to begin:

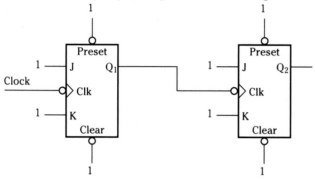

Section 7.7

25. If the following J-K flip-flop has a setup time of t_{su} = 8 nsec and a hold time of t_h = 1 nsec, will the flip-flop operate correctly with the signals given? The square wave clock frequency is 62.5 MHz:

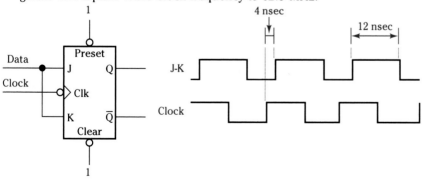

---

* See Appendix F: Answers to Selected Problems.

Section 7.8

26. Identify if the following waveforms exhibit static zero, static one, or dynamic logic hazards:

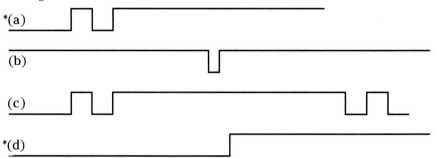

*(a)

(b)

(c)

*(d)

*27. Design a hazard-free clocking circuit for the following flip-flop. Clocking follows the truth table given:

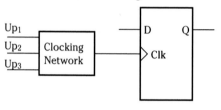

Up3	Up2	Up1	Clock
0	0	0	1
0	0	1	0
0	1	0	1
0	1	1	1
1	0	0	0
1	0	1	0
1	1	0	0
1	1	1	1

28. Using problem 27, assume every gate in the clocking network has a 5 nsec propagation delay time. Draw timing diagrams to:
(a) Show the hazard created from the basic truth table.
(b) Show how the hazard-free network eliminates the hazard.

*29. Joe Tech is having a problem with the following circuit. Explain the problem to him:

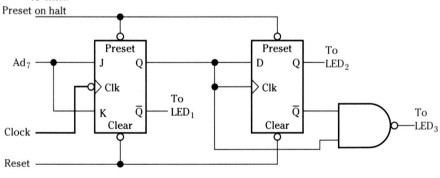

_____

* See Appendix F: Answers to Selected Problems.

Section 7.9

30. Determine the resistor sizes required to obtain the following output pulses from a 74121 one-shot using a 0.01 μfd capacitor:
    *(a) 1 msec   (b) 150 μsec   (c) 950 nsec

31. Determine how a 74HC123 one-shot can be triggered with a negative going edge to produce a 496 μsec pulse. Use a resistor value of 500 ohms for the timing network.

32. If a retriggerable one-shot is connected to provide a 25 nsec pulse when triggered with a positive going edge, how will the output respond to the following triggering waveform?

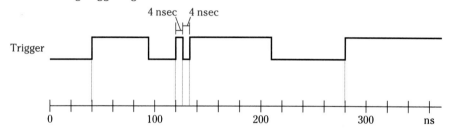

33. Determine the output waveforms for Q_2 and Y using the following circuit and input data:

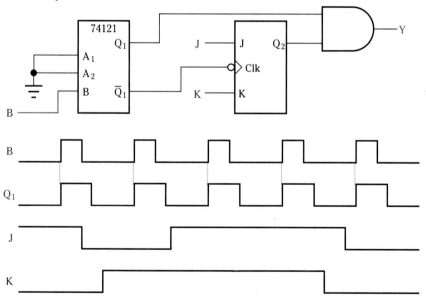

34. Show how a 555 timer can be used to replace the one-shot in problem 33. Use a resistor value of 10k to produce a pulse width of 15 μsec.

---

* See Appendix F: Answers to Selected Problems.

CHAPTER 8

SEQUENTIAL CIRCUITS— COUNTERS AND SHIFT REGISTERS

OBJECTIVES

To explain the function of digital counters and the differences between asynchronous and synchronous counters.

To illustrate how natural binary asynchronous counters are designed using the flip-flop toggling function.

To explain how the natural counting sequence of an asynchronous counter may be altered.

To show several useful applications for digital counters.

To explain how modifications to counter designs, such as strobing, help to eliminate glitches in output signals.

To explain the operation and advantages of synchronous counters.

To illustrate techniques used to truncate the count sequence of synchronous counters.

To show how to design up, down, and presettable counters.

To illustrate the operation of shift registers and to identify common shift register configurations.

To explain how shift registers can be utilized as ring counters and Johnson counters.

PREVIEW

The study of flip-flops introduced the concept of sequential circuits. In this chapter we examine frequently used sequential circuits, specifically the counter and shift register. The many counter and shift register designs that exist form the basis for more complex sequential circuit designs. In addition to basic sequential circuit operation, we discuss some of the design problems and solutions that are often encountered with these circuits.

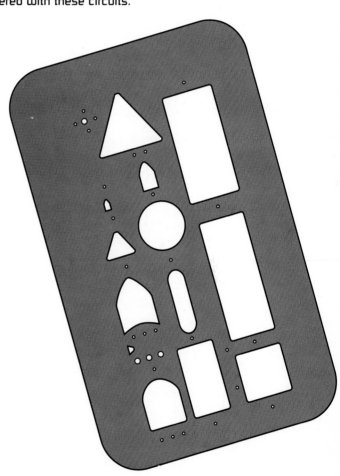

8.1 ■ DIGITAL COUNTERS

The need for digital counting circuits is fundamental in logic design. We can all appreciate the need for this circuitry in a manufacturing environment where items must be counted as they are processed along an assembly line. It is much easier and much less tedious to have an automated circuit tracking the process rather than assigning a person to do the same boring job. Similar needs for counters abound in industry, military, and consumer products. But counters are also fundamental elements in sequential circuit design, forming the basic building blocks for many complex circuits. You will not be surprised to learn, based on our other discussions on sequential circuitry, that there are two classes of counters—**asynchronous** and **synchronous**. Asynchronous counters are relatively easy to design, but they lack speed. In addition, they suffer from glitch-related problems. Synchronous counters are faster and less susceptible to glitches than asynchronous counters, but they require a little more effort to design.

8.2 ■ ASYNCHRONOUS COUNTERS

Asynchronous counters are designed around the flip-flop toggling function. This is illustrated in Figure 8.1 where three D flip-flops are wired for toggling operation. The counter is considered asynchronous because every flip-flop receives a different clocking signal. The flip-flop on the right is the least significant and is the only one to receive the external clocking signal; all other flip-flops are clocked by adjacent flip-flop outputs. (Most schematics show signal flow from left to right or from top to bottom. This counter is drawn from right to left so that the count sequence can be easily analyzed.) The external clock may be either a continuous system clock signal or a signal corresponding to some external process. For instance, if the counter is designed to count the number of cars passing over a section of highway, the car count could be obtained by converting the pressure of a pneumatic hose into an electrical signal, which would clock the counter. Since cars do not pass over a section of road in a regular timing pattern, the electrical pulses clocking the counter would also occur at an irregular rate. This is not a problem, it just shows how nonperiodic events can be detected.

This least significant flip-flop in Figure 8.1 toggles every time a positive edge clock signal is received. Assuming that all flip-flops are initially reset, this means that the second flip-flop does not toggle from zero to one based on the first external clocking signal, but rather, it toggles after the first flip-flop changes from one to zero. Examine the circuit and you will see why. The second flip-flop does not receive a positive going edge from the $\overline{Q}$ output until the first flip-flop Q output makes the transition from one to zero. $\overline{Q}$, of course, changes in a complement fashion from zero to one, and this output provides the necessary positive going edge.

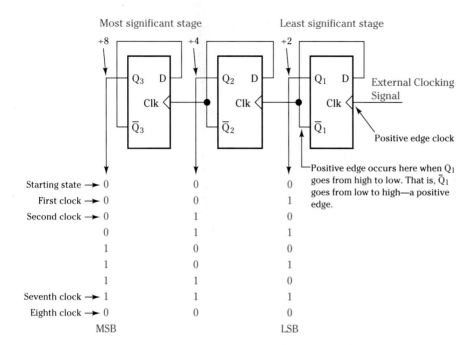

Most significant stage		Least significant stage	
Starting state → 0	0	0	
First clock → 0	0	1	
Second clock → 0	1	0	
0	1	1	
1	0	0	
1	0	1	
1	1	0	
Seventh clock → 1	1	1	
Eighth clock → 0	0	0	
MSB		LSB	

Positive edge occurs here when Q_1 goes from high to low. That is, $\bar{Q}_1$ goes from low to high—a positive edge.

Figure 8.1
Asynchronous Counter

Since the first flip-flop toggles during every clock cycle, its output varies in a zero, one, zero, one, and so on pattern. The second flip-flop changes at one-half of this rate or one-quarter of the original clock rate, varying only when the first flip-flop changes from one to zero. The third flip-flop also changes based on a one to zero transition from its controlling device's output, the second flip-flop. The third flip-flop varies at a rate equal to one-half of the second flip-flop toggling rate or one-eighth of the original clock rate. When the changes in output are written down as shown in Figure 8.1, a familar binary count is obtained. Since this is a 3-bit counter (three flip-flops), eight states (0–7) are possible. Also, since eight states are possible, eight clock pulses are required to advance the counter through all states (000 through 111). When the counter receives the ninth pulse, the counter **rolls-over** or **recycles** back to its initial starting state 000.

Basic decoding hardware can be tacked on to the counter's output to determine when any particular state is reached. The size of the counter can be extended simply by adding additional toggling flip-flops to the existing circuit. An increased count and corresponding reduction in output frequency is obtained.

In general, the output frequency for any flip-flop in a binary counter is $1/2^n$ of the input clock frequency, where "n" = the bit position of the flip-flop. Thus, the frequency division rate for the eighth flip-flop of an 8-bit counter is $1/2^8 = 1/256$. If the input clock frequency is 512 kHz, the output for the eighth flip-flop is $1/256*512$ kHz = 2000 Hz.

**DESIGN
EXAMPLE 8.1**

Draw the timing diagram for the counter depicted in Figure 8.1. Identify the states and frequency division.

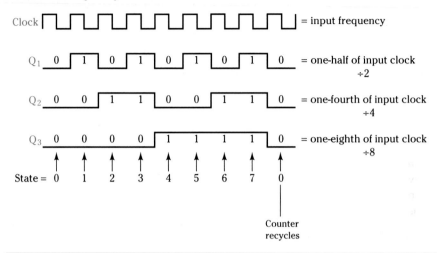

Typically, counters are clocked by continuous clock pulses when the frequency division capabilities of the counter are utilized. When used to divide frequency, the counters are often referred to as **divide by n** counters, where n refers to the frequency division possible. For example, a divide by 5 counter output is one-fifth of the input clock frequency. As an additional example, common digital alarm clocks used a divide by 60 counter to convert a 60-cycle AC line frequency into seconds.

Another interesting use of divide by n counters is cascading them. For instance, by connecting the output of a divide by 12 counter to the input of a

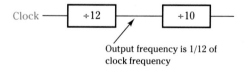

Output frequency is 1/10 of the counter input frequency, or 1/120 of the original clock frequency.

That is:

$$1/12 \times 1/10 = 1/120$$

That is, if the original clock frequency = 75 kHz,

the ÷ 12 output frequency = $1/12 \times 75$ kHz = 6.25 kHz
the ÷ 10 output frequency = $1/10 \times 6.25$ kHz = 625 Hz

**Figure 8.2
Frequency Division
Using Counters**

or

75 kHz × 1/120 = 625 Hz

divide by 10 counter, a divide by 120 counter is created. Since the divide by 12 reduces the frequency of the input clock by a factor of 12, the divide by 10 counter receives a clock input that is already one-twelfth of the original clock signal. This clock is reduced in frequency by a factor of 10, or 120th of the original clock signal. Figure 8.2 summarizes this example.

Show with block diagrams how a 60 Hz AC input signal can be converted into a 1 Hz digital waveform. Assume that divide by 6 and divide by 10 counters are available.

DESIGN EXAMPLE 8.2

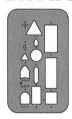

Solution   First, the AC signal, a sine wave, must be converted into a digital waveform by a wave-shaping circuit. This involves reducing the voltage (transformer), clipping off the lower half of the wave (diode), and converting the sine shape into a rectangular shape (a device known as a Schmitt trigger can do this). The resulting 60 Hz digital waveform must be divided down by a factor of 60 to produce the 1 Hz signal.

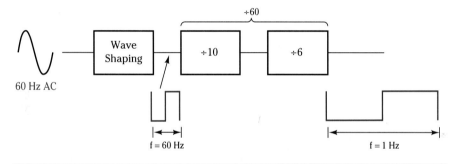

The counter just described is an **asynchronous** or **ripple counter.** The clock pulse changing the counter state must propagate or ripple through each flip-flop in succession before the final steady state is reached. The first flip-flop receives the initial clock pulse and begins to toggle. But toggling is not complete until after the propagation delay time of the flip-flop has passed, and not until this delay is over does the second flip-flop in the cascade receive its signal to toggle. This flip-flop cannot signal the next in line until its delay time has elapsed, a situation faced by each flip-flop in the counter. This limits counting speed and can create other design problems as we will see. However, ripple counters are satisfactory for many basic counting applications because they are very easy to design. Frequently, the number of states in a counter design are designated as the **modulus** of the counter, where modulus is a mathematical term referring to a constant number. The 3-bit counter example described earlier has eight states. Thus, the counter discussed can be called a modulus 8 or, more typically, a MOD 8 counter.

Asynchronous Counter Design Summary

To design a natural length binary asynchronous counter:

a. Select the correct number of flip-flops such that the number of counter states is equal to 2^n, where n equals the number of flip-flops required.
b. Configure each flip-flop to toggle upon receipt of a clock pulse.
c. Connect the clock or source of clocking signals directly to the clock input of the counter's least significant flip-flop.
d. Connect all other flip-flops together so that the output of one flip-flop connects to the clock input of the next most significant flip-flop in the cascade.
e. For positive edge-triggered flip-flops, connect $\overline{Q}$ to the next stage clock; for negative edge-triggered flip-flops, connect Q to the next stage clock.

DESIGN
EXAMPLE 8.3

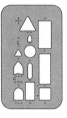

Design a MOD 16 asynchronous counter using J-K flip-flops.

Solution   A MOD 16 counter requires four flip-flops since $2^4 = 16$. Each flip-flop will have the J and K inputs tied high so that the flip-flops toggle whenever a "clock" signal is received. Since an asynchronous design is desired, the least significant flip-flop receives the counting pulses directly; all other flip-flops receive their clocking signal from an adjacent flip-flop. Assuming negative edge-triggered J-Ks, the Q output of one flip-flop provides the clock signal for its next most significant neighbor since we expect the more significant flip-flops to change when the less significant neighboring flip-flop changes from high to low. This negative going signal occurs on the Q outputs during this transition.

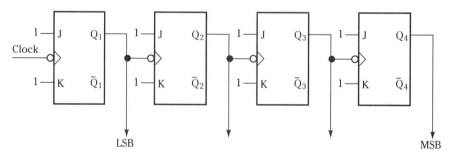

8.3 ■ COUNTER APPLICATIONS

In our brief look at counters we have already identified several counter applications. The following section shows how additional circuitry or simple circuit modifications can enhance the usefulness of counters.

Decoding the State of a Counter

A decoder, designed from a fundamental product, is attached to the 3-bit counter shown in Figure 8.3. In this case the binary state for 5 (101) is detected by attaching Q_1, $\overline{Q}_2$, and Q_3 to an AND gate. The AND gate output will go high only when the counter has reached the appropriate state (101), otherwise the AND output is low. Simple decoders like this allow designers to detect when sequences of events have transpired. However, this decoder-ripple counter combination has the potential to create circuit hazards because of the flip-flop propagation delay, a problem we will investigate shortly.

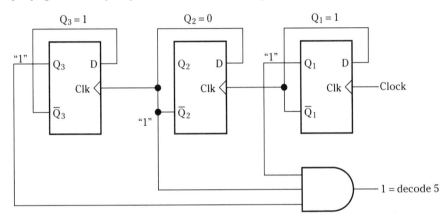

Figure 8.3
Decoding the State of a Counter

A MOD 4 counter is used to control the sequence of operations on a conveyor line. When the counter reaches the 00 state, a start-up signal is provided as a warning indication; when state 01 is reached, a signal is needed to start the line; when state 10 is reached, a signal is required to apply pneumatic pressure; when state 11 is reached, a halt signal is provided.

Assuming that the interfaces between the digital signals and the conveyor equipment are in place, design the digital circuits required to provide the necessary control signals.

Solution Each signal is simply a decode of a counter state. For instance, state 00 can be identified when $Q_1Q_2 = 00$, or $\overline{Q}_1\overline{Q}_2$. The decoder output can then initiate the desired action. The decoders can be designed individually as shown here, or in this example a 2-to-4 line decoder IC could be used.

DESIGN EXAMPLE 8.4

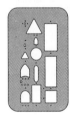

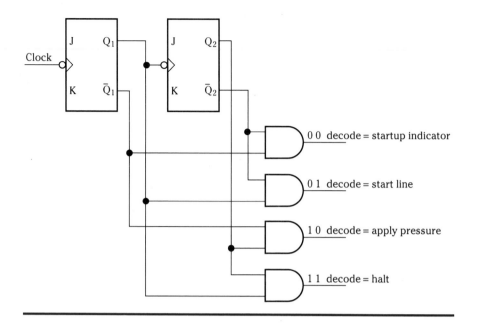

Stopping a Counter

It may be necessary to force a counter to halt at a specific state in some applications if, for instance, the state must be maintained for additional processing to take place. This is easily accomplished as shown in Figure 8.4. The binary counter shown is designed with J-K flip-flops and makes use of the flip-flops' ability to toggle. All stages of the counter, except the first, have their J and K inputs tied to a high logic level, placing them permanently in the toggle condition. Since the J-K flip-flops in this design respond to negative clock edges, the clock inputs for all flip-flops except the first are obtained from the Q outputs of adjacent flip-flops. This is directly opposite the D flip-flop design used earlier where the $\overline{Q}$ outputs fed the positive edge sensitive clock inputs.

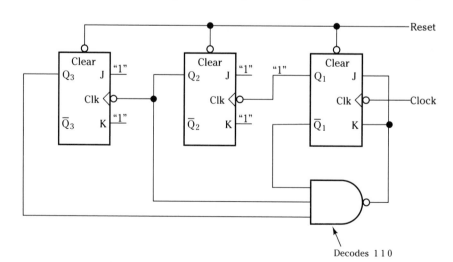

Figure 8.4
Self-Stopping Counter

The decoder attached to the counter is formed from a NAND gate, and in this example it decodes the count 110. When the counter is not in this state, the NAND gate output is high, which keeps the first flip-flop J-K inputs also high. This allows the flip-flop to be free to toggle on every negative going clock edge. As the counter proceeds through its counting sequence, it will eventually reach the count 110. At this point the NAND output goes low, forcing the first flip-flop into a previous state condition. Any additional clock pulses are ignored by the counter and the count freezes at 110. Restarting the count can be accomplished by resetting the counter using the flip-flop asynchronous clear inputs.

Up/Down Counters

Another useful counter variation is to count from a specified number down to zero. Timing circuits can be derived from "count down" operation since periodic clock pulses and the counter state can be related. Usually, counters of this kind are designed as up/down counters with a "mode select" line determining whether the count proceeds from zero (up) or from a specified number toward zero (down).

Forcing an asynchronous counter to count down is relatively simple. In the J-K counter described earlier the flip-flops obtained their clock signals from the Q outputs of adjacent flip-flops. If the clock signal were obtained from the $\overline{Q}$ outputs instead, the counter would count in the opposite direction. Changing the direction of the count becomes a simple matter of altering the clock signal polarity.

Design a MOD 16 asynchronous down counter. Use J-K flip-flops.

Solution An asynchronous down counter is designed by connecting each $\overline{Q}$ output to the next flip-flop stage's clock. The count sequence shown indicates that a flip-flop will toggle when its less significant neighboring flip-flop changes from low to high. The flip-flop $\overline{Q}$ output provides a negative going edge when this occurs. Four flip-flops are required for 16 states.

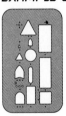

DESIGN EXAMPLE 8.5

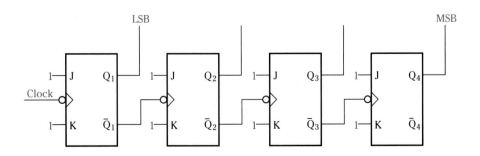

Count Sequence			
Q_4	Q_3	Q_2	Q_1
1	1	1	1
1	1	1	0
1	1	0	1
1	1	0	0
1	0	1	1
1	0	1	0
1	0	0	1
1	0	0	0
0	1	1	1
0	1	1	0
0	1	0	1
0	1	0	0
0	0	1	1
0	0	1	0
0	0	0	1
0	0	0	0

Typically, synchronous counters are used as the basis of an up/down counter, eliminating the possibility of false counts that can occur when an asynchronous counter switches modes. Details on synchronous counters will be given later.

DESIGN EXAMPLE 8.6   Analyze and determine why the following MOD 8 asynchronous "up/down" counter will not function properly:

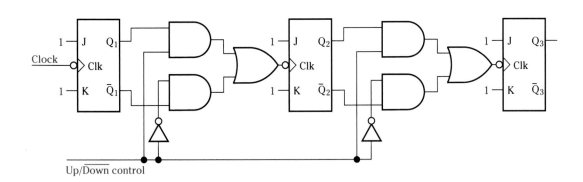

Solution   The counter appears functional because the AND-OR network inserted between stages will steer either the Q output to the clock input of the next stage (up count) or steer the $\overline{Q}$ output to the clock input of the next stage (down count). The choice of an up or down count is determined by the level on the up/down control line.

The control line is the root of the problem. The counter should change state only when a clock pulse occurs, but changing the up/down control line level will inadvertently cause an edge to occur on the clock input. This hazard will force the counter into the incorrect state. To identify whether the hazard is correctable (a logic hazard), or insolvable (a function hazard), we can create a truth table and K-map of the function.

A = Q output
B = $\overline{Q}$ output
C = UP/DOWN control
A high output occurs when Q
and the control line are high
or when Q and the control
line are low.

A	B	C	OUTPUT
0	0	0	0
0	0	1	0
0	1	0	1
0	1	1	0
1	0	0	0
1	0	1	1
1	1	0	0
1	1	1	0

	$\overline{C}$	C
$\overline{A}\,\overline{B}$	0	0
$\overline{A}B$	1	0
AB	0	0
$A\overline{B}$	0	1

$$\text{OUTPUT} = \overline{A}B\overline{C} + A\overline{B}C$$

Since more than one variable changes from one prime implicant to another, this is a function hazard and cannot be corrected.

The equation developed from the truth table clearly shows that a function hazard is present in this design. However, the circuit depicted does not match the equation. This is because the circuit shown was designed using some don't care conditions. Since Q and $\overline{Q}$ cannot equal one another, the input conditions on the truth table where this occurs can be considered as don't care states.

A = Q output
B = $\overline{Q}$ output
C = UP/DOWN control

A	B	C	OUTPUT
0	0	0	d
0	0	1	d
0	1	0	1
0	1	1	0
1	0	0	0
1	0	1	1
1	1	0	d
1	1	1	d

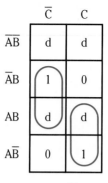

$$\text{OUTPUT} = B\overline{C} + AC$$

It appears that only a logic hazard is present since two K-map pairs are next to each other. Usually, by overlapping the two essential prime implicant pairs, we can eliminate the hazard. Why won't that approach work in this design? Consider the K-map again.

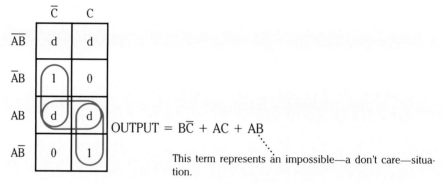

$$\text{OUTPUT} = B\overline{C} + AC + AB$$

This term represents an impossible—a don't care—situation.

The AB term encircled to correct the hazard supposedly will not help because it is created strictly from two don't care states. Since A and B (Q and $\overline{Q}$) can never equal one during normal operation, the AND function represented by AB will never have a high output to correct the hazard.

Summary of Basic Counter Applications

a. Any counter state can be decoded simply by attaching an AND/NAND gate to the appropriate individual flip-flop outputs (Q or $\overline{Q}$) such that the state desired provides a unique input combination (fundamental product) to the AND/NAND gate.

b. Counters may be stopped at any count by decoding a specific state and using the decoder output to inhibit further clocking.

c. Up counters may be converted into down counters simply by altering the output to clock connections between each flip-flop stage.

That is, if counting up occurs by connecting Qs to clock inputs, then counting down can be achieved by connecting the $\overline{Q}$s to the clock inputs.

Changing the Counter Modulus

The modulus of the ripple counter is based on powers of 2. That is, the modulus can be 2, 4, 8, 16, and so on, with the determining factor being the number of flip-flops included in the design. These counters are also called **pure binary counters** or **natural binary counters** since they count in a straightforward binary sequence with the number of states exactly equal to a power of 2. However, sometimes a modulus not exactly related to a power of 2 is required. For instance, how can a MOD 5 counter be constructed? Five states cannot be obtained directly from the counters we have studied (two flip-flops have four states, three flip-flops have eight), so we must modify the counting sequence so that only five states exist. Typically, the states will be 0, 1, 2, 3, and 4 for a MOD 5 counter, but, strictly speaking, the modulus of a counter identifies only the number of states, not the sequence of states. The basic counting sequence is used because it makes the counter design much easier. Altering or reducing the normal counting sequence of a counter is known as **truncating** the count sequence.

We design the MOD 5 counter by first selecting the appropriate number of flip-flops. Three are required because three can provide the required number of states. Two flip-flops only provide four states maximum—not enough for a MOD 5—whereas three flip-flops provide eight states maximum—more than enough. A MOD 5 design must prevent three of the eight states from being used.

Using J-K flip-flops for the counter design simplifies our work since the toggling feature is easily utilized. In addition, we will use flip-flops with an asynchronous clear input, allowing us to reset the flip-flops with an active low signal. Perhaps you can now visualize the counter design. The counter counts through a normal binary sequence starting at zero and ending when an attached decoder detects that the appropriate number of states has occurred. Upon detecting the last state, the decoder then provides a low level signal to all flip-flop clear inputs, forcing the counter back to its initial zero starting state. The counter recycles from this point. However, the one question to answer is, What state do we decode to determine when the reset should occur?

Figure 8.5 shows the basic configuration for the MOD 5 counter. All J and K inputs are tied high to allow toggling, the clock inputs are properly connected, and the NAND gate decoder output feeds all clear inputs. Since the counter must progress through five states, which all exist for a full clock period, we could decode the sixth state to force the reset. The sixth state would only persist for a short time as compared to the time of one full clock cycle since the decoder would immediately reset the counter back to the zero state once the sixth state is detected. Thus, the counter is effectively a MOD 5. This **decode and reset** method works, but it creates the problem shown in Figure 8.6.

Because of propagation delay times associated with the flip-flops and decoder, the counter will actually be in a sixth state (101) momentarily before the reset occurs. As can be seen in Figure 8.6, the NAND gate decoder output goes low shortly after the negative clock edge that moves the counter into state 101. Since the NAND output is connected directly to the active level low clear input of each flip-flop, a reset occurs almost simultaneously with the NAND

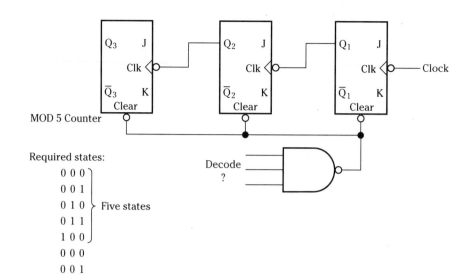

Required states:

$$\left.\begin{array}{l}0\ 0\ 0 \\ 0\ 0\ 1 \\ 0\ 1\ 0 \\ 0\ 1\ 1 \\ 1\ 0\ 0 \end{array}\right\} \text{Five states}$$

0 0 0
0 0 1

Figure 8.5
Implementing the
Decode and Reset
Method

going low. For most of the clock period during which the reset condition occurs, the state of the counter is 000. However, the counter is briefly in the 101 or sixth counter state during the beginning of the clock period, which is not entirely appropriate for a MOD 5 counter. The end result is a glitch being generated at the Q_1 flip-flop output, an undesirable situation in most circuits.

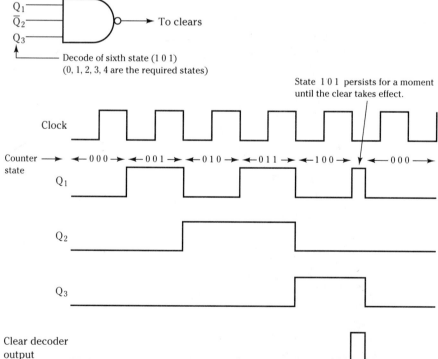

Figure 8.6
Asynchronous Counter
Glitch Problem

Other digital circuits connected to the Q_1 line, particularly high speed circuits, may inadvertently respond to this false state.

   To prevent the creation of an inadvertent sixth state, we can decode the fifth state instead. However, simply decoding the fifth state does not alleviate the problem because the fifth state would only persist for a brief instant, essentially creating a glitchy MOD 4 counter. We need to maintain the fifth state for a duration of time equal to the duration of all other states. This necessitates delaying the reset until the clock period following the fifth state occurs. This is shown in Figure 8.7.

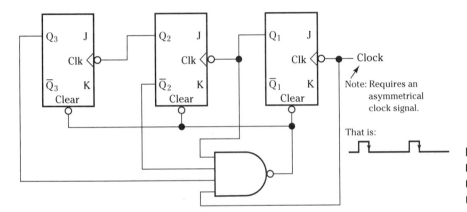

Figure 8.7
Using the Counter
Clock to Prevent
Glitching

   The counter clock signal is fed to the NAND gate decoder circuit. When the clock edge advances the counter to the fifth state, the clock also degates the NAND gate since it is going low at the time. The NAND gate is degated before the fifth state actually occurs because of the counter ripple delay characteristic. In order to maintain the fifth state for approximately the same duration as the other states, the clock must be asymmetrical (a rectangular waveform with a duty cycle not equal to 50%), so that the NAND gate remains degated for almost a full clock cycle. The NAND gate clear signal does not occur until shortly before the counter tries to attain the sixth state. At this time the clear signal resets all flip-flops back to the zero state and the normal counting sequence restarts. The reset occurs before the sixth state appears because the NAND gate decoder already has the proper reset condition decoded. The NAND only waits for the enabling clock signal to carry out the reset, which is accomplished in a far shorter time than propagation delay of the counter. Therefore, the counter never reaches the sixth state and five states of nearly equal duration occur. A word of caution, however; this solution places a constraint on the clocking signal that may not be tolerable in all situations. If the duration of each state must be so precise that either variation of a decode and reset design is impracticable, then a synchronous counter would be used (discussed later in this chapter).

   Although the techniques shown for altering the modulus of an asynchronous counter may not produce an ideal response, they nevertheless provide a relatively simple method to create counters of any modulus desired. The delays through each flip-flop accounting for some of the nonideal characteristics also create other problems that will be covered in the next section.

DESIGN EXAMPLE 8.7

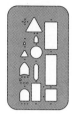

Design a MOD 10 asynchronous counter using the decode and reset method.

Solution   MOD 10 counters are very common and are often called **decade counters**. If the count progresses in a sequence from 0000 to 1001, then the counter is also called a **BCD counter**.

Four flip-flops are necessary to create ten states. To truncate the normal count sequence, decode the eleventh state (1010) to create a reset pulse; that is, $1010 = Q_4\overline{Q}_3Q_2\overline{Q}_1$.

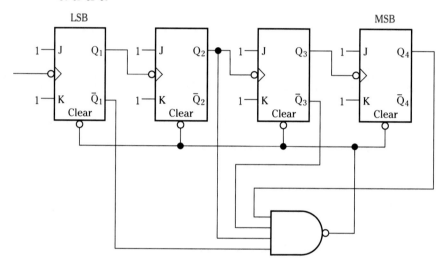

In this example the decoder can be simplified to:

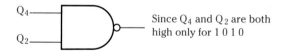

Since Q_4 and Q_2 are both high only for 1 0 1 0

DESIGN EXAMPLE 8.8

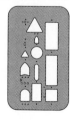

Draw the timing diagram for the MOD 10 decade counter designed in Design Example 8.7.

Solution   Counting proceeds normally until the eleventh state—1010—truncates the sequence.

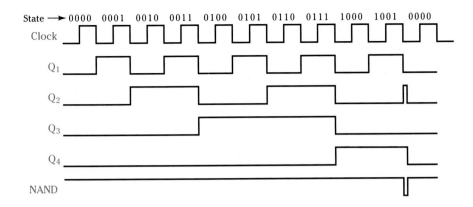

State →0000 0001 0010 0011 0100 0101 0110 0111 1000 1001 0000

Clock

Q_1

Q_2

Q_3

Q_4

NAND

A Quick Way to Approximate Counter Waveforms

a. Write down the counting sequence in truth table order, leaving space between the individual columns.
b. Next to each zero, draw a vertical line. Next to each one, draw a vertical line displaced a bit to the right of the lines for the zeros.
c. Connect all vertical lines with horizontal lines.
d. Turn the paper 90°.

That is, MOD 10 sequence

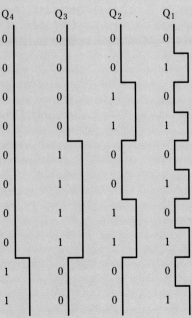

Q_4	Q_3	Q_2	Q_1
0	0	0	0
0	0	0	1
0	0	1	0
0	0	1	1
0	1	0	0
0	1	0	1
0	1	1	0
0	1	1	1
1	0	0	0
1	0	0	1

FOR YOUR INFORMATION

Design Summary for Truncating Asynchronous Counter Count Sequence

a. Begin by designing a basic natural length asynchronous counter with a number of states greater than the truncated modulus desired.

b. Using the decode and reset method, attach to the counter outputs a decoder to decode the counter state equal to the desired counter modulus, for example, decode state five for a MOD 5 counter.

c. Connect the decoder output to the clear inputs of each flip-flop in the counter, for example, use a NAND gate decoder if the flip-flop clear inputs are active low.

d. When the decoded state occurs, the decoder forces each flip-flop to clear, resetting the counter.

e. If the glitches created by this method are a problem, then alternate designs controlling clock pulse width are required. A more practical solution may be to use a synchronous counter.

DESIGN EXAMPLE 8.9

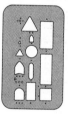

Using block diagrams, show how the states of a BCD counter can be displayed on a seven segment display device.

Solution   Converting BCD to a seven segment code requires a decoder. Therefore, a seven segment decoder is driven by the BCD counter outputs. In order for the display information to be meaningful, the counter clock rate must be slow enough for our eyes to perceive the number on the display. If the clock rate is too fast, all seven segments will appear to be on.

Note: The clock could be the output of a sensor or a switch. Switches used in this application must be "debounced" to prevent false counts. Debouncing is covered in Chapter 11.

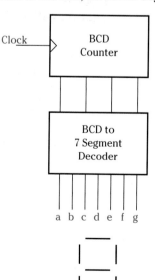

8.4 ■ PRACTICAL ASYNCHRONOUS COUNTER CONSIDERATIONS

Counter Speed

The speed of a counter is a function of the counter delay, which limits the maximum clocking rate of the design. This is true in counters of all kinds. The delay is not restricted to the counter design itself, for any circuitry connected to the counter must be given enough time to sense the counter state and react to it. This implies that the counter speed must also account for the delay of associated circuitry.

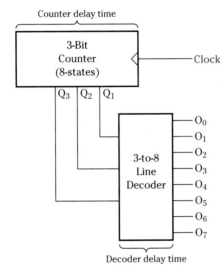

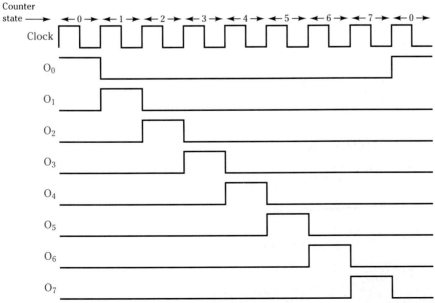

Figure 8.8
Fully Decoded 3-Bit Counter

One of the most typical counter applications is attaching decoder circuits. We have used decoders to alter the state of a counter and to detect the various counter states. Figure 8.8 shows how useful timing signals can be generated from a counter and decoder. We must now consider how circuit delay affects the usefulness of these signals.

As shown in the figure, delay exists through the counter and through the decoder circuitry. Assume that the counter flip-flops each provide a worst case propagation delay of 18 nsec and the decoder provides a 22 nsec delay. The total delay is 3*18 nsec + 22 nsec = 76 nsec. This means that 76 nsec elapse between the active clock transition and the availability of a valid decoder signal. Furthermore, this overall delay restricts the maximum counter clock rate to 1/76 nsec = 13.16 MHz. Obviously, not every counter is required to clock at a maximum rate, but in performance based systems this is a real concern. Considerable time is spent engineering solutions that minimize delays or spent specifying faster logic families so that the system performance goals can be met.

DESIGN EXAMPLE 8.10

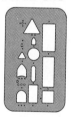

A MOD 32 asynchronous counter is constructed using flip-flops with a 10 nsec propagation delay. Decoders attached to the counter add 16 nsec of delay. What is the maximum clock rate for this counting system?

Solution   A MOD 32 counter has five flip-flops ($2^5 = 32$). In an asynchronous design each flip-flop clocks an adjacent flip-flop, so the propagation delay is the product of the individual delay times the number of flip-flops. The decoder delay is added to the flip-flop delay since the decoder output is not valid until the counter state is stable. Thus:

$$\text{System delay} = (5 \times 10 \text{ nsec}) + 16 \text{ nsec}$$
$$\text{System delay} = 50 \text{ nsec} + 16 \text{ nsec}$$
$$\text{System delay} = 66 \text{ nsec}$$

$$\text{Maximum clock frequency} = 1/66 \text{ nsec} = 15.152 \text{ MHz}$$

Counter Glitch Problems

When we take a closer look at the ripple delay characteristics of the asynchronous counter, we will discover other potential problems. During an active clock transition we expect the counter to move to its next state. However, the change in state takes place one flip-flop at a time, beginning with the least significant stage. The first flip-flop responds to the clock signal, and after its propagation delay time it settles down. The new level in the first flip-flop may trigger a change in the second flip-flop, but the second flip-flop does not change until its propagation delay time has expired. As you can see, the counter goes through several "false states" on its way to the correct state. This is illustrated in Figure 8.9.

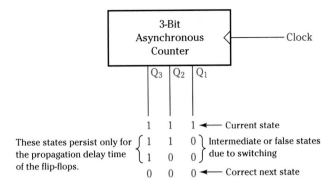

These states persist only for the propagation delay time of the flip-flops.

1 1 1 ← Current state

1 1 0 ⎱ Intermediate or false states
1 0 0 ⎰ due to switching

0 0 0 ← Correct next state

Figure 8.9
False States in an Asynchronous Count Sequence

As is evident from the timing diagram in Figure 8.10, three pulses occur at the decoder output—one is the correct decode for the state 100, whereas

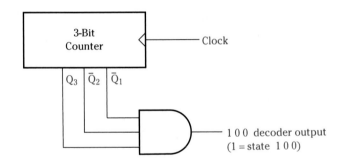

1 0 0 decoder output
(1 = state 1 0 0)

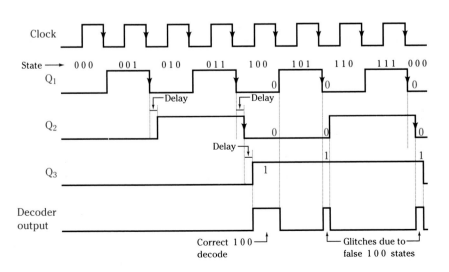

Correct 1 0 0 decode

Glitches due to false 1 0 0 states

Figure 8.10
How Glitches Are Created in an Asynchronous Counter

the other two are very short glitches caused by the counter delay times as the counter moves from state 101 to 110 and from 111 to 000. The delays from stage to stage cause the state 100 to occur temporarily on two different occasions. Obviously, this is incorrect since any circuitry responding to the decoder output will be fooled twice into believing that state 100 is present in the counter. The glitches are also very short in duration and may be difficult to debug. Furthermore, since the glitches also contain edges, they can be particularly troublesome in sequential circuits.

There are several ways to overcome the glitch problem. The counter clock signal can be used as a gating signal for the decoder, as shown in Figure 8.11. Since the clock initiates a change in state on a negative going transition, the low clock level occurring after the negative edge degates the decoder during the time that glitches are generated. Therefore, the decoder cannot produce a false response. This technique shortens the valid decoder response since the decoder is always degated for a portion of each clock cycle, even during a decode of the correct state. This is usually not a problem because many decoders are designed to indicate the occurrence of a specific counter state, which is accomplished with a shortened pulse as easily as it is with a long pulse. However, the technique may constrain the duty cycle characteristics of the clock signal since it must remain low longer than the ripple delay time of the counter. Fortunately, other alternatives are possible.

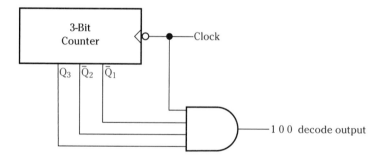

Figure 8.11
Using the Counter
Clock to Enable
Decoding

Figure 8.12 shows the same counter circuit and decoder as before except for the **strobe** line attached to the decoder input. Strobe signals enable portions of the logic circuitry and bear a timing relationship to the circuit they are controlling. In this example the strobe line degates the decoder briefly during the beginning of each clock cycle, effectively inhibiting any glitches occurring during this time. During the remainder of each clock cycle the strobe line enables the decoder so that its output signal is present. Therefore, when the correct state for decoding occurs, the decoder output will be active only after the strobe signal becomes active. The strobe pulse can easily be generated using a one-shot periodically triggered by the counter clock.

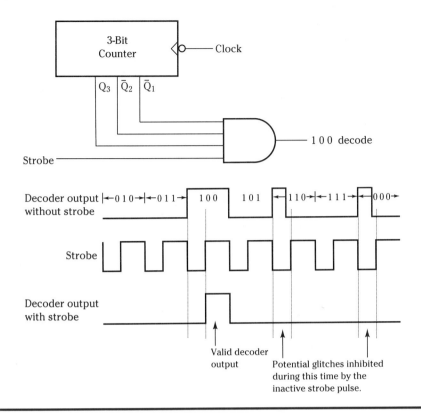

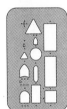

Figure 8.12
Controlling Decoding
Time with a Strobe
Signal

Design the strobe pulse required to eliminate glitching on the decoder outputs of a MOD 16 asynchronous counter. The decoders decode states 0010, 0110, 1000, and 1100. The counter advances on a positive going edge and requires 50 nsec before a state is established. The clock rate is 5 MHz.

DESIGN EXAMPLE 8.11

Solution   The strobe signal must enable the decoder only after the counter has firmly established a state. Therefore, the strobe should not be active until at least 50 nsec after a positive clock edge occurs. To design some safety margin into the strobe pulse timing, design the pulse to occur 70 nsec after the clock edge. A 5 MHz clock has a 200 nsec period, so 70 nsec allows plenty of time for the counter to settle down to a state and still allow for adequate decoding time.

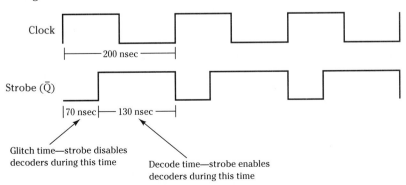

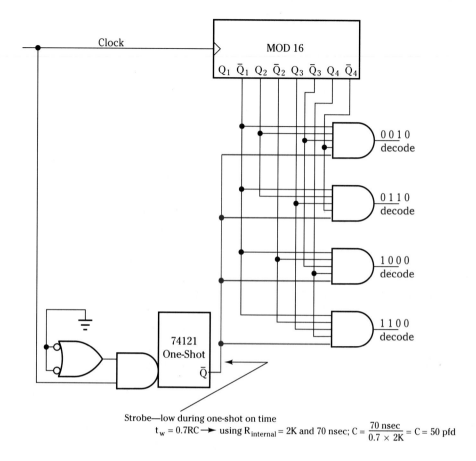

Strobe—low during one-shot on time

$t_w = 0.7RC \longrightarrow$ using $R_{internal} = 2K$ and 70 nsec; $C = \dfrac{70\ nsec}{0.7 \times 2K} = C = 50$ pfd

8.5 ■ SYNCHRONOUS COUNTERS

Synchronous or **parallel** counters have several advantages over their asynchronous counterparts. They can operate at faster speeds because all flip-flops are clocked simultaneously, thus eliminating the slowing effects of ripple delay. The reduction in ripple delay also minimizes glitch problems to a great extent, although not completely. The overall counter delay, neglecting attached decoders, is approximately equal to one flip-flop delay. However, no two flip-flops switch at exactly the same rate nor do they switch from zero to one or one to zero in the same amount of time. There are also some delays caused by intervening circuitry. Therefore, glitches are still possible. The techniques previously covered to eliminate glitches in asynchronous counters are also effective for synchronous counters.

We will examine two methods for designing a synchronous counter. In this chapter we look at counters from a practical perspective and design the counter based on operational characteristics. In a later chapter we will formalize sequential machine design techniques to include not only counters but other sequential circuits as well.

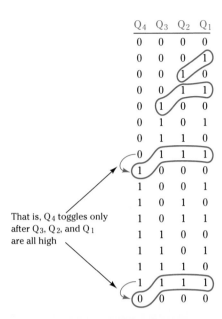

$$\begin{array}{cccc} Q_4 & Q_3 & Q_2 & Q_1 \\ \hline 0 & 0 & 0 & 0 \\ 0 & 0 & 0 & 1 \\ 0 & 0 & 1 & 0 \\ 0 & 0 & 1 & 1 \\ 0 & 1 & 0 & 0 \\ 0 & 1 & 0 & 1 \\ 0 & 1 & 1 & 0 \\ 0 & 1 & 1 & 1 \\ 1 & 0 & 0 & 0 \\ 1 & 0 & 0 & 1 \\ 1 & 0 & 1 & 0 \\ 1 & 0 & 1 & 1 \\ 1 & 1 & 0 & 0 \\ 1 & 1 & 0 & 1 \\ 1 & 1 & 1 & 0 \\ 1 & 1 & 1 & 1 \\ 0 & 0 & 0 & 0 \end{array}$$

That is, Q_4 toggles only after Q_3, Q_2, and Q_1 are all high

Figure 8.13
4-Bit Count Sequence Illustrating Flip-Flop State Change

The counting sequence illustrated in Figure 8.13 is a useful starting point for synchronous counter design, for it shows when the various flip-flop outputs should change. Several examples are circled on the diagram. Whenever a flip-flop output changes from zero to one or from one to zero, it is evident that all less significant stages are ones prior to the change in counter state. For example, when output Q_4 is clocked to initiate a change from zero to one, outputs Q_3, Q_2, and Q_1 are in the one state prior to the change. Q_4 will not change unless this is the case. Therefore, in synchronous counter design, lower order flip-flop states determine when the higher order flip-flops change. Q_4 changes when Q_3 AND Q_2 AND Q_1 are high. Similar equations can be stated for the other flip-flops. For instance, Q_3 changes when Q_2 AND Q_1 are high. Of course, the change in any flip-flop occurs after a clock transition. The ANDing of flip-flop outputs merely sets the conditions under which a change can occur. A 4-bit synchronous counter is illustrated in Figure 8.14.

It is clear from this figure that we are dealing with a synchronous counter since the clock signal is connected directly to each flip-flop. All flip-flops have the potential to undergo a change at the same moment in time. The AND gates inserted into the circuit enable succeeding flip-flops (in terms of bit significance) to change during a clock transition. Notice that by change we are referring to a toggling action. The AND gates set the condition by which a flip-flop may or may not toggle. Remember that all flip-flops receive a clock signal every clock period, but this does not mean that each flip-flop will change level. The AND gates determine exactly when a change in level is appropriate by applying high levels to both J and K when a toggle should occur. Otherwise J and K are held low in the previous state condition.

The modulus of natural synchronous counters can be extended by adding more flip-flops and AND gates. The AND gates must combine the signals from all less significant stages. AND gates with as many inputs as necessary are used to keep the AND gate delay down to one gate per stage.

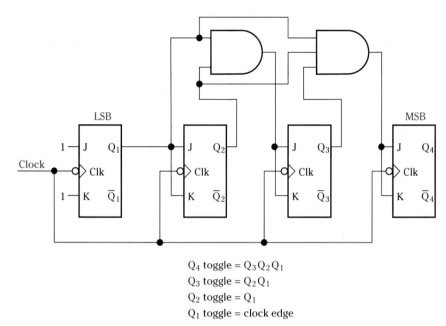

Figure 8.14
MOD 16 Synchronous
Counter

Q_4 toggle $= Q_3 Q_2 Q_1$
Q_3 toggle $= Q_2 Q_1$
Q_2 toggle $= Q_1$
Q_1 toggle $=$ clock edge

The worse case delay for the counter is determined by the flip-flops controlled by AND gates. Since the flip-flop change in state is subject to its own propagation delay plus the delay experienced by signals passing through an AND gate, the total delay is equal to that of a single flip-flop plus the delay of a single AND gate. This, of course, affects the maximum clocking rate. (Any decoders or other circuitry added to the counter outputs also affect clocking speed.)

DESIGN
EXAMPLE 8.12

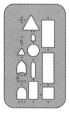

What is the delay for the counter in Figure 8.14 if the AND gate propagation delay is 11 nsec and the flip-flop delay is 14 nsec? What is the maximum clocking rate?

Solution Synchronous counter delay is equal to the delay of one flip-flop plus the delay of one AND gate.

Total delay $= 11$ nsec $+ 14$ nsec $= 25$ nsec
Maximum clock rate $= 1/25$ nsec $= 40$ Mhz

Altering the Modulus of Synchronous Counters

The synchronous counter design discussed earlier is a **natural length binary counter**. That is, the count sequence is a normal binary count sequence based of 2^n. Naturally, there exists a need for count sequences that are not related to a power of 2. We can truncate the count sequence of a synchronous counter in a manner similar to asynchronous counters by eliminating unnecessary

FOR YOUR
INFORMATION

Natural Length Synchronous Counter Design Summary

a. Choose the number of flip-flops (n) required so that 2^n = number of states.

b. Connect the clock line to the clock input of each flip-flop.

c. Attach to each J-K input the ANDed outputs of all less significant stages, assuming that J-K flip-flops are used.

d. Treat the counter as a state machine, an alternate design approach, as discussed in Chapter 13.

states. However, we will not force a reset as we did using the asynchronous decode and reset method because we are not interested in generating glitches. Remember, one advantage of the synchronous design is the fact that glitches are minimized. We would not be doing ourselves any favor by introducing glitches into the design. To truncate the count sequence in a synchronous counter we use a method called **decode and steer.** Rather than abruptly re-setting the counter, we "steer" the counter into the next desired state during a normal clock period. This alters the counter sequence but eliminates glitches because the transition from one state to the next will occur smoothly. Typically, the counter is steered to the all zeros state after the maximum count has been reached. Although the all zeros state is the same state produced by a reset, we will only allow this state to occur synchronized with the clock.

Suppose that we need a MOD 5 synchronous counter. (Refer to Figure 8.15.) It is readily apparent that three flip-flops are required to provide five distinct states. To begin, design a natural binary counter first. In this example a MOD 8 synchronous counter is designed in the fashion discussed earlier. This design will then be modified so that only five states are present. Assuming that the states 000, 001, 010, 011, and 100 are the five desired states, add to this counter a low active level decoder to detect the last state 100. The decoder output will drive the counter steering logic. Unless the counter is in the decoded state, the counter will count in the prescribed normal sequence.

Notice the AND gates in Figure 8.15. Their primary function is to present the state of the lesser significant flip-flops to the J-K inputs of the more sig-nificant flip-flops. This is exactly what a natural binary synchronous counter must do and is exactly what our previous synchronous counter design did. As long as the decoder output is high, the AND gates and OR gates are enabled for normal counting operation. The various J-K inputs either receive signals to toggle or remain in a previous state. That is, the AND gates feeding the J inputs are enabled so that the less significant flip-flop information is presented to the appropriate J inputs. The OR gates feeding the K inputs are also enabled by the complement of the decoder signal so that the AND gate information is presented to the K input through an OR gate. For example, flip-flop Q_3 receives signals from flip-flops Q_1 and Q_2 through the AND gate connected to its J input. When the 100 decoder output is high, the AND gate is enabled and the less

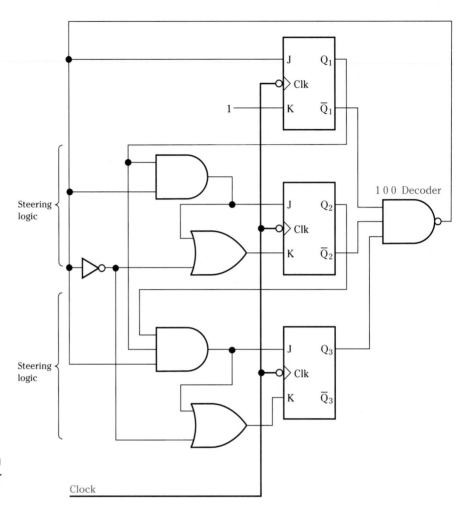

Figure 8.15
MOD 5 Synchronous Counter Design Using the Decode and Steer Method

significant flip-flops control the change in state of flip-flop Q_3. The K inputs are similarly controlled by the OR gates.

Only when the decoder output goes low is counter operation altered. When this happens, the decoder disables all AND gates so that every J input is low. The OR output goes high during this time because the complement of the decoder output is an input signal to each OR gate. This forces every K input high. Therefore, each flip-flop has J = 0 and K = 1, so that on the following clock edge every flip-flop resets and the initial counter state 000 is reestablished. It is important to note that the change back to the 000 state is synchronized with a clock edge.

The steering logic remains the same from stage to stage (except for the least significant stage). That is, an AND gate is required to present the values of all least significant flip-flop stages to the next most significant J input. The AND gate size (number of inputs) varies depending on the position of the flip-flop in the counter. The more significant the flip-flop is in the counter, the greater is the number of inputs because there are more flip-flop stages preceding it. The AND gate also includes the decoder signal. Each stage also has a 2-input OR gate feeding the K input. The OR inputs are always the AND gate output and the complement of the decoder output. With these facts in mind, any size counter can be developed by cascading this basic design.

However, it may be possible to simplify the design further by comparing the last counter state to the all zeros state. As the counter recycles back to zero, many of the flip-flops may be in the zero state just by following their natural counting sequence. For instance, a MOD 13 counter's last state is 1100. Upon recycling to 0000, only three of the four flip-flops actually have to be forced to zero (Q_4, Q_3, and Q_1 because the next natural count is 1101). The remaining flip-flop (Q_2) will go to zero anyway. Noting this, some of the steering logic can be removed from the Q_2 flip-flop.

Figure 8.16 shows how the MOD 5 synchronous counter we designed can be modified by considering the count sequence. Since the last state in the counter is 100, only the first and last stage need to be forced to zero. The second flip-flop stage will normally stay at zero (101) after the 100 count passes.

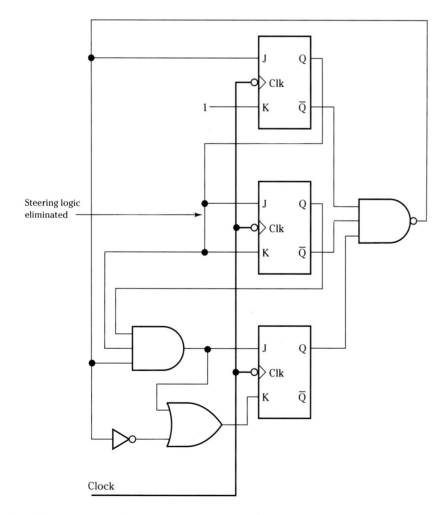

Figure 8.16
Simplifying a
Truncated Count
Synchronous Counter
Design

Notice how the second stage receives its information only from Q_1. There is no need for the OR gate or the decoder signal. The least significant stage never needs to be altered since the design shown for that stage is simple and always works.

DESIGN EXAMPLE 8.13

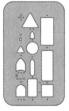

Design a MOD 10 (decade) synchronous counter using the decode and steer method.

Solution   If the counter is designed so that the counting sequence is from 0000 to 1001, a natural binary synchronous counter can be utilized as the basis for the design. A decoder can be created to decode the last state (1001) and to steer the counter back to the 0000 state. Steering logic can be used on each flip-flop to alter the count. However, logic circuitry may be minimized by examining when the count rolls over. Normally, the counter would proceed from $1001 \rightarrow 1010$. Noting that the third bit will remain zero whether or not we force the counter from $1001 \rightarrow 0000$, we can eliminate the steering logic from the third stage of the counter. Only the AND gate combining all lower order Q outputs needs to be included.

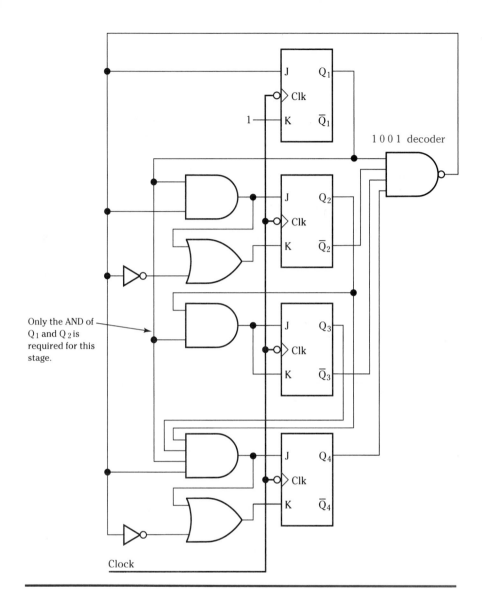

Only the AND of Q$_1$ and Q$_2$ is required for this stage.

1 0 0 1 decoder

Clock

8.6 ■ SYNCHRONOUS COUNTER APPLICATIONS

Combining Counters

Frequently, synchronous counters are combined with others to create larger counters. Doing this usually destroys any semblance of a normal counting sequence, but it is very useful for frequency division purposes. Figure 8.17 shows how two MOD 3 counters are combined to produce a MOD 9 counter. The Q_2 output frequency of the first MOD 3 is one-third that of the input clock,

Figure 8.17
Combining Two MOD
3 Counters to Create a
MOD 9 Counter

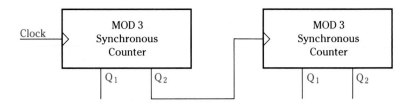

whereas the Q_2 output frequency of the second MOD 3 is one-ninth that of the original clock.

Strictly speaking, this MOD 9 counter is no longer a synchronous counter, but rather, a "hybrid counter." Each MOD 3 counter is synchronous in design; however, the clock signal driving the first counter does not drive the second. The second MOD 3 is clocked by the output of the first counter, adding an asynchronous element to the circuit. This reduces speed somewhat, although it is still very useful. Many counter designs readily available as integrated circuits are designed this way and can be combined to build larger counters such as those shown in Figure 8.18.

Figure 8.18
Divide by 1600
Counter Fabricated by
Cascading Counters

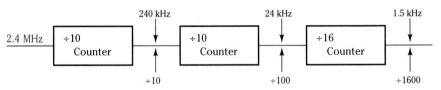

Synchronous Down Counter

All our synchronous counter designs discussed to this point have followed a count sequence from zero to the maximum count of the counter. Therefore, these counters are designed as up counters. By examining the following up and down count sequences, we can determine how to construct a synchronous down counter:

Up Count	Down Count
0000	1111
0001	1110
0010	1101
0011	1100
0100	1011
0101	1010
0110	1001
0111	1000
1000	0111
1001	0110
1010	0101
1011	0100
1100	0011
1101	0010
1110	0001
1111	0000

We have already determined that for an up count a flip-flop toggles only when all lesser significant stages are ones. The down count sequence shows just the opposite: A flip-flop toggles only when all lesser significant stages are zeros. Designing a down counter based on this premise is rather easy considering that we can use flip-flops with Q and $\overline{Q}$ outputs. Since Q and $\overline{Q}$ are complementary outputs, whenever the counter Q outputs are displaying the count down sequence, the $\overline{Q}$ outputs will have the required levels needed to toggle the flip-flops. Rather than ANDing Q outputs, we can AND the $\overline{Q}$ outputs to create the signals necessary for the count down sequence. For instance, when the down counter is poised to move from state 1000 to state 0111, outputs Q_3, Q_2, and Q_1 are all zeros. At this time the $\overline{Q}_3$, $\overline{Q}_2$, and $\overline{Q}_1$ outputs are all ones. Therefore, we can AND these three $\overline{Q}$ outputs as the toggling signal for the fourth flip-flop. A MOD 16 synchronous down counter following this principle is shown in Figure 8.19.

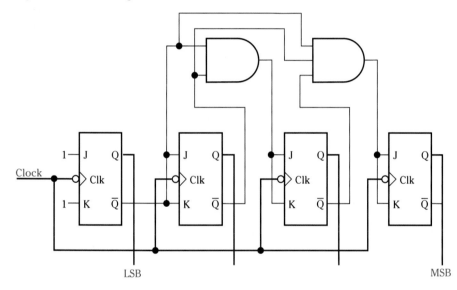

Figure 8.19
MOD 16 Synchronous Down Counter

Design a MOD 5 synchronous down counter.

Solution A down counter can be designed by ANDing the various $\overline{Q}$ outputs. These in turn drive the J-K inputs to determine when the flip-flops should change state. However, since this counter has a truncated count sequence, we will have to include decode and steering logic as well. The steering logic remains the same as in an up counter except that the $\overline{Q}$ rather than the Q outputs are used.

The decoder can be designed by examining the count sequence:

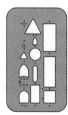

DESIGN EXAMPLE 8.14

100   Since the counter is counting down, once
011   the 000 state is reached, the counter must be
010   forced back to 100. The decoder must decode 000
001   to indicate when this should happen.
000

To determine where the steering logic goes, we examine how the normal count would proceed once the 000 state is reached. Normally, in a down counter the next state would be 111. This state is one of those skipped in this design. This MOD 5 must go from 000 to 100.

$$\begin{array}{ll} 000 & \text{previous state} \\ 111 & \text{next "normal" state} \\ 100 & \text{desired next state} \end{array}$$

The least significant bit (LSB) is not a problem because our truncated synchronous counter design already takes care of this bit. The second bit must be forced to zero since it normally would go to one. Therefore, steering logic is required on the second flip-flop. The third most significant bit (MSB) flip-flop will go to one naturally, so that steering logic is not required for this flip-flop. Only AND gate logic is necessary.

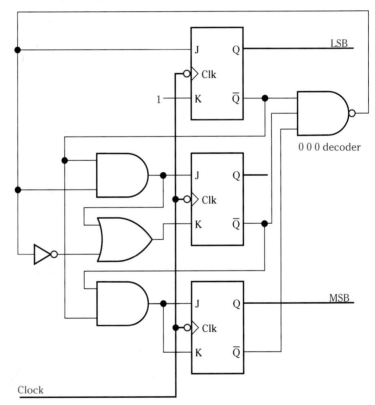

Synchronous Up/Down Counter

It is possible and also very useful to combine the up and down count functions into one counting system. A synchronous up/down counter can be created by devising an AND-OR network that determines whether the Q or $\overline{Q}$ outputs of one flip-flop are routed to the J-K inputs of the following flip-flop in the counter.

A control line is set aside to indicate when up or down counting operation will take place.

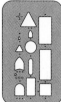

Design a MOD 8 synchronous up/down counter.

Solution Three flip-flops are required along with a control line to distinguish between count up and count down operating modes. AND gates combining Q outputs and AND gates combining $\overline{Q}$ outputs are ORed together to present the correct state to the J-K inputs of each flip-flop. Naturally, all lesser significant flip-flop outputs are ANDed together as well to determine when a toggle should occur.

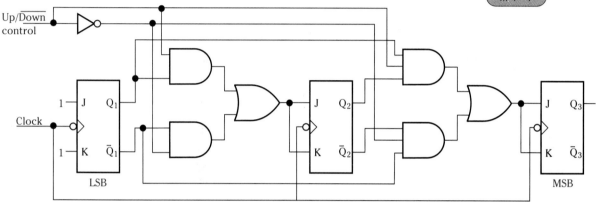

Presettable Counters

Presettable or **programmable** counters are designed so that they may be "pre-loaded" with any number. After preloading the normal count sequence begins at the count of the preloaded number. For instance, assume that a presettable MOD 16 up counter is preloaded with the state 0101. The counter then counts from 0101 to 0110, 0111, 1000, and so on, normally. The counter will roll over to 0000 at the appropriate time. Presettable counters, and other counters, are often used as timing elements since the counting rate is related to clock frequency. Presettable counters can also be used as programmable modulus counters. That is, a general application presettable counter can be wired to operate as a counter of any modulus desired.

Figure 8.20 shows a presettable MOD 8 counter. This particular counter is a synchronous counter, but preloading a count is accomplished in an asynchronous fashion. Since the data input information is presented to the counter stages via the preset and clear inputs, preloading occurs without any synchronization to the clock. This counter has three data input bits, which are connected to the flip-flop preset and clear inputs through NAND gates. For example, if D_0 is high and the data load control line is high, then the D_0 preset NAND gate output goes low and presets the first flip-flop. If D_0 was low, the inverter

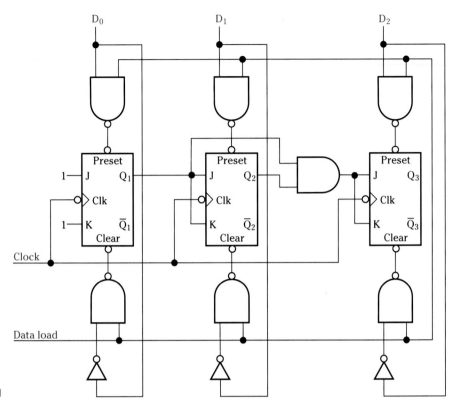

Figure 8.20
Presettable
Synchronous
Counter—
Asynchronous Preload

on the clear input NAND gate allows a clear operation to take place. Once the data load control line is made inactive (low), counting proceeds in a normal fashion. Synchronous presettable counters are available as well. These counters' load operations are synchronized to the counter clock using logic similar to that used to alter the modulus of a counter. That is, preload data is presented to the flip-flop J-K inputs through a logic network and is loaded into the flip-flops on a clock edge. The load network is disabled once preloading is complete and normal counting operation begins.

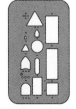

DESIGN
EXAMPLE 8.16

Draw the timing diagram for the following circuit to determine how the MOD 8 presettable counter modulus was modified. Determine the count sequence:

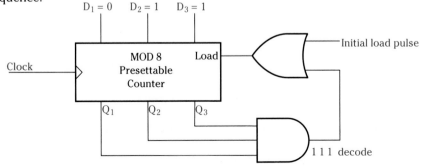

Solution This presettable counter modulus is altered by decoding the 111 state and forcing a preload to the 011 state. Using the asynchronous preload design discussed earlier, the 111 state will not persist for long since this condition will immediately preload the counter with 110. Therefore, the counting sequence will be 011, 100, 101, 110—a MOD 4. If a synchronous preload were used, the decoder signal would not initiate the preload until the following clock edge. That is, state 111 would persist for a full clock cycle. This count sequence would be 011, 100, 101, 110, 111—a MOD 5. The asynchronous preload timing is:

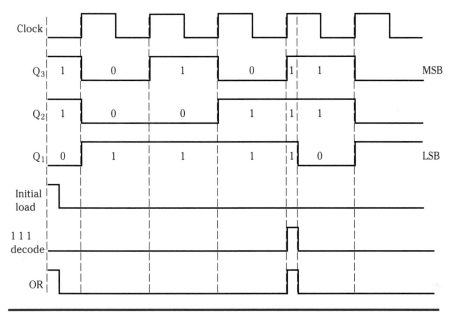

Some Common Counter ICs

A quick glance through any manufacturer's TTL, CMOS, or ECL data book will instantly verify that many of the counter designs we have discussed are readily available in integrated circuit form. You may wonder why we go through all the design details when ready-made counters are so easy to obtain. Good question. First of all, understanding how a counter works gives you plenty of insight into how counters can be used in real applications. You do not learn to use a counter to sequence events in an automated factory simply by knowing that a decade counter can be purchased. You have to understand the operation, advantages, and limitations of counters in order to apply them.

Second, predesigned integrated circuits are somebody else's design and may never meet all your design specifications. Furthermore, modern design is moving more and more toward programmable logic and custom logic. You cannot use off-the-shelf parts for these applications; you design you own. Nevertheless, a familiarity with common TTL, CMOS, and ECL parts is beneficial since they are perfect for many applications, they are excellent learning devices, and many of their part numbers have been adopted for more advanced logic technologies.

Most common IC counters are designed to be flexible. That is, each chip can be wired to provide several functions. A case in point is counters such as the 7490 decade counter, the 7492 divide by 12 counter, and the 7493 4-bit binary counter. Each counter chip is actually multiple counters as Figure 8.21 indicates.

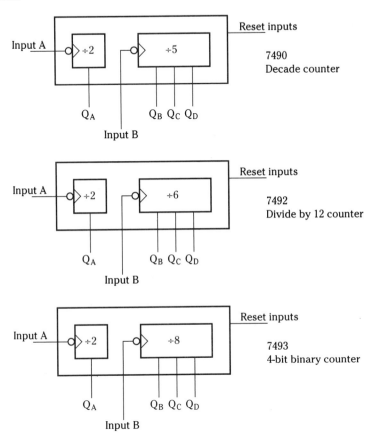

Figure 8.21
Common TTL Counters

The designer can configure the counters to operate any way required. For instance, only a portion of the 7492 needs to be used if a divide by 6 is needed. Multiple reset lines provide some logic design flexibility as well. However, some of this flexibility may be a disadvantage. For instance, any unused gates in the chip will be consuming power from your system power supply—a costly item in any design. Reset pins, even though you may not need them, require connections to appropriate logic levels. The pins require placement on the printed circuit board. These disadvantages are mentioned because they affect your system design and layout even if they do not affect your counter design.

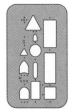

Design a divide by 480 counter using 7490s, 7492s, and 7493s.

Solution   Large divide by n counters can be created by cascading smaller divide by n counters together. Using the divide by 2, 5, 6, and 8 functions available in the 7490, 7492, and 7493 counters, a divide by 480 can be created.

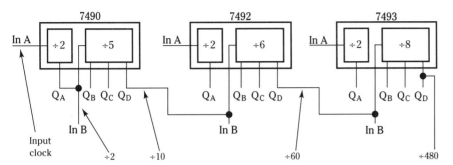

Analyze the application of the following counter.

Solution   The following circuit uses a MOD 4 counter to create a timing pulse equal in duration to four clock pulses after a short "Begin" pulse arrives. Counter operation is inhibited until the Begin pulse arrives. Once Begin arrives, the J-K flip-flop is set on the following clock. Then counting starts on subsequent clocks. When the counter reaches the 11 state, a high is applied to the flip-flop K input (J is already low). The flip-flop resets on the next clock edge and the timing pulse ends. The timing pulse duration is equal to the time the J-K flip-flop remains set.

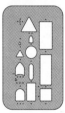

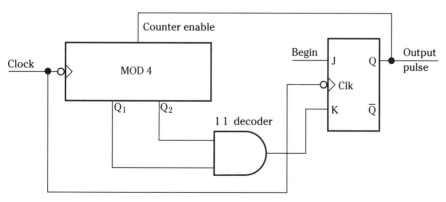

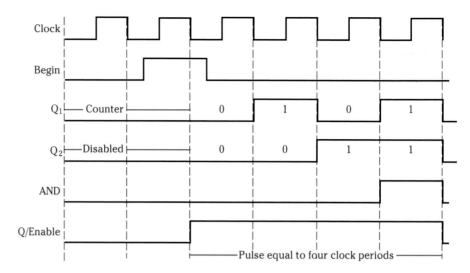

Different pulse lengths can be created using a different modulus counter.

8.7 ■ SHIFT REGISTERS

A group of flip-flops used for a common purpose are often called a **register**. For instance, when an arithmetic calculation takes place in a computer's arithmetic logic unit (ALU), the result must be saved. Each bit of the result is stored in a distinct flip-flop within the register. Since the flip-flops storing the result are all serving a common purpose, it makes sense that they be grouped and designated with a common name. Registers are prevalent in microprocessor and associated peripheral chips.

A certain class of registers called **shift registers** store data and also allow the data to be shifted left or right within the register, 1-bit at a time. Data can also be transferred in **serial** fashion—1-bit at a time—or in **parallel**—multiple bits at a time. The four basic shift register designs, illustrated in Figure 8.22, are defined as follows:

- **Serial-In, Serial-Out (SISO).** Data is presented to the SISO 1-bit at a time, and on each clock pulse data is shifted into the register. One clock pulse is required for each data bit shifted into the register. When the register is full, additional clock pulses shift the data out of the register, also 1-bit at a time.
- **Serial-In, Parallel-Out (SIPO).** Data is loaded into this shift register on a 1-bit per clock pulse basis. When the register is completely loaded, all data bits are immediately available at the register outputs.
- **Parallel-In, Serial-Out (PISO).** Data is loaded into a PISO with a single clock pulse because of the parallel nature of the inputs. Data is clocked out of the register 1-bit at a time.

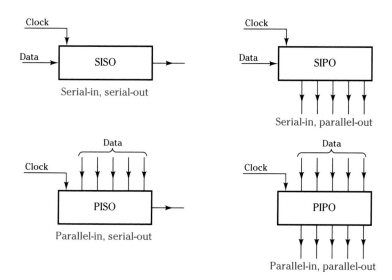

Figure 8.22
Shift Register Designs

■ **Parallel-In, Parallel-Out (PIPO).** Data is loaded and unloaded in a single clock cycle using a PIPO. While data is present in the shift register, it may be shifted left or right with additional clocking signals. A basic computer register performs the same data transfer function as a PIPO except for the shifting capability.

The major differences between serial and parallel shift registers is design complexity and speed. Parallel registers require one wire per data bit to allow data to be presented to or obtained from the shift register hardware. This requirement goes beyond just the shift register design since any other hardware attached to the device will require similar parallel connections. There is a cost penalty for parallel operation as compared to serial operation because additional hardware is required, but enhanced operational speed is the advantage gained. Since all data bits are transferred at once, data transfer operations can occur in one clock cycle regardless of the number of data bits involved. For instance, if 16-bits of data are to be transferred into a register in a serial fashion at 1 μsec per bit, it takes 16 μsec for the complete transfer. The same 16-bits require only a single microsecond when transferred in parallel. Although the costs of parallel connections may not seem obvious, they represent a significant portion of the printed circuit board, assembly, testing, and rework costs incurred by industry. Sometimes the designer has no choice between serial or parallel operation since many applications already dictate the mode of data transfer required. We will examine some shift register applications in the following sections.

Serial Input

Figure 8.23 shows a fundamental shift register design based on D flip-flops. All flip-flops receive a common clock signal and the first flip-flop also receives the input data bit. In this case only 1-bit per clock pulse is entered into the

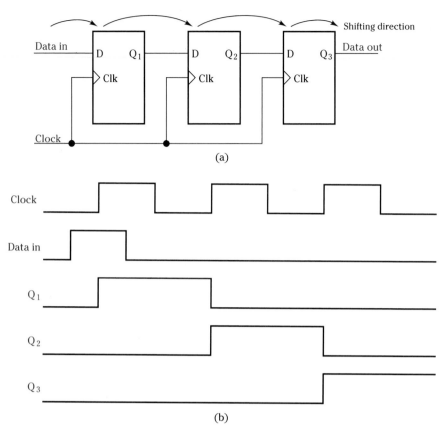

Figure 8.23
(a) D Flip-Flop Shift
Register Design (b)
Shift Register Timing
Waveforms

shift register since it has a serial input. The data present at the D input of the first flip-flop is stored on the first clock pulse. All other flip-flops store the levels present at their inputs—the Q output of adjacent flip-flops. After the first clock pulse the data stored in the first flip-flop is present on the D input of the second flip-flop, and on the second clock pulse the data is transferred into the second flip-flop. As you can see, data shifts 1-bit to the right on each clock pulse. The number of pulses required to shift the data through the register depends on the size or "width" of the register. Furthermore, as the first bit is entered into the register, subsequent bits are applied to the data input line and follow along. The order of the bits and the timing of their presentation to the data input line are important considerations for shift register applications.

DESIGN
EXAMPLE 8.19

Show how a 4-bit SISO shift register can be converted into a 4-bit SIPO shift register. Identify how many clock cycles are required to right shift 1101 into the register; identify how many clock cycles are required to output the data. Which data bit must be entered first?

Solution   Parallel output means that all data bits are available at one time. Therefore, each flip-flop output must provide an output signal. The shift register will then be able to function as a SIPO and a SISO.

Since there are four flip-flops, four clock cycles are required to shift the data into the register. Once entered, all data is immediately available with parallel outputs. Three additional clocks would be required to output the data with a serial output.

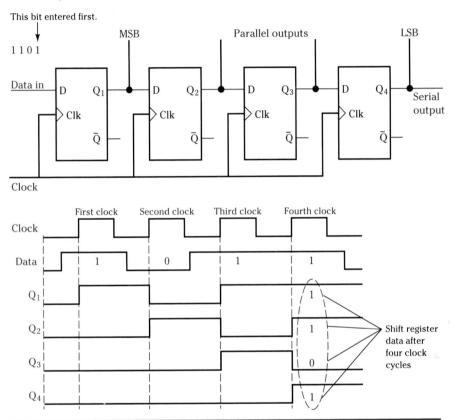

Parallel Input

Shift registers with parallel inputs receive all data bits simultaneously prior to shifting. Parallel loading or **broadside loading** a shift register can be done in two different ways—asynchronously or synchronously. In fact, the loading circuits and methods are quite similar to the way presettable counters are loaded with data. Asynchronous or **jam loading** involves using flip-flop preset and clear inputs to initialize the shift register flip-flop values. Since the flip-flop preset and clear inputs override the flip-flop clocking function, the parallel loading operation is not synchronized to the shift register clock. Figure 8.24 shows a typical method used for asynchronous loading. Initially, every flip-flop is cleared to zero. Each input data bit feeds a NAND gate, which in turn is connected to the flip-flop preset input. When the load enable input is high, any data bits equal to one preset their respective flip-flop through the NAND

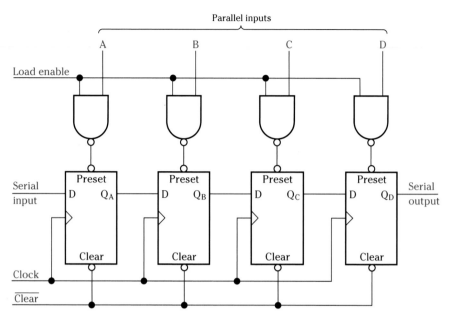

Figure 8.24
Parallel-In, Serial-Out
Shift Register with
Asynchronous Load

Operation:
- Load enable = 0
- Clear = 0
- Clear = 1
- Set data value
- Load enable = 1
- Load enable = 0
- Clock to shift

circuitry. Data input bits at the low level cannot cause a preset to occur; their associated flip-flops are already low because of the initial clear operation. Shifting can occur when both load enable and clear are inactive. Notice that a serial input line is available, allowing this shift register to function as a SISO or a PISO.

Synchronous loading coordinates data transfer to the shift register using the shift register clock. Thus, shift register loading and shifting operations are all keyed to the clock. Figure 8.25 illustrates a synchronous load PISO. An AND-OR network controls each flip-flop D input. When the load/shift line is low, the parallel data input information is presented to each D input. This parallel data is stored in each flip-flop on a clock transition. When load/shift is high, normal right shifting takes place. The presence of a serial input line allows this PISO to also function as an SISO.

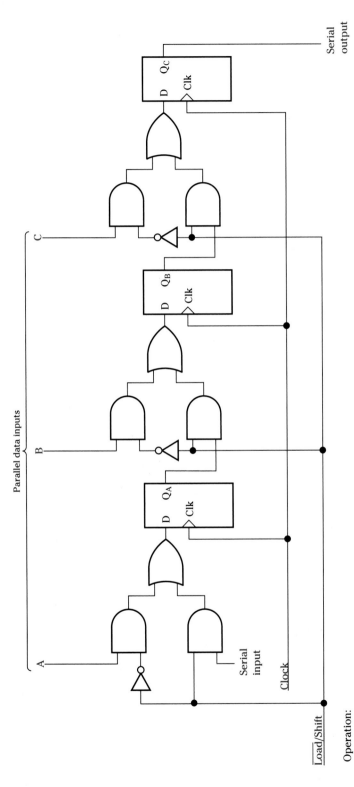

Parallel data inputs

Serial
output

Serial
input

Clock

$\overline{\text{Load}}$/Shift

Operation:
- Set input data
- $\overline{\text{Load}}$/Shift = 0 to enable loading
- Clock
- Load/Shift = 1 to enable shifting
- Clock to shift

Figure 8.25
Parallel-In, Serial-Out
Shift Register with
Synchronous Load

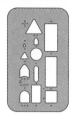

DESIGN
EXAMPLE 8.20

Modify the synchronous load PISO to allow either right or left shifting to take place.

Solution   Right shifting occurs as data moves through the shift register from left to right. Therefore, left shifting is the opposite operation—movement of data from right to left. Modifying the AND-OR network to include the necessary data path for left shifting is possible provided that an additional right/left control line is included to determine the shifting direction. A left shift and a right shift data input will also be needed.

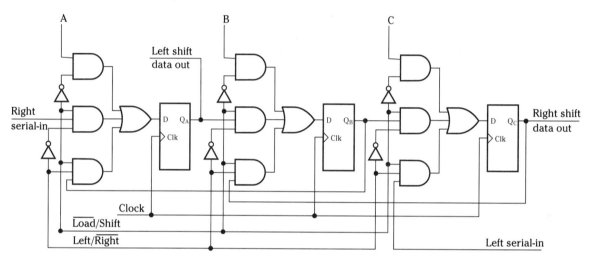

Shift Register Applications

The most typical application for a shift register is converting serial data into parallel data or vice versa. For example, computer systems process information in a parallel form because of the speed advantages of the parallel format. However, when one computer system requires information from another computer system, a serial transfer is typically used for reasons of economy. When information is transferred using a **modem** (modulate-demodulate), serial information transfer is a necessity because the information is transmitted along a telephone wire, a distinctly serial medium. Data sent between computers in a **local area network (LAN)** are also sent in serial form even though the computers may be relatively close to each other. The cost of installing parallel wiring in a building is prohibitive.

Shift registers are important elements in many complex integrated circuits. Specialty devices known as an **UART** (universal asynchronous receiver transmitter) or an **USART** (universal synchronous/asynchronous receiver transmitter) employ parallel-to-serial and serial-to-parallel shift registers to convert data from one form to another. The UART and USART also provide the circuitry to support data transmission for a range of speeds and formats. The shift register makes the basic transmission process possible.

Shift registers are also used to provide mathematical capability. To illustrate a simple example, shift the bits in the number 0110 one position to the left. The result is 1100. The left shift operation provides multiplication by 2 since the number changed from a binary 6 to a binary 12. Shifting to the right will divide a number by 2. More complex multiplication requires a combination of shifting and addition, with the shift register playing an important part in this kind of high speed math hardware.

Explain the operation of the shift registers in the following serial adder:

DESIGN
EXAMPLE 8.21

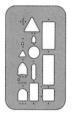

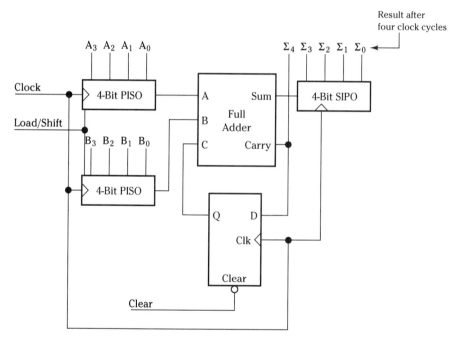

Solution The two PISOs are loaded with the numbers to be added. Initially, the LSBs are added by the full adder. On the first clock pulse the sum of the two LSBs is stored in the SIPO, whereas the carry is stored in the D flip-flop (cleared to start). On each succeeding clock pulse another set of bits is added with the previous carry and then stored in the SIPO on the following clock. After four clock cycles the MSB of the result is at the carry output. The SIPO contains SUM 3 through SUM 0.

8.8 ■ RING COUNTERS

Figure 8.26 shows two interesting applications for shift registers. When the last flip-flop output is fed back to the first flip-flop input, a **ring counter** is created. Any information within the ring counter is circulated throughout the device. A **Johnson counter** is similar to the ring counter except that the last flip-flop $\overline{Q}$ output is fed back to the input.

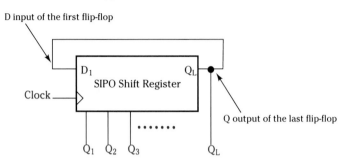

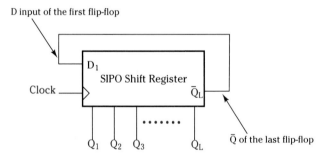

Figure 8.26
(a) Ring Counter Block Diagram (b) Johnson Counter Block Diagram

Using a ring counter, the last flip-flop Q output feeds the input of the first flip-flop. On every clock pulse the ring counter data is shifted one position to the right. After a series of clock pulses the data reaches the last flip-flop in the ring and is recirculated through the ring counter on subsequent clock pulses. Typically, a data value with only a single high bit, such as 1000, is parallel loaded into the register before clocking begins. As the information is circulated through the ring counter, each flip-flop receives the high bit once for every n clock pulses, where n equals the number of flip-flops. For example, using a 4-bit ring counter, each flip-flop is high once for every four clock pulses. Since the data pattern in each flip-flop is predictable and related to the clock frequency, ring counters are often used for the generation of system timing signals as Figure 8.27 illustrates.

The ring counter timing diagram may look familiar because it can also be generated by decoding every state of a continuously clocked 2-bit counter. That is, if a 2-to-4 line decoder is attached to a 2-bit counter, the same set of timing pulses with the same timing relationship will be created. However, the

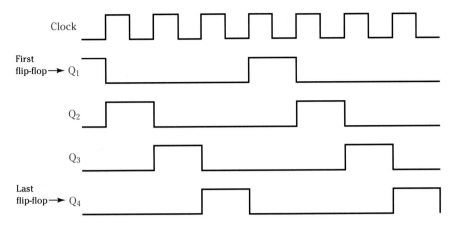

Figure 8.27
4-Bit Ring Counter
Timing Diagram

ring counter does not require any decoding circuits because the number of states in the ring counter equals the number of flip-flops. This is not an efficient use of flip-flops because the ring counter requires more than an equivalent counter, but it can be useful nonetheless since eliminating the decoders eliminates the delays associated with the decoding circuits.

Design a 5-bit ring counter using D flip-flops. Show how the counter can be preloaded with 10000.

**DESIGN
EXAMPLE 8.22**

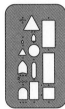

Solution The ring counter design can be extended by adding additional flip-flops to the basic design. Using preset and clear inputs, we can load the initial starting value.

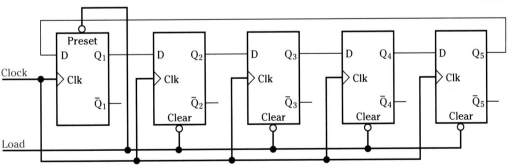

The Johnson counter provides a number of states or a modulus equal to two times the number of flip-flops, doubling the number of states possible with a ring counter. For instance, a six flip-flop Johnson counter design has 12 distinct states (MOD 12). However, to obtain all possible Johnson counter output states for timing purposes, we need decoders. Fortunately, the decoders do not have to decode each state completely. For the 4-bit Johnson counter timing illustrated in Figure 8.28 and accompanying circuit in Figure 8.29 only 2-input AND gates are necessary.

The Johnson counter is initially placed in the all zeros state by a clear operation. No other loading of data is required since the $\overline{Q}$ output of the last flip-flop will provide high bits for the shifting pattern. The all zeros condition can be decoded by ANDing $\overline{Q}_1$ and $\overline{Q}_4$ since these two outputs are only high simultaneously when the all zeros state exists. Similarly, the all ones state is detected by decoding Q_1 AND Q_4. All other Johnson counter states are detected by ANDing the output of a flip-flop with the complement of the next flip-flop in the counter or by decoding the complement of a flip-flop and the true output of the next flip-flop. For instance, the state $Q_1, Q_2, Q_3, Q_4 = 1100$ is detected by ANDing Q_2 with $\overline{Q}_3$. State $Q_1, Q_2, Q_3, Q_4 = 0011$ is detected by ANDing $\overline{Q}_2$ with Q_3. The chart in Figure 8.29 defines all the AND gate inputs required for full decoding.

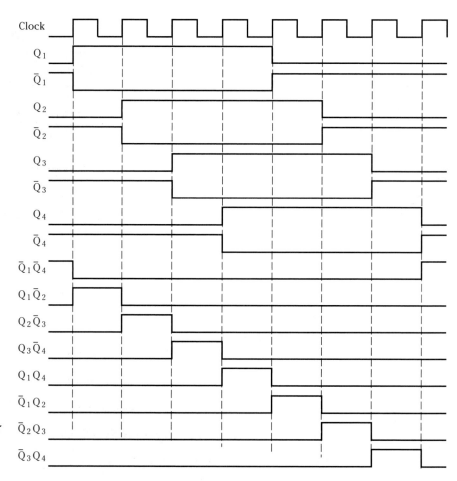

Figure 8.28
4-Bit Johnson Counter
and Decoder Timing
Diagram

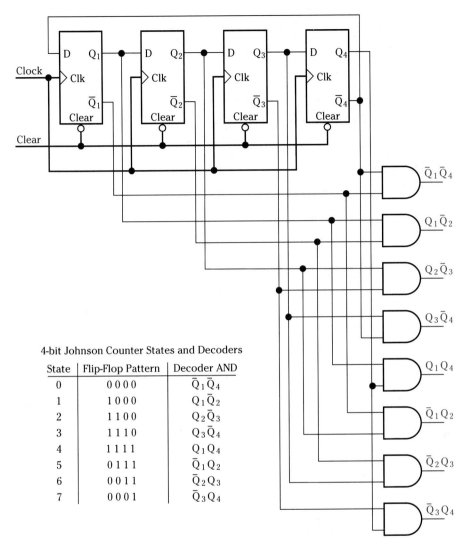

4-bit Johnson Counter States and Decoders

State	Flip-Flop Pattern	Decoder AND
0	0 0 0 0	$\bar{Q}_1\bar{Q}_4$
1	1 0 0 0	$Q_1\bar{Q}_2$
2	1 1 0 0	$Q_2\bar{Q}_3$
3	1 1 1 0	$Q_3\bar{Q}_4$
4	1 1 1 1	Q_1Q_4
5	0 1 1 1	$\bar{Q}_1Q_2$
6	0 0 1 1	$\bar{Q}_2Q_3$
7	0 0 0 1	$\bar{Q}_3Q_4$

Figure 8.29
4-Bit Johnson Counter
Circuit Design

SUMMARY

- Digital counters may be broadly classified as asynchronous or synchronous.

- The various stages in an asynchronous counter change at differing times. Although simple to design, counting speed decreases as the counter length increases.

- Synchronous counters are generally more complex than asynchronous counters, but they are much faster because every counter stage is clocked simultaneously.

- A truncated asynchronous counter sequence is easily designed by modifying a basic counter using the decode and reset method.

■ A truncated synchronous counter sequence may be designed using the decode and steer method.

■ The modulus of a counter refers to the number of states through which the counter progresses.

■ Shift registers are groups of flip-flops used to move data 1-bit at a time from right to left or left to right. Four basic designs exist: serial-in, serial-out; serial-in, parallel-out; parallel-in, serial-out; and parallel-in, parallel-out.

■ Simple modifications made to a basic shift register design can convert the circuit into a ring counter or a Johnson counter.

PROBLEMS

Section 8.2

1. Design a MOD 32 asynchronous counter out of J-K flip-flops. Include provisions for an asynchronous manual reset and manual preload to state 01010. Draw the output waveforms for all flip-flops and their relationship to the input clock.

*2. Design a MOD 4 asynchronous counter using D flip-flops. Assume that each flip-flop has a 15 nsec propagation delay. Draw the output waveforms for the counter based on this delay.

*3. How many flip-flops are required to build an asynchronous counter with divide by 512 capability?

4. Is there any difference in the output signal of a MOD 16 asynchronous counter using clock waveform a or b shown here? Why?

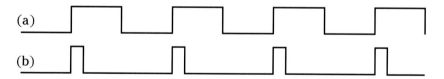

(a)

(b)

Sections 8.3—8.4

5. Design a MOD 8 asynchronous counter with individual decoders that decode states 010, 100, and 110.

*6. Draw the timing waveforms for the counter and decoder designed in problem 5.

*7. What is the modulus of a binary counter created with ten flip-flops?

8. Design a MOD 8 asynchronous counter with a single decoder that decodes for states 100, 101, and 110. Draw the timing diagram for the counter and decoder outputs.

---

* See Appendix F: Answers to Selected Problems.

9. (a) What are the output frequencies at each stage in the following circuit?

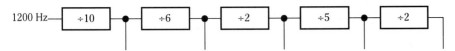

1200 Hz ——— ÷10 ——— ÷6 ——— ÷2 ——— ÷5 ——— ÷2

*(b) What is the frequency division at each point in the circuit with respect to the original input frequency?

10. Demonstrate when glitches occur for a 3-bit asynchronous counter decoding state 110.

*11. Design a self-stopping counter to stop at the count of 10. Include the provisions for a reset.

12. Design a counting system to count the number of balls allowed per player in a pinball game. The system should allow five balls to be played. After five are used the system is not allowed to operate further. A signal should also be provided to warn the player that he is using the last ball. Assume that a signal is available in the system every time a ball is taken out of play.

13. Design a MOD 32 asynchronous down counter using D flip-flops.

*14. Design a MOD 6 asynchronous counter using the detect and reset method.

15. Draw ideal waveforms for the MOD 6 counter designed in problem 14. Assume that the clock is a perfect square wave (50% duty cycle). How will the duty cycle of the individual flip-flop outputs change with respect to the clock duty cycle. What is the frequency division ratio of each flip-flop output as compared to the clock frequency if the clock frequency is 5.1 MHz?

*16. Draw the waveforms for the MOD 6 counter designed in problem 14 to determine when glitches occur on the flip-flop output lines. Assume that the clock rate is 10 MHz and that each flip-flop and logic gate has a 10 nsec propagation delay time.

17. Add decoders to the MOD 6 counter from problem 14 to decode each state of the counter. Use the flip-flop output waveforms found in problem 16 to determine when the 100 decoder output will create glitches.

18. What is the maximum clocking rate for the following ripple counter designs?
    *(a) MOD 24—14 nsec flip-flops and 10 nsec decoding gates
    (b) Divide by 12—8 nsec flip-flops
    (c) MOD 1024—5 nsec flip-flops

19. Design a MOD 32 asynchronous counter with decoders for states 00101, 01001, 01101, 10010, and 11011. Using a clock frequency of 3 MHz, also design a decoder strobe signal using a 74121 one-shot to account for the 50 nsec ripple counter delay. Draw a portion of the timing diagram for the 01001 through 01101 states to illustrate the strobing operation.

_____

* See Appendix F: Answers to Selected Problems.

Sections 8.5–8.6

20. Design a MOD 32 synchronous counter. Draw a timing diagram showing the flip-flop outputs and the outputs of the controlling AND gates for the first ten clock pulses starting at a counter state of 01100.

*21. Design a MOD 12 synchronous counter using the decode and steer method. Eliminate steering logic wherever possible.

22. Draw the timing diagrams for the MOD 12 counter designed in problem 21. Compare the clock to the flip-flop outputs as well as to the steering logic AND and OR outputs.

*23. Design a divide by 1000 counter by cascading decade counters together. How many flip-flops are required for this approach? How many flip-flops are required to design the counter directly as a MOD 1000?

24. Design a synchronous MOD 32 down counter. Compare timing waveforms for the clock to the flip-flop outputs as well as to the controlling AND gate outputs for states 10100 through 01100.

*25. If a MOD 64 down counter is continuously clocked at a 2.5 MHz rate, how often will a decoder decoding the all zeros state produce a pulse? How often will a decoder decoding 101101 produce a pulse?

26. Design a MOD 13 synchronous down counter using J-K flip-flops.

27. Design a MOD 32 synchronous up/down counter using J-K flip-flops.

28. Design a synchronous presettable MOD 16 counter. Design the preloading logic so that preloading is a synchronous operation. A control line may be used to distinguish between preload and normal counting operations.

29. Modify a MOD 8 presettable counter to function as a MOD 6 with the counting sequence 010, 011, 100, 101, 110, 111. Assume asynchronous preloading such as that used in Design Example 8.16.

30. Determine how a 74160 presettable BCD decade counter can be preloaded with 0110. Determine how a reset can take place. What happens to the counter if it is preloaded with an illegal BCD state?

*31. The following divide by 3 counter was designed using state machine techniques, the subject of Chapter 13. Starting with the counter in the clear state, analyze and draw the timing diagram for the flip-flop outputs to verify operation.

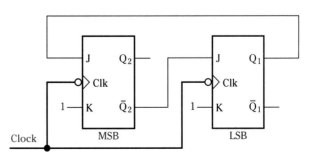

---

* See Appendix F: Answers to Selected Problems.

32. Using the pulse generator circuit illustrated in Design Example 8.18, design a timing circuit capable of creating a pulse equal in duration to five clock periods. Show how the counter enable circuitry is created.

33. Using 7490s, 7492s, and 7493s create the following divide by n counters:
    *(a) n = 6400   (b) n = 14000   (c) n = 40960   (d) n = 10000

34. Using block diagrams, show how a BCD keyboard could be used to load a value into a synchronous down counter. A load-run switch should be included. When low, the keyboard data is loaded into the counter. When high, keyboard operation is inhibited and the counter counts down.

35. Modify the circuit in problem 34 to include a seven segment display and also a zero state decoder, which will inhibit count operation once the zero state is reached.

36. Using 7490, 7492, and 7493 counter functions, show (using block diagrams) how the seconds, minutes, and hours circuitry of a digital clock are created. Also include seven segment displays as part of the circuitry. Assume that a one pulse per second clock is available. (*Hint:* Frequency division can be accomplished by cascading counters together.)

Section 8.7

37. How many clock pulses are required to load the data pattern 11001101 into the following shift registers? How many pulses are required to unload the information?
    *(a) SISO   (b) PIPO   (c) SIPO   (d) PISO

38. Draw the timing diagram for a 4-bit SIPO shift register right shifting the data pattern 0110. Assume the register is cleared to start.

*39. Design a 6-bit SIPO shift register using J-K flip-flops. Include a reset capability.

40. Design a 5-bit asynchronous loading PIPO shift register. The load should be able to take place whether or not a reset has occurred. Use J-K flip-flops in the design.

*41. Modify the PIPO designed in problem 40 to allow synchronous loading of the data.

42. Design a 4-bit SISO left/right shifting shift register. Include all necessary control lines for proper operation.

*43. Using block diagrams, show how data may be transferred back and forth from a parallel data system across a serial line to an adjoining parallel data system. Determine how a common clock can be used. Include any other control lines that may be necessary.

Section 8.8

44. Draw the timing diagram for the 5-bit ring counter in Design Example 8.22. Illustrate the timing pattern for each flip-flop Q output over ten clock periods.

---

* See Appendix F: Answers to Selected Problems.

*45. How many states are possible with ring counters constructed with 10, 12, and 18 flip-flops? How many states would a Johnson counter provide if built with the same number of flip-flops?

*46. Design a 5-bit Johnson counter using J-K flip-flops. Draw the timing diagram of each flip-flop Q and $\overline{Q}$ output.

47. Using the 5-bit Johnson counter design from problem 46, design the decoding network to decode completely all ten counter states.

48. Draw the decoder output timing diagram utilizing the information from problems 46 and 47.

CHAPTER 9

MEMORY

OBJECTIVES

To explain how a bus structure allows multiple devices to be connected together and to discuss associated interface chips used for this purpose.

To describe the differences and similarities between RAM and ROM.

To illustrate and define memory read and write cycles.

To illustrate and define the concepts of addressing and word size.

To explain the operating principles of static and dynamic RAM.

To explain the operating principles of EPROMs and PROMs.

To explore how memory system address space and word size can be enlarged.

To investigate innovations in semiconductor memory technology.

To explain the purpose of semiconductor memory testing.

PREVIEW

Memory is an important element in many digital systems. From small memories storing a few bytes of data to large, mass storage systems storing megabyte quantities, there is a memory technology available to meet most needs. This chapter explores the semiconductor memory devices commonly used in modern digital systems and discusses how they function and how they fit into a logical memory structure.

9.1 ■ THE NEED FOR MEMORY

We have seen a need for basic storage devices such as flip-flops and registers in practical digital hardware. Naturally, storage capability is required in larger quantities beyond that provided by the flip-flop and register to store programs, tables of information, word-processing files, and so on. Most people are generally familiar with the memory requirements of personal computers, although they might be surprised to learn how many kinds of memory are actually used. The "main storage" or "main memory" in a personal computer consists of memory chips that are capable of accessing stored data in a relatively short time. These chips must be **dense**, that is, large quantities of storage per chip, to minimize the number of chips required. A personal computer also has some memory that is **nonvolatile**. This means that data remains within the memory storage device even when electrical power is removed. Memory devices exhibiting this property are used for computer or digital system start-up and initialization. For instance, when a computer system is first turned on, the nonvolatile memory is the first portion of the overall memory system accessed. Flip-flops and registers can be set to meaningful levels from the information stored in the nonvolatile memory so that additional processing may begin.

Other lesser known memory "subsystems" make up a portion of personal computer storage. For instance, internal to the **microprocessor** chip controlling the entire system there is memory containing **microcode**. Microcode programs determine how the computer functions at the logic level, thereby controlling the sequence of events occurring as the microprocessor processes information. The memory holding the microcode must be very fast to maintain high performance. The microprocessor may also have very fast **scratchpad memory** assisting with high speed calculations. Larger computer systems have a **cache**, **DLAT (Directory Look Aside Table)**, and other associated hardware to increase the operational speed and size of the memory system. Peripheral chips, such as floppy disk controllers, CRT display controllers, and timers have their own internal memory requirements as well. Taking into account nonsemiconductor storage devices such as floppy disks, hard disks, tape, and CD disks, it is apparent that memory requirements are extensive. Many of these same memory components are necessary elements in digital systems not characterized as personal computers since storage is a fundamental requirement for most systems.

Since this chapter is concerned with semiconductor memory devices, we must first study the electrical/logical methods used to connect memory chips to other memory chips and other digital system components. Usually, a single memory chip is not sufficient to meet the storage requirements of a digital system. Since more than one chip will be used, some provisions must be made so that all the different "physical" memory devices appear as a single "logical" memory system. This is typically accomplished utilizing a **system bus** architecture that allows multiple devices to share common wires. The bus structure is set up so that only one memory chip is operational at a time since it makes no sense to have multiple chips providing information to one destination simultaneously. Although many bus architectures are in use, they share similar electrical/logical properties. The **tristate** or **three-state** bus is the most common interconnection.

9.2 ■ TRISTATE DEVICES

Tristate devices have three possible logical outputs—high, low, and Hi-Z. Electrically, this implies that a tristate device can produce a high level or a low level output voltage such as normal logic gates. In the Hi-Z condition the output looks electrically like an extremely high resistance—an open circuit. This means that tristate outputs can be wired together because a tristate device placed into the Hi-Z state will not be electrically connected to any other circuitry. As long as only one tristate bus is active (high or low output) at a time, many devices can share a common wire. A connection of this type is called a **bus.**

Figure 9.1 shows a typical tristate inverter. Compared to the "normal" inverter symbol, an additional control line labeled "enable" is shown. When

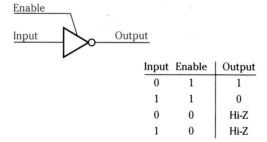

Input	Enable	Output
0	1	1
1	1	0
0	0	Hi-Z
1	0	Hi-Z

Figure 9.1
Tristate Inverter

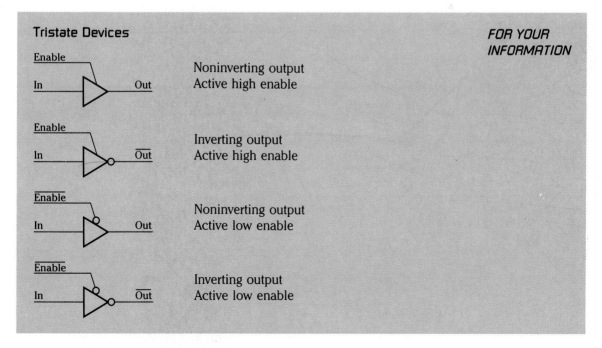

Tristate Devices

FOR YOUR
INFORMATION

Noninverting output
Active high enable

Inverting output
Active high enable

Noninverting output
Active low enable

Inverting output
Active low enable

enable is high, the inverter functions as a typical inverter, but when enable is low, the device output is placed into the Hi-Z condition. Logic levels at the device input are ignored under this condition. Many basic logic gates, flip-flops, memory chips, and microprocessor related chips are available with tristate outputs.

Figure 9.2 illustrates how tristate devices are wired to form a common bus. All outputs are connected together, but to avoid electrical problems, only one device can be enabled at a time. In other words, all devices except one must be in the Hi-Z condition at any time. However, it is all right to have all devices in the Hi-Z state at the same time, provided logical problems are not created, since the bus will not have any valid logic level on it under this condition. When every device connected to a bus is in the Hi-Z state, the bus is said to be "floating."

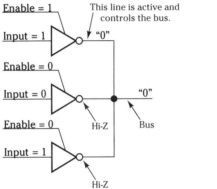
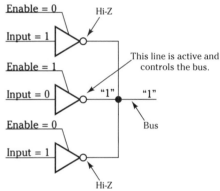

Figure 9.2
How a Tristate Bus Is Controlled

DESIGN EXAMPLE 9.1

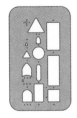

Draw the timing diagram for the following tristate inverter:

Device functions as an inverter except when disabled.

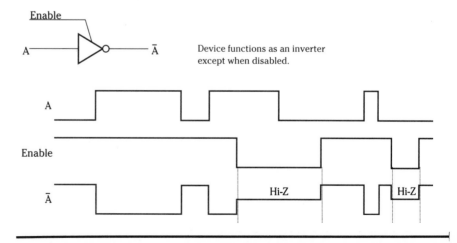

Show how eight D flip-flops can be connected to function as a tristate register.

Solution A register is a group of flip-flops acting together for a common purpose. This implies that all eight flip-flops will be clocked simultaneously. Connecting each flip-flop $\overline{Q}$ to a tristate inverter will provide tristate capability. One enable control line can control the tristate operation of the register.

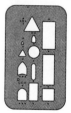

DESIGN EXAMPLE 9.2

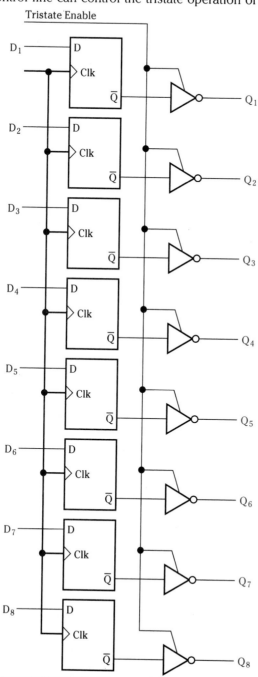

9.3 ■ MEMORY CLASSIFICATION

Memory devices fit into one of two broad classifications, although the distinction between the two blurs with ongoing technological advances in memory device fabrication. **RAM** or **random access memory** characterizes memory devices or systems that access any of the stored information in the same amount of time. This accessibility can be better appreciated by considering information stored on magnetic tape. Information stored at the end of a tape cannot be obtained as quickly as information stored at the beginning because of the serial nature of the storage. Random access devices can retrieve or store any piece of information in the same amount of time regardless of the internal location of the information.

ROM or **read only memory** devices have the provision to retrieve information, but generally do not allow information to be stored within the device, other than for initial loading of information. ROM devices are **nonvolatile storage** devices and are very useful for storing constant data. Although the two memory classifications make it appear that only RAM data can be accessed randomly, this is not the case. Data in semiconductor ROM chips are also accessed in the same amount of time, regardless of their storage location.

9.4 ■ MEMORY TERMINOLOGY

Several points mentioned earlier can be formally defined as they apply to memory components. Storing information within a memory chip is called a **write operation.** Under these circumstances information is placed into a storage area on the chip. Generally, any previously written information residing at the same location is replaced by the new information.

A **read operation** occurs when data is retrieved from a memory device. A read operation is usually nondestructive; that is, only a copy of the information is retrieved from the memory.

Both read and write operations require a certain amount of time to take place. This is referred to as the **access time** and is roughly analogous to propagation delay time. When a memory device is commanded to read or write, the operation is not valid until the device access time has transpired.

Internally, a memory device is an arrangement of **storage cells**, with each cell storing 1-bit of data. Some memory chips access data on a cell basis. This means 1-bit of information is read or written at a time. Other chips organize cells into nibble or byte quantities so that a group of data bits is read or written at once. In all cases a mechanism exists to point to the particular data of interest. This is called **memory addressing**. Each group of data has its own address within the chip, similar to each house on a city block having its own unique address. Information written to a certain address can be read back at a later time by referring to the same address.

From an organizational point of view a memory chip is a uniform array of storage cells arranged in rows and columns as shown in Figure 9.3. The number

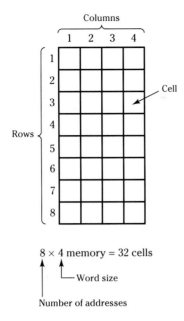

Columns

Rows

8 × 4 memory = 32 cells

Word size

Number of addresses

Figure 9.3
Memory Organization

of rows determines the number of addresses, whereas the number of columns determines how many bits of information are read or written at a time. The number of bits accessed at a time is also referred to as the memory **word size**. In this example the word size is 4-bits and the number of addresses is eight; thus, this memory is referred to as an 8 × 4 memory. Of course, real memory chips are more dense than this. Typical memory sizes are 1024 × 4 or 1K × 4, 65,536 × 1 or 64K × 1, and 262,144 × 4 or 256K × 4. [In the computer field the kilo prefix (K) is equal to 1024 or 2^{10}. Therefore, 256K = 256*1024 = 262,144.]

The number of addresses in a memory chip is related to the size of the memory array. Each group of bits accessed by the chip requires a unique address, implying that each group of bits is individually controlled by addressing circuitry. In a chip with 1K addresses, this means 1024 addressing wires are needed to access the individual data groups. Since this is an inordinate amount of wiring, even for a relatively small memory, the bulk of the wiring is done on the integrated circuit (IC) itself. Figure 9.4 illustrates how an internal **address decoder** selects the memory cells and provides external encoded address lines for system use. For instance, the 1K × 4 chip illustrated requires ten address lines (A_0–A_9) since 2^{10} = 1024. An external address of A_0–A_9 = 0000000000 addresses the first storage address on the chip; A_0–A_9 = 0000000001 addresses the second chip address; A_0–A_9 = 1111111111 refers to the last chip address (1023).

In addition to addressing requirements, other control lines are necessary to distinguish between read and write operations and to allow for memory system expansion. Naturally, data input and output lines are also needed. We will examine the purpose of these lines in the following sections.

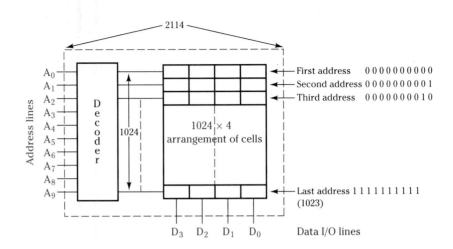

Figure 9.4
Logical Operation of
Memory Address
Lines

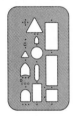

DESIGN
EXAMPLE 9.3

How many address lines are required for the following memory devices; what is the word size of each?

(a) 64K × 1   (b) 512 × 9   (c) 1024K × 1   (d) 16K × 4

Solution   The first number represents the number of addresses (remember, K = 1024). The number of address lines present on the chip is found by 2^n = addresses, where n = number of address lines. The second number refers to the memory chip word size.

(a) 64K × 1—64K or 65,536 addresses require 16 address lines. The word size is 1-bit.
(b) 512 × 9—512 addresses require 9 address lines. The word size is 9-bits.
(c) 1024K × 1—1024K or 1 meg or 1,048,576 addresses require 20 address lines. The word size is 1-bit.
(d) 16K × 4—16K or 16,384 addresses require 14 addresses lines. The word size is 4-bits.

9.5 ■ RANDOM ACCESS MEMORY DEVICES

Static RAM memory chips are one of two basic RAM memory technologies. Static RAMs are volatile memory devices, but they operate at relatively fast speeds. The density of a static RAM is not so great as other chips; however, the static RAM speed, combined with its straightforward circuit interface, makes it an ideal device for high speed, low density applications.

Dynamic RAM's primary advantage for memory system design is its density. One megabyte dynamic RAMs are common with current technology. Comparing this to maximum static RAM densities of 64K to 256K, you can understand why dynamic RAM chips are the mainstay of large memory system designs.

Static RAM

Figure 9.5 shows the logic symbol for a typical static RAM device, the 2114 1024×4 static RAM. (The complete data sheet is shown in the back of the textbook.) The 1K chip address space necessitates ten address lines shown as inputs A_0 through A_9. The 4-bit word size requires four lines for data input and four lines for data output. However, data input and output functions are combined into the four lines listed as I/O_1 through I/O_4. These data lines are **bidirectional**, meaning that data can flow in either direction. Naturally, if a read operation is taking place, data appears on the I/O (input/output) lines from the 2114; during a write operation, data is presented to the I/O lines from an external source. The I/O lines are also tristate lines. When the chip is not involved in a read or write operation, the I/O lines are in the Hi-Z condition. This takes the 2114 off the system bus electrically so that another device may use the bus.

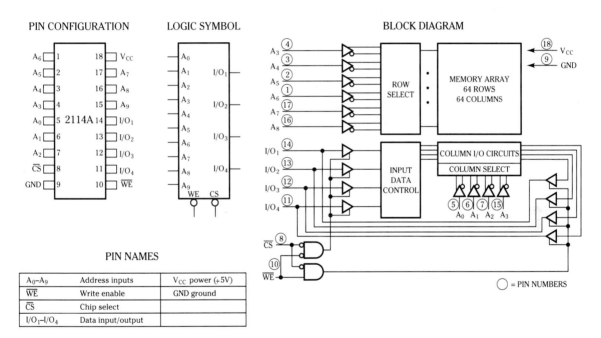

PIN CONFIGURATION   LOGIC SYMBOL   BLOCK DIAGRAM

PIN NAMES

A_0–A_9	Address inputs	V_{CC} power (+5V)
$\overline{WE}$	Write enable	GND ground
$\overline{CS}$	Chip select	
I/O_1–I/O_4	Data input/output	

Figure 9.5
2114 Data Sheet

Two additional control lines affect the operation of the 2114. The write enable line ($\overline{WE}$) distinguishes between read and write operations. When low, the chip is set for writing; a high on this line places the chip in read mode. The actual read or write operation does not take place until the chip is selected for operation by the chip select ($\overline{CS}$) line. Chip select is essentially a chip enable line. When it is high, read and write operations are prohibited and the I/O lines are in a Hi-Z state. Chip select is often used as an address line when more than one memory chip is used and is the control signal determining whether the chip has access to the system bus. In addition, much of the chip timing is based on an active chip select signal.

Reading or writing operations occur as a predefined sequence of events. Address, data, and control lines must have their levels stable in a prescribed timing relationship to one another for proper operation. Furthermore, inadvertently changing levels on any line may prevent a read or write operation from taking place, or worse, corrupt data stored in the chip.

The basic sequence of events for a 2114 read operation are:

- Set the write enable line high, indicating read mode.
- Set the memory address lines to point to the location where the data is stored.
- Bring the chip select line low. Shortly after this occurs the addressed data will appear on the data I/O lines.
- Bring the chip select line high to deactivate the chip and return the I/O lines to a Hi-Z condition.

The basic sequence of events for a 2114 write operation are:

- Begin with write enable high. Holding this line high before the address lines are changed prevents an accidental write operation. This is not necessary for all memory chips, but it is recommended for the 2114.
- Place the data destined for storage onto the I/O lines.
- Set the address lines to point to the location where the data will be stored.
- Bring the write enable line low, enabling the chip for a write operation.
- Bring chip select low to initiate the write. Shortly after activating this line data is stored in the addressed location.
- Remove the active chip select signal to disable the chip.

The preceding sequence of events is similar for most memory chips. The manufacturers' data sheets for the memory device must be utilized for precise information on a particular chip's operation.

DESIGN EXAMPLE 9.4

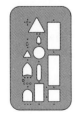

A 2114 memory chip has the following signals present. What operations are taking place?
(a) A_9-A_0 = 1011101101, $I/O_4-I/O_1$ = 1011, $\overline{CS}$ = 0, $\overline{WE}$ = 0
(b) A_9-A_0 = 0010100011, $I/O_4-I/O_1$ = 1100, $\overline{CS}$ = 0, $\overline{WE}$ = 1

Solution   Chip select and write enable determine which operation takes place. The address and data lines determine the address location and data values.

(a) Write operation since both $\overline{CS}$ and $\overline{WE}$ are low. The data 1011 will be stored in address 1011101101.
(b) Read operation since $\overline{CS}$ = 0 and $\overline{WE}$ = 1. The data 1100 is being read from address 0010100011.

The timing diagrams for 2114 read and write cycles are shown in Figure 9.6 and provide more details on the exact timing relationship for the previously described sequence of events.

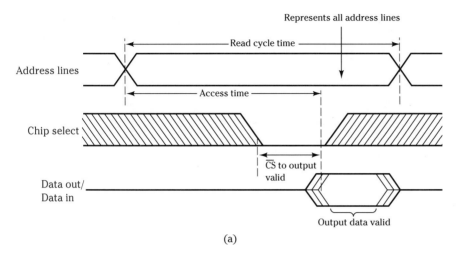

(a)

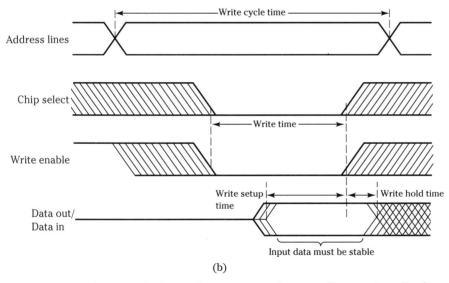

(b)

Figure 9.6
(a) 2114 Read Cycle (WE Is High)
(b) 2114 Write Cycle

The timing diagrams indicate the sequence of events discussed earlier for 2114 read and write operations. In both diagrams the overall length of time for the **read cycle** and the **write cycle** is determined by the time it takes to change from one address to another. This time is at least as long as the chip access time but may be longer. For a read operation, access time is the time that elapses between a valid memory address signal and the appearance of stable data on the I/O lines. For a write operation, the write time determines the minimum amount of time required for a reliable write to take place and is

Waveform Notation for Timing Diagrams

WAVEFORM	INPUT MEANING	OUTPUT MEANING
	High or low level must be stable; can represent several lines	High or low level will be stable
	Don't care. Okay to change input level	Output unknown; signal is changing
	Input permitted to change from high to low	Output is changing from high to low during the time interval specified
	Input permitted to change from low to high	Output is changing from low to high during the time interval specified
	Does not apply	Center line represents the Hi-Z condition

dictated by the active chip select signal. Naturally, the memory chip timing values and the timing considerations of the system in which it will be used should be comparable. Using very fast memory chips with a slow digital system is uneconomical. On the other hand, slow memory chips degrade the performance of fast digital systems. Determining the best trade-offs among speed, power, and density is a dilemma constantly facing the digital designer.

DESIGN
EXAMPLE 9.5

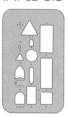

Using the pin diagram for the CY7C164/CY7C166-25 static RAM (pin diagram given; additional information in the appendix), determine the following:

The active power
The standby power
The number of address lines
The purpose of $\overline{OE}$
The read access time

Solution

- **The active power**—read directly from the spec sheet as 385 mW. This is the power consumed during normal operation.
- **The standby power**—the power consumed by the chip when chip select is inactive is 110 mW.
- **The number of address lines**—16K addresses require 14 address lines $2^{14} = 16384$. The address lines are listed as A_0 through A_{13}.
- **The purpose of $\overline{OE}$.** Output enable places data on the I/O lines during a read cycle. This information is found in the 7C166 truth table and timing diagrams.

■ The read access time—depending on the mode of operation, two access times, both 25 nsec, are given: t_{AA} is the address to data valid time for continuously selected operation; t_{ACE} is the $\overline{CS}$ to valid data time.

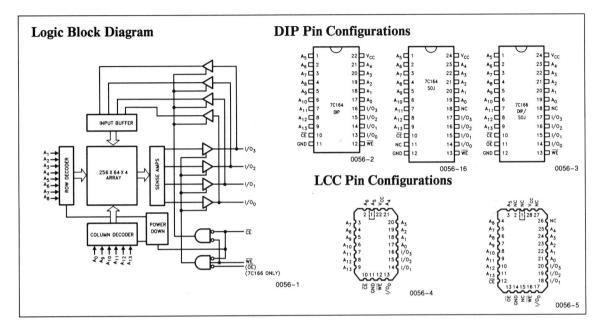

Datasheet courtesy of Cypress Semiconductor.

Dynamic RAM

Dynamic RAM (DRAM) chips have much greater storage capacity than static chips due to the internal construction of the device. While the cell structure of a static RAM chip is of flip-flop design and requires at least four transistors, a basic dynamic RAM cell consists of a single transistor and capacitor. This construction takes up considerably less space or "real estate" on the silicon "die," allowing many more storage cells to be packed together. Some trade-offs are made for this increase in storage capacity—dynamic RAM access times are longer than static RAMs'. In addition, the storage capacitors in the dynamic RAM cells are leaky. The presence of charge in a capacitor is equivalent to a one state, whereas a lack of charge is equivalent to a zero state. However, if the charge leaks off, then all cells eventually revert to the zero state. In dynamic RAMs a fully charged cell will lose its charge in approximately 2 msec to 4 msec. In order to prevent this loss of data, dynamic RAM chips require a **refresh cycle**. During a refresh cycle stored data are periodically read and rewritten to recharge the storage cells. Every cell must be refreshed within a 2 msec to 4 msec interval. Although 2 msec to 4 msec may not seem very long, when considering the much shorter access time of a dynamic RAM chip, we find that plenty of read/write operations can take place between the refresh cycles. For example, assume a dynamic RAM has a 120 nsec access time. Then the number of potential read/write cycles that can occur between a 2 msec refresh interval is:

$$\frac{2 \text{ msec}}{120 \text{ nsec}} = 16{,}666$$

Quite a few!

The dynamic RAM refresh cycle requires special support circuitry to ensure that all storage cells are refreshed on a periodic basis. This circuity is rather complex, but the difficulties encountered in dynamic RAM memory system design are alleviated by the use of **dynamic RAM controller** chips. These large-scale integrated circuits contain all required refresh circuitry and simplify using dynamic RAM chips. Naturally, understanding dynamic RAM operation is a necessary prerequisite to understanding the functions of a dynamic RAM controller.

Dynamic RAM Addressing

Since dynamic RAMs are very dense, they contain a significantly greater number of addresses than static RAMs. This would mean that more physical address pins are required on the dynamic RAM package to accommodate the increased address space. In order to minimize the number of address pins, and consequently the size of the integrated circuit package, we multiplex the address pins so that the dynamic RAM chip receives the address in two segments—a **row address** and a **column address**. The row address (low order address bits) and column address (high order address bits) are presented to the chip address lines one after another. When the row address is valid, the dynamic RAM controller chip or other controlling circuitry brings the **RAS (row address strobe)** line low. At this time the row address is stored in an internal dynamic RAM register. Next, the column address is presented to the RAM chip and the **CAS (column address strobe)** line is brought low; the dynamic RAM chip stores this portion of the address in another on-board register. This sequence places a complete address into the dynamic RAM chip, allowing a read or write cycle to take place. RAS and CAS also function as chip select lines when multiple DRAM chips are used. (See Figure 9.7 for an illustration of typical DRAM write cycle timing.)

The 4256 256K $\times$ 1 dynamic RAM shown in Figure 9.7 shows the small size of the chip due to address multiplexing. A 256K (262,144) address chip usually needs 18 address lines, but in a multiplexed chip only 9 address lines are required. The reduction in the number of address pins allow 256K worth of memory to be packaged on a small 16-pin IC.

TMS4256, TMS4257
262,144-BIT DYNAMIC RANDOM-ACCESS MEMORIES

MAY 1983—REVISED JANUARY 1988

- **262,144 × 1 Organization**
- **Single 5-V Power Supply**
  - **5% Tolerance Required for TMS4256-8**
  - **10% Tolerance Required for TMS4256-10, -12, -15, and TMS4257-10, -12, -15**
- **JEDEC Standardized Pinouts**
- **Performance Ranges:**

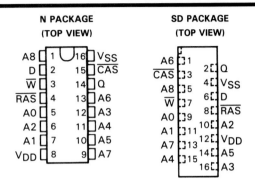

DEVICE	ACCESS TIME ROW ADDRESS (MAX)	ACCESS TIME COLUMN ADDRESS (MAX)	READ OR WRITE CYCLE (MIN)	V_{DD} TOLERANCE
'4256-8	80 ns	40 ns	160 ns	± 5%
'4256-10 '4257-10	100 ns	50 ns	200 ns	± 10%
'4256-12 '4257-12	120 ns	60 ns	220 ns	± 10%
'4256-15 '4257-15	150 ns	75 ns	260 ns	± 10%

- **Long Refresh Period . . . 4 ms (Max)**
- **Operations of the TMS4256/TMS4257 Can Be Controlled by TI's SN74ALS2967, SN74ALS2968, and THCT4502 Dynamic RAM Controllers**
- **All Inputs, Outputs, and Clocks Fully TTL Compatible**
- **3-State Unlatched Outputs**
- **Common I/O Capability with "Early Write" Feature**
- **Page Mode ('4256) or Nibble-Mode ('4257)**
- **Low Power Dissipation**
- **$\overline{RAS}$-Only Refresh Mode**
- **Hidden Refresh Mode**
- **$\overline{CAS}$-Before-$\overline{RAS}$ Refresh Mode**
- **Available with MIL-STD-883B Processing and L(0 °C to 70 °C), E(− 40 °C to 85 °C), or S(− 55 °C to 100 °C) Temperature Ranges (SMJ4256, with 10% Power Supply)**

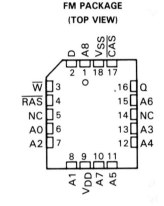

PIN NOMENCLATURE	
A0-A8	Address Inputs
$\overline{CAS}$	Column-Address Strobe
D	Data In
NC	No Connection
Q	Data Out
$\overline{RAS}$	Row-Address Strobe
V_{DD}	5-V Power Supply
V_{SS}	Ground
$\overline{W}$	Write Enable

Figure 9.7(a)
4256 Data Sheet

TMS4256, TMS4257
262,144-BIT DYNAMIC RANDOM-ACCESS MEMORIES

write cycle timing

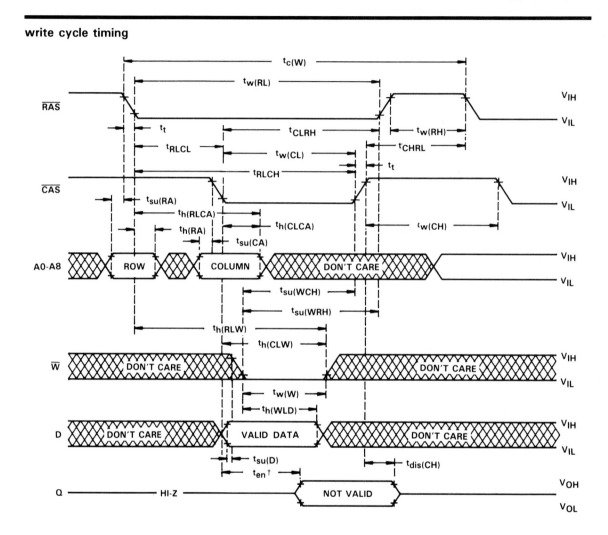

†The enable time (t_{en}) for a write cycle is equal in duration to the access time from $\overline{CAS}$ ($t_{a(C)}$) in a read cycle; but the active levels at the output are invalid.

Figure 9.7(b)
4256 Write Cycle
Timing

Printed by permission of the copyright holder,
Texas Instruments Incorporated.

Dynamic RAM Data

Notice also that the chip is 256K × 1. Every one of the 256K addresses accesses a single data bit. There is one data in line (D) and one data out line (Q) for this purpose. Naturally, several memory chips are required to extend the word size for a practical memory system. For instance, an 8-bit word size requires eight DRAM chips. Details of memory expansion are discussed later. The dedicated data in and data out pins may be connected directly to a system data bus because data out is internally tristated during write operations. Frequently, **bidirectional bus transceivers** are used on each bus wire to control the transfer of data in either direction (to or from memory) and to provide sufficient current driving capacity. The bus transceiver is composed of two tristate buffers wired so that only one at a time is enabled as shown in Figure 9.8. Thus, data can flow either to the left or the right under control of the enable lines.

Typical Schematic

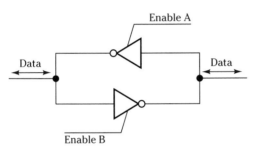

To move data left to right:
    Enable A = 0
    Enable B = 1

To move data right to left:
    Enable A = 1
    Enable B = 0

Figure 9.8
Bidirectional Bus
Transceiver

The only other DRAM control line not yet mentioned is write enable, which determines whether a read or write operation is to take place. A low level enables writing, whereas a high level enables reading.

Summary of DRAM read and write operations:

Read

1. Set system address lines.
2. Write enable = 1.
3. When $\overline{RAS}$ is activated, the low order half of the system address is stored in the DRAM as the row address.
4. When $\overline{CAS}$ is activated, the high order half of the system address is obtained by the DRAM as the column address.
5. Data from the addressed location appears at data out after the access time of the chip.

Write
1. Set system address lines.
2. When $\overline{RAS}$ is activated, the low order half of the system address is stored in the DRAM as the row address.
3. When $\overline{CAS}$ is activated, the high order half of the system address is obtained by the DRAM as the column address.
4. Apply data to the data in line.
5. Activate write enable.

DRAM Refresh

Refresh operations are carried out by holding the $\overline{CAS}$ line inactive while pulsing the $\overline{RAS}$ line 256 times as 256 row addresses are cycled on the A_0 through A_7 address lines. This is shown in Figure 9.9. (A_8 is not used in this operation, making the chip compatible with older DRAM controllers.) Internally, the 4256 is organized so every $\overline{RAS}$ address refreshes 1000 cells. 256 row addresses times 1000 cells equal 256K storage locations, so that 256 $\overline{RAS}$ pulses refresh the entire chip. Remember, when refresh operations are being carried out, normal memory functions are halted. Using a dynamic RAM controller, the refresh cycle can be "hidden" during normal system processing so that refresh delays are minimized. In addition, sophisticated controller chips have several refresh modes to meet the needs of a variety of memory applications.

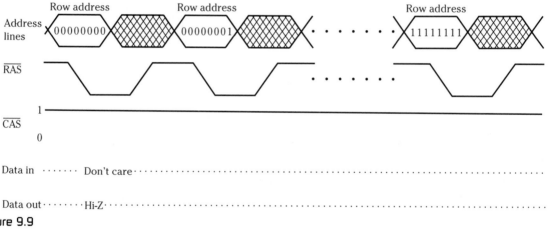

Figure 9.9
Typical DRAM Refresh Timing

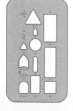

DESIGN EXAMPLE 9.6

Draw the block diagram of the interface required to connect the address and data lines to a 16K × 1 DRAM. Use the $\overline{CAS}$ line to distinguish between the row and column addresses.

Solution   Multiplexers are required to present half of the address lines to the DRAM for the row address. Using $\overline{CAS}$, the remaining half of the address lines can be presented to the DRAM as a column address. A bus transceiver is required to interface the data in and data out lines to the system data bus.

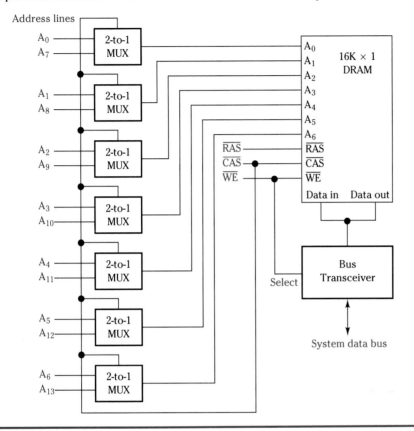

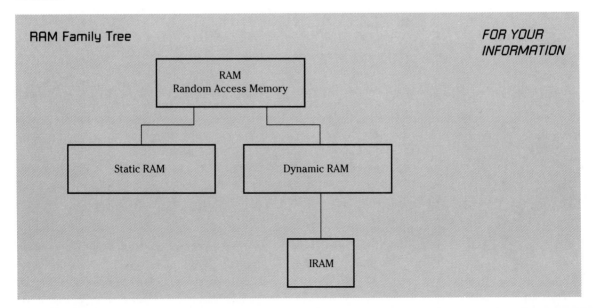

RAM Family Tree

FOR YOUR INFORMATION

DESIGN
EXAMPLE 9.7

Sketch a 4256 DRAM timing diagram for the following operation. Write a one to storage location 28 (decimal):

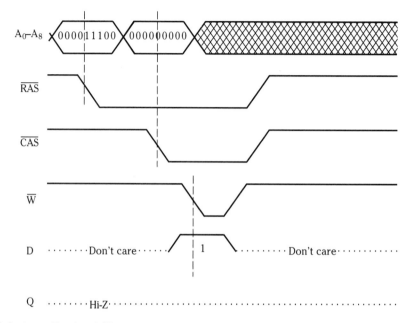

Solution   Decimal 28 must be represented as an 18-bit binary address, which is divided into row and column segments. 28 = 000000000000011100. The input data must be stable on the timing diagram when write enable is active.

9.6 ■ READ ONLY MEMORY

Any computer or digital system requires power to run. This immediately presents a problem since nothing is 100% reliable in the "real world"—our computer power sources included. Batteries do wear out, and simply inserting a plug in the wall does not guarantee an uninterrupted flow of electricity. In addition, power-on switches are placed on our computing equipment because we do have the desire to turn it off now and then. Unfortunately, any information stored in volatile memory devices is lost the instant power is removed. Therefore, we need backup storage systems, such as floppy disks, to retain our vital information. However, there is also a need for nonvolatile storage that reacts at digital circuit speeds. This is where ROM is used.

ROM storage devices are similar to RAM devices because they must interface to the same digital circuitry. Therefore, we can expect to see addressing lines, data lines, and control signal lines present on these chips. Since most ROM chips are truly read only, no read operations take place, and so there is really no need to distinguish between read and write operations. However, this characteristic is slowly changing as manufacturers strive to develop nonvolatile, fast, dense, read and write storage devices.

At the present, several kinds of ROM storage are heavily used.

EPROM

The **EPROM** or **Erasable PROM** is a reusable PROM. That is, the programmed data can be erased and the chip can be reprogrammed. EPROM is like any other ROM device in operation and serves as a good introductory device to illustrate the signal and control lines affecting chip operation. The primary distinction between any ROM technology is the method used to induce the chip's initial data information.

EPROMs store ones and zeros by the presence or absence of electric charge. The charge is stored in a **floating gate,** an electrically isolated region within the basic EPROM storage cell. A suitable programming voltage forces electric charge into the floating gate where it becomes trapped. Internal sensing circuits can determine whether or not a selected cell contains charge.

EPROM chips are purchased as **blank** or **virgin** devices. In this unprogrammed state every address provides the same output pattern. The designer specifies the information required for each address and takes the appropriate action to see that correct chip programming takes place. This is typically accomplished using a computer software development system. The software providing this capability runs on personal computing systems and allows the designer to specify and verify easily all data placed into the EPROM. When the designer is satisfied that the information is correct, a blank device is inserted into an **EPROM programmer.** Under control of the software, commands and data are sent from the computer to the EPROM programmer to "burn" data into the device.

Erasing an EPROM is accomplished by exposing the chip to ultraviolet light. In fact, an identifying characteristic of an erasable device is the presence of a small transparent window on the IC package. It is worth your time to examine one of these chips since the actual chip, not the chip package, is clearly visible under the window. Shining ultraviolet light onto the chip induces enough energy to allow the charge to escape from the floating gate, returning the EPROM to an unprogrammed state. The whole EPROM is erased during this process so that complete reprogramming is necessary.

A quality ultraviolet light source can completely erase an EPROM in 15 to 20 minutes. As a precautionary measure, most EPROM windows are covered with opaque tape after programming to prevent accidental erasure. This is important since sunlight and fluorescent lighting contain ultraviolet wavelengths that can affect the level of charge over a period of time. Direct sunlight can cause erasure after approximately one week; fluorescent lighting can erase a device after approximately 3 years exposure.

EPROMs are used extensively to store programs, games, and data permanently. In addition, EPROMs are crucial in the designer's software development effort. Although PROMs (see next section) are less expensive than EPROMs, they cannot be erased. Any software or data errors result in a PROM destined for the scrap pile. Therefore, in the early stages of software development, ERPOMs are used when changes in data or programming code are frequent. Once the program is correct, less expensive PROMs are substituted to bring down production costs.

Figure 9.10 shows a 2732A 4K × 8 EPROM. A 4K address space requires 12 address lines, shown as A_0 through A_{11}. In addition, eight outputs (Q_0–Q_7)

TMS2732A
32,768-BIT UV ERASABLE PROGRAMMABLE READ-ONLY MEMORY

AUGUST 1983 – REVISED FEBRUARY 1988

- **Organization . . . 4096 × 8**

- **Single 5-V Power Supply**

- **All Inputs/Outputs Fully TTL Compatible**

- **Max Access/Min Cycle Times**
  TMS2732A-17   170 ns
  TMS2732A-20   200 ns
  TMS2732A-25   250 ns
  TMS2732A-45   450 ns

- **Low Standby Power Dissipation . . .**
  158 mW (Maximum)

- **JEDEC Approved Pinout . . . Industry Standard**

- **21-V Power Supply Required for Programming**

- **N-Channel Silicon-Gate Technology**

- **PEP4 Version Available with 168 Hour Burn-In, and Extended Guaranteed Operating Temperature Range from −10 °C to 85 °C (TMS2732A-_ _JP4)**

J PACKAGE
(TOP VIEW)

```
        ┌───∪───┐
 A7  ☐ 1      24 ☐ VCC
 A6  ☐ 2      23 ☐ A8
 A5  ☐ 3      22 ☐ A9
 A4  ☐ 4      21 ☐ A11
 A3  ☐ 5      20 ☐ G̅/VPP
 A2  ☐ 6      19 ☐ A10
 A1  ☐ 7      18 ☐ E̅
 A0  ☐ 8      17 ☐ Q8
 Q1  ☐ 9      16 ☐ Q7
 Q2  ☐ 10     15 ☐ Q6
 Q3  ☐ 11     14 ☐ Q5
 GND ☐ 12     13 ☐ Q4
        └──────┘
```

PIN NOMENCLATURE	
A0-A11	Address Inputs
E̅	Chip Enable
G̅/VPP	Output Enable/21 V
GND	Ground
Q1-Q8	Outputs
VCC	5-V Power Supply

description

The TMS2732A is an ultraviolet light-erasable, electrically programmable read-only memory. It has 32,768 bits organized as 4,096 words of 8-bit length. The TMS2732A only requires a single 5-volt power supply with a tolerance of ±5%.

The TMS2732A provides two output control lines: Output Enable (G̅/VPP) and Chip Enable (E̅). This feature allows the G̅/VPP control line to eliminate bus contention in multibus microprocessor systems. The TMS2732A has a power-down mode that reduces maximum power dissipation from 657 mW to 158 mW when the device is placed on standby.

This EPROM is supplied in a 24-pin dual-in-line ceramic package and is designed for operation from 0 °C to 70 °C. The TMS2732A is also offered in the PEP4 version with an extended guaranteed operating temperature range of −10 °C to 85 °C and 168 hour burn-in (TMS2732A-_ _JP4).

Figure 9.10
2732A Data Sheet

Printed by permission of the copyright holder,
Texas Instruments Incorporated.

are required for the 8-bit word size. These lines are listed as outputs because, other than during programming, data is only read from the EPROM. Also shown are two control lines, chip enable and output enable. Chip enable activates the chip for operation and is used to select a particular chip when multiple memory chips are utilized. This aspect of memory expansion is discussed later in the chapter. Chip enable, when inactive, also places the PROM into a standby power mode, reducing the power dissipation requirements of the chip. The active current for this chip is 125 mA, whereas the standby current is only 30 mA. This current reduction is significant in systems with many memory chips and very beneficial considering that a memory chip is inactive far more than it is active in typical systems. Standby features are becoming more widespread in memory and programmable device technologies.

The output enable line also has two functions. For normal operation output enable controls the tristate nature of the output lines, allowing the chip to be connected to a computer bus. During programming output enable is referred to as V_{pp}—the programming voltage input. Programming is accomplished by applying 21 V to this pin while the correct data and addresses are supplied. Pulsing chip enable briefly while a specific data word and address are presented to the chip completes the programming for the addressed location.

The following waveform illustrates typical EPROM read cycle timing. Determine:
(a) When is the address valid?
(b) When is the chip in standby mode?
(c) When is data valid?
(d) What determines the access time?

DESIGN
EXAMPLE 9.8

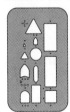

Solution Refer to the diagram.

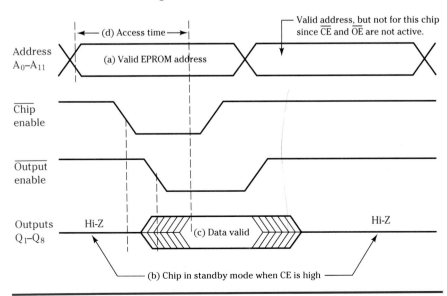

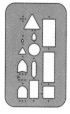

DESIGN EXAMPLE 9.9

Code conversion is a typical application for ROMs. Using the address lines as BCD inputs, show how seven segment code can be obtained from the outputs. What is the minimum sized ROM required?

Solution   Since there are ten BCD codes, each 4-bits, four address lines are necessary. Seven segment code consists of ten 7-bit codes, so that seven output lines are needed. The minimum sized ROM is 4×7.

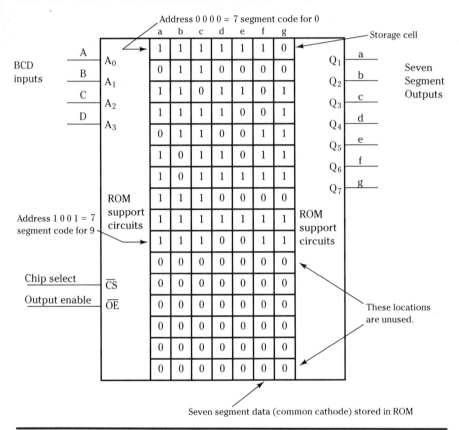

Seven segment data (common cathode) stored in ROM

PROM

PROM (**programmable read only memory**) are permanently programmed ROMs, many of which are available in the same sizes, pin outs, and part number designations as EPROMs. For instance, the 2732 EPROM is identical to the 27P32 PROM except for the erasing window. PROMs tend to be less expensive than EPROMs and are used when there is little chance that data will need to be altered.

A PROM is programmed by blowing a set of internal microscopic fuses called **fusable links**. Each address has one fuse per data bit, so that a 256 × 8 PROM has 2048 fuses. An intact fuse passes an electrical current, whereas a blown fuse will not—a simple system making the presence of a one or zero possible. During programming a PROM programmer provides an address and the correct data pattern. A **programming voltage** is applied at this time to blow the internal fuses. Typical programming voltages range from 12 to 21 volts and are sufficiently greater than the typical 5 V operating voltage. Once programmed, the PROM functions as a read only device with address, data, and control lines operating exactly like that of an EPROM.

Masked ROMs

Masked ROMs are the least expensive ROM available, but they are not acceptable for all applications. During the IC manufacturing process the IC is fabricated by growing layer upon layer of silicon, silicon dioxide (glass), and metal. This is accomplished in a series of **masking** steps, which lay down the features for each layer of the chip. The basic layout for any ROM is the same except for the programming information. In a masked ROM the manufacturer builds and stocks ROM chips that are complete except for the final masking steps that program the device. A designer sends programming information to the manufacturer specifying how the ROM should be masked. The information may be provided to the manufacturer in the form of a programmed EPROM, a floppy disk file, or a data sheet. In any case the ROM manufacturer completes the integrated circuit process by customizing the ROM to the designer's specifications. Masked ROMs are only inexpensive when purchased in large quantities since masking costs are high. The cost per ROM is reduced because of the volume purchase of the devices. For instance, an electronic game company may use masked ROMs if it plans to sell thousands of the same electronic game. The masked ROM is cost-effective since all of the games contain the same software and the ROM programming task is carried by the ROM manufacturer. The disadvantage to the masked ROM is the time required to obtain the devices because masking is a lengthy as well as expensive process.

EEPROMs

EEPROMs or E^{2}PROMs are **electrically erasable PROMs** that have two advantages—they are nonvolatile and possess the ability to be read or written in circuit. However, the write capability is limited in terms of speed and frequency, although the read capability is the same as that for other ROM technologies. For this reason the EEPROM is also known as **read-mostly memory.**

The EEPROM can be read simply by providing addressing and control signals. Access times are typically in the 200 nsec range, similar to other ROM devices. Writing, on the other hand, takes much longer, approximately 9 msec. This long length of time makes the device impractical for microsystem use because it is too slow to keep pace with the read/write cycles normally associated with static and dynamic RAM chips. In addition, the number of write cycles is limited. After approximately 10,000 write operations (to the same storage cell) cell reliability diminishes due to charge retention problems. The number of write operations for the whole chip is limited to that of the worst

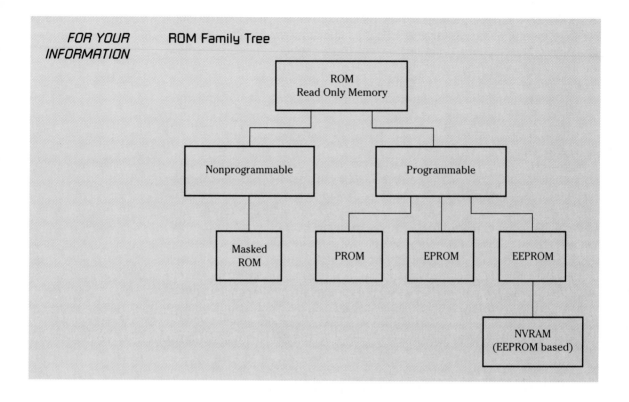

FOR YOUR INFORMATION

ROM Family Tree

case cell. Read operations are unlimited, however. Even though write operations are restricted, write operations can take place in circuit and at the byte level, making EEPROMs an extremely useful device for many applications.

9.7 ■ INNOVATIVE MEMORY TECHNOLOGIES

Many innovations in memory devices occur as manufacturers strive to produce nonvolatile, fast RAM chips. Other innovations occur through the efforts of manufacturers to reduce the cost of memory system design. The following section discusses some of these innovative devices.

IRAM

IRAM or Integrated RAM combines the density of dynamic RAM storage and complex dynamic RAM refresh circuitry on one chip. Internally, the memory undergoes periodic refresh cycles, but to the designer the IRAM functions as a static RAM chip. This means that memory system development is simplified because the designer does not have to contend with the design of refresh control circuitry nor the timing of refresh operations. In addition, the parts count of the memory system is reduced. On the downside, integrated RAM does not have the massive storage density of standard dynamic RAM chips since a portion of the chip space is relegated to the refresh circuitry. Thus, most applications for IRAM are in smaller memory systems.

NVRAM

NVRAMs or Nonvolatile RAMs are formed by combining static RAM with EEPROMs. Each static RAM cell has a corresponding EEPROM cell that provides a backup or "shadow" of the static RAM contents. Under normal memory operation the static RAM portion of the NVRAM chip carries out read and write operations at typical system speeds. Under the control of a store signal the entire contents of the static RAM are copied to the EEPROM portion of the chip in approximately 10 msec. This provides an ideal backup storage system in the event of a power failure. EEPROM data is copied back to the static RAM with a recall signal.

Since the NVRAM contains both EEPROM and static RAM circuitry on the same chip, storage density is restricted. Typical sizes are 128×8 and 512×8. Figure 9.11 shows the pin diagram and functional block diagram for a 2201 1024×1 NVRAM.

PIN CONFIGURATION

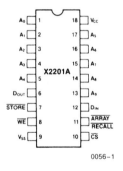

0056-1

PIN NAMES

A_0-A_9	Address Inputs
D_{IN}	Data Input
D_{OUT}	Data Out
$\overline{WE}$	Write Enable
$\overline{CS}$	Chip Select
$\overline{ARRAY\ RECALL}$	Array Recall
$\overline{STORE}$	Store
V_{CC}	+5V
V_{SS}	Ground

Printed by permission of Xicor, Inc.

FUNCTIONAL DIAGRAM

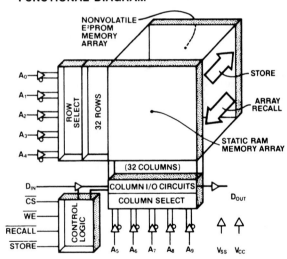

Figure 9.11
2210A NVRAM Data
Sheet

9.8 ■ MEMORY SYSTEM EXPANSION

In a typical memory system a single memory chip will not provide enough storage capacity. Therefore, multiple chips must be employed to expand the memory system. Expansion can take place in two directions. The word size or number of data bits per address can be expanded. Typical word sizes are 8-, 16-, and 32-bits. The address range can also be expanded. A basic 8-bit microprocesser chip typically addresses 64K storage locations. More sophisticated microprocessors can address 4G (giga) storage locations. This range of address space can only be met with multiple chips.

Increasing Word Size

We will use the 2114 1K × 4 memory chip to illustrate memory expansion techniques. Assume that a 1K × 8 memory system is required. Using 2114s, two chips are required to increase the word size from 4-bits to 8-bits. A simple calculation verifies this. The number of storage cells in a 1K × 8 is 1024 × 8 = 8192 cells. A single 2114 has 1024 × 4 = 4096 cells; 8192/4096 = 2 chips. This calculation is relatively obvious on this small scale, but it is useful for more complex expansions.

DESIGN EXAMPLE 9.10

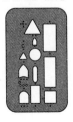

How many 2114 chips are needed to build a 16K × 8 memory system?

Solution   Determine the number of cells in a 16K × 8 system. Divide this number by 4096, the number of cells in a single 2114, to obtain the required number of chips. Thus:

$$16K = 16 \times 1024 = 16{,}384$$
$$16{,}384 \times 8 = 131{,}072 \text{ cells}$$
$$\frac{131{,}072}{4096} = 32 \text{ chips}$$

When expanding a memory system, we must keep in mind what quantity is actually being expanded. In this case we are increasing the word size only. Therefore, the address space will remain the same. A 1K address space, whether for a 4-bit or 8-bit word size will always require ten address lines ($2^{10} = 1024$). Since each 2114 has ten address lines, all of the lines must be connected together. That is, A_0 of the first chip is connected to A_0 of the second chip, A_1 to A_1, and so on. This makes sense since we are accessing 8-bits at a time, so both chips must be simultaneously active (4-bits per chip).

In addition to address lines and I/O lines, the memory control lines take part in the expansion. In this example both chips must be read and written together, so the write enable lines are tied together. In a similar manner, the chip enable lines are tied together to enable each chip simultaneously. Figure 9.12 illustrates the 2114 word size expansion.

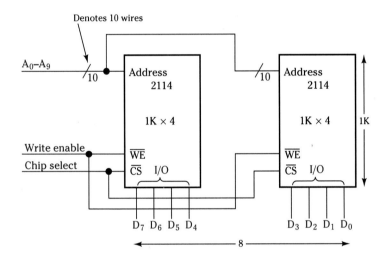

Figure 9.12
2114 Word Size
Expansion

Design a 256K × 4 memory using 4256 dynamic RAMs.

DESIGN EXAMPLE 9.11

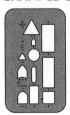

Solution Since a 4256 is a 256K × 1 DRAM, we must expand the word size. Four chips are necessary for the word size expansion.

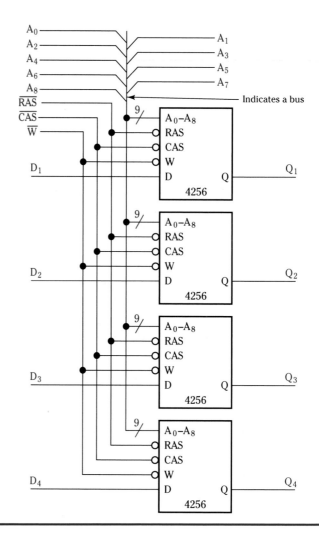

Increasing Address Space

Using the 2114 again, we will now create a 2K × 4 memory. Two chips are required since $2 \times 1024 \times 4 = 8192$ and $8192/4096 = 2$. However, this time the addressing lines must be modified to accommodate the increased address range. Figure 9.13 shows how the expansion is accomplished.

Since two chips are used, but only four I/O lines are required (word size is 4-bits), the I/O lines are connected together to maintain four overall. This is exactly how a computer bus structure is formed—multiple devices sharing a common set of wires. The tristate nature of the 2114 is useful here, since only one of the two chips can place data on the bus at any time.

The write enable line is still common to both chips because this line merely selects the mode of operation. If the chip is not selected by its chip select line, then write enable is ignored.

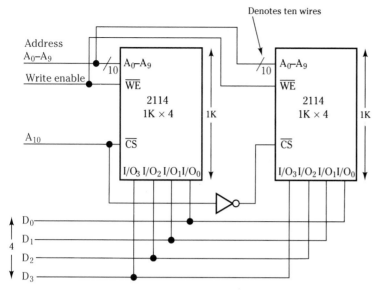

Figure 9.13
2114 Address
Expansion

$A_{10} = 0$ First 2114 selects 1024 addresses

$A_{10} = 1$ Second 2114 selects an additional 1024 addresses

The ten address lines on each chip can still be tied together since the 1024 addresses within each 2114 must be uniquely accessible. It always takes ten address bits to access a 1024 address space. However, if a certain address is desired in one of the 2114s, it will also be accessed in the other unless some additional circuitry prevents both chips from becoming simultaneously active. This is the function of the chip select line.

Since we are designing a 2K (2048) memory, we need 11 (2^{11} = 2048) address lines. Ten of the lines are already physically present on the 2114, so we need to create the eleventh. Since chip select can have either a high or low level and controls device selection, it becomes the eleventh address bit. The inverter added to the circuitry allows only one of the 2114s to be selected at any time. Therefore, when one 2114 is selected, 1024 unique addresses can be accessed within the chip. Changing the chip select level to activate the other 2114 provides access to an additional 1024 addresses. Together, the two 2114s provide 2048 unique addressable storage locations. Chip select is used as part of the addressing structure for most memory systems. The terms "chip select circuitry" or "address decoding circuitry" generally refer to the addressing circuits connecting multiple memory chips in microcomputer systems.

Design a 512K × 4 memory using 4256 DRAMs.

Solution Both word size and address space need to be increased since a single 4256 is 256K × 1 in size.

$$256K = 256 \times 1024 = 262{,}144$$
$$262{,}144 \times 1 = 262{,}144 \text{ cells per } 4256$$
$$512K = 512 \times 1024 = 524{,}288$$
$$524{,}288 \times 4 = 2{,}097{,}152 \text{ total cells}$$
$$\frac{2{,}097{,}152}{262{,}144} = 8 \text{ chips}$$

DESIGN EXAMPLE 9.12

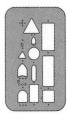

Multiple RAS lines are used to distinguish between the first 256K of memory and the second 256K of memory.

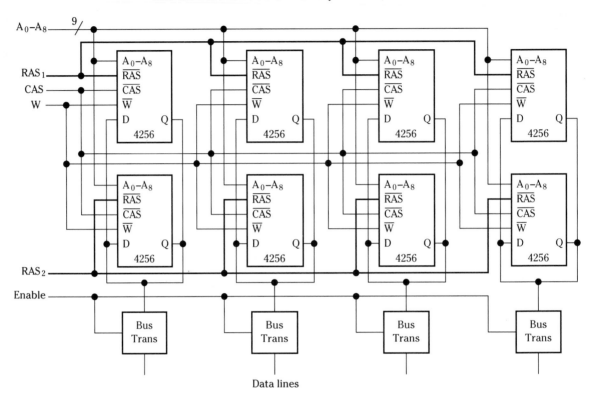

Data lines

DESIGN EXAMPLE 9.13

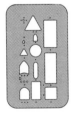

A computerized point of sale terminal is being designed for retail store use. The terminal requires memory for the following uses (all data is 8-bits wide).

> Initial start-up program—2K ROM
> Current prices—57K of data loaded every day
> Permanent system diagnostic program—1K
> Sales programs—6K

The following memory chips meet the performance specifications of the machine. Which chips are the best to use?

Available: 1K × 4 static RAM, 128K × 1 dynamic RAM, 4K × 4 EPROM, 64K × 8 static RAM, 128K × 8 static RAM, 2K × 8 EPROM, 8K × 8 EPROM

Solution Even though the total memory requirement is 66K, the needs are divided between RAM and ROM storage. Since the start-up, sales application, and diagnostic functions are all permanent on line operations, an EPROM is suitable. The total memory requirements for these functions is 2K + 1K + 6K = 9K. Use one 8K × 8 EPROM and one 2K × 8. The current prices data changes daily and requires RAM storage. A 64K × 8 static RAM is the best fit for 57K worth of information.

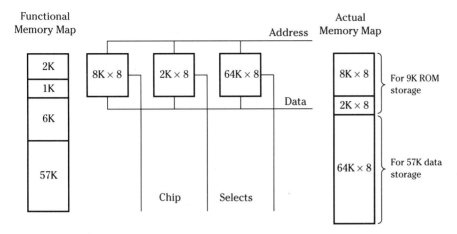

Determine the address range for the memory expansions in Figures 9.12 and 9.13 and in Design Examples 9.11, 9.12, and 9.13.

DESIGN
EXAMPLE 9.14

Solution Memory address ranges are typically expressed in hexadecimal notation. The address range is determined by the number of address bits and is equal to 2^n. The range of addresses is from zero to $2^n - 1$, assuming that each of the preceding memory systems begins at the all zeros address.

- **Figure 9.12** 1K × 8 address range is from 0000000000 to 1111111111 or 000 to 3FF hex.
- **Figure 9.13** 2K × 4 address range is from 00000000000 to 11111111111 or 0000 to 7FF.
- **Design Example 9.11** 256K × 4 address range is from 000000000000000000 to 111111111111111111 or 00000 to 3FFFF.
- **Design Example 9.12** 512K × 4 address range is from 0000000000000000000 to 1111111111111111111 or 00000 to 7FFFF.
- **Design Example 9.13** 74K × 8 address range is from 00000000000000000 to 10010100000000000 or 00000 to 12800.

$$74K = 74 \times 1024 = 75776 \text{ addresses}$$

9.9 ■ MEMORY TESTING

The sheer number of storage cells in a memory device makes it imperative that some memory testing be carried out. Naturally, the chip manufacturer has tested the chips during production to ensure reliability, but in-system tests are often useful diagnostic tools. Many of the testing procedures used by the manufacturer can be utilized for memory system designs as well.

Testing a memory chip is unlike testing a simple gate or flip-flop since logic levels cannot be obtained directly from a memory cell. All signals are available at the memory chip outputs, so a deliberate sequence of reads and writes is required to obtain internal values. In addition, a knowledge of memory device **failure modes** is necessary for effective testing. Typical internal failures

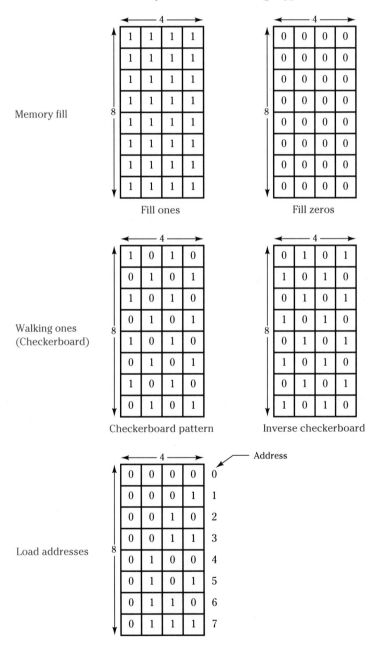

Figure 9.14
Memory Test Patterns
for an 8 × 4 Memory

occur because of cell to cell interaction, where the logic level in one cell affects that of another. Wiring errors can also alter proper addressing and memory chip selection in a memory system.

Testing memory systems involves writing specific test patterns to the individual chip or complete memory system and then reading the patterns back. If the patterns are the same as those written, then the system checks out. If the patterns read back incorrectly, then the nature of the error may be determined by reviewing the incorrect test pattern result. Some common test patterns are shown in Figure 9.14.

The most common test is completely filling the memory with all ones or zeros and then reading the data back. This is called a **memory fill** and isolates cells stuck at certain logic levels. For instance, when all zeros are written to memory, any ones showing in the pattern after reading indicates a cell "stuck at one." Conversely, cells "stuck at zero" can be found by filling memory with all ones. Stuck faults indicate inoperative cells and prevent the cells' associated address from being used. Sophisticated memory systems can ignore the address and use the rest of the chip, while simpler computer systems will need the chip replaced.

The **walking ones—walking zero** or **checkerboard** test pattern is useful for finding adjacent cell interactions. If the test pattern is not read back as written, it indicates that one memory cell is affecting the logic levels of its neighbors. This test is particularly useful for the memory chip designer since it identifies potential manufacturing problems.

Writing addresses and reading them back is another useful test at the system level because it identifies possible wiring errors. For instance, if address 25 is loaded with its address (25) but another number is read back, then an address wiring problem is indicated.

Memory tests are common in computer systems. Many systems perform a series of memory tests every time the system is turned on to check and verify proper memory system operation.

Four address locations in a memory are loaded with the ASCII numbers for A, B, C, and D. When read back, the data reads C, B, C, F. What memory problem is likely? Would parity checking catch this error? Would EEC logic correct it?

DESIGN EXAMPLE 9.15

Solution   The ASCII codes for the letters are:

A—1000001   B—1000010   C—1000011   D—1000100
C—1000011   B—1000010   C—1000011   F—1000110

    ↑   ↑   ↑   ↑

└ This bit is stuck at one since it incorrectly forces the code for A to C and the code for D to F.

Parity checking would detect that an error has occurred during a memory read since the fault creates a single bit error. The resulting parity error would halt system operation.

Error correction code logic would correct the single bit error when data is read. The system could still continue to function.

SUMMARY

■ Traditionally, memory technologies have been classified as RAM and ROM. A more appropriate classification for current semiconductor memory technology might be volatile or nonvolatile memory.

■ Static RAM devices are volatile memory components. Data may be read or written at high rates of speed using relatively simple addressing signals.

■ Dynamic RAM requires a DRAM controller to manage the DRAM addressing process. DRAMs also require a periodic refresh cycle to maintain data.

■ PROMs are user programmable storage devices. PROM technologies include PROM (programmable read only memory), EPROM (erasable programmable read only memory), and EEPROM (electrically erasable programmable read only memory).

■ Semiconductor memory is organized as an array of storage cells defined as an arrangement of M × N, where M is the address space and N is the word size.

■ Most memory systems require more storage capacity than is possible with a single integrated circuit memory device. Therefore, memory expansion using multiple memory chips is a common memory system design practice.

■ Memory systems are typically expanded in two ways—by increasing address space and/or by increasing word size.

PROBLEMS

Section 9.2

1. Write the truth tables for the following tristate devices:

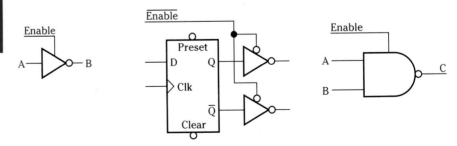

2. Draw the timing diagram for the following circuit:

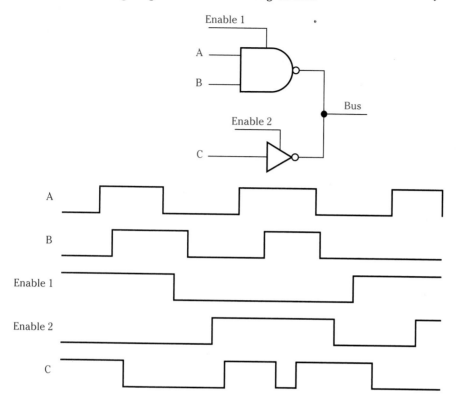

Sections 9.3–9.4

3. Identify for the following memory sizes:
  *(a) the word size
   (b) the address space
   (c) the number of address lines required
   Memory sizes: (1) 128K × 4, (2) 1M × 1, (3) 32K × 8
4. How many storage cells are in each of the memory chips identified in problem 3?
  *(a) (b) (c)

Section 9.5

5. Identify the operations taking place when the 2114 receives the following signals:
   (a) CS = 0   WE = 1 A_9–A_0 = 0000110000 I/O_4–I/O_1 = 0010
   (b) CS = 1   WE = 1 A_9–A_0 = 0000110000 I/O_4–I/O_1 = 0010
  *(c) CS = 0   WE = 0 A_9–A_0 = 0011010011 I/O_4–I/O_1 = 1010

---

6. Show how a 2114 static RAM would be connected to two separate registers. One register, the data in register, provides 4-bits of data to the 2114 during a write operation. The other register, the data out register, receives data from the 2114 during a read operation. Show how the data out register will store the data supplied from the memory. Each register should be attached to the memory bus and allowed to function without interfering with each other or the memory chip.

7. Draw read and write timing diagrams for the memory design in problem 6. Indicate the relationship among data in, data out, chip select, and write enable. Show when the data out register clock should be activated as well with respect to the memory waveforms. Assume that address information is available and will not constrain the timing.

*8. Determine the row address and the column address required to read a data bit from chip address 3F2D5 of a 256K × 1 dynamic RAM.

Section 9.6

9. What minimum sized ROM would be required to convert BCD numbers into equivalent ASCII numbers? (i.e., BCD 4(0100) → ASCII 4(0110100) Complete a table listing how BCD addressing information can be converted into ASCII output information.

*10. A microprocessor is connected to an EPROM as illustrated below. If the microprocessor signals the EPROM for data every 220 nsec, which versions of the 2732 EPROM will be suitable for this application? Why?

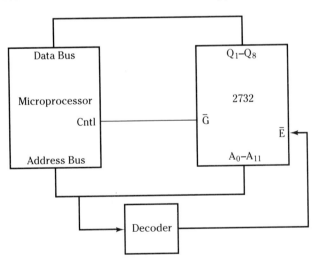

11. Using the preceding diagram, if the microprocessor provides an address for 180 nsec and the decoder provides 25 nsec delay, what is the minimum access time required for the EPROM?

---

* See Appendix F: Answers to Selected Problems.

Section 9.8

12. How many 2114s are required to accomplish the following memory expansions?
    (a) 4K × 8 *(b) 256K × 16 (c) 64K × 4 (d) 1M × 8
13. How many 2732s are required for the following memory expansions?
    *(a) 8K × 8 (b) 256K × 16 (c) 64K × 8 (d) 1M × 8
14. Determine the quantity of 4256 DRAM chips required to create the following memory expansions:
    (a) 256K × 16 (b) 512K × 8 (c) 1M × 8 (d) 4M × 32
15. Design a 1K × 12 memory system using 2114 static RAMs.
*16. Design a 4K × 4 memory system using 2114 static RAMs.
17. Draw the circuitry required to create a 2K × 8 memory system using 2114 static RAM chips. Include the details of the chip selection circuitry.
18. Show how four 2732 EPROMs can be combined to produce a 16K × 8 ROM storage system. What is the address range for this system?
*19. Using the same number of 2732s as in the previous system, modify the design to form an 8K × 16 ROM storage system. What is the address range for this system?
20. Draw the circuitry to show how a 2732 EPROM and two 2114s can be connected to form a 5K × 8 memory. The EPROM should occupy the address range from 0000 to 0FFF, whereas the RAM chip occupies the address range from 1000 to 13FF. Which chip determines the overall operating speed? Why?
21. Design the circuitry to create a 512K × 8 DRAM memory system using 4256 DRAMs.

Section 9.9

22. What problems are indicated by the following memory fill patterns? Which addresses are involved?

*(a)

1	1	1	1
1	1	1	1
1	0	1	1
1	1	1	1
1	1	1	1
1	1	1	1
1	1	1	1

(b)

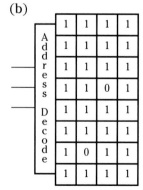

1	1	1	1
1	1	1	1
1	1	1	1
1	1	0	1
1	1	1	1
1	0	1	1
1	1	1	1

(c)

1	0	1	0
0	1	0	1
1	0	1	1
0	1	0	1
1	0	1	0
0	1	0	1
1	0	1	0
0	1	0	1

---

* See Appendix F: Answers to Selected Problems.

CHAPTER 10

PROGRAMMABLE LOGIC

OBJECTIVES

To explain how programmable logic differs from and compares with typical logic devices.

To investigate the many forms of programmable logic including the PROM, PAL, and PLA.

To describe how programmable logic schematics are drawn.

To discuss the benefits obtained using programmable logic.

To describe how programmable logic design takes place.

To discuss how programmable logic development software is used to design and simulate programmable logic.

To explain the processes leading from a basic design to a finished, programmed PAL device.

PREVIEW

Programmable logic is one of the newest frontiers in the logic design evolutionary process. Although programmable logic in one form or another has been in use for quite some time, its initial use was confined to companies with massive financial and computing resources. With the creation of inexpensive development software and personal computing, programmable logic design is easily accomplished.

This chapter describes the various forms of programmable logic hardware and concentrates on the most commonly used programmable logic devices.

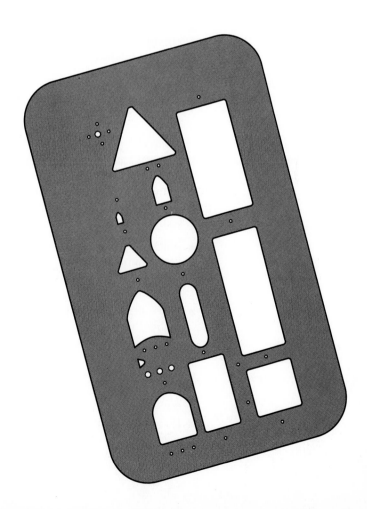

10.1 ■ WHY PROGRAMMABLE LOGIC?

Consider this. You have designed a logic circuit board containing 14 TTL parts. This design performs at a certain speed and consumes a certain amount of power. Shortly thereafter you hear that your design can be replaced by two or three chips, operate much faster, and consume considerably less power using a new technology. What would you do? Of course, you would change to the new technology if the specifications given are true and the reliability of the technology can be proven. This is the same scenario faced by many logic designers over the past years as programmable logic proved its usefulness and cost-effectiveness.

Programmable logic designs can be found in more and more designs for some of the following reasons:

- Fewer parts are required—increases reliability of systems.
- Propagation delay times are reduced.
- Power consumption is reduced.
- Programmable logic designs are inexpensive.
- Fewer parts need to be in stock.
- Designs can be prototyped in a matter of seconds once design work is complete.
- Proprietary design work can be protected.

10.2 ■ PROGRAMMABLE LOGIC DEVICES

Programmable logic devices come in a wide variety of forms, which tends to complicate the initial understanding of this technology. To make matters worse, we are flooded with programmable logic acronyms that are very similar. The following glossary briefly defines programmable logic and other integrated circuit names currently in use. The terms in this list will be followed throughout the text for clarity.

- **ASIC—application specific integrated circuit**   This general term refers to programmable chips that are customized for a specific purpose. ASIC chips with the same part number do not necessarily perform the same function because of their programmable nature.
- **PLD—programmable logic device**   This acronym refers to any programmable logic device.
- **PROM—programmable read only memory**   PROMs are simple programmable logic devices as well as memory devices.
- **PAL—programmable array logic**   This is one of the most widely used programmable logic devices, consisting of a programmable AND array and a fixed OR array (defined later).
- **PLA—programmable logic array**   A device consisting of programmable AND and OR arrays.
- **EPLD—erasable programmable logic device**   This term generally refers

to a highly complex programmable device that allows a designer to inter-connect predesigned "macrocells" and other logical building blocks.

- **Full custom**   This term describes an integrated circuit (IC) that was de-signed completely at the electronic, logic, and chip fabrication levels. Full custom design is a complex, time-consuming process offering maximum design flexibility but long design and development times.
- **Gate arrays**   Gate arrays are ICs that are partially completed by the chip manufacturer. A designer using a gate array takes a chip full of basic gates (300 to 50,000 on a chip) and combines them to produce a circuit design. The chip manufacturer then finishes fabricating the chip to the designer's specifications. This approach is inexpensive for volume chip production. Programmable gate array technology, the latest development in this tech-nology, will make this a more practical design approach.
- **Standard cells**   Using standard cells, a logic designer chooses from prede-signed logic functions, similiar to basic TTL and CMOS functions, and connects them together on a single chip. These chips are very dense and take a significant amount of time to design and fabricate. However, stan-dard cells make available to the logic designer everything from basic gates to microprocessor functions. In addition, the basic standard cells often are equivalent to TTL parts and carry the same part numbers. This means that many of the design techniques are similar to those of well-known discrete logic design and reflect the historic nature of TTL part numbers.
- **Masterslice**   A masterslice is another name for a gate array.
- **Microsequencers**   This is a class of programmable logic used in state machine design (discussed in later chapters).
- **Standard product**   A standard product is a commercially available logic device such as a TTL or CMOS part. Although not referring to a program-mable device, standard products are used to compare and evaluate alter-native logic technologies.

As you can see, there is an abundance of programmable logic devices and their complexity necessitates special design methods (discussed in Appendix C). Figure 10.1, which illustrates the cost and development trade-offs encoun-tered with the various technologies, indicates that programmable logic is a relatively inexpensive and timely design medium. A discussion of many of the devices listed above is far beyond the scope of this text. These devices require engineering expertise and experience to obtain full benefit of the technology. However, your eventual understanding of the more complex devices is directly related to your understanding of the logic design fundamentals discussed throughout this book, and your experience at this point is more than sufficient to begin tackling programmable logic designs. Figure 10.2 shows the typical development cycle for a programmable logic design.

As we progress through this chapter and the next, we will see how each of these steps is applied while designing a programmable logic part. At this point, though, it is worth noting the processing steps involved. For the re-mainder of this chapter we will cover important aspects of the more common programmable devices, particularly the PAL.

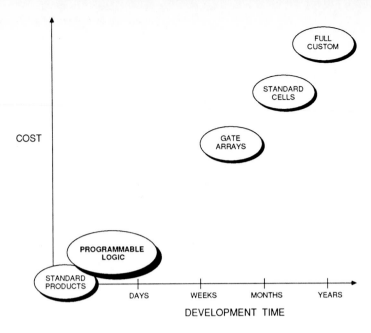

Figure 10.1
Cost Versus Time
Comparison for
Differing Logic
Technologies

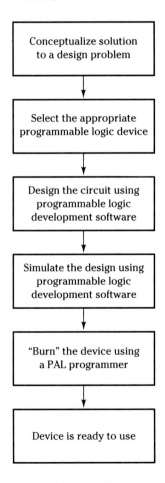

Figure 10.2
Programmable Logic
Design Process

10.3 ■ PROGRAMMABLE LOGIC ARCHITECTURE

The PROM, PAL, and PLA all share a similar internal structure as illustrated in Figures 10.3, 10.4, and 10.5. Each consists of an AND-OR gate structure, allowing for easy implementation of sum of product (SOP) expressions. All the devices have inverters on the inputs, so that true and complement input levels may be generated. Outputs are taken directly off the OR gate in simple devices. The AND gates are referred to as the **AND gate array**, whereas the OR gates are called the **OR gate array**. The differences between the various device technologies are determined by what elements can be programmed.

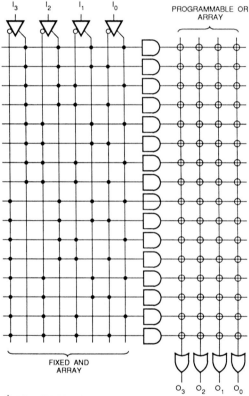

$\oplus$ INDICATES PROGRAMMABLE CONNECTION

$+$ INDICATES FIXED CONNECTION

Figure 10.3
PROM Array Structure

PROMs have a fixed AND array and a programmable OR array. View the PROM for a moment as a memory device and you can understand why. PROM address inputs feed an internal address decoder—the AND array—which re-

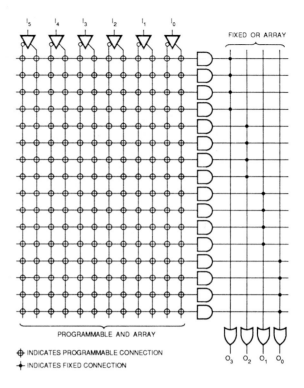

Figure 10.4
PAL Device Array
Structure

Copyright © Advanced Micro Devices, Inc., 1989.
Reprinted with permission of copyright owner.
All Rights Reserved.

mains fixed. The data programmed into the PROM is actually the interconnections, or lack of interconnections, between the AND and OR gates. As you may recall, PROMs are programmed by blowing fuses. When an address is presented to the PROM, the selected AND gate output provides a high level to the OR gates with fuses still intact (to provide a one output level); the OR gate inputs with blown fuses provide an opposite output level. The fixed AND array limits the number of input pins, a serious constraint. For instance, eight input pins require 256 AND gates. Increasing the number of input pins to nine doubles the number of AND gates required, 512. Even though this creates 512 product terms, of nine inputs each, they are all fixed and generally far more than are necessary for normal logic implementation.

PALs have a programmable AND array that allows for the creation of varied product terms. PLAs have programmable AND and OR arrays that provide the most flexibility of the three. However, from a practical standpoint the programmable OR array is not so useful as the programmable AND array for most applications because most logic can be implemented in SOP form. SOP expressions are easily generated with only a programmable AND array. Hence, PALs are the preferred logic device.

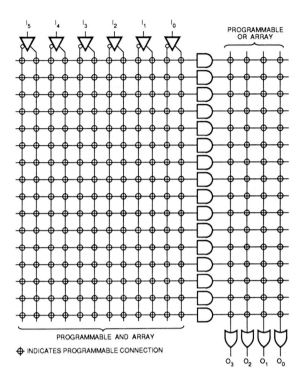

Figure 10.5
PLA Device Array
Structure

10.4 ■ PROGRAMMABLE LOGIC SCHEMATICS

Before delving further into programmable logic design, we must fully understand the notation used to represent programmable logic schematics. Since so much function is packed onto a chip, it becomes impossible, or at the very

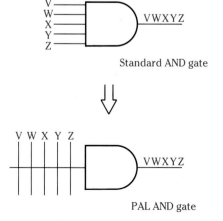

Figure 10.6
PAL Schematic
Notation

least, impractical, to draw every interconnection. A shorthand notation is used instead, which is illustrated in Figure 10.6.

In the example shown the five inputs to the PAL AND gate are represented by a single line. The fact that five inputs are present is made clear by the intersection of the input lines with the single line feeding the AND gate. The intersection of a variable input line with the AND gate input line represents a connection between the variable and the gate. It is this connection that is programmable in the PAL structure.

DESIGN EXAMPLE 10.1

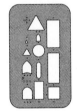

Using PAL schematic notation, determine the expressions for the following circuits:

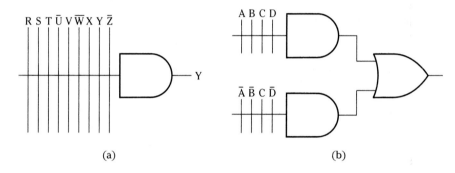

(a)   (b)

Solution   Each intersection of a variable with the AND gate input is the equivalent of a dedicated AND gate input line.
(a) $Y = RST\bar{U}V\bar{W}XY\bar{Z}$
(b) $Y = ABCD + \bar{A}\bar{B}CD$

Figure 10.7 shows a more typical PAL structure. Each input passes through an inverter and the resulting true and complement signals feed each AND gate

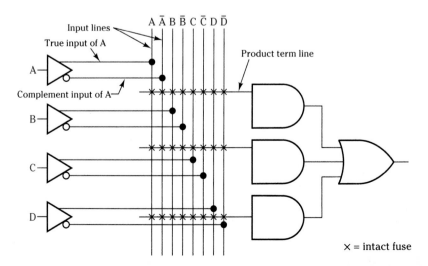

Figure 10.7
Typical PAL AND-OR Structure

in the array. Initially, all fuses are intact so that each input signal goes to each AND gate. Programming the array involves blowing selected fuses to create the required logic structure. This is accomplished with a **PAL programmer,** similar to a PROM programmer. When working at the schematic level, as opposed to the software level, we place a small "x" at each intersection to indicate an intact fuse. The lack of an x indicates a blown or programmed fuse.

What is the output expression for the PAL circuit shown in Figure 10.7?

Solution Each AND gate is an 8-input gate. Since both true and complement levels of each variable are ANDed and then ORed together, logic simplification shows that the output level is zero. Any PAL AND gate with both true and complement values that are present as inputs is disabled. This is done frequently to enable OR gates in the PAL structure.

$$Y = A\overline{A}B\overline{B}C\overline{C}D\overline{D} + A\overline{A}B\overline{B}C\overline{C}D\overline{D} + A\overline{A}B\overline{B}C\overline{C}D\overline{D}$$
$$Y = \quad 0 \quad + \quad 0 \quad + \quad 0 \qquad (X\overline{X} = 0)$$
$$Y = \quad 0$$

DESIGN EXAMPLE 10.2

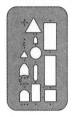

Figure 10.8 illustrates how the PAL structure is utilized to create a logic expression. Basically, all undesired input signals are disconnected from AND gates by blowing their fuses. All required input signals retain their fuses. For example, the top AND gate retains fuses only for inputs A and $\overline{B}$. Therefore, this AND gate produces the product term $A\overline{B}$. Similarly, the intact fuses for the middle AND gate provides the product term BCD while the lower AND gate produces AD. The programmed AND gates feed the OR gates to create a SOP expression. As we know, any truth table function can be implemented in the

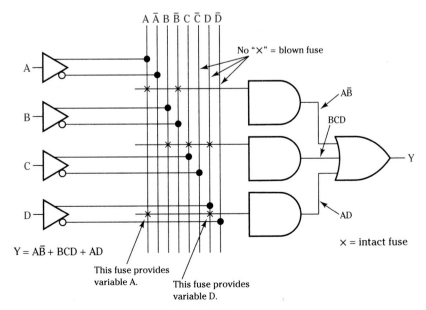

Figure 10.8
Creating a Logic Expression Using a PAL Device

SOP form, implying that the PAL is an ideal logic structure for AND-OR networks. Typical PAL chips have 16 inputs, so sizable SOP expressions are possible. Another advantage of the PAL and the SOP form is that all equations are produced in only two levels of logic. This minimizes propagation delays since there are fewer gates that signals pass through.

DESIGN EXAMPLE 10.3

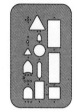

Produce $Y = \overline{A}BC + \overline{B}D + \overline{C}\overline{D}$ using the PAL structure shown in Figure 10.7.

Solution   The fuses representing the correct level of the variables for each product term must be left intact. Mark these with an x on the diagram. For example, the first product term—$\overline{A}BC$—necessitates that the fuses for $\overline{A}$, B, and C be left in place. All other fuses associated with that product term must be blown. These are represented on the diagram by the lack of an x.

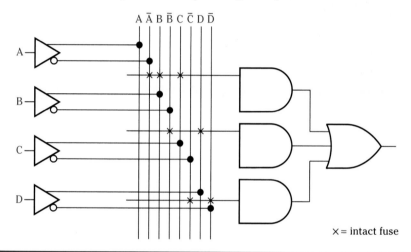

10.5 ■ PAL PART NUMBERING SYSTEM

PAL part numbers convey a considerable amount of information about the capabilities of a particular device. As seen in Figure 10.9, the part number consists of several entities. The letters PAL identify the part as a PAL device as opposed to some other programmable part. Usually, these letters are preceded by letters representing the manufacturer of the part. After the letters PAL, there may or may not be a letter or number/letter combination representing the device technology. PAL devices are manufactured in TTL, CMOS, and ECL technologies, providing a sizable range of power and speed specifications.

The next three sections of the part number tell how the PAL device is organized. For instance, in a 16R8, the number 16 identifies the number of inputs pins, the R refers to the logical treatment or "architecture" of the outputs, and the 8 refers to the number of outputs. Following this essential information there are indications of the part's speed and power.

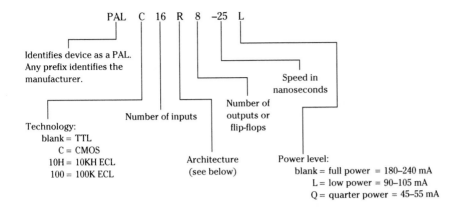

PAL   C   16   R   8   –25   L

Identifies device as a PAL.
Any prefix identifies the
manufacturer.

Technology:
  blank = TTL
  C = CMOS
  10H = 10KH ECL
  100 = 100K ECL

Number of inputs

Architecture
(see below)

Number of
outputs or
flip-flops

Speed in
nanoseconds

Power level:
  blank = full power = 180–240 mA
  L = low power = 90–105 mA
  Q = quarter power = 45–55 mA

Architecture—Combinatorial Devices

Code Letter	Meaning
H	Active high outputs
L	Active low outputs
P	Programmable output polarity
C	Complementary outputs
XP	Exclusive-OR gate, programmable
S	Product term steering

Architecture—Registered Devices

Code Letter	Meaning
R	Registered outputs
X	Exclusive-OR gates
RP	Registered—polarity programmable
RS	Registered—term steering
V	Versatile—varied product terms
RX	Registered Ex-OR
MA	Macrocell

Figure 10.9
PAL Part Numbering
System

10.6 ■ PRACTICAL PAL DEVICES

Figure 10.10 shows the data sheet for a 16L8 PAL. This part contains eight AND-OR networks, each consisting of seven 32-input AND gates feeding a 7-input OR gate. Each OR gate feeds a tristate inverter, which is enabled by its own 32-input AND gate. The figure shows how physical pins on the chip are sometimes shared to function as input or output lines. The eight output pins are fairly obvious—they are all from a tristate inverter output. The part number indicates that there are 16 inputs. Where are they?

Pins 1 through 9 and 11 are dedicated inputs. Pins 12 through 18 are output pins, but they may also function as inputs. If the tristate inverter connected to each one of these outputs is disabled, then the inverter output is in the Hi-Z state. Since no electrical signal is present from the inverter at this point, an input signal may be applied here. Notice how these pins feed around through an inverter/buffer into the AND gate array. When the associated output is disabled, the pin functions as an input feeding the AND array. If the output is used, the output signal still feeds the AND array, allowing the signal to be fed back into the AND array logic. Now the output signal can be included in other AND-OR networks on the chip or fed back to its own AND-OR network if this

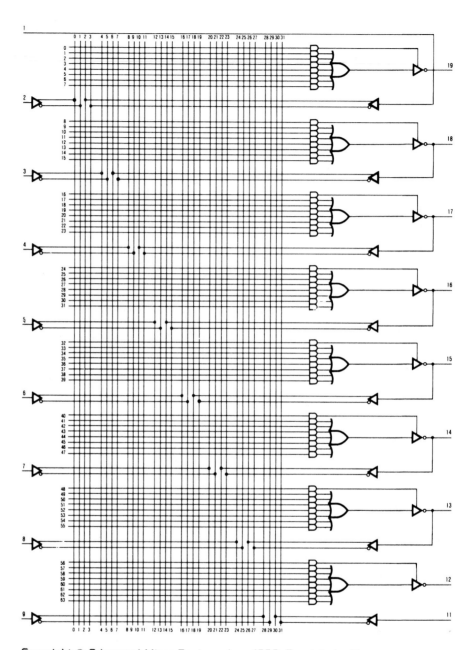

Figure 10.10
16L8 Data Sheet

is desirable (be careful doing this—latches may be formed). Designating pins as inputs or outputs is a decision made by the designer using the PAL device. This designation is generally accomplished with the PAL software development system (Section 10.8). Sharing physical pins is important in large-scale IC development because pins cannot be added without increasing the size of the chip. In order to maintain standard sized chips, or **chip footprints**, the number of pins is fixed. It is a lot easier for the IC developer to squeeze in an extra logic gate than to add an extra physical pin.

Notice also that since the output pins in the 16L8 are connected to tristate inverters, the output signals are active low (an AND-OR-invert structure is formed).

How must the PAL be programmed to allow pin 14 to function as an output? To function as an input?

DESIGN
EXAMPLE 10.4

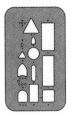

Solution Enabling the tristate inverter controlling pin 14 allows the pin to function as an output. An input pin is set aside as the enable line for this function.

To allow pin 14 to function as an input, you must permanently disable the tristate inverter. This is accomplished by leaving every pin to the controlling AND gate intact. Since the AND receives the true and complement levels from all inputs, the AND gate output will always be low regardless of the input combinations.

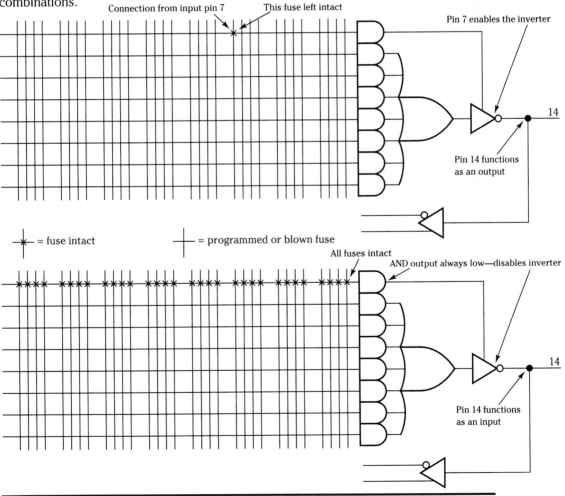

DESIGN
EXAMPLE 10.5

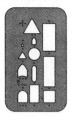

Design the circuit represented by the following truth table using a 16L8 PAL:

A	B	C	D	E	Y	
0	0	0	0	0	1	
0	0	0	0	1	1	
0	0	0	1	0	1	
0	0	0	1	1	1	
0	0	1	0	0	0	$-\overline{A}\overline{B}C\overline{D}\overline{E}$
0	0	1	0	1	1	
0	0	1	1	0	1	
0	0	1	1	1	1	
0	1	0	0	0	1	
0	1	0	0	1	1	
0	1	0	1	0	0	$-\overline{A}B\overline{C}D\overline{E}$
0	1	0	1	1	1	
0	1	1	0	0	1	
0	1	1	0	1	1	
0	1	1	1	0	1	
0	1	1	1	1	0	$-\overline{A}BCDE$
1	0	0	0	0	1	
1	0	0	0	1	1	
1	0	0	1	0	1	
1	0	0	1	1	1	
1	0	1	0	0	1	
1	0	1	0	1	1	
1	0	1	1	0	0	$-A\overline{B}CD\overline{E}$
1	0	1	1	1	1	
1	1	0	0	0	0	$-AB\overline{C}\overline{D}\overline{E}$
1	1	0	0	1	1	
1	1	0	1	0	1	
1	1	0	1	1	1	
1	1	1	0	0	1	
1	1	1	0	1	1	
1	1	1	1	0	1	
1	1	1	1	1	1	

Solution   Since the 16L8 has active low outputs (inverter following the OR gate), design the circuit using the zero outputs from the truth table—the complement method. The circuit may be designed directly from the truth table without reducing since circuit simplification takes place automatically when using computers for the design work (see Appendix C).

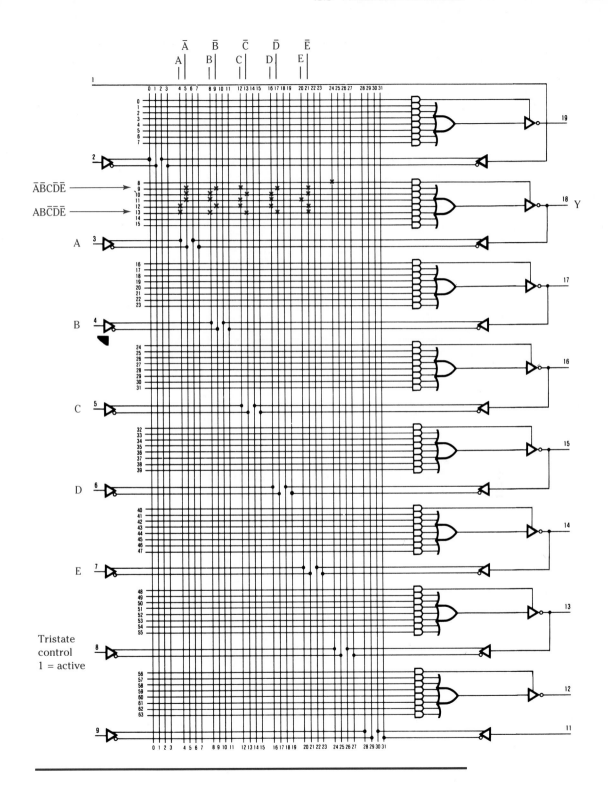

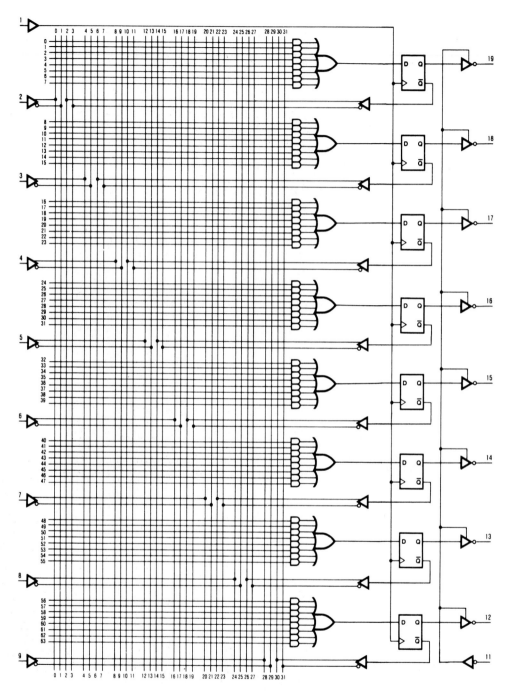

Figure 10.11
16R8 Data Sheet

Registered Devices

The 16L8 is a combinatorial device because there are no flip-flops in the chip. However, many flip-flop devices, referred to as **registered devices**, are available. The 16R8 is an example (Figure 10.11).

The basic logic structure for the 16R8 is the same as for any other PAL—an AND array feeding an OR array. Registered devices also have a flip-flop, and often other circuitry, inserted before the output. The 16R8 contains eight D flip-flops whose data input pins are fed by the OR gate outputs. Pin 1 is dedicated as a clock signal to feed the clock input of each D flip-flop. This makes the chip a synchronous device since all flip-flops are clocked simultaneously. In addition, each flip-flop $\overline{Q}$ output feeds back into the AND array. Since these $\overline{Q}$ lines feed an inverter/buffer gate, it is possible to have either the Q or $\overline{Q}$ value from the flip-flop feed back into the logic. **Registered feedback** is common in PAL devices and means that data stored in a flip-flop can be used to control the states of other flip-flops in the PAL—very useful for complex sequential circuits.

Design a 4-bit PISO shift register using a 16R8 PAL.

DESIGN EXAMPLE 10.6

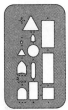

Solution Use the shift register schematic that follows as the basis for the design. Since output signals from flip-flops will be fed back into the PAL array, the schematic shows outputs taken from the flip-flops' $\overline{Q}$ pins. Note that this manual design effort is greatly simplified using a computerized software development system (see Appendix C).

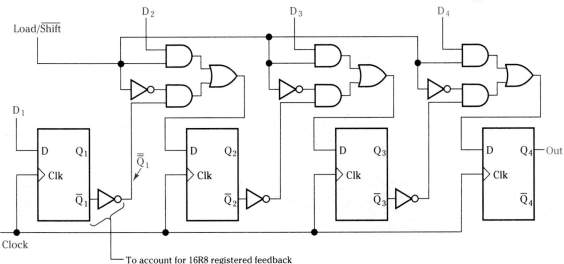

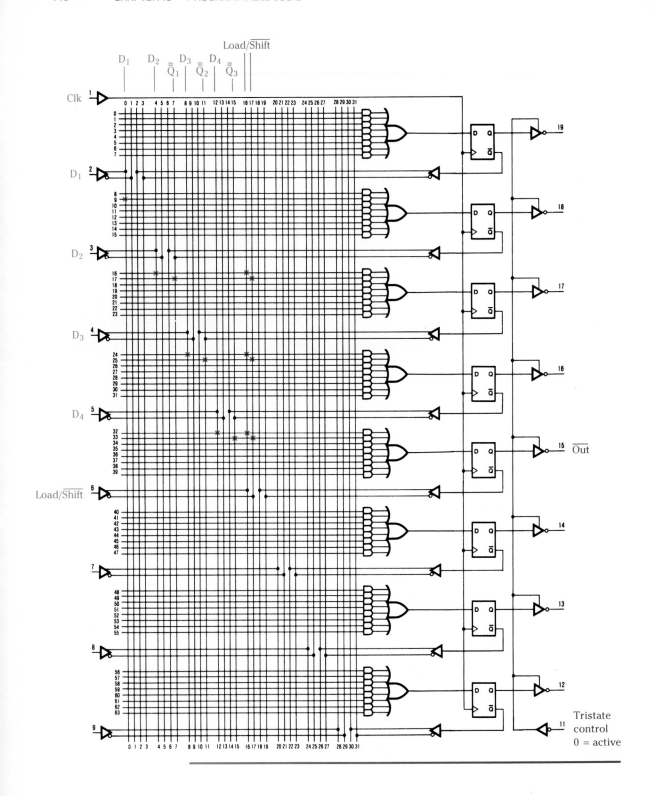

Macrocells

PAL devices with increasing complexity are designed by adding more and more programmable functions to the structure beyond the OR array. The increased number of functional elements generally include flip-flops, multiplexers, and

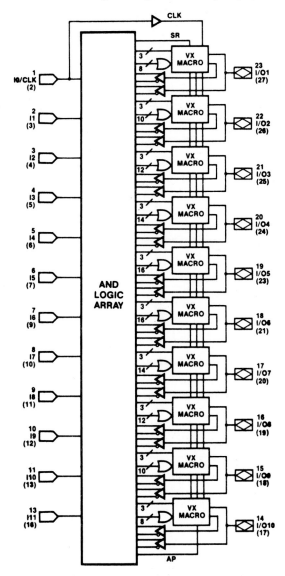

Note: PLCC pin numbers are indicated in parentheses.
PLCC pins 1, 8, 15, and 22 are not connected.

Figure 10.12
32VX10 Organization

Exclusive-OR gates. All these elements are connected by programmable junctions so that a wide variety of logic configurations is possible. An arrangement of elements as described is called a **macrocell**, a structure replicated throughout the chip. The 32VX10 shown in Figure 10.12 is an example. The majority of the complete chip is still an AND array, feeding a versatile output structure. A macrocell for this device consists of a portion of the AND-OR array and the associated output circuitry. This is illustrated in more detail in Figure 10.13. A macrocell such as this allows the designer to configure the PAL to include the following capabilities: J-K, S-R, D flip-flops, by modifying the flip-flop input lines; programmable output polarity, by programming the $Q/\overline{Q}$ multiplexer; and asynchronous or synchronous presets and clears, by modifying the output multiplexers. In addition, this PAL, as well as many others, includes a programmable **security fuse**. Once blown, the device cannot be read by a PAL programmer, so proprietary designs may be protected.

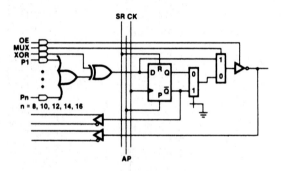

Figure 10.13
32VX10 Macrocell

DESIGN
EXAMPLE 10.7

Show how the 32VX10 macrocell can be configured for:
(a) AND-OR operation
(b) AND-OR-invert operation
(c) D flip-flop with $\overline{Q}$ feeding output

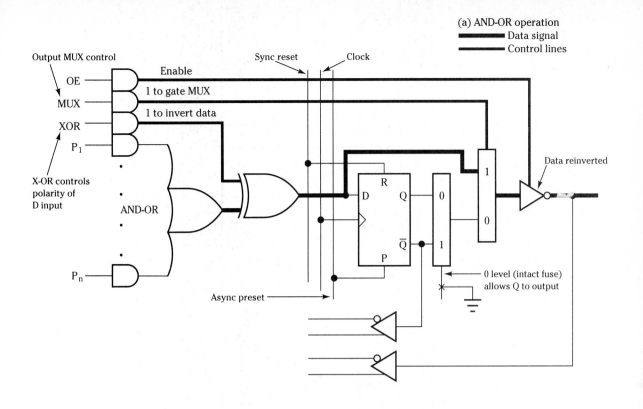

(a) AND-OR operation
■■■ Data signal
■ Control lines

Output MUX control

OE

MUX

XOR

P_1

X-OR controls
polarity of
D input

AND-OR

P_n

Enable

1 to gate MUX

1 to invert data

Sync reset   Clock

R

D   Q

$\overline{Q}$

P

Async preset

1

0

0

1

Data reinverted

0 level (intact fuse)
allows Q to output

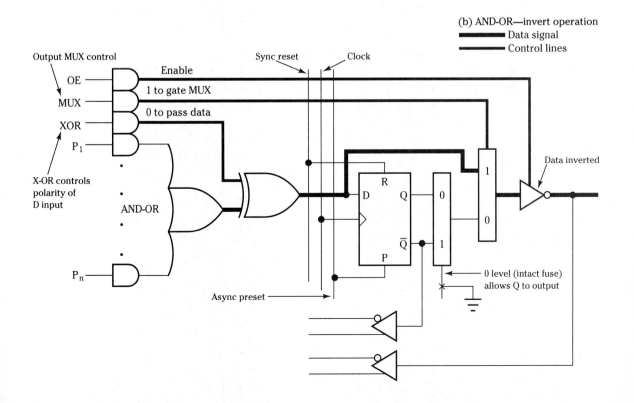

(b) AND-OR—invert operation
■■■ Data signal
■■■ Control lines

Output MUX control

OE

MUX

XOR

P_1

X-OR controls
polarity of
D input

AND-OR

P_n

Enable

1 to gate MUX

0 to pass data

Sync reset   Clock

R

D   Q

$\overline{Q}$

P

Async preset

1

0

0

1

Data inverted

0 level (intact fuse)
allows Q to output

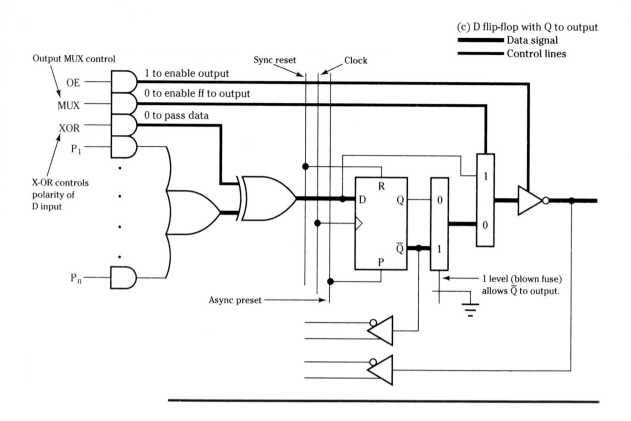

(c) D flip-flop with Q to output
■ Data signal
■ Control lines

Output MUX control

OE — 1 to enable output

MUX — 0 to enable ff to output

XOR — 0 to pass data

P_1

X-OR controls
polarity of
D input

P_n

Sync reset   Clock

Async preset

R
D   Q   0

Q̄   1

P

1

0

1 level (blown fuse)
allows Q̄ to output.

10.7 ■ THE PROGRAMMABLE FUSE

Before concluding this section it is worth spending a moment to understand the character of the "fuses" in programmable devices because they are the portion of the structure that characterizes our designs. There are two technologies used to create programmable devices—fuses and floating gates.

If these two terms sound familiar it is not surprising since these are the same devices used in PROMs and EPROMs. Fuses are commonly made from platinum-silicide and are truly fuses in every sense of the word. Figure 10.14 shows how the fuse structure is made narrower than the surrounding material. During programming a voltage pulse melts the narrow section of material, thus "blowing" the fuse. Surface tension pulls the material away from the region,

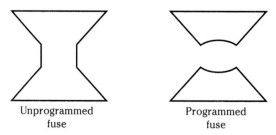

Figure 10.14
Programmable Logic
Fuse

Unprogrammed
fuse

Programmed
fuse

preventing the fuse from regrowing. Figure 10.15 shows the nature of the schematic interconnection. The crossing of wires on the schematic is actually
connected via a fuse in series with a diode. The connection persists if the fuse
remains intact, otherwise the connection is broken.

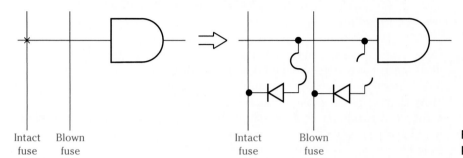

Intact   Blown   Intact   Blown
fuse   fuse   fuse   fuse

Figure 10.15
PAL Fuse Connections

    Floating gates are areas within the semiconductor structure that are con-
structed to trap electric charge. The floating gate actually controls the on-off
activity of the transistors comprising the AND gate transistors. An advantage
of this kind of device is that the PAL can be reprogrammed after exposing the
chip to ultraviolet light. As you recall, ultraviolet light releases the charge stored
in the floating gate, rendering the device suitable for programming again.

10.8 ■ PROGRAMMABLE LOGIC—SOFTWARE

The discussion in previous sections centered on the many forms of programmable logic and showed how programmable AND-OR arrays can be used to
implement logical functions. The design examples used to illustrate the ver-
satility of programmable logic were manually created by placing an x on the
PAL schematic wherever an AND gate connection was desired. Manual creation
of PAL circuits in this manner is time-consuming, error-prone, and somewhat
tedious. Programmable logic design is much more easily accomplished using
programmable logic development software. This section introduces programmable logic design using standard development software. Included in Appendices C and D are specific details regarding a commonly used PAL programming
software package called PALASM2. This development software, or any other
appropriate development software, is essential if you wish to program your
own PAL parts. The next few sections also serve as an introduction to computer
simulation and computer generated timing diagrams.

Programmable Logic Software Systems

Software for programmable logic development is readily available from many
sources. Manufacturers of programmable logic devices often have software
specifically tailored to program the parts they produce. Other independent
software companies offer software products designed to program most of the

programmable parts on the market. In any case the designer can obtain development software easily.

Programmable logic software varies in complexity, depending on the nature of the programmable logic device used. For instance, the software required to support the development of a gate array is significantly different from that required for a PAL.

The development software provides several benefits for the designer. Design time is greatly reduced because the software is optimized to assist the designer in the development work. Initially, designs are entered into the system as Boolean equations, truth tables, state equations, or schematics. Word-processing software is often utilized to assist with design entry. The development software minimizes the equations and produces a **fuse map** representing the blown and intact fuses in the device. Before burning the part, simulation tests are run to ensure the functionality of the design. When the designer is satisfied with the results, the software communicates the fuse data to a PAL programmer or other appropriate programming device to actually blow the fuses.

The software runs on most personal computing systems, although very powerful and sophisticated software packages may require the computing power of workstations or mainframe computers. The PAL programmer is usually a separate attachment that communicates with the computer through an interface card. Figure 10.16 shows the computing requirements for programmable logic development.

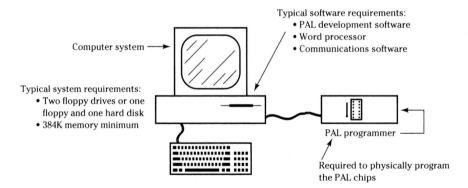

Typical software requirements:
• PAL development software
• Word processor
• Communications software

Computer system

Typical system requirements:
• Two floppy drives or one floppy and one hard disk
• 384K memory minimum

PAL programmer

Required to physically program the PAL chips

**Figure 10.16
PAL Programming
Requirements**

The Programmable Logic Development Flow

Several steps take place between the conception of a design idea and the finished product, as shown in Figure 10.17. This is true in any design undertaking, but using programmable logic, the software dictates the sequence of steps to take. This is not restrictive, though, since the software is written to ensure the production of quality designs.

Creation of the logic equations that represent the circuit under design is the first step in the process. Typically, a standard word-processing program is used to create the **input design file**. The programmable logic software accepts the input file as the design information source for further processing. The format

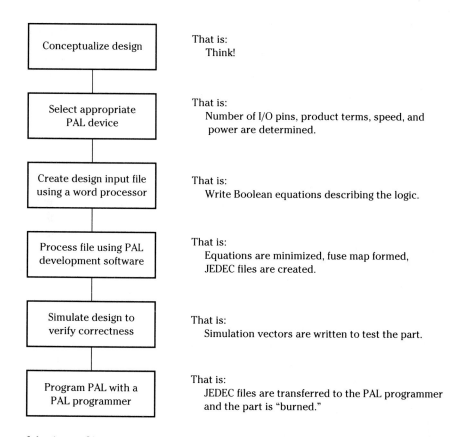

| Conceptualize design | That is:
    Think! |

| Select appropriate
PAL device | That is:
    Number of I/O pins, product terms, speed, and
        power are determined. |

| Create design input file
using a word processor | That is:
    Write Boolean equations describing the logic. |

| Process file using PAL
development software | That is:
    Equations are minimized, fuse map formed,
    JEDEC files are created. |

| Simulate design to
verify correctness | That is:
    Simulation vectors are written to test the part. |

| Program PAL with a
PAL programmer | That is:
    JEDEC files are transferred to the PAL programmer
    and the part is "burned." |

Figure 10.17
PAL Processing Steps
Using Development
Software

of the input file varies depending on the development software in use, but most are quite similar.

Once an input file is created, the development software reads it and checks for proper command syntax. Any errors are flagged with messages so that the errors can be corrected. Once the file is error-free, further processing continues.

Typical programmable logic development software will examine the logic equations and attempt to minimize the logic structure. Some software packages have several minimizing options for this purpose; others have only one. For most designs the minimizing software eliminates nonessential variables, shortening product terms. A big advantage to the designer is that minimizing software eliminates manual reduction normally accomplished with K-maps or other methods.

After the reduction steps are completed, the development software creates a fuse map indicating which fuses are to be blown. (Some PAL devices do not have fuses, but rather, have charge storage gates similar to EPROMs. However, the programming concepts are the same.) Typically, several computer files are generated at this point. The fuse map shows a pattern of "x" and "–" symbols, indicating where the fuses are blown. (An x indicates an intact or unprogrammed fuse, a – indicates a blown or programmed fuse.) The map is usually used by the designer only to check the design. A more important file created for the actual programming task is the **JEDEC fuse data file.** The data in this file also represents the device fuse pattern using a standard format utilized by PAL programmers. This file is downloaded to the PAL programmer from the computer when physical programming takes place. (JEDEC stands for Joint Electronic Device Engineering Council, a group setting standards for use in the electronics industry.)

Simulation tests are run following fuse map creation to verify the functional operation of the design. **Simulation vectors**, indicating the expected output states occurring for specific input combinations, are written as a special simulation file or included with the initial input data file. The designer writes test vectors to verify that the part will perform as expected. Frequently, several simulation runs will be required as errors are uncovered and corrected. An error usually means that the input design file requires a change that necessitates rerunning all previous steps. Fortunately, these steps run in a matter of seconds, so that there is little impact on the designer's time to rerun the software. (A major impact may occur if the design itself requires a significant change since the designer will have to reevaluate the actual circuit design.) Simulation should take place before the PAL is actually programmed to eliminate creating scrap parts.

When the design has been verified as functional, the programmable logic device is programmed. A blank or virgin device of the appropriate part number is placed into a PAL programmer and then the JEDEC fuse file is transferred by the software to the programmer. Naturally, the PAL programmer is connected to the computer system for this operation. The PAL programmer interprets the

FOR YOUR INFORMATION

Design Methods Comparison

	Unit Cost Range	NREs/ Development Costs	Purchase Volume For Best Economics	Design Flexibility	Gate Count per Device	Engineering Design Time	Design Turnaround Time	Cost of Each Design Change
PLDs: BLANK	Low–Med	None	5K–200K	Med	200–2K	1/2 Week	Short (1–10 Days)	Very Low
PLDs: FACTORY PROGRAMMED	Low–Med	Low	10K–200K	Med	200–2K	3–10 Weeks	Moderate (8–10 Wks.)	Med
DISCRETE LOGIC	Low–Med	None	1K–10K	Low	10–30	1 Week	Short (1–10 Days)	Med
GATE ARRAYS	Low	Med–High	10K–200K	Med–High	1K–10K	12–40 Weeks	Long (3–9 Mos.)	High
STANDARD CELLS	Low	High	100K–300K	Med–High	1K–10K	26–52 Weeks	Long (6–12 Mos.)	High
FULL CUSTOM	Very Low	High	200K and Up	High	1K–60K	1–2 Yrs.	Long (6–12 Mos.)	High

fuse map data and applies the proper programming voltages to the pins of the PAL, blowing designated fuses. This is accomplished in a few seconds. Once complete, the part is ready for use.

- Programmable logic devices such as PALs represent a portion of the ASIC (application specific integrated circuit) design approach currently used by the digital logic design industry.

- PALs offer an easy approach to implementing common AND-OR circuits on a single chip with the added benefit of reduced propagation delay time.

- Due to the large number of gates and gate inputs in a typical PAL, special schematic notation is used.

- Programmable logic is customized by blowing fuses within the chip's AND array.

- Programmable logic development is typically carried out using computer software written to simplify the design, testing, and production of the PAL part.

- PALs range in complexity from combinatorial parts to registered macrocell parts.

Section 10.3

*1. A PROM has 12 address input lines. How many AND gates are required in the device's AND array? What is the minimum number of AND gates that would be required in a PAL or PLA with the same number of inputs?

*2. Determine the output expression for the following PAL schematics. Assume that every fuse is intact.

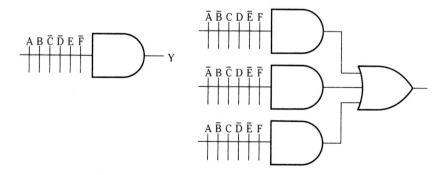

---

* See Appendix F: Answers to Selected Problems.

3. Determine the output expression for the following PAL devices: * = intact fuse

*(a)

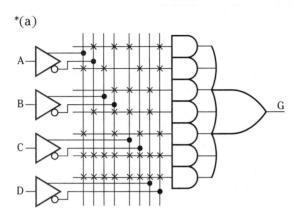

(b)

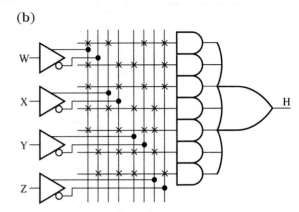

4. Using the following PAL structure, indicate where fuses should or should not be blown to create the following logic expressions: + = blown fuse; * = intact fuse

*(a) $Y = \overline{A}\overline{B}\overline{C} + \overline{A}BC + A\overline{B}C$
(b) $Y = A\overline{B}\overline{C} + A\overline{B}C + AB\overline{C}$
(c) $Y = \overline{A}\overline{B}C + \overline{A}BC + ABC$

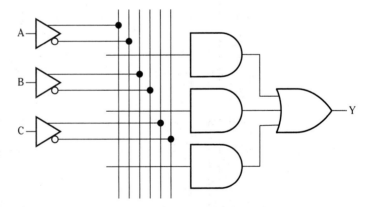

5. Use the PAL structure in problem 4 to implement the following truth tables:

*(a) A	B	C	Y
0	0	0	1
0	0	1	1
0	1	0	0
0	1	1	0
1	0	0	0
1	0	1	1
1	1	0	0
1	1	1	0

(b) A	B	C	Y
0	0	0	0
0	0	1	0
0	1	0	1
0	1	1	0
1	0	0	0
1	0	1	1
1	1	0	0
1	1	1	1

(c) A	B	C	Y
0	0	0	0
0	0	1	0
0	1	0	0
0	1	1	1
1	0	0	1
1	0	1	1
1	1	0	0
1	1	1	0

* See Appendix F: Answers to Selected Problems.

Sections 10.5–10.6

6. Determine the number of inputs, outputs, and the architecture for the following PAL part numbers:
   *(a) PAL 16R6   (b) PAL 10H8   (c) PAL 18P8   (d) PAL22XP10
*7. If a 16L8 is programmed to contain 14 inputs, how many outputs are available?

Section 10.6

8. Sketch the connections required for a 16L8 to implement the following equations:

$$Y = \overline{A}\overline{B}C + A\overline{B}\overline{C} + AB\overline{C} + ABC$$
$$W = \overline{S}TUV + \overline{S}T\overline{U}\overline{V} + S$$

*9. Design an 8-bit tristate bus using a 16L8. How can any one of the eight outputs control the bus at a time? The bus should take the following form:

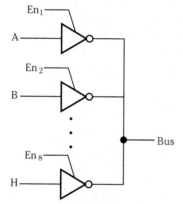

10. Use a 16L8 to implement the following truth table:

A	B	C	D	Y
0	0	0	0	0
0	0	0	1	1
0	0	1	0	1
0	0	1	1	0
0	1	0	0	0
0	1	0	1	1
0	1	1	0	0
0	1	1	1	0
1	0	0	0	1
1	0	0	1	0
1	0	1	0	0
1	0	1	1	1
1	1	0	0	0
1	1	0	1	1
1	1	1	0	0
1	1	1	1	1

---

* See Appendix F: Answers to Selected Problems.

11. Use a programmable logic device to create the following functions:
    (a) inverter
    (b) 4-input AND gate
    *(c) 3-input OR gate
    (d) 6-input NAND gate
    (e) 7-input NOR gate
    (f) Exclusive-OR gate
12. Using a PAL data book, determine the following specifications for a 16R6 PAL:
    (a) the number of outputs
    (b) the number of inputs
    (c) the number of product terms
    (d) the output polarity
13. Repeat problem 12 for a 22V10 PAL.
14. Design a MOD 8 synchronous counter using a 16R8. Draw the schematic and mark all PAL pins on it. On a 16R8 data sheet, mark all blown and intact fuses for the product terms used.
15. Sketch the design of a 6-bit PISO shift register using a 16R8. Draw the logic schematic and mark all PAL pins on it. On a 16R8 data sheet, mark all blown and intact fuses for the product terms used.
*16. Show how the 32VX10 macrocell shown here can be programmed so that an AND-OR-invert function supplies the D input of the flip-flop with data. In addition, what must be done so that the Q output data feeds back into the AND-OR array? How can the output pin be utilized as an input pin?

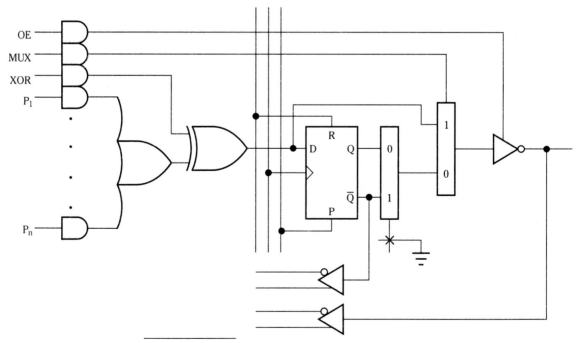

---

* See Appendix F: Answers to Selected Problems.

CHAPTER 11

COMMUNICATING WITH THE ANALOG WORLD

OBJECTIVES

To explain how analog signals are converted into digital signals.

To explain how digital signals are converted into analog signals.

To illustrate practical circuits used to carry out digital signal interfacing.

PREVIEW

The utility of digital logic circuits becomes apparent when we establish the means to send and receive analog information with the circuitry. Since we live in an analog environment, communication with digital circuits is a matter of interfacing analog devices and associated actions with the binary nature of the digital system. This chapter shows many ways to accomplish interfacing, ranging from simple switches and lights to more complex analog-to-digital converters. This chapter also delves into a satisfying aspect of design work that is using digital systems to control practical, real-life devices.

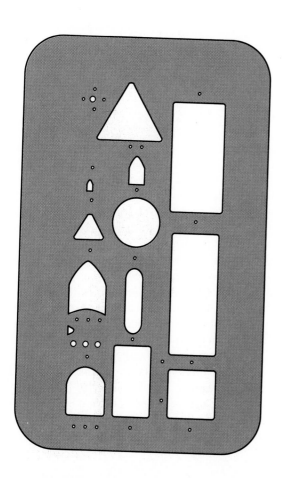

11.1 ■ BASIC INPUT AND OUTPUT INTERFACING

Digital input and output signals can be connected in a wide variety of ways to enable communication between analog and digital environments. The most common method is using switches as input devices and LEDs (light emitting diodes) as output indicators.

Switches

Figure 11.1 shows how a switch (single pole, single throw) is connected to provide solid logic one and zero levels for digital circuit use. The switch symbol shown represents a wide variety of common switches such as push button, toggle, and rocker switches.

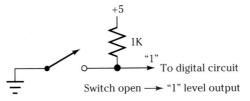

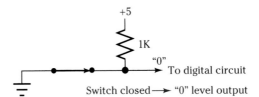

Figure 11.1
Wiring a Switch to
Provide Logic Levels

When the switch is closed, a low logic level is presented to the digital circuit through the switch's connection to ground. When the switch is open, a high logic level is presented to the digital circuit via the "pull-up" resistor connection to the power supply. The pull-up resistor is necessary for several reasons. First, it does provide the voltage necessary to represent the one logic level. Theoretically, the switch alone could provide a "floating one" level when opened. Since some digital technologies interpret the lack of any voltage on an input pin as a high level, the open switch contact could be intrepreted as a high level input. However, floating one inputs are not recommended because an unconnected pin easily picks up noise signals that can disrupt circuit operation or result in excessive power dissipation. Therefore, the pull-up resistor is included to provide a meaningful logic one.

The pull-up resistor also prevents the power supply positive lead from becoming shorted to the ground lead when the switch is closed. Shorted voltage terminals are very upsetting to power supplies. The 1K value for the pull-up resistor is a typical value. Using too small a resistance will draw unnecessarily large amounts of current from the power supply; too large a pull-up resistance can slow down the rate of change of the input signal.

Show how three waterproof SPST switches activated by a float can be used to detect whether a tank is full, half full, or empty.

DESIGN
EXAMPLE 11.1

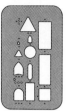

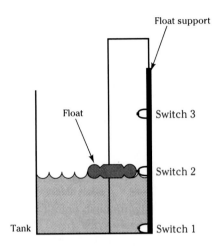

Float support

Float

Switch 3

Switch 2

Tank

Switch 1

Solution The switches should each be connected with a pull-up resistor so that one terminal supplies a logic high level. The other terminals are connected to ground to supply a low indication.

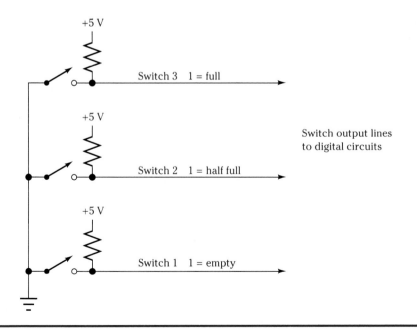

+5 V

Switch 3   1 = full

+5 V

Switch output lines
to digital circuits

Switch 2   1 = half full

+5 V

Switch 1   1 = empty

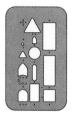

DESIGN EXAMPLE 11.2

Show how 16 switches can be utilized to form a hexadecimal keypad.

Solution   Sixteen switches can be used with each switch providing one hexadecimal signal. For instance, if switch 12 is depressed, an active signal on the switch wire is interpreted as a Hex C. This method requires 16 individual output wires. To conserve wires, switches are usually connected as an array. The array input lines are constantly scanned by a series of digital pulses. The array output is constantly monitored. When a switch closure is detected, the scan code (input signal) and return code (output signal) at the moment determine which key was closed.

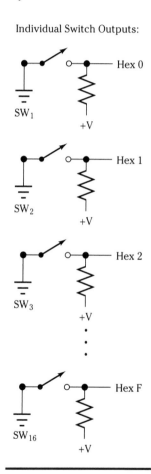

Individual Switch Outputs:

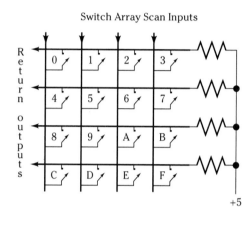

Switch Array Scan Inputs

Switch Bounce

Mechanical switching devices have an annoying digital property called **switch bounce**. When the switch is activated, causing the switch contact and wiper either to connect or disconnect, the motion and associated momentum of the contact elements prevent a clean transition from taking place. For instance, assume that the switch is being closed. Ideally, the moving parts of the switch would settle into position against the stationary parts of the switch and hold still. Instead, the moving parts of the switch (wiper) hit against the stationary switch parts (contacts) and bounce on and off the contact for a duration of several milliseconds. Bouncing also occurs when the switch is opened because the breaking apart of the wiper and contact does not occur cleanly nor instantaneously. The net effect of switch bounce is the production of many one-zero or zero-one logic level transitions when there should only be one. Since these transitions are actually positive and negative going edges, switch bounce is particularly troublesome in sequential circuitry. Figure 11.2 summarizes the switch bounce problem.

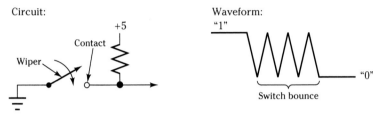

Circuit:

Waveform:

As the switch wiper makes a connection with the switch contact, the wiper bounces up and down due to the mechanical action and momentum.

**Figure 11.2
Switch Bounce**

Switch bounce problems are overcome in two ways using hardware and software solutions. Most computer systems "debounce" switches through programming. The computer program simply waits a few milliseconds before accepting any switch level. This gives any undesirable switch activity time to subside before the computer accepts the switch data.

Some hardware debouncing solutions are shown in Figure 11.3, the actual solution depending on the switch used. NAND-NAND Set–Reset latches (or the

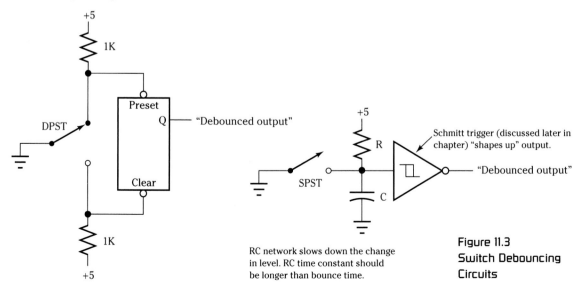

RC network slows down the change in level. RC time constant should be longer than bounce time.

**Figure 11.3
Switch Debouncing
Circuits**

preset and clears on D or J-K flip-flops) are often used for debouncing. The switch is connected to the preset and clear pins while the actual digital level is extracted from the flip-flop Q output. Since preset and clear are only sensitive to logic levels (not edges), they respond to a change in switch level while ignoring subsequent bouncing. Meanwhile, the Q output produces a nice clean output transition that is suitable for digital circuitry.

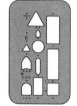

DESIGN EXAMPLE 11.3

Draw the timing diagram for the bouncy switches and the NAND-NAND latch debouncing circuit to demonstrate how bounces are controlled.

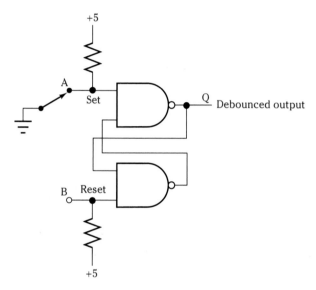

Solution   Both switch contacts (A and B) will exhibit bouncing as the contacts are moved. Since the NAND-NAND latch responds only to low levels, Q will equal zero the moment the latch detects a low on the reset input. Conversely, Q will equal 1 the moment the latch detects a low on the reset input. All bounces will be ignored since they are primarily edges. However, all subsequent low levels, after the first detected one, will also be ignored by the latch since it will already be set or reset.

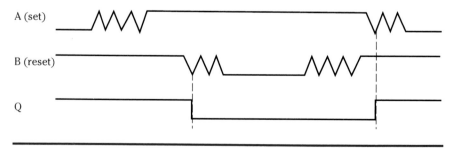

LEDs

Light emitting diodes (LEDs) are physically small solid state devices that provide illumination when activated by low voltages. LEDs do require moderate amounts of current flow, but the low voltage characteristic of the LED, combined with low cost, makes an LED a primary digital display device.

Connecting an LED to a digital gate requires knowledge of both logic family and LED characteristics. LEDs must operate within a specified current range or they will burn out. The logic gate must be able to supply the necessary current or the LED will not light. Figure 11.4 shows a typical LED-logic gate arrangement.

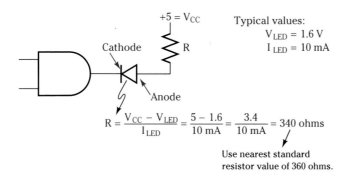

Typical values:
$$V_{LED} = 1.6 \text{ V}$$
$$I_{LED} = 10 \text{ mA}$$

$$R = \frac{V_{CC} - V_{LED}}{I_{LED}} = \frac{5 - 1.6}{10 \text{ mA}} = \frac{3.4}{10 \text{ mA}} = 340 \text{ ohms}$$

Use nearest standard
resistor value of 360 ohms.

Figure 11.4
Using a Logic Gate to Drive an LED

LEDs act like normal diodes in that current can flow only in one direction. An LED has to be connected such that the voltage polarity across the device makes the "anode" more positive than the "cathode." In addition, the magnitude of this "forward bias potential" must be great enough to allow LED current flow to exist. In the figure, notice how the diode is connected so that the anode is connected toward the +5 V supply and the cathode is connected to the AND gate output. In this configuration the LED lights when the AND gate output is low since a typical low logic level is approximately 0 V. Under this condition current flows (sink current) through the LED from the supply into the AND gate. A high logic level, on the other hand, keeps the LED off because the LED will not have sufficient forward bias to conduct. This configuration is typically used since most logic gates provide greater amounts of current flow with low outputs than with high outputs. Naturally, the logic gate used must be rated to supply the necessary LED current. The value of the resistor depends on the LED voltage and current requirements. The calculations in Figure 11.4 show typical values. LEDs can be purchased with many different characteristics. Some have low current requirements making the interface to any logic gate easy. Most LEDs increase light output as current flow increases. For large current requirements other driving interface circuits may be required. In addition, LEDs are available as individual lamps, as seven segment displays, as bar graphs, and in other useful forms.

DESIGN
EXAMPLE 11.4

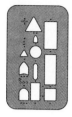

Devise a circuit to illuminate a red LED if one switch is opened or a yellow LED if a second switch is opened. A green LED should light if both switches are opened. The red and yellow should be off under this condition. Typical LED values are shown as follows:

MV50B standard red LED	MV53B yellow LED	MV5474C high efficiency green
$I_f = 20$ mA	$I_f = 20$ mA	$I_f = 20$ mA
$V_f = 1.65$ V	$V_f = 2.1$ V	$V_f = 2.2$ V

Solution Each switch may drive the required logic gates to produce a low output for the required function. Calculate appropriate values of current limiting resistance to restrict current to rated values. Set up a truth table to design the logic circuit.

SW_1	SW_2	RED	YELLOW	GREEN	
0	0	0	0	0	switch levels: 0 − switch closed
0	1	0	1	0	1 − switch opened
1	0	1	0	0	output levels: 0 − LED off
1	1	0	0	1	1 − LED on

$$\text{Red} = SW_1 \, \overline{SW_2}$$
$$\text{Yellow} = \overline{SW_1} \, SW_2$$
$$\text{Green} = SW_1 \, SW_2$$

Since low levels are needed to light the LEDs, complement all output levels. Thus:

$$\text{Red} = \overline{SW_1\overline{SW_2}}$$
$$\text{Yellow} = \overline{\overline{SW_1}SW_2}$$
$$\text{Green} = \overline{SW_1SW_2}$$

$$R_{LED} = \frac{V_{CC} - V_f}{I_f}$$

$$R_1 = \frac{5 - 1.65}{20 \text{ mA}} = 167 \text{ ohms}$$

$$R_2 = \frac{5 - 2.1}{20 \text{ mA}} = 145 \text{ ohms}$$

$$R_3 = \frac{5 - 2.2}{20 \text{ mA}} = 140 \text{ ohms}$$

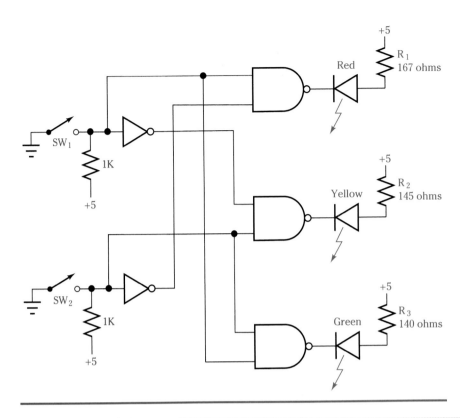

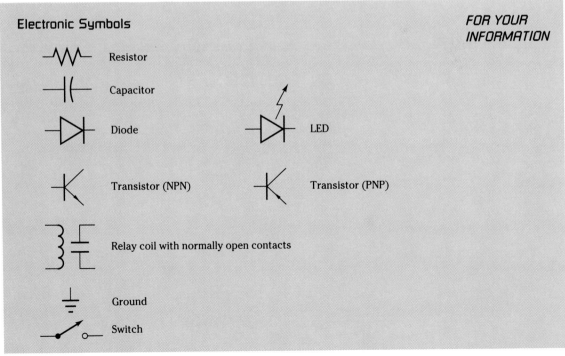

Electronic Symbols

FOR YOUR
INFORMATION

—⩗⩗— Resistor

—‖— Capacitor

—▷⊢ Diode   ⫸ LED

—ⱪ Transistor (NPN)   ⱪ Transistor (PNP)

⫶‖ Relay coil with normally open contacts

⟂ Ground

⟋° Switch

11.2 ■ DIGITAL DEVICES USED FOR INTERFACING

Many logic families contain devices that are useful for interfacing applications. These devices either provide current levels greater than that available from standard devices or fit a particular applications niche. We have covered some already in this book, such as tristate devices, but others also deserve our attention.

Open Collector/Open Drain

Open collector (TTL) and open drain (CMOS) devices are similar to standard logic devices except for the IC's (integrated circuit's) output circuitry. The output of these devices must be connected through a pull-up resistor or other load to the power supply in order to provide a logical high output. Although this reduces switching speed, decreases noise margin, and increases power dissipation, the open collector/open drain configuration has some advantages.

Open collector/open drain circuits can provide greater current handling capability than standard logic gates, making them very useful as an interface to high current devices. In addition, open collector/open drain circuits can

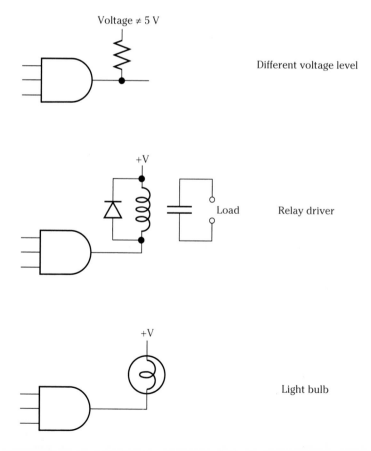

Voltage ≠ 5 V

Different voltage level

+V

Load   Relay driver

+V

Light bulb

Figure 11.5
Open Collector/Drain
Applications

have their outputs pulled-up to power supply voltages greater than that of the logic power supply. This allows the higher current and voltage requirements of nonlogic devices to be met by the digital logic family. Open collector devices like this are usually called **buffer/drivers**. Naturally, you should consult the device data sheets to determine maximum current and voltage ratings since some open collector devices cannot handle any more current than a standard logic gate. Figure 11.5 shows examples of open collector interfaces.

Another interesting aspect of open collector/open drain devices is **wired logic**. The term wired logic means that a logical function is created simply by connecting the outputs of gates together. This is a logic function in addition to that provided by the open collector gate function. This can only be done with open collector/open drain circuitry since tying standard logic gate outputs together would cause the chips to fail.

Figure 11.6 shows how TTL open collector devices produce a **wired-AND** or **dot-AND** function. The individual open collector gate outputs are tied together, but the level that appears at this connection is dependent on the AND function of the three device outputs. In other words, the wired connection produces a one level only when all three gate outputs are also producing one outputs. This "free" level of logic is used extensively in some technologies and is also used in some memory expansion schemes.

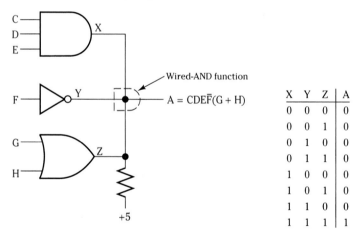

Wired-AND function

$A = CDE\overline{F}(G + H)$

X	Y	Z	A
0	0	0	0
0	0	1	0
0	1	0	0
0	1	1	0
1	0	0	0
1	0	1	0
1	1	0	0
1	1	1	1

Figure 11.6
Wired-AND Logic

Design the interface allowing the following truth table function to run a 117 V AC, 3.5 A fan. Use a 12 V relay as the interface device:

DESIGN EXAMPLE 11.5

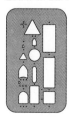

A	B	C	Fan_on
0	0	0	0
0	0	1	1
0	1	0	0
0	1	1	0
1	0	0	1
1	0	1	1
1	1	0	0
1	1	1	0

Solution   Determine the logic equation from the truth table. Interface the logic to the relay using an open collector device.

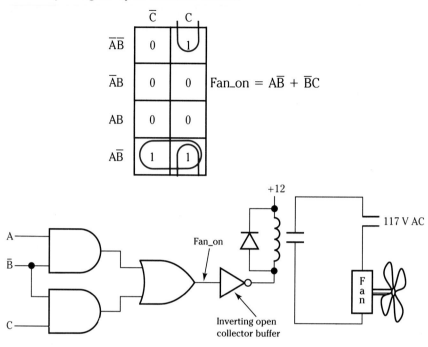

$$Fan\_on = A\overline{B} + \overline{B}C$$

Inverting open collector buffer

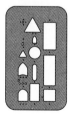

**DESIGN
EXAMPLE 11.6**

Determine if the relay and open collector devices used in Design Example 11.5 are adequate for the circuit design. The relay is a 12 volt 650 ohm relay. The 7406 open collector buffer has the following specifications: maximum $V_{OH} = 30$ V; and a maximum $I_{OL} = 40$ mA.

Solution   Since the supply voltage in Design Example 11.5 is listed as 12 volts, the relay voltage specification as 12 V will be present across the relay coil when the buffer output is low (0 V). The buffer voltage specification is also safe because the chip can withstand a maximum voltage of 30 V.

The relay will draw current when the buffer output is low. The magnitude of the current I is

$$I = \frac{V_{CC}}{R_{relay}} \qquad \text{or} \qquad I = \frac{12}{650} = 18.5 \text{ mA}$$

This current is safe with the buffer's current rating limit of 40 mA.

Schmitt Trigger

Schmitt triggers are used to shape up slowly changing digital signals. Most digital circuits require input signals that switch rapidly between the one and zero levels since input voltages between the actual one and zero voltage ranges do not represent any logic level. These "forbidden" voltages can cause a normal digital circuit to malfunction or exhibit strange behavior. When a digital circuit is driven by another logic device, fast transitions are assured. But when the input signal is derived from an analog source, the signal's rate of change may be much too slow for a digital circuit to process reliably. Schmitt triggers form a good interface between the slowly changing signal and the digital circuit because the Schmitt trigger can respond to slowly changing signals yet provide a rapid change in output level. Figure 11.7 shows this "signal conditioning" interface.

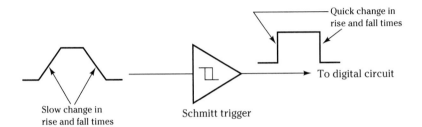

Quick change in rise and fall times

To digital circuit

Slow change in rise and fall times

Schmitt trigger

Figure 11.7
Using a Schmitt Trigger to Improve the Edges of a Waveform

Although standard logic gates change state when the input voltage approaches approximately the midpoint voltage between a logic high and low, a Schmitt trigger has two distinct switching or threshold voltages. One threshold voltage determines when the Schmitt trigger output switches from low to high and is called the **upper trip point** or **upper threshold voltage**. Another voltage called the **lower trip point** or **lower threshold voltage** determines when the output switches from high to low. These two threshold voltages are not the same, so the device is said to exhibit **hysteresis**. The hysteresis symbol is typically drawn inside any logic device that has this property. Figures 11.8a, b, and c show the effects of hysteresis.

The shaping of slowly changing input signals is not due so much to the upper and lower threshold voltages as it is to internal feedback, which causes a fast output transition to occur once the output changes in state begins. However, the differing threshold voltages are beneficial. Figures 11.8a, b show how signals with slow rise and fall times are converted to normal digital signals. Note that with positive going input signals, the output response does not occur until the upper trip voltage is reached. Conversely, the Schmitt trigger's output does not change for negative going inputs until the lower trip point voltage is reached. Despite the input signal's rate of change, nothing happens until the input signal passes through the appropriate trip voltage.

Figure 11.8c shows how two distinctly different threshold voltages can aid in noise rejection. The input signal shown has noise superimposed upon it. A standard logic gate responds to any signal that passes through the gate's single threshold voltage. As shown, a glitch is created since the noise signal would bring a standard logic device through its threshold. But the Schmitt trigger totally ignores the large noise spike because the lower threshold voltage is much lower than the minimum input voltage brought about by the noise. The noise rejection mechanism works with noise present on a low level input as well.

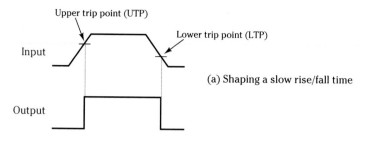

Upper trip point (UTP)

Lower trip point (LTP)

Input

Output

(a) Shaping a slow rise/fall time

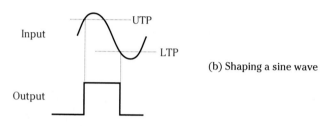

Input

UTP

LTP

Output

(b) Shaping a sine wave

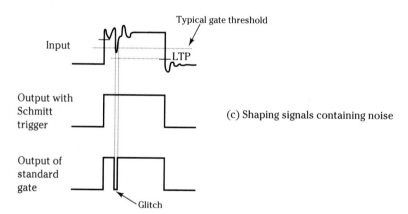

Typical gate threshold

Input

LTP

Output with
Schmitt
trigger

(c) Shaping signals containing noise

Output of
standard
gate

Glitch

**Figure 11.8
Schmitt Trigger
Applications**

Voltage Comparators

A **voltage comparator** compares two input analog voltages and provides a high output logic level indicating when one input voltage is greater than another. Comparators can detect a voltage difference in the millivolt range and are useful for detecting when an analog property (temperature, pressure, etc.) has reached some predetermined level. For instance, a temperature-sensing probe can easily provide a voltage proportional to the measured temperature. If the probe voltage is connected to one of the two comparator's inputs and a reference voltage representing a desired temperature is connected to the other input, the comparator's output logic level will indicate when the probe temperature has reached or exceeded the reference temperature. (See Figure 11.9.)

Temperature probe

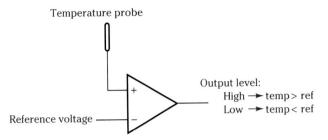

Output level:
High → temp > ref
Low → temp < ref

Figure 11.9
Voltage Comparator
Application

A comparator is a specialized form of an **op-amp** (**operational amplifier**), another beneficial interfacing device.

Use voltage comparators to determine when a 0 V to 5 V input signal is producing a voltage between 1.6 V and 3.3 V.

Solution Use two voltage comparators. One detects when the input is above 3.3 V and the other determines when the input voltage is below 1.6 V. To set the voltage limits, 3.3 V and 1.6 V references are required.

DESIGN
EXAMPLE 11.7

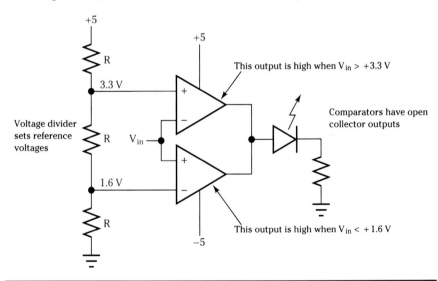

+5

R

3.3 V

Voltage divider
sets reference
voltages

R V_{in}

1.6 V

R

−5

+5

This output is high when $V_{in} > +3.3$ V

Comparators have open
collector outputs

This output is high when $V_{in} < +1.6$ V

Analog Switches-Multiplexers

An **analog switch** is a solid state circuit that switches an analog signal on or off under digital logic control. A typical analog switch is shown in Figure 11.10.

An analog signal placed at input S_1 appears at output D_1 when the digital control input line is high. When the control line is low, the switch is open and no signal appears at the output. Since the control line can be exercised by logic gates, digital control of analog signals is possible. Some analog switch

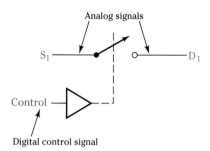

Figure 11.10
Analog Switch
Schematic

parameters of interest are the **on resistance (Ron)** and the **break before make** characteristics.

The on resistance is the amount of resistance the switch presents to an analog signal when the switch is closed. Ideally, a closed switch presents zero resistance to a signal, but a semiconductor switch will have some resistance that must be considered to prevent excessive signal loss. Typical on resistance values are 50 ohms to 150 ohms.

Break before make is an important parameter when using multiple switches because, as the term implies, a closed analog switch will open before other switches close. This property prevents signals from inadvertently being connected together when switches share common signal lines. Break before make times are typically in the 100 nsec range. Other timing specifications also exists for switch turn-on, switch turn-off, and enabling delay times.

An **analog multiplexer** consists of several analog switches connected with a common output line. Similar to a digital multiplexer, a decoder controlled by digital levels determines which switch's analog input signal appears at the output.

DESIGN
EXAMPLE 11.8

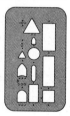

Draw the output waveform for the following analog switch:

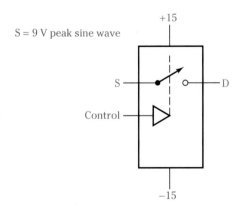

Solution   The D output will follow the analog input waveform when the digital control line is high. When the digital input signal is low, the output will be zero. The maximum peak analog input voltage recommended is $V_{CC} - 4$ V or 11 V peak.

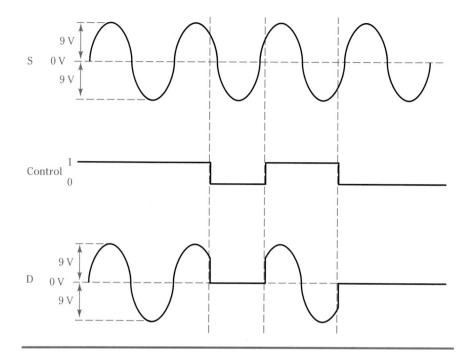

Line Drivers and Line Receivers

In previous chapters we discussed the problems encountered when sending digital signals over long lengths of wire. In most cases the best advice is keeping wires as short as possible. But what can be done when signals must be transmitted over a significant distance, such as from a computer terminal to a computer separated by a distance of several hundred feet? The solution is boosting the signal voltage and current levels so that the transmission can be carried out. This is the function of **line drivers** and **line receivers**.

There are a significant assortment of line drivers and matching line receivers available to meet the applications of the many different bus and communications standards used in the computing industry. For instance, **RS-232-C** is a well-known communications standard used to connect many kinds of peripheral equipment to computer systems. The standard covers many aspects of data communications including the logic voltage requirements. Using RS-232-C, we define a high logic level as a voltage between -3 V and -25 V, whereas a low logic level is a voltage between $+3$ V and $+25$ V. Obviously, these voltage levels are incompatible with those of most logic families, so that an interface is necessary. Figure 11.11 shows how a typical RS-232-C interface is formed using line drivers and line receivers.

Digital logic levels are converted into levels compatible with the transmission standard using a specially designed line driver. This IC must be powered with the appropriate voltages required for transmission. The line driver sends the reconverted digital data over the transmission line to a line receiver. Here the signals are reconverted by the receiver IC back to normal logic levels.

Figure 11.11
Using Line Drivers and
Line Receivers to
Drive a Transmission
Line

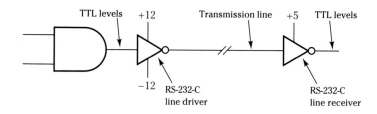

DESIGN
EXAMPLE 11.9

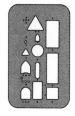

Show how a MC1488 line driver and a MC1489 line receiver can be used to connect a RS-232 computer terminal to the serial input and output data lines of a microprocessor chip.

Solution   Both interface chips contain four interface circuits. Only one is needed from each since the microprocessor has a single data in and data out line. The line driver requires a +12 V and a −12 V supply to be compatible with the RS-232 standard. Assume that the terminal supplies correct RS-232 levels.

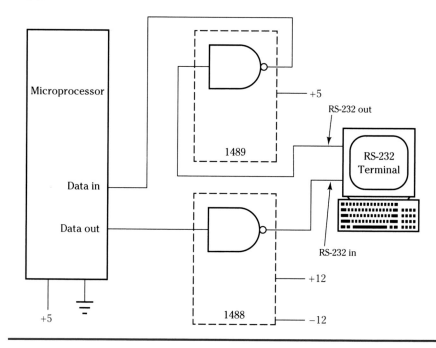

11.3 ■ ELECTRONIC INTERFACING ALTERNATIVES

Often a digital/analog interface cannot be structured with the relatively low powered digital devices available. This is particularly true when using digital systems to control power hungry equipment such as motors and lights. Some useful electronic interfacing circuits are discussed in the following sections.

Transistor Circuits

The transistor's current amplification characteristics make it a useful interfacing component, allowing the relatively low digital device current levels to be converted into the higher current levels required for electrical equipment. In addition, greater voltage levels can also be utilized than is possible with digital components. In a typical circuit low values of current drive the transistor's **base** lead. Transistor operation multiplies the level of base current by a factor of the transistor's **beta**. The increased current flows through the transistor's **collector** and **emitter** leads where the high current load is placed. There are many ways to use a transistor as an interfacing device; Figure 11.12 shows a basic transistor interface.

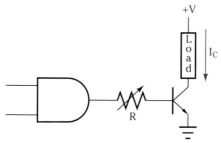

+V

I_C

R

Adjust R to control current through the load.

Figure 11.12
Basic Transistor
Interface for Digital
Logic

Create the interface circuit required for a 7408 AND gate to drive a 5 V 60 mA incandescent light bulb using a transistor interface. The bulb should go on when the AND gate output is high.

*DESIGN
EXAMPLE 11.10*

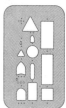

Solution The following steps are used to determine the appropriate transistor and limiting resistance required:

1. Find the maximum high level output current (I_{OH}) for the digital gate.
2. Find the minimum high level output voltage (V_{OH}) for the digital gate.
3. Compute the value of base limiting resistance with these specifications:

$$R = \frac{V_{OH} - 0.7}{I_{OH}}$$

4. Compute the transistor beta needed to supply adequate current to the load:

$$B = \frac{Ic}{Ib}$$

where Ic = load current and Ib = I_{OH}.

7408 Specs:
$V_{OH\ min} = 2.4\ V$
$I_{OH\ max} = 800\ \mu A$

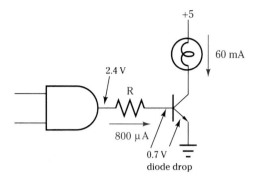

$$R = \frac{2.4 - 0.7}{800\ \mu A} =$$

$$R = 2125\ ohms$$

$$beta = \frac{60\ mA}{800\ \mu A}$$

$$beta = 75$$

An NPN transistor is used since the active on signal is a logic high.

Solid State Relays

Solid state relays are semiconductor based circuits possessing many of the advantages of standard relays. They can both switch large voltage and current loads using small control voltages and currents, allowing digital levels to be the controlling inputs. In addition, both relays provide **electrical isolation** so that the digital controlling circuits are not electrically connected to the more powerful load circuitry. In case of a failure on the high powered electrical end the digital controlling circuits will not fail as well. Solid state relays have an advantage over mechanical relays—they operate faster. Because mechanical relays rely on magnetic operation to move the relay contacts physically, inertia and other physical effects prevent fast switching. Typical relay switching rates are in the milliseconds range, whereas a solid state relay can switch on or off in microseconds.

Solid state relays are constructed using **opto-coupling** circuits. Opto-coupler operation involves transmitting light between an LED and a photosensitive transistor. While the LED is controlled by low current digital signals, the phototransistor output is used to activate the high voltage/current circuitry connected to the load. Since the LED and phototransistor are only connected by a light path, they are electrically isolated from one another. Using this system, any failure on the load side of the circuit will not affect the controlling low voltage, sensitive, and expensive digital circuitry. Solid state relays are designed to operate on normal TTL levels yet are capable of controlling many amperes of current. Using a solid state relay can be as simple as inserting one between a digital gate and a higher power load.

Shaft Encoders

Measuring the rotational speed or position of a motor shaft is a typical digital system industrial application. Typically, a **shaft encoder** is fixed on the shaft of rotating machinery (motors, generators) for this purpose. The encoder contains a disk encoded with a suitable binary code, such as the Gray code, that spins as the motor turns. A pattern of clear and dark areas on the disk represents the Gray code. LEDs and light sensors are fixed on either side of the semi-transparent disk to shine through and detect the actual disk/motor shaft position. Output pins attached to the encoder transmit the encoded data to a digital system for processing.

Integrated Circuit Drivers

Since interfacing is common in many design applications, it is not surprising that IC manufacturers have created specialty chips for interfacing purposes. Special interface ICs can be purchased for almost any interfacing task. Some typical functions available include LED display drivers, electroluminescent display drivers, AC/DC plasma display drivers, motor drivers, and general purpose drivers and actuators. Manufacturers' data books are great sources of information for these specialty circuits. In addition, interface devices are available to convert logic signals from one family to those of another. For instance, an ECL to TTL converter allows an ECL gate to drive a TTL device.

11.4 ■ DIGITAL-TO-ANALOG AND ANALOG-TO-DIGITAL CONVERSION

Digital-to-Analog (D/A) conversion is the process of converting digital data into a representative analog voltage or current. **Analog-to-Digital** (A/D) conversion is the direct opposite, converting an analog voltage or current into an equivalent binary code. Both A/D and D/A conversion are sophisticated ways to bring the analog and digital worlds together. Our previous interfacing techniques and examples have simply allowed the two worlds to communicate in an elementary fashion. For example, we used a comparator to detect when a temperature probe reached a certain temperature. This circuit only detects a single temperature and cannot detect multiple temperatures. How can multiple temperature values be detected? We could include a comparator circuit for each temperature of interest, but if the specific temperatures required for detection are numerous, the number of comparators becomes unwieldy. An A/D converter solves the problem by producing unique binary codes that represent the unique temperature values. That is, each analog input voltage representing a specific temperature is converted into a corresponding unique binary code. The number of temperatures represented is only limited by the number of binary bits created during the A/D process.

In a similar manner, digital data can be translated into an equivalent analog voltage. Again, the number of output voltages possible is only limited by the number of input digital bits. Figure 11.13 outlines the basic A/D and D/A processes.

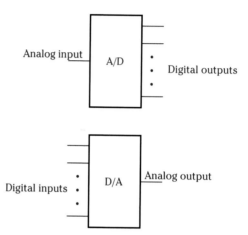

Figure 11.13
Block Diagram View of
A/D and D/A Functions

The Digital-to-Analog Conversion Process

The typical digital-to-analog converter (DAC) consists of a R-2R resistor network, a voltage or current reference, electronic switching and level shifting circuits, and a summing circuit (Figure 11.14). This represents a substantial amount of circuitry that must be fabricated to close tolerances if accurate results are to be obtained. Fortunately, DAC are available in IC form, making the digital designer's job one of applying D/A technology into a digital design rather than one of designing a DAC.

D/A Voltage Reference

The essential tasks for a DAC are to accept a number of digital data bits from logic gates or a computer and convert that information into a representative DC voltage. If the digital information changes over time, then the analog signal also changes, forming an AC output voltage from many individual DC values. One complicating factor is that the digital data bits presented to a DAC vary in voltage level since logic high and low values are represented by a range of voltages rather than a specific voltage. That is, one digital gate at the high logic level may be producing an output voltage of 3.3 V while another high voltage output may be 3.8 V. Since these voltages will actually be translated into a meaningful analog voltage, you can understand why level shifting is a necessary

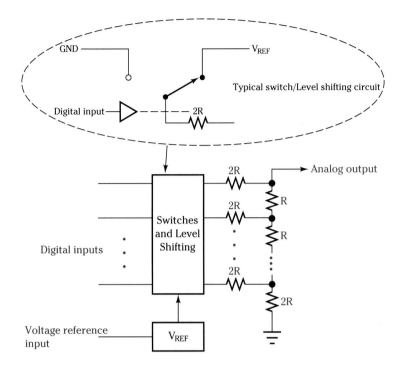

Figure 11.14
Digital-to-Analog
Converter Block
Diagram

requirement. All digital voltages must be adjusted to a predefined level in order for the translation circuitry to convert effectively a high logic level into an appropriate analog response. The D/A voltage reference circuitry controls the level shifting and also sets the range of output analog voltages. For instance, it may be required that the binary input codes produce an output voltage varying between 0 V and +10 V. Another application may require a voltage range from 0 V to +20 V. The voltage reference controls this adjustment. The actual shifting is easily accomplished by allowing the binary input signals to be the controlling inputs to a series of analog switches. When an input bit is low, the switch passes a 0 V voltage level to the resistive network. When the input bit is high, the switch passes the reference voltage along to the resistive network.

D/A Resistor Network

Each "rung" of the R-2R resistor network receives a properly adjusted digital input voltage, which is translated into a voltage proportional to the significance of its corresponding binary input. The relative size of the resistors with respect to each other produces a voltage proportional to each bit of the digital input code. The sum total of these individual voltages is the output analog voltage. The R-2R resistor configuration is the predominate network used to accomplish D/A conversion. (Some D/As use a weighted resistor network—many resistors whose values are proportional to each other—to accomplish the translation, but the network is not so practical since many more precision resistor values are required than with the R-2R network.)

D/A Resolution

Of primary concern in DAC specifications is the number of digital bits. The number of bits determines the **resolution** of the system. The resolution determines how finely an analog voltage may be represented. For example, an 8-bit DAC has the capability to produce $2^8 = 256$ analog voltages. In other words, a voltage range may be divided into 256 specific voltages—one voltage per binary combination. A 12-bit DAC can provide 4096 analog voltages for the same voltage range—a significantly greater resolution. Figure 11.15 illustrates how the number of bits affects the output signal.

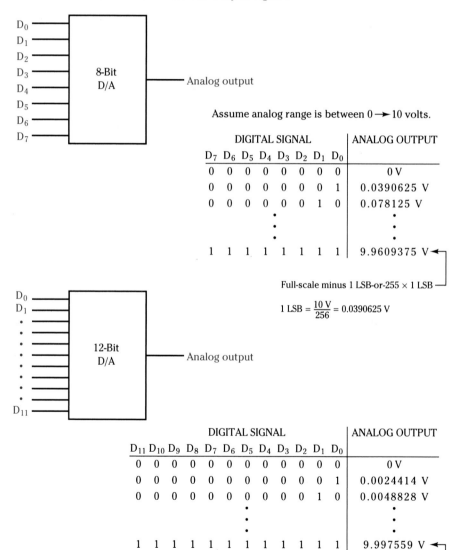

Assume analog range is between 0 → 10 volts.

D_7	D_6	D_5	D_4	D_3	D_2	D_1	D_0	ANALOG OUTPUT
0	0	0	0	0	0	0	0	0 V
0	0	0	0	0	0	0	1	0.0390625 V
0	0	0	0	0	0	1	0	0.078125 V
⋮								⋮
1	1	1	1	1	1	1	1	9.9609375 V

Full-scale minus 1 LSB-or-255 × 1 LSB

$$1\ \text{LSB} = \frac{10\ \text{V}}{256} = 0.0390625\ \text{V}$$

DIGITAL SIGNAL | ANALOG OUTPUT

D_{11}	D_{10}	D_9	D_8	D_7	D_6	D_5	D_4	D_3	D_2	D_1	D_0	ANALOG OUTPUT
0	0	0	0	0	0	0	0	0	0	0	0	0 V
0	0	0	0	0	0	0	0	0	0	0	1	0.0024414 V
0	0	0	0	0	0	0	0	0	0	1	0	0.0048828 V
⋮												⋮
1	1	1	1	1	1	1	1	1	1	1	1	9.997559 V

Full-scale minus 1 LSB-or-4095 × 1 LSB

$$1\ \text{LSB} = \frac{10\ \text{V}}{4096} = 0.0024414\ \text{V}$$

Figure 11.15
Comparison of 8- and 12-Bit D/A Converter Resolution

If an 8-bit resolution is desired for a voltage range between 0 V and 10 V, the resolution is determined by computing the output voltage change resulting when moving from one binary input code to the next, or the change of the LSB (least significant bit). This is equal to the maximum voltage divided by the number of input combinations or 10 V/256 = 0.0390625 V. In other words, each binary combination produces a change in the analog output voltage equal to 0.0390625 V. Increased resolution means that a given analog range can be divided into finer incremental voltage changes. The same 0 V to 10 V span can be divided into 10 V/4096 = 0.0024414 V increments with a 12-bit D/A. Note that the full-scale output voltage (maximum D/A output voltage) in all cases is equal to the analog voltage represented by 1 LSB times $2^n - 1$. In the 12-bit example this is equal to 0.0024414 V $\times$ 4095 = 9.997559 V. The D/A output does not quite reach the full 10 V output. Adjustment of the voltage reference may be necessary to attain a specific full-scale output voltage. Also, an amplifier (op-amp) is often attached to the output of the D/A to provide adequate output voltage or to scale the output voltage.

What is the resolution required, in number of bits, for a DAC to provide an analog output voltage in 1024 increments?

DESIGN EXAMPLE 11.11

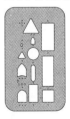

Solution   Resolution is dependent on the number of digital input bits. Consequently, the range of analog output voltages is constrained to the number of input combinations (2^n) allowed by the digital inputs. $2^{10} = 1024$; therefore, a 10-bit DAC is needed.

If the reference voltage for a 10-bit DAC is +15 V, what is the analog output voltage represented by the LSB? What is the full-scale analog output voltage? What is the output voltage represented by the input code 0011011101?

DESIGN EXAMPLE 11.12

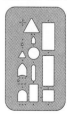

Solution   The output voltage represented by the LSB is determined by dividing the reference voltage by the number of DAC input combinations. Thus:

$$\frac{15 \text{ V}}{1024} = 0.014648437 \text{ V}$$

In other words, a change in the LSB creates a 0.014648437 V change in the analog output.

The full-scale output voltage is the analog voltage occurring when the digital input is all ones. This is 1111111111 for a 10-bit DAC and represents a decimal input code of 1023 ($2^n - 1$). The full-scale output is obtained by multiplying this value by the analog voltage produced by the LSB. Thus:

$$1023 \times 0.014648437 \text{ V} = 14.98535156 \text{ V}$$

To find the analog voltage for any input combination, multiply the decimal equivalent value of the combination by the LSB voltage. The output voltage for 0011011101 (221 decimal) is

$$221 \times 0.014648437 \text{ V} = 3.237304577 \text{ V}$$

D/A Accuracy

The number of bits available is of little importance if the DAC is inaccurate. DAC **accuracy** determines how close the actual analog output voltage is to the expected output voltage. Accuracy is usually expressed as a variation in the ideal output voltage in terms of LSB voltage. A typical specification may place accuracy within the range of one-half the LSB voltage, meaning that the analog output voltage for a specific digital input is near the expected value by an amount equal to the voltage change produced by one-half the LSB change. For instance, if the voltage change produced by changing the LSB is 0.05 V, then the output voltage can be expected to be within $\pm\frac{1}{2}$ LSB or ± 0.025 V of the ideal voltage.

D/A Speed

Speed or the **conversion rate** is another important DAC consideration. High speed applications, such as music reproduction, are limited by the conversion rate because the analog values are obtained from digital data only after the DAC has completed the conversion process. This DAC delay is equivalent to digital circuit propagation delay time and is called **settling time.** Typical settling times range from 135 μsec to 5000 μsec. If a DAC has a settling time of 300 μsec, the maximum rate of conversion is 1/300 μsec = 3333 conversions per second. The conversion rate is particularly important in A/D conversion because low sampling (conversion) rates can lead to false analog signal representation.

Monotonicity

Testing a DAC is relatively straightforward. A **monotonicity test** shows if the DAC output changes in a prescribed manner. A DAC is monotonic if the analog output either increases or remains constant for an increasing digital input. A free-running counter is attached to the DAC binary input bits. As the counter progresses through its states, the DAC produces an increasing output voltage until the counter recycles back to zero. At this time the DAC output returns to its minimum output voltage condition as the process begins anew. Figure 11.16 shows how a monotonicity test is carried out along with the resulting oscilloscope staircase waveform.

The waveform indicates several things about the DAC's operation. The number of steps should be equal to the number of states in the counter and the steps should progress as shown. If the progression of steps does not match

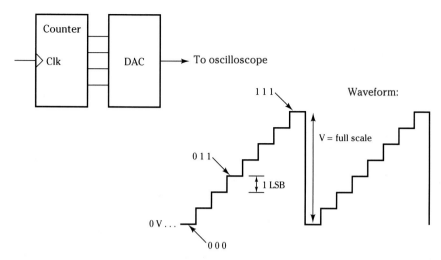

Figure 11.16
Monotonicity Test

the waveform shown, it is most likely the result of an incorrect digital input level. If the input bits were connected improperly with regards to significance or if inputs were stuck at high or low levels, an erratic pattern would result. The voltage increase for each step should conform with that expected for the change of the LSB. Measuring the voltage change for each step will verify the accuracy of the DAC. Furthermore, the last step should agree with the expected full-scale output voltage. Variations in voltage levels from expected values point to an incorrect voltage reference or poor quality components drifting in value.

A 3-bit DAC is connected to a reference voltage of 8 V. Determine the output voltages for every input binary code. Draw and label values on the staircase waveform that results from continuous clocking of the DAC.

DESIGN
EXAMPLE 11.13

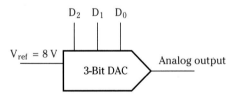

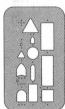

Solution Three-input bits produce eight input combinations. The output voltage change due to an increase of the LSB is 8 V/8 = 1 volt per combination. Thus:

D_2	D_1	D_0	ANALOG OUTPUT
0	0	0	0 V
0	0	1	1 V
0	1	0	2 V
0	1	1	3 V
1	0	0	4 V
1	0	1	5 V
1	1	0	6 V
1	1	1	7 V

Monotonicity Test:

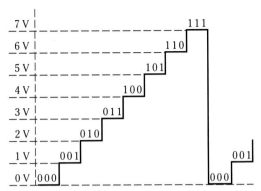

***

DESIGN EXAMPLE 11.14

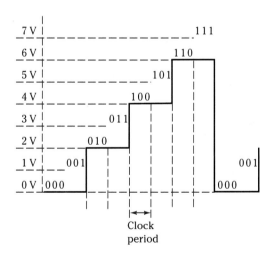

Using the 3-bit DAC from Design Example 14.13, troubleshoot the problem evident by the monotonicity test waveform shown here.

Solution   Compare voltages and states of the errant waveform to that of the expected waveform. A discrepancy in values will pinpoint the problem.

Missing Codes	Missing Voltages
0 0 1	1 V
0 1 1	3 V
1 0 1	5 V
1 1 1	7 V

The least significant bit is stuck at zero since the DAC output remains in 0, 2, 4, and 6 conditions for two clock periods.

Draw the output analog waveform for the 3-bit DAC used in Design Examples 14.13 and 14.14 if the binary inputs undergo the following changes:

DESIGN EXAMPLE 11.15

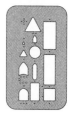

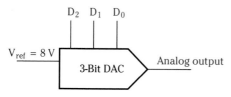

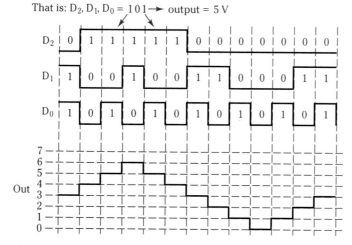

That is: $D_2, D_1, D_0 = 101 \rightarrow$ output = 5 V

This output waveform approximates a sine wave.

Solution During each increment of time the binary input code produces a specific output voltage. Determine the input code's corresponding analog voltage to produce the analog output signal.

D/A Current Outputs

Many DACs provide an output current flow that is proportional to the binary input value. The output current represents the sum of the currents flowing through the individual resistors in the D/A's resistive network. To avoid loading down this network by direct connection to a load, we use an op-amp as a current-to-voltage converter as shown in Figure 11.17. The op-amp presents

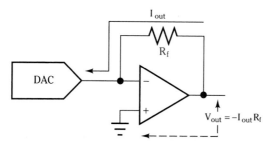

Figure 11.17 Op-Amp Current-to-Voltage Converter

8-Bit Monolithic
Multipling D/A Converter

AD1408/AD1508*

PRELIMINARY TECHNICAL DATA

FEATURES
Improved Replacement for Industry Standard 1408/1508
Improved Settling Time: 250ns typ
Improved Linearity: ±0.1% Accuracy Guaranteed Over
Temperature Range (-9 Grade)
High Output Voltage Compliance: +0.5V to -5.0V
Low Power Consumption: 157mW typ
High Speed 2-Quadrant Multiplying Input: 4.0mA/μs
Slew Rate
Single Chip Monolithic Construction
Hermetic 16-Pin Ceramic DIP

AD1408/AD1508 FUNCTIONAL BLOCK DIAGRAM

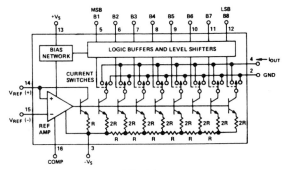

TO-116

PRODUCT DESCRIPTION

The AD1408 and AD1508 are low cost monolithic integrated circuit 8-bit multiplying digital-to-analog converters, consisting of matched bipolar switches, a precision resistor network and a control amplifier. The single chip is mounted in a hermetically sealed ceramic 16 lead dual-in-line package.

Advanced circuit design and precision processing techniques result in significant performance advantages over older industry standard 1408/1508 devices. The maximum linearity error over the specified operating temperature range is guaranteed to be less than ±¼LSB (-9 grade) while settling time to ±½LSB is reduced to 250ns typ. The temperature coefficient of gain is typically 20ppm/°C and monotonicity is guaranteed over the entire operating temperature range.

The AD1408/AD1508 is recommended for all low-cost 8-bit DAC requirements; it is also suitable for upgrading overall performance where older, less accurate and slower 1408/1508 devices have been designed in. The AD1408 series is specified for operation over the 0 to +75°C temperature range, the AD1508 series for operation over the extended temperature range of -55°C to +125°C.

PRODUCT HIGHLIGHTS

1. Monolithic IC construction makes the AD1408/AD1508 an optimum choice for applications where low cost is a major consideration.

2. The AD1408/AD1508 directly replaces other devices of this type.

3. Versatile design configuration allows voltage or current outputs, variable or fixed reference inputs, CMOS or TTL logic compatibility and a wide choice of accuracy and temperature range specifications.

4. Accuracies within ±¼LSB allow performance improvement of older applications without redesign.

5. Faster settling time (250ns typ) permits use in higher speed applications.

6. Low power consumption improves stability and reduces warm-up time.

7. The AD1408/AD1508 multiplies in two quadrants when a varying reference voltage is applied. When multiplication is not required, a fixed reference is used.

*Covered by Patent Numbers 3,961,326; 4,141,004

Figure 11.18
AD1408 Data Sheet. Courtesy of Analog Devices.

such a high impedance to the DAC's output pin, that all current flow from the converter passes through the op-amp feedback resistor. The voltage developed across this resistor ($I_{OUT} \times R_f$) is the output voltage of the circuit. This voltage is also proportional to the D/A binary input code.

A practical DAC, such as the 8-bit AD1408 as shown in Figure 11.18 (see Appendix A for a complete data sheet) utilizes this output circuitry. The range of output current is set by the input voltage reference (pin 14), which is converted into a reference current determined by the value of resistor R14. For instance, if the reference voltage is 2.5 V and R14 is 1.25K ohms, then the reference current is 2 mA (V/R). The maximum D/A output current is one LSB less than 2 mA or 2 mA × 255/256 = 1.9921 mA. This current flows into the op-amp feedback resistor to determine the maximum value of output voltage. For example, if a maximum output voltage of 9.961 V is desired, then R_f (R_0 in Figure 11.18) is found by

$$R_f = \frac{V_{out}}{I_{out}} = \frac{9.961 \ V}{1.9921 \ mA} = 5K \ ohms$$

The output voltage will vary between 0 V and 9.961 V.

Show how an AD1408 DAC can be utilized to convert 8-bits of binary information into a 0 V to 5 V output voltage range.

DESIGN EXAMPLE 11.16

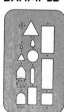

Solution   Using the AD1408 data sheet from Appendix A, the positive reference connection is followed.

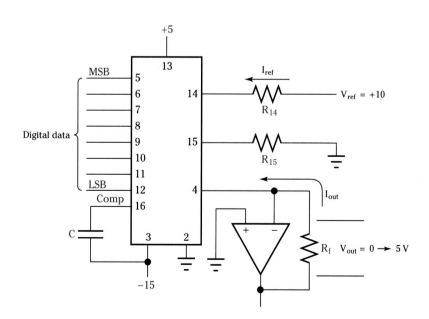

Specs:

$$I_0 = \frac{1}{256} \text{ less than } I_{ref}$$

$$R_{14} = R_{15}$$

$$C \approx 15 \text{ pfd per } 1K \text{ of } R_{14}$$

(a) If $V_{out\ max} = 5$ V and R_f is chosen as 2K, then

$$I_0 = \frac{V}{R_L} = \frac{5}{2K} = 2.5 \text{ mA}$$

(b) Since

$$I_0 = I_{ref} - \frac{I_{ref}}{256} = I_{ref}\left[1 - \frac{1}{256}\right] = I_{ref}\left[\frac{255}{256}\right]$$

then $I_0 = I_{ref}\ 0.996$ or

$$I_{ref} = \frac{I_0}{0.996} = \frac{2.5 \text{ mA}}{0.996} = 2.51 \text{ mA}$$

(c) With $V_{ref} = +10$

$$R_{14} = \frac{V_{ref}}{I_{ref}} = \frac{10}{2.51 \text{ mA}} = 3984 \text{ ohms}$$

$$R_{14} = R_{15} = 3984 \text{ ohms}$$

$$C \cong 15 \text{ pfd} \times 4 = 60 \text{ pfd}$$

11.5 ■ THE ANALOG-TO-DIGITAL CONVERSION PROCESS

Analog-to-digital converters (ADC) change an analog signal into an equivalent binary number. ADCs are built in a variety of different ways, supplying the designer with many options in terms of ADC complexity and speed. Many of the same parameters of interest to DAC selection are also important to ADC applications. Resolution, accuracy, and conversion rate all impact the performance of an ADC system. The resolution of the ADC is determined by the number of digital output bits. As with DACs, the more bits supported by the chip, the more an analog signal may be subdivided. However, the accuracy of any digital signal is only as good as the accuracy of the ADC. Quality designs, both the ADC and the applications circuitry, determine how meaningful a signal is produced. As may be expected, ADCs do not produce an output instantaneously. Once the command is given to start an A/D conversion, a valid digital signal is not produced until the conversion time elapses.

Flash Converters

Flash converters or **parallel converters** are the fastest ADCs available. Constructed using multiple comparators and voltage references, these converters function similarly to the temperature probe-comparator circuit description given earlier. Although fast, the amount of circuitry required doubles for each bit of increased resolution, making the flash converter quite expensive. Flash converters are often fabricated using ECL technology for maximum speed.

The flash converter shown in Figure 11.19 shows the primary components required—a resistive network, voltage comparators, and an encoder. The resistors form a voltage divider string that divides the input reference voltage into individual reference voltages for each comparator. The reference voltage for any comparator is at least one LSB greater than the reference voltage of the comparator below it. Using this approach, every comparator with a reference voltage less than the analog input signal (which is the other input to every comparator) will have a high output. All comparator outputs feed a digital encoder, which converts the comparator data into an appropriate digital output code.

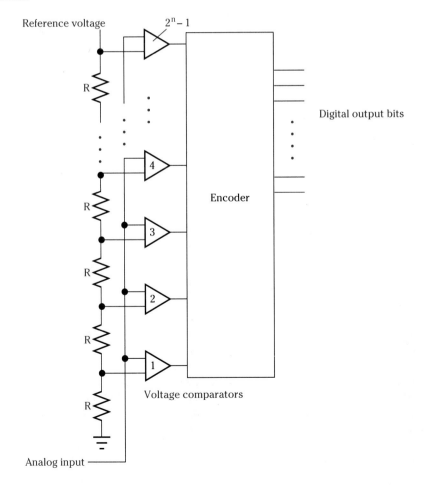

Figure 11.19
Flash Converter

Unfortunately, high resolution flash converters are impractical because of the number of comparators required. The number of comparators is equal to $2^n - 1$. Therefore, if a 4-bit resolution is needed, 15 comparators are required. An 8-bit ADC necessitates 255 comparators, whereas a 12-bit flash ADC would use 4095 comparators.

Integrating Converters

Other conversion techniques use **integrating** methods that count clock pulses for a length of time proportional to the input voltage. The circuit relys on an integrator—a resistor-capacitor op-amp network—to convert a voltage magnitude into a function of time. A typical integrating device is a **dual slope** converter, which is used frequently in digital voltmeters. The block diagram shown in Figure 11.20 will be used to explain the process.

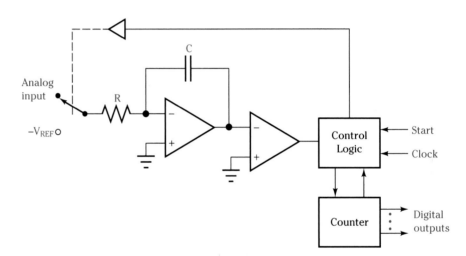

Figure 11.20
Block Diagram of a
Dual-Slope ADC

1. An analog input voltage is applied to the ADC.
2. A start signal is received by the ADC, which accomplishes the following: The capacitor is allowed to charge from zero by the analog input signal and the counter begins counting at a rate set by the clock.
3. The capacitor will charge in a linear fashion until a specified amount of time has elapsed. The charging time, which is determined by the counter, ends when the counter reaches a predetermined value.
4. Charging is halted at the end of the fixed time interval as the counter and control logic switch the analog voltage signal out (via an analog switch) and insert an opposite polarity reference voltage. The counter is also reset at this time. The charge on the capacitor at this point is not the analog voltage signal magnitude but, rather, is proportional to the analog voltage signal magnitude. This is shown in Figure 11.21.
5. The capacitor now discharges (charging in the opposite direction) as it begins the process of acquiring the potential of the opposite polarity reference voltage. The counter begins counting anew at this time as well. The capacitor acquires the reference potential in a predetermined pattern since the voltage

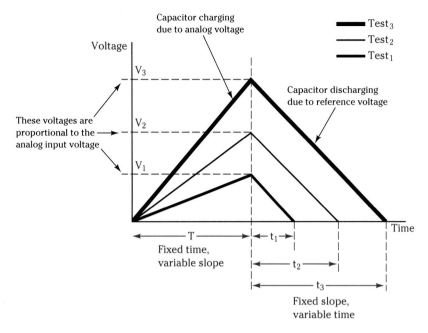

The voltage magnitude acquired by the capacitor during the fixed time charge interval is proportional to the time for discharge during the fixed slope interval.

Therefore: $V_1 \propto t_1$, $V_2 \propto t_2$, $V_3 \propto t_3$

Since times t_1, t_2, and t_3 control the count time and the state of the counter, the digital counter state is related to the input analog voltage.

Figure 11.21
Dual-Slope Converter Operation

as well as the resistor and capacitor values in the charging circuit are constants. Therefore, the discharge slope is fixed (Figure 11.21).

   6. The capacitor never reaches the potential of the reference voltage because the instant that the capacitor voltage reaches 0 V, on its way from the analog potential to the reference potential, the comparator stops the counter.

   7. The state of the counter reflects the time it took the capacitor to reach 0 V since the counter was clocked at a steady rate as the capacitor discharged. If the capacitor discharged from a large analog potential (Test$_3$ in Figure 11.21), then the counter state will be relatively high (a high binary number) because the time for discharge is long. This allows the counter time to receive many clock pulses. If the capacitor discharged from a lower analog potential (Test$_1$ in Figure 11.21), then the counter state is lower since the capacitor reached zero quicker and the counter is not allowed to receive as many clock pulses.

   In any case the binary number in the counter after it stops is proportional to the time the counter was allowed to receive clock pulses. This time period for counting was determined by the time it took the capacitor to discharge to zero. That time period (to discharge) was determined by the charge rate of the original analog voltage signal since the analog signal determined the voltage magnitude from which the discharge began. Therefore, the counter state is proportional to the analog input signal and constitutes the digital output signal.

   This kind of ADC is insensitive to the clock rate and the capacitor and resistor values because these elements affect both the charging and discharging operations. This is an advantage since external component values do not affect ADC accuracy. One disadvantage of this method is that conversion time depends on the magnitude of the analog signal. This means that systems using an integrating ADC may have to be designed using the worse case conversion rate, which slows overall operation.

Successive Approximation Converters

A **successive approximation** converter has a constant, fairly rapid conversion time and is perhaps the most widely used A/D conversion technique. The technique involves successive comparisons of the input analog voltage to an internally generated reference voltage. The comparisons are done on a binary weighted basis. That is, the analog signal is compared to a reference voltage equal to the voltage represented by a single binary output bit. If the analog voltage exceeds this weighted value, the bit is retained as part of the digital output; otherwise the bit is reset. There is a comparison made for every digital output bit. For instance, an 8-bit DAC of this type requires eight comparisons.

A successive approximation conversion begins by comparing (using a comparator) the analog input voltage with the voltage produced by an internal DAC as shown in Figure 11.22. The internal DAC processes a digital signal stored within a **successive approximation register** (SAR). The SAR value is the output of the ADC system as well as the value feeding the internal DAC. Comparisons begin with the MSB (most significant bit). If the analog input voltage is greater than the equivalent voltage of the MSB of the SAR when it is set, then the MSB is kept high. This means that the analog input voltage will eventually yield an output code with the MSB and other bits set to one. As an example, assume that the analog input voltage is 9 V. Assume further that the SAR MSB = 1 represents a voltage of 8 V. The SAR MSB bit is retained high in this case since the analog magnitude is greater than 8 V. This bit and at least one additional bit will be set to digitally represent 9 V. However, if the analog signal was 5 V, the SAR MSB would be reset (since 5 is less than 8) and other less significant bits would be set via additional comparisons to represent 5 V digitally.

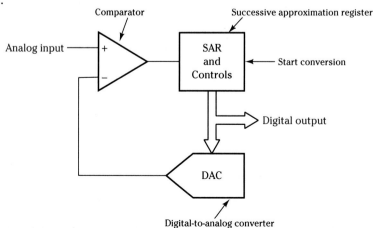

**Figure 11.22
Successive
Approximation ADC
Block Diagram**

Comparisons continue like this for each digital bit in the SAR until all bits are compared to the analog input voltage. As more and more SAR bits become set, the internal DAC output voltage increases and approaches or approximates the magnitude of the input voltage. The result remaining in the SAR after all comparisons are complete are the final digital bits in the ADC process. It is interesting that the comparison of analog voltage to equivalent digital voltage requires a DAC. The DAC provides the means to produce a variable reference voltage that is compared to the incoming analog signal. Since the number of comparisons always equals the number of digital bits, the conversion rate is fixed as well as being relatively fast. This is a strong advantage in data acquisition systems where conversions must occur on a periodic basis. Many successive approximation ADCs are made with microprocessor compatible signal and data lines, allowing microprocessor control of the ADC.

Using block diagrams, outline how a microprocessor and an ADC can control the heating and cooling of a greenhouse. Assume that the microprocessor can be programmed to accept data and transmit control signals.

DESIGN EXAMPLE 11.17

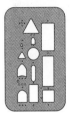

Solution Temperature can be determined using a temperature probe. The voltage produced by the probe is the analog input signal to an ADC. The ADC output is processed by the microprocessor, which determines whether heating or cooling should take place.

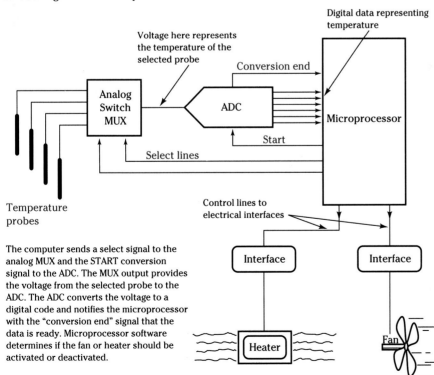

The computer sends a select signal to the analog MUX and the START conversion signal to the ADC. The MUX output provides the voltage from the selected probe to the ADC. The ADC converts the voltage to a digital code and notifies the microprocessor with the "conversion end" signal that the data is ready. Microprocessor software determines if the fan or heater should be activated or deactivated.

Sampling Rate

Figure 11.23 shows the significance of A/D conversion and the number of "conversion samples." The number of samples is determined by the conversion rate. If too few samples are taken, the "digitized" signal may not bear any resemblance to the actual signal. The signal shown resembles the analog signal but could be improved with additional sampling points. The **Nyquist Sampling Theorem** states that accurate reproduction without sampling errors occurs only if the sampling occurs at a rate equal to at least twice the highest frequency of the analog signal. In actual practice, sampling may occur at three or four times the highest frequency. With this in mind, it is apparent why fast conversion is required for high frequency signals. In addition, slow sampling produces **aliasing** or the production of low frequencies related to the actual frequency. The alias frequencies cannot be distinguished from the actual frequency and, of course, produce incorrect results.

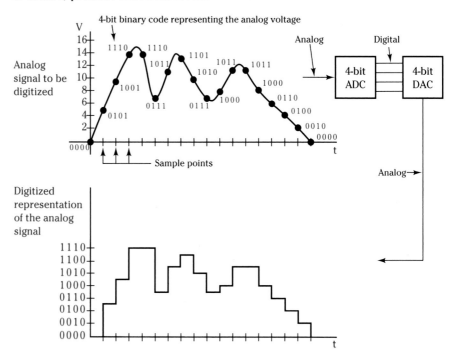

Figure 11.23
Digitizing an Analog Signal

DESIGN EXAMPLE 11.18

What is the required sampling frequency to meet the Nyquist sampling criteria for a 250 kHz signal? At what minimum speed must the ADC operate to handle this frequency?

Solution   The minimum sampling frequency is equal to twice the signal frequency or 2×250 kHz $= 500$ kHz. The ADC minimum speed is $1/500$ kHz $= 2 \ \mu\text{sec}$.

■ Digital systems "talk" to the analog world through translation circuits referred to as interfaces.

■ Converting digital signal on-off information into an appropriate analog response is accomplished with a variety of devices including LEDs, open collector gates, and line drivers.

■ Converting analog information into basic digital signals is accomplished using devices such as comparators, Schmitt triggers, and switches.

■ Converting quantities of digital data into equivalent analog data is carried out using a digital-to-analog converter (DAC); quantities of analog data are converted into digital data using an analog-to-digital converter (ADC).

■ The quality of an analog-to-digital or a digital-to-analog conversion is determined by the converting device's accuracy, speed, and resolution specifications.

■ Analog-to-digital conversion is carried out using several differing processes including flash conversion, integrating methods, and successive approximation.

Section 11.1

*1. Show how four switches can be connected to a 2114 memory chip's I/O lines to provide an input data signal for write operations.
 2. In the following schematic for a common anode seven segment display each LED segment draws 20 mA at a forward voltage drop of 2 V. Assuming that the anodes will be connected to a +24 V supply and the segment cathodes will be driven by a suitable source (0 V indicates segment on), what value of resistance is required to limit current to acceptable levels?

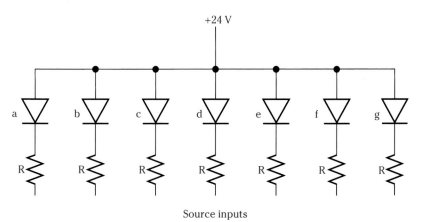

Source inputs

---

* See Appendix F: Answers to Selected Problems.

Section 11.2

*3. Determine how the three open collector NAND gates on a 7412 can be used in a wired logic system to produce the function $\overline{ABC}\,\overline{DEF}\,\overline{GHI}$.

*4. The waveform given here is the input to a Schmitt trigger. Draw the output

Output

5. Three temperature probes are connected to voltage comparators to sense a temperature range between +5°C and +25°C. One comparator goes high when the temperature is above 5°C; a second comparator goes high when the temperature is above 15°C; and a third comparator goes high when the temperature is above 25°C. Assuming that appropriate reference voltages are available and the temperature probes are set to produce meaningful voltages with respect to the reference voltages, sketch the voltage comparator system.

6. Design a logic network to attach to the voltage comparator system designed in problem 5 to indicate the following:
   (a) a green LED lights if the temperature is between 5°C and 15°C
   (b) a red LED lights if the temperature is between 15°C and 25°C
   (c) a clear LED light, if the temperature is over 25°C
   (d) a yellow LED light, if the temperature is less than 5°C. Only one LED may be lit at a time.

*7. Draw the output waveform for the following analog switch:

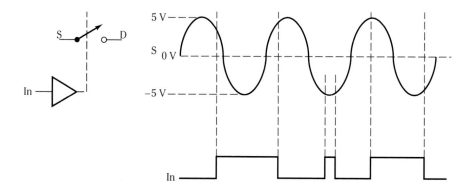

---

8. Draw the output waveform for the following analog switch multiplexer:

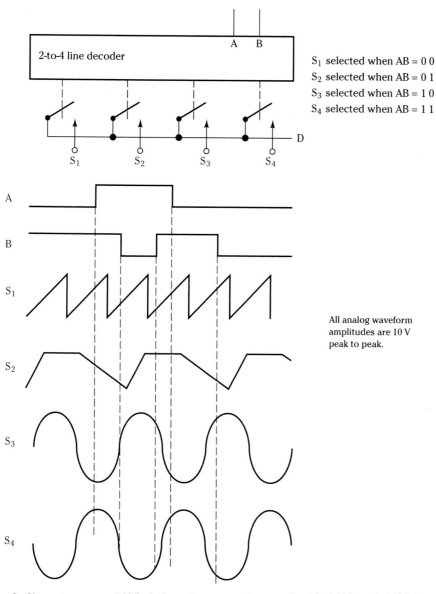

S_1 selected when AB = 0 0
S_2 selected when AB = 0 1
S_3 selected when AB = 1 0
S_4 selected when AB = 1 1

All analog waveform amplitudes are 10 V peak to peak.

9. Show how two 7495 shift registers can be used with 1488 and 1489 line drivers and line receivers to left shift information between the two 7495s in a serial fashion over 50 feet of twisted pair cable. Also include provisions so that a common clock may be used to control the shifting between the two registers.

Section 11.3

*10. Show how the following solid state relay can be used to allow an AND gate to control the on-off operation of an electric heater. The heater draws 10 A at 110 V.

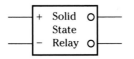

Relay specs:

Control signal voltages 3-32 V DC

Load current 25 A max

Line voltage 45-140 V @ 60 Hz

Sections 11.4–11.5

11. What is the resolution of a digital-to-analog converter (DAC) if the number of digital bits is:

*(a) 6   (b) 10   (c) 16

12. What is the analog voltage represented by the LSB of a DAC whose reference voltage is +12 V if the DAC resolution is:

*(a) 6-bits   (b) 10-bits   (c) 16-bits

*13. Using the DAC data from problem 12, determine the full-scale output voltage for each DAC.

14. A 12-bit DAC is designed to produce an output voltage based on a 5-V reference. What is the output voltage for the following input codes?

*(a) 000000100000

(b) 111100001111

(c) 101010101010

15. An 8-bit DAC has an accuracy of one-half the LSB. If the DAC's reference voltage is +10 V, what is the voltage output range that can be expected for the following digital inputs?

(a) 00000000   (b) 11111111

(c) 00011100   (d) 01100100

16. Determine the conversion rate for DACs with the following settling times:

*(a) 1 μsec   (b) 0.150 μsec   (c) 0.600 μsec   (d) 5 μsec

17. Determine the resolution, LSB analog voltage, and conversion rate for the DAC producing the following monotonic waveform:

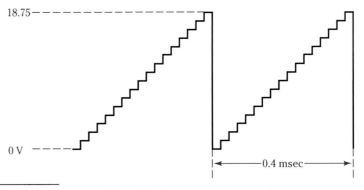

---

* See Appendix F: Answers to Selected Problems.

*18. The following waveform for all digital input codes to a 4-bit DAC indicates a problem. What is wrong?

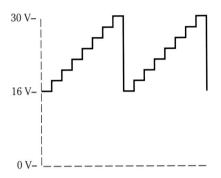

*19. The following flash converter converts an analog voltage ranging from 0 V to 4 V into a 2-bit analog code. All comparators are low when the input voltage is less than 1 V. The lower comparator is high for voltages less than 2 V but greater than or equal to 1 V. The other comparators go high in a similar fashion for every 1 V increment in the input analog voltage. Design the encoder network to produce a 2-bit binary code representing the analog input voltage.

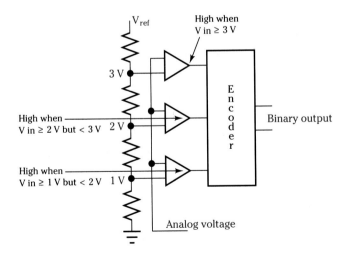

20. Redesign the flash converter encoder network from problem 19 to produce a Gray code output corresponding to the analog input voltage.

---

*21. The following analog waveform is to be digitized by a 4-bit analog-to-digital converter (ADC). Using the sample points shown here, determine the digital codes representing the analog voltage at each sample point.

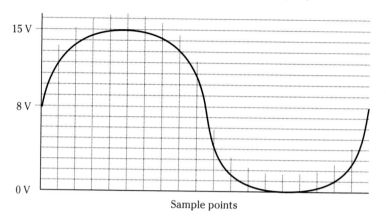

Sample points

22. Using the same analog waveform as shown in problem 21 and the same 4-bit ADC, determine the binary codes resulting if the sampling rate is doubled.

23. Using the digital data gathered from the analog waveform in problem 21, draw the reconstructed analog waveform resulting if the digital data is presented to a 4-bit DAC.

24. Repeat problem 23 using the data gathered from problem 22. Compare this reconstructed waveform to that from problem 23. Why is one a better representation of the original waveform than the other?

---

* See Appendix F: Answers to Selected Problems.

CHAPTER 12

PRACTICAL DIGITAL SYSTEM DESIGN CONSIDERATIONS

OBJECTIVES

To explain the factors affecting the selection of one logic family or technology over another.

To explain the meaning of fan-out, speed, power, and noise margin and their impact on circuit performance

To discuss the concept of a transmission line and the nature of transmission line materials.

To illustrate potential problems due to transmission line effects on high speed signals.

To explain the utility of line drivers and line receivers on long distance signal transmission.

To discuss the details necessary to ensure proper printed circuit board construction.

PREVIEW

Eventually, all the efforts placed into the design, simplification, simulation, and testing of a logic network lead to the construction of an actual working circuit. The physical layout of the chip or printed circuit board connecting the logic demands as much attention as does the logic itself to ensure that your terrific high speed circuit design performs as expected. A number of construction design details are considered at design inception and are modified and reconsidered at all stages of the product development cycle. As an introduction to construction of a working circuit, this chapter focuses on the many, nonlogic design tasks facing the logic designer.

12.1 ■ INTEGRATED CIRCUIT LOGIC FAMILIES

Without a doubt, virtually all modern digital systems are constructed with **integrated circuits** (ICs) or **chips.** ICs are semiconductor devices that are typically fabricated from silicon and may contain several thousand transistor circuits. Commercially available ICs designed to contain logic circuits with similar electrical characteristics are referred to as **logic families.** A logic family consists of many IC types, each performing a specific logical function. Each IC has a part number designation that identifies the function of the part. Typical logic families contain anywhere from tens to hundreds of unique part numbers.

An advantage of logic family use is ease of design. The ICs are designed and tested to perform a specific logic function that the logic designer can use as building blocks to create complex logical circuits. Many electronic design concerns are alleviated by using a logic family since the bulk of the electronic design is complete. The logic designer merely has to interconnect devices in a logical manner to create the desired logic function. However, electronics is not tossed out the window altogether. The designer must design logic circuits following the **design rules** of the particular logic family. The design rules dictate how many devices can be interconnected, the length of interconnecting wires, power supply levels, and a host of other concerns. Adherence to the logic family design rules is crucial for reliable operation. Many of these factors are discussed later in this chapter.

Each chip requires attachment to the system power supply ground and voltage terminals. Each IC has pins designated for this purpose, which are typically labeled as GND (ground) and V_{CC} or V_{DD} (voltage).

Several logic families are in use today that are designed and fabricated using a variety of semiconductor technologies. Each technology/logic family offers advantages and disadvantages that the logic designer must consider. Typically, selecting a logic family is a trade-off between different device characteristics, although as technology improves the choices get easier to make.

Logic families gained widespread use when the IC fabrication process became reliable. The earliest logic families were based on logic circuit designs originally developed for individual transistor circuits. Initially, **chip density,** that is, the number of devices on a chip, was low. One or two logic circuits were all that could be placed on a single IC at the time. As the IC process evolved and requirements for additional chip performance were made, logic families also improved. In addition to the several logic families presently in use, some other IC technologies are also used for newer developments. Generally, logic families and/or technologies are distinguished by performance and density factors.

12.2 ■ TTL

Transistor-transistor logic (TTL) is a logic family based on **bipolar transistor** technology. (TTL is so named because it describes the circuit arrangement of transistors in the device—transistors for inputs and transistors for outputs.)

This widely used family has grown over 20 years to include hundreds of parts. Even the TTL part numbers have attained historical significance. The TTL part numbering system is as familiar to the digital designer as is his or her own name. Many newer logic families use variations of the TTL part numbering system to maintain consistency and familiarity with older parts. The TTL family itself also includes many subfamilies that offer variations in performance characteristics. The most predominant characteristics of concern in any logic family are speed and power. (These parameters are discussed more fully in later sections of this chapter.) In addition, the term "TTL compatible" is used in many technologies to indicate that non-TTL parts can be interconnected with TTL parts and that communication between the two can take place.

Figure 12.1 shows how logic parts are identified in the TTL part numbering system. Often the TTL parts are referred to as 7400 parts or the 7400 series since the "74" identifies commercially available parts. Parts preceded with a "54," such as 5400, are military TTL parts that are functionally equivalent.

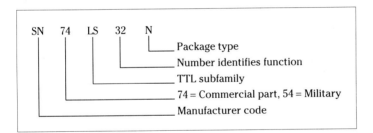

Figure 12.1
TTL Part Numbering
System

When using TTL or any other logic family, you need the manufacturer's data book. Each part is listed in the book along with pin assignments, logic function, and other important specifications. For example, Figure 12.2 shows the data sheet for a 74ALS00 quadruple 2-input positive-NAND gate. Quadruple means that four 2-input NAND gates are present on the chip. The pin diagram identifies which pins are inputs and which are outputs. Other information shows how the part reacts to logic signals. Naturally, this information is essential when connecting the devices together.

TTL Logic Levels

A TTL device is considered to be in the high or one state when the output voltage is between 2.4 volts and 5 volts (V) and in the low or zero state when the output voltage is in the 0 V to 0.4 V range. TTL inputs, when connected to other TTL device outputs, will respond to these voltage levels and correctly interpret the logic level. TTL inputs can actually respond to a wider range of voltages than the device output is capable of producing. Any input voltage between 2 V and 5 V is correctly interpreted as a high level, whereas any voltage between 0 V and 0.8 V is correctly interpreted as a low logic level. The 0.4 V difference between input and output levels is referred to as **noise margin** and is indicative of the logic family's ability to reject noise signals. (Noise margin is discussed more completely later in this chapter.) As long as TTL devices

SN54ALS00A, SN54AS00, SN74ALS00A, SN74AS00
QUADRUPLE 2-INPUT POSITIVE-NAND GATES

D2661, APRIL 1982−REVISED SEPTEMBER 1987

- Package Options Include Plastic "Small Outline" Packages, Ceramic Chip Carriers, and Standard Plastic and Ceramic 300-mil DIPs

- Dependable Texas Instruments Quality and Reliability

description

These devices contain four independent 2-input NAND gates. They perform the Boolean functions $Y = \overline{A \cdot B}$ or $Y = \overline{A} + \overline{B}$ in positive logic.

The SN54ALS00A and SN54AS00 are characterized for operation over the full military temperature range of −55°C to 125°C. The SN74ALS00A and SN74AS00 are characterized for operation from 0°C to 70°C.

SN54ALS00A, SN54AS00 . . . J PACKAGE
SN74ALS00A, SN74AS00 . . . D OR N PACKAGE
(TOP VIEW)

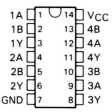

SN54ALS00A, SN54AS00 . . . FK PACKAGE
(TOP VIEW)

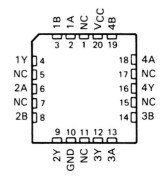

NC−No internal connection

FUNCTION TABLE (each gate)

INPUTS		OUTPUT
A	B	Y
H	H	L
L	X	H
X	L	H

logic symbol[†]

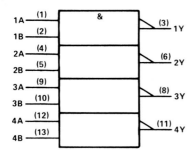

[†]This symbol is in accordance with ANSI/IEEE Std 91-1984 and IEC Publication 617-12.
Pin numbers shown are for D, J, and N packages.

logic diagram (positive logic)

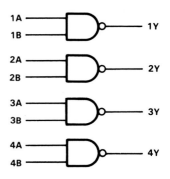

Figure 12.2
74ALS00 Data Sheet. Courtesy Texas Instruments.

SN54ALS00A, SN74ALS00A
QUADRUPLE 2-INPUT POSITIVE-NAND GATES

absolute maximum ratings over operating free-air temperature range (unless otherwise noted)

Supply voltage, V_{CC} . . . . 7 V
Input voltage . . . . 7 V
Operating free-air temperature range: SN54ALS00A . . . . $-55\,°C$ to $125\,°C$
                                        SN74ALS00A . . . . $0\,°C$ to $70\,°C$
Storage temperature range . . . . $-65\,°C$ to $150\,°C$

recommended operating conditions

		SN54ALS00A			SN74ALS00A			UNIT
		MIN	NOM	MAX	MIN	NOM	MAX	
V_{CC}	Supply voltage	4.5	5	5.5	4.5	5	5.5	V
V_{IH}	High-level input voltage	2			2			V
V_{IL}	Low-level input voltage			$0.8^{\dagger}$			0.8	V
				$0.7^{\ddagger}$				
I_{OH}	High-level output current			-0.4			-0.4	mA
I_{OL}	Low--level output current			4			8	mA
T_A	Operating free-air temperature	-55		125	0		70	°C

†Tested at $-55\,°C$ to $70\,°C$.
‡Tested at $70\,°C$ to $125\,°C$, per MIL-STD-833, method 5005, sub-group 1, 2, and 3. Static test is performed at $25\,°C$, $125\,°C$, and $-55\,°C$.

electrical characteristics over recommended operating free-air temperature range (unless otherwise noted)

PARAMETER	TEST CONDITIONS		SN54ALS00A			SN74ALS00A			UNIT
			MIN	TYP§	MAX	MIN	TYP§	MAX	
V_{IK}	$V_{CC} = 4.5$ V,	$I_I = -18$ mA			-1.5			-1.5	V
V_{OH}	$V_{CC} = 4.5$ V to 5.5 V,	$I_{OH} = -0.4$ mA	$V_{CC}-2$			$V_{CC}-2$			V
V_{OL}	$V_{CC} = 4.5$ V,	$I_{OL} = 4$ mA		0.25	0.4		0.25	0.4	V
	$V_{CC} = 4.5$ V,	$I_{OL} = 8$ mA					0.35	0.5	
I_I	$V_{CC} = 5.5$ V,	$V_I = 7$ V			0.1			0.1	mA
I_{IH}	$V_{CC} = 5.5$ V,	$V_I = 2.7$ V			20			20	μA
I_{IL}	$V_{CC} = 5.5$ V,	$V_I = 0.4$ V			-0.1			-0.1	mA
$I_O{}^{\P}$	$V_{CC} = 5.5$ V,	$V_O = 2.25$ V	-30		-112	-30		-112	mA
I_{CCH}	$V_{CC} = 5.5$ V,	$V_I = 0$ V		0.5	0.85		0.5	0.85	mA
I_{CCL}	$V_{CC} = 5.5$ V,	$V_I = 4.5$ V		1.5	3		1.5	3	mA

§All typical values are at $V_{CC} = 5$ V, $T_A = 25\,°C$.
¶The output conditions have been chosen to produce a current that closely approximates one half of the true short-circuit output current, I_{OS}.

switching characteristics (see Note 1)

PARAMETER	FROM (INPUT)	TO (OUTPUT)	$V_{CC} = 5$ V, $C_L = 50$ pF, $R_L = 500\ \Omega$, $T_A = 25\,°C$	$V_{CC} = 4.5$ V to 5.5 V, $C_L = 50$ pF, $R_L = 500\ \Omega$, T_A = MIN to MAX				UNIT
			'ALS00A		SN54ALS00A		SN74ALS00A	
			TYP	MIN	MAX	MIN	MAX	
t_{PLH}	A or B	Y	7	3	15	3	11	ns
t_{PHL}	A or B	Y	5	2	9	2	8	ns

NOTE 1: Load circuit and voltage waveforms are shown in Section 1 of the ALS/AS Logic Data Book, 1986.

Figure 12.2
74ALS00 Data Sheet. Courtesy Texas Instruments.

SN54AS00, SN74AS00
QUADRUPLE 2-INPUT POSITIVE-NAND GATES

absolute maximum ratings over operating free-air temperature range (unless otherwise noted)

Supply voltage, V_{CC} . . . . 7 V
Input voltage . . . . 7 V
Operating free-air temperature range: SN54AS00 . . . . −55 °C to 125 °C
SN74AS00 . . . . 0 °C to 70 °C
Storage temperature range . . . . −65 °C to 150 °C

recommended operating conditions

		SN54AS00			SN74AS00			UNIT
		MIN	NOM	MAX	MIN	NOM	MAX	
V_{CC}	Supply voltage	4.5	5	5.5	4.5	5	5.5	V
V_{IH}	High-level input voltage	2			2			V
V_{IL}	Low-level input voltage			0.8			0.8	V
I_{OH}	High-level output current			−2			−2	mA
I_{OL}	Low-level output current			20			20	mA
T_A	Operating free-air temperature	−55		125	0		70	°C

electrical characteristics over recommended operating free-air temperature range (unless otherwise noted)

PARAMETER	TEST CONDITIONS		SN54AS00			SN74AS00			UNIT
			MIN	TYP[†]	MAX	MIN	TYP[†]	MAX	
V_{IK}	V_{CC} = 4.5 V,	I_I = −18 mA			−1.2			−1.2	V
V_{OH}	V_{CC} = 4.5 V to 5.5 V,	I_{OH} = −2 mA	V_{CC}−2			V_{CC}−2			V
V_{OL}	V_{CC} = 4.5 V,	I_{OL} = 20 mA		0.35	0.5		0.35	0.5	V
I_I	V_{CC} = 5.5 V,	V_I = 7 V			0.1			0.1	mA
I_{IH}	V_{CC} = 5.5 V,	V_I = 2.7 V			20			20	μA
I_{IL}	V_{CC} = 5.5 V,	V_I = 0.4 V			−0.5			−0.5	mA
I_O[‡]	V_{CC} = 5.5 V,	V_O = 2.25 V	−30		−112	−30		−112	mA
I_{CCH}	V_{CC} = 5.5 V,	V_I = 0 V		2	3.2		2	3.2	mA
I_{CCL}	V_{CC} = 5.5 V,	V_I = 4.5 V		10.8	17.4		10.8	17.4	mA

[†]All typical values are at V_{CC} = 5 V, T_A = 25 °C.
[‡]The output conditions have been chosen to produce a current that closely approximates one half of the true short-circuit output current, I_{OS}.

switching characteristics (see Note 1)

PARAMETER	FROM (INPUT)	TO (OUTPUT)	V_{CC} = 4.5 V to 5.5 V, C_L = 50 pF, R_L = 50 Ω, T_A = MIN to MAX				UNIT
			SN54AS00		SN74AS00		
			MIN	MAX	MIN	MAX	
t_{PLH}	A or B	Y	1	5	1	4.5	ns
t_{PHL}	A or B	Y	1	5	1	4	

NOTE 1: Load circuit and voltage waveforms are shown in Section 1 of the ALS/AS Logic Data Book, 1986.

Figure 12.2
74ALS00 Data Sheet. Courtesy Texas Instruments.

are connected properly, the actual high or low voltage is unimportant since the voltages will be interpreted as valid logic levels. This allows the designer to step back from electronic design and concentrate on logic design. However, any voltages between the 0.8 V and 2 V levels are "forbidden" and indicate a circuit or device problem. Figure 12.3 summarizes the TTL logic levels.

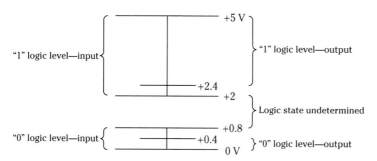

Figure 12.3
TTL Voltage Levels

TTL Subfamilies

Many varieties or subfamilies of TTL devices have been created as technology improvements or as manufacturers' answers to competition in the IC market. Some of the earlier TTL subfamilies are no longer used, having been supplanted by better performing subfamilies. In general, any TTL subfamily can be inter-connected with any other TTL subfamily when proper design guidelines are followed. Figure 12.4 contains information about the various TTL subfamilies naming convention.

7408   Standard TTL
74L08   Low power TTL
74H08   High speed TTL
74S08   Schottky TTL
74LS08   Low power Schottky TTL
74AS08   Advanced Schottky TTL
74ALS08   Advanced low power Schottky TTL
74F08   Fast TTL
Other variations are possible depending on the manufacturer.

Figure 12.4
TTL Subfamily
Identification

12.3 ■ CMOS

Another major logic family and logic technology is **CMOS**. CMOS stands for complementary metal oxide semiconductor, which describes the transistor construction used in the IC fabrication process. CMOS technology has under-gone tremendous growth and is rapidly overtaking the TTL logic family in speed, one of CMOS's early shortcomings. In addition, CMOS has very low power consumption, another beneficial attribute.

Several CMOS part numbering systems are in use, although newer CMOS devices have adopted the TTL part numbering system. The presence of a "C" in the part number, such as 74C00, indicates a CMOS part. Figure 12.5 indicates many of the CMOS families in existence.

4000 Series CMOS
29FCTxx—Fast CMOS-TTL compatible
74C08   CMOS part using TTL numbering system
74HC08   High speed CMOS using TTL numbering system
74HCT08   High speed CMOS-TTL compatible
74FCT08   Fast CMOS TTL compatible logic

Figure 12.5
CMOS Logic Families

CMOS Logic Levels

Whereas TTL devices are based on 5 V systems (V_{CC}), this is not true of CMOS. Many CMOS parts can operate with power supply voltages between 2 V and 15 V. Of course, logic level outputs are related to the power supply voltage chosen. Typical high and low logic level voltages for CMOS parts are 30% of V_{CC}. That is, the low level voltage is 30% of V_{CC}, whereas the high level voltage is V_{CC} minus 30% of V_{CC}. For a CMOS device to be TTL compatible, the supply voltage must equal 5 V.

The advantage of variable power supply levels is the ability to reduce power consumption. Since power can be related by the equation $P = VI$, reducing voltage directly reduces power consumption. Power dissipation specifications are of major importance in digital systems design.

12.4 ■ ECL

Emitter Coupled Logic (ECL) is very fast and, unlike other logic families, is not constructed around transistors that operate as simple on-off devices. Rather, the ECL transistors are designed to operate at two different "on" and closely spaced voltage levels. Consequently, device speed is very good because switching from one logic level to the other only requires a small voltage swing and some of the delays associated with on-off transistor operation are gone. On the other hand, power dissipation tends to be high because the transistors are never off and are always dissipating power.

Logic levels in the ECL family are defined with negative voltages. An output high logic level occurs for voltages between -0.81 V and -0.96 V, whereas a low logic level is defined as a voltage between -1.65 V and -1.85 V. Again, the beauty of a logic family is that the actual voltage is not critical as long as it falls within the specified range. Therefore, compatible devices will logically understand each other. Figure 12.6 outlines ECL logic levels.

The high speed of ECL (and other fast technologies) brings with it special design considerations. Wiring and other interconnections must be short and properly laid out to avoid noise problems that can hamper logic circuit performance. Design constraints of this nature are discussed in detail later in this chapter.

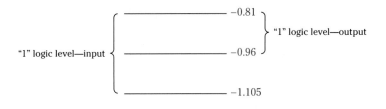

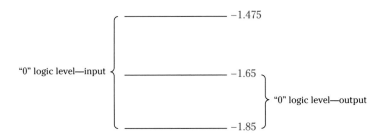

Figure 12.6
10K ECL Voltage Levels

12.5 ■ ALTERNATIVE LOGIC DEVICES

For a number of reasons not all logic circuit design is done using standard logic families. The convenience of logic family devices to some applications can be a constraint to others. For instance, a digital system designed with a large number of TTL chips can easily be redesigned with one chip using an alternate technology. Reducing the number of chips in a system increases system reliability as well as performance. However, changing from the tried and true logic family methods may introduce additional design methodology problems. Many of these aspects of design have been discussed throughout this book.

Some of the alternate design approaches include **programmable logic, gate arrays,** and **standard cells.** Each alternative has advantages and disadvantages that must be sorted out by the designer or design team.

12.6 ■ LOGIC TECHNOLOGIES

During the early days of IC based logic design (not that long ago), selecting a logic family generally meant choosing between the well-established TTL, CMOS, or ECL families. Choices such as these still face logic designers today, but more and more the selection process is one of technology rather than of logic family. Not surprisingly, the technologies predominating are still TTL, CMOS, and ECL. Does this sound confusing? Actually it is not. The three logic families mentioned have been around for quite awhile and are continually being improved. However, choosing a standard logic family, that is, using predesigned parts, will in many ways constrain your design because you will be forced to design using only the standard logic building blocks of the selected family. This may or may not be a good tactic and is one of the design decisions facing a logic system development team. It is common in digital design to combine standard logic family parts and programmable or ASIC (application specific IC—custom parts) parts to produce the best combination of performance and economics.

As newer logic design alternatives are created, such as programmable logic, they are produced using the three proven technologies because a proven technology reduces costs and increases device reliability. This also gives the designer plenty of flexibility to choose a technology that is appropriate for specific design applications. For example, we might choose to use one of the various logic families for convenience or use a newer logic design approach for increased design power. In any case we will still be choosing among a TTL, CMOS, or ECL technology [although gallium arsenide (GaAs) shows promise as a new high speed technology]. For instance, if erasable devices are required, then a CMOS based technology is needed since other technologies do not provide erasable devices. Extremely high speed circuit requirements may dictate that the designer investigate ECL technology. These decisions will become even more interesting as other technology choices become commercially available.

12.7 ■ SPEED AND POWER

Two of the most commonly used criteria for component selection are **speed (propagation delay time)** and **power dissipation**. Speed determines how fast digital systems can process information—faster chips beget faster systems. Power dissipation affects the design of many other costly system elements. Increased power dissipation means that larger power supplies are required, that cooling systems must be used or increased in size, and that battery operation is out of the question. Unfortunately, speed and power considerations generally mean design trade-offs—increase speed and power consumption increases as well.

Figure 12.7 shows how power dissipation values and propagation delay specifications compare for many of the commonly used logic families. In general, high speed is obtained at the expense of power consumption. However, several recent logic innovations that are apparent exceptions to this rule are also shown on the chart. The advanced CMOS logic family (ACL) combines CMOS's small power dissipation with the propagation delays normally associated with faster TTL parts. Trends toward better overall circuit performance are likely to continue in this fashion. Currently, CMOS technology is providing the best performance in the speed-power categories (unimaginable several years ago). For another view of this subject, Figure 12.8 shows a chart comparing speed and power characteristics of common logic families.

One final note regarding CMOS power dissipation specifications. Although the CMOS power dissipation is very low, this is only true for components that are not switching at high rates of speed. When CMOS parts are fixed at a logic

Technology	Propagation Delay Time	Power Dissipation (per gate)
TTL		
74XX	10 ns	10 mW
74LXX	33 ns	1 mW
74HXX	6 ns	22 mW
74SXX	3 ns	19 mW
74LSXX	9.5 ns	2 mW
74ASXX	1.7 ns	8 mW
74ALSXX	4 ns	1.2 mW
74FXX	3 ns	3.5 mW
PALs (AND-OR)	10–25 ns	150 mW
CMOS		
4000	60 ns	0.001 mW
74HCXX	8 ns	0.025 mW
74HCTXX	8 ns	0.025 mW
74FCTXX	4.5 ns	10 mW
74ACTXX	3 ns	10 mW
29FCTXX	6.5 ns	10 mW
PALs (AND-OR)	25 ns	0.5 mW standby
ECL		
10K	2 ns	25 mW
100K	0.75 ns	40 mW
PALs (AND-OR)	6 ns	1.1 W
GaAS		
UBG7XX	400 ps	400 mW

Figure 12.7
Logic Family Speed-
Power Comparison

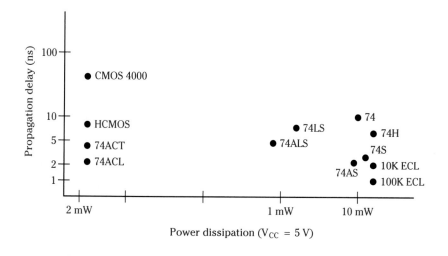

Figure 12.8
Speed vs. Power
Characteristics of
Common Logic
Families

level, they do not conduct current. Hence, negligible power is consumed. (This is not true of other logic families; TTL, for example, consumes power when in the high or low logic state.) When a CMOS device is switching from one logic state to another, current flows briefly and power is consumed. The power consumption is proportional to the rate of switching and approaches that of other logic families at high switching speeds. However, in many digital designs it is unlikely that most of the logic circuitry will be switching at top speed, so power consumption can be kept very low using a CMOS technology. CMOS can also operate at voltages lower than other logic families, providing additional power savings. [$P = EI$, so reducing E (voltage) reduces P (power).]

It should be noted that power consumption is also dependent on load capacitance as well as switching frequency and supply voltage. A true picture of CMOS power dissipation requires a precise knowledge of the circuit and circuit layout.

DESIGN EXAMPLE 12.1

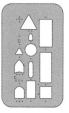

Several logic families (standard TTL, TTLALS, HCT CMOS, and 10K ECL) are under consideration for a design project. All meet the minimum speed and power requirements of the design. Which should be used?

Solution   This choice becomes a matter of economy. Generally, the technology providing the best speed with the lowest power consumption is preferred. A **speed-power product** comparison is often used to make such a selection. The speed-power product for a logic family is found by multiplying the propagation delay time (tpd) by the power dissipation. The family with the lowest speed-power product is the best choice if speed and power are the only overwhelming design considerations. Thus

Speed-Power Product = tpd × Power Dissipation

Since speed is in nanoseconds and power is expressed in milli-watts (mW), the product is in watt-seconds. However, since a watt equals a joule per second and typical values of speed and power are in nanoseconds (ns) (10^{-9}) and milli-watts (mW) (10^{-3}), the speed-power product is typically expressed as a number of picojoules (pj) (10^{-12}).

For the logic families under consideration:

TTL	10 ns × 10 mW = 100 × 10^{-12} = 100 pj
TTLALS	4 ns × 1.2 mW = 4.8 pj
HCT CMOS	8 ns × 0.025 mW = 0.2 pj
10K ECL	2 ns × 25 mW = 50 pj

Based on this comparison alone, the HCT CMOS parts would be the best choice.

12.8 ■ OTHER IMPORTANT TECHNOLOGY PARAMETERS

With the impact of speed and power specifications understood, we can investigate other important logic family specifications that weigh heavily in logic design.

Unused Inputs

In many designs not all component input pins are used. For example, a 3-input NAND function is possible using a 4-input NAND gate assuming that the fourth unnecessary input is left in a condition to enable the gate. In TTL logic this can be accomplished (although incorrectly) by simply leaving the pin disconnected. The TTL chip interprets the lack of any valid logic level as a high logic level. Since this would enable the NAND gate, the method should work. However, unconnected inputs easily pick up noise signals and this can cause false switching to occur. Since this could prove disastrous, most logic families do not recommend unconnected or "floating" inputs. Some general procedures to handle the problem properly include:

- Connect unused inputs to an independent power supply.
- Connect unused inputs to another driven input on the same gate.
- Connect unused inputs to the chip's source of power through a 1K ohm "pull-up" or "pull-down" resistor. (A pull-up resistor brings the input pin up to the supply voltage. A pull-down resistor brings the pin down to the supply voltage.) Typically, up to 25 unused inputs may be connected to a single pull-up resistor.
- Connect unused inputs to an output of a device that has its output level fixed at the proper level. For example, an inverter with an input tied low has a fixed high output voltage.
- Connect unused inputs directly to ground if a low logic level is desired.

These recommendations are satisfactory for both TTL and CMOS families, although it is recommended that unused gates on a CMOS chip also be connected to reduce power consumption. ECL chips may have their inputs left unconnected since they have internal pull-down resistors fabricated along with the chip.

Show how to tie properly the inputs of a 5-input NAND gate and a 5-input OR gate to correct levels so that both gates can be used as 4-input devices.

Solution AND and NAND gates must be tied to a high level through a pull-up resistor (or other appropriate manner) to enable the gates. OR gates and NOR gates must be tied low (to ground or to another driven input) to allow the gates to be enabled.

DESIGN
EXAMPLE 12.2

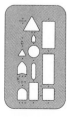

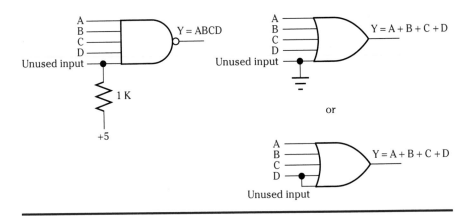

Fan-out

Fan-out is a specification describing the drive capability of a logic technology. **Fan-out** is defined as the number of loads (chip inputs) a gate output can drive reliably. For instance, standard TTL gates have a specified fan-out of ten, implying that one TTL gate can drive up to ten TTL inputs and still maintain valid high and low logic levels.

Fan-outs vary considerably from logic family to logic family. In addition, the fan-outs within logic subfamilies depend on the subfamilies used. Refer to manufacturers' specifications for fan-out information particularly when mixing logic families.

Fan-out is restricted by the amount of current flow a device can handle. For example, TTL devices **sink** current when outputs are low. This means that a low TTL output accepts current from the device it is driving. As more devices are attached to the driving gate output, the amount of sink current increases. An increase in sink current can reduce the logic device's ability to reject noise and can also increase the power dissipation of the driving chip.

Not all technologies' fan-outs are limited by sink current. CMOS devices are not affected since the technology does not produce significant current flow into the driving device because of the nature of the MOS (metal oxide semiconductor) transistors used. However, each load attached to a CMOS output adds capacitance to the output line. This is true in any logic family, but it is very significant in CMOS. When a logic gate changes level, the capacitance associated with the loads must be charged and discharged by the signal energy, a process that takes longer and longer as capacitance increases. In this case fan-out is limited by increased capacitance, which affects signal rise and fall times. ECL logic is also affected by added capacitance. Fan-outs no greater than ten are recommended for ECL. Figure 12.9 shows some of the factors affecting fan-out.

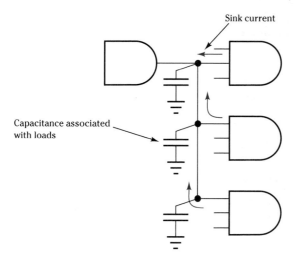

Sink current

Capacitance associated with loads

Figure 12.9
Chip Fan-Out

What is the maximum fan-out for the following interconnections?
(a) LSTTL feeding LSTTL
(b) ALSTTL feeding ALSTTL
(c) ALSTTL feeding LSTTL

DESIGN
EXAMPLE 12.3

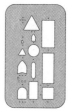

Solution   Feedback is determined by the current handling capabilities of the driving gate. Each gate has a maximum output current rating for the high and low states. In addition, each gate has an input current rating for the high and low states. The ratio of output to input currents for each state determines the fan-out.

The following terms and definitions are commonly found in data books:

- **Low level output current** (I_{OL})   The current (sink current) through a device output when the output is in the low state.
- **High level output current** (I_{OH})   The current through a device output when the output is in the high state.
- **Low level input current** (I_{IL})   The current through a device input when the input is held at a low logic level.
- **High level input current** (I_{IH})   The current through a device input when the input is held at a high logic level.

From these definitions fan-out can be computed as

$$\text{Fan-out (high)} = \frac{I_{OH}}{I_{IH}} \qquad \text{Fan-out (low)} = \frac{I_{OL}}{I_{IL}}$$

(a)
From 74LS08 data sheet:

$$I_{OL} = 4 \text{ mA}$$
$$I_{OH} = 400 \text{ } \mu\text{A}$$
$$I_{IL} = 0.4 \text{ mA}$$
$$I_{IH} = 20 \text{ } \mu\text{A}$$

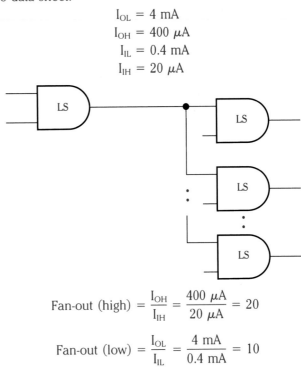

$$\text{Fan-out (high)} = \frac{I_{OH}}{I_{IH}} = \frac{400 \text{ } \mu\text{A}}{20 \text{ } \mu\text{A}} = 20$$

$$\text{Fan-out (low)} = \frac{I_{OL}}{I_{IL}} = \frac{4 \text{ mA}}{0.4 \text{ mA}} = 10$$

Use the lowest fan-out figure → 10

(b)
From the 74ALS08 data sheet:

$$I_{OL} = 8 \text{ mA}$$
$$I_{OH} = 0.4 \text{ mA}$$
$$I_{IL} = 0.1 \text{ mA}$$
$$I_{IH} = 20 \text{ } \mu\text{A}$$

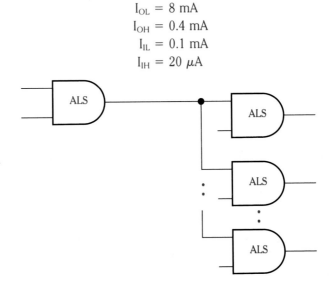

$$\text{Fan-out (high)} = \frac{I_{OH}}{I_{IH}} = \frac{0.4 \text{ mA}}{20 \text{ } \mu\text{A}} = 20$$

$$\text{Fan-out (low)} = \frac{I_{OL}}{I_{IL}} = \frac{8 \text{ mA}}{0.1 \text{ mA}} = 80$$

Use the lowest fan-out figure → 20

(c)

$$I_{OL} = 8 \text{ mA} \text{ ----> ALS}$$
$$I_{OH} = 0.4 \text{ mA} \text{ ----> ALS}$$
$$I_{IL} = 0.4 \text{ mA} \text{ ----> LS}$$
$$I_{IH} = 20 \text{ } \mu\text{A} \text{ ----> LS}$$

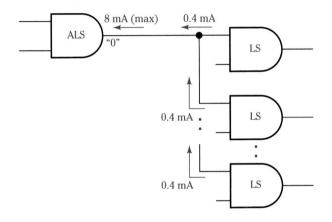

$$\text{Fan-out (high)} = \frac{I_{OH}}{I_{IH}} = \frac{0.4 \text{ mA}}{20 \text{ } \mu\text{A}} = 20$$

$$\text{Fan-out (low)} = \frac{I_{OL}}{I_{IL}} = \frac{8 \text{ mA}}{0.4 \text{ mA}} = 20$$

Use the lowest fan-out figure → 20

The signal from a 74ALS08 is required by 45 loads in a particular design. Since 45 exceeds the ALS fan-out limit of 20 loads, what can be done?

DESIGN EXAMPLE 12.4

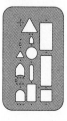

Solution   The AND gate signal can feed three other ALS devices. These three then fan out to the required loads. Typically, a buffer (noninverter) is used. Two possible solutions are given here:

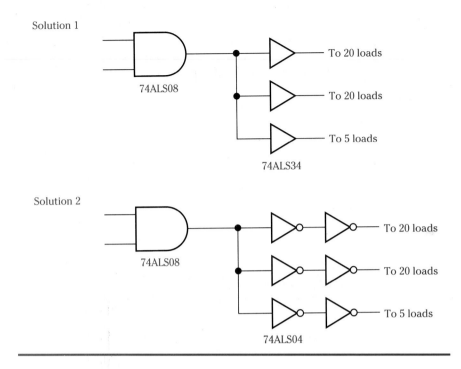

Noise Margin

Noise is a serious problem in virtually all digital systems. External noise can cause false switching as easily as internally generated circuit noise. Factors contributing to electrical noise are discussed in the following sections. **Noise margin** is a specification detailing a logic device's ability to reject noise impulses when they occur. Investigating noise margin requires an understanding of a typical logic device **transfer characteristic**.

The transfer characteristic for a logic device relates an input voltage change to the corresponding output response. Thus, the complete switching characteristics of a particular logic family can be noted from a representative transfer curve. The high and low level input voltages referenced on the diagram in Figure 12.10 are defined as follows:

- V_{IL}—The low level input voltage required for guaranteed operation. Since logic levels represent a range of voltages, the worst case, low level input voltage is defined as $V_{IL(max)}$. Any voltage greater than this value is no longer treated as a low logic level by a receiving gate.
- V_{IH}—The high level input voltage required for guaranteed operation. The worst case, high input voltage is defined as $V_{IH(min)}$. A logic gate will not respond to any input voltage less than this value as a high logic level.
- V_{OL}—The maximum guaranteed low level output voltage. The worst case, low output voltage, is defined as $V_{OL(max)}$. Output voltages greater than this value are not considered as low logic levels.

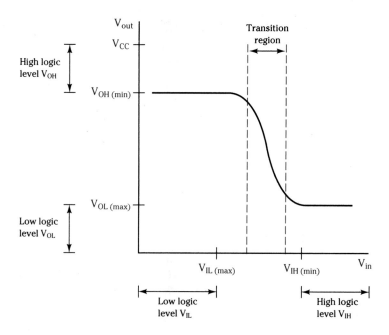

Figure 12.10
Logic Device Transfer
Curve

■ V_{OH}—The minimum guaranteed high level output voltage. The worst case, high output voltage is defined as $V_{OH(min)}$. Output voltages less than this value are not considered as high logic levels.

It is worth mentioning at this point the difference between a "guaranteed" specification and a "typical" specification. Manufacturers' guaranteed specifications are very conservative to ensure that marginal components are not sold. However, the typical component performs better than the minimum guaranteed specification. Generally, you can expect better performance than that stated on the data sheet. It is advisable to design to the guaranteed specifications since, from a statistical point of view, variations within the operating range can be expected. When many systems are manufactured from a set of parts, the statistical differences can cause problems if the manufacturers' specifications are not adhered to. Therefore, for reliability reasons most logic systems are designed using the manufacturers' specifications.

An ideal transfer curve switches from one state to another instantly when a "threshold" voltage is reached. This takes place as the input voltage changes from either a high to low level or from a low to high level. However, many real logic devices do not have an ideal characteristic. Figure 12.10 clearly shows that the actual change in state is more gradual, occurring around a small voltage range. This means that one threshold voltage occurs for low level inputs while another threshold occurs for high level inputs. For example, assume that a logic gate's typical low level input voltage is 0.1 V and the threshold voltage where the device senses a high input voltage is 1.1 V. The 1 V difference between these two levels determines when the device switches state. When the device switches under normal conditions, the input voltage (supplied by a driving gate) will increase up through the threshold level and force a corresponding output change. This occurs because the device controlling the input voltage undergoes a normal transition of its own from a low to a high logic level.

Now with the same low level input and threshold voltage conditions, assume that the input level supplied by a driving gate remains constant at 0.1 V. The output should also remain constant since there is no variation in the input. Again, this is normal. An exception to this seemingly steady state condition takes place when a noise pulse appears on the input line. Noise occurs all the time and has the potential to cause false switching. If a 1 V noise spike (glitch) occurs on the input in question, the threshold voltage of 1.1 V (1 + 0.1) will be reached and the gate will momentarily change level. This creates a glitch on the output, which finds its way into the rest of the logic attached to that output with equally unfavorable results. If the input noise spike only had an amplitude of 0.9 V, then the threshold would not be reached, the output would not switch, no glitch would be produced, and the system operation would remain normal. Summing up this example: The gate is immune from noise voltages less than 1 V in amplitude or, stated differently, the noise margin is 1 V. The noise margin specifies the ability of a logic technology to reject noise voltages. Figure 12.11 illustrates the previous example.

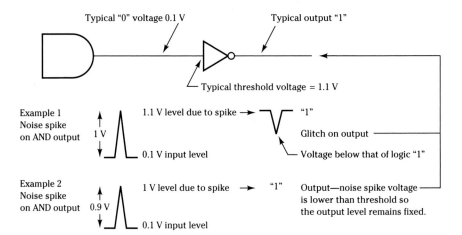

Figure 12.11 How Signal Noise Affects Circuit Response

It is important to realize that the threshold voltage just described and the manufacturer's minimum and maximum input/output voltages are not the same. The manufacturers' published specifications are not even close to the threshold voltage for a good reason—designing a system to the threshold levels would almost guarantee an unreliable system. No system should be designed exactly to the margins of operation. Some safety margin—and that is what the manufacturer's specifications guarantee—should be designed into the system. So, even though the actual noise margins may be quite good, the published safe margins are computed as follows:

$$\text{Noise Margin High} = V_{OH(min)} - V_{IH(min)}$$
$$\text{Noise Margin Low} = V_{IL(max)} - V_{OL(max)}$$

Review the definitions of the voltages in these equations and you will see why this makes sense. All noise margins are computed with the worst case manufacturers' specifications so that a worst case noise margin is obtained. If the published specification says that the noise margin high is 400 mV, then, theoretically, any negative going noise glitch greater than 400 mV would cause an accidental change in level at the output of the device receiving the glitch. Figure 12.12 shows how noise margin information is obtained from a transfer curve.

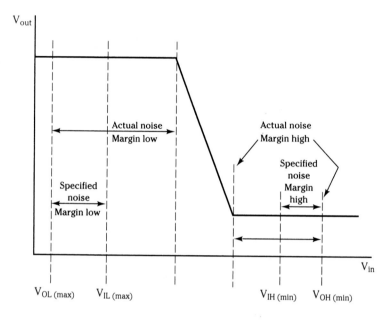

$$\text{Specified noise margin low } = V_{IL\,(max)} - V_{OL\,(max)}$$
$$\text{Specified noise margin high} = V_{OH\,(min)} - V_{IH\,(min)}$$

Figure 12.12
Obtaining Noise
Margin Specifications
from a Transfer Curve

You may wonder how a noise spike gets onto a signal line in the first place. That is another story altogether and is discussed in the next section.

Determine the high and low level noise immunity for the ACT CMOS logic family. Use the transfer curve and data supplied here.

DESIGN
EXAMPLE 12.5

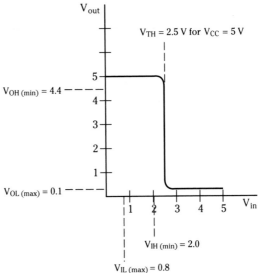

ACT Transfer Curve:

Noise Margin Low = $V_{IL\,(max)} - V_{OL\,(max)} = 0.8 - 0.1 = 0.7$ V
Noise Margin High = $V_{OH\,(min)} - V_{IH\,(min)} = 4.4 - 2.0 = 2.4$ V

Solution Noise immunity is the difference in high/low level input and output voltages. Notice that for CMOS logic the transition between high and low levels is abrupt and the transition voltage is equal to 50% of the supply voltage.

12.9 ■ NOISE IN DIGITAL SYSTEMS

Electrical noise can turn an otherwise superb digital system into a flailing, nervous pile of silicon. Noise can cause anything from erratic operation to permanent damage in systems that are not protected from its effects. In order to prevent noise problems from becoming bothersome, we must be aware of the potential sources of noise. Two basic sources are possible—external noise generated near the digital system and internal noise generated by the digital system itself.

Externally Generated Noise

Line Transients

Whenever an elevator, soda machine, conveyor belt, or any piece of electrical equipment is turned on or off, electrical noise is generated. The noise results from the sudden demands for current from the electrical devices or from the arcing that occurs when switches and contacts are opened and closed. The fleeting noise voltages caused by these phenomena are called **transient volt-**

ages and can find their way into digital systems. Often transient voltages appear on the electrical power lines running throughout a building. Since digital system power supplies obtain their source of electricity from the same electrical power lines and convert it into DC current for each chip, the transient noise could affect every component in a system. Digital logic levels are related to the power supply voltages so that a temporary change in voltage supply level affects the output logic levels of the chips within a digital system. If the logic level change is severe, then false states occur.

Since we cannot prevent external noise of this sort, we must design our system to reject power supply noise. Several things can be done. Filters and surge suppressors placed on power supply lines can minimize the levels of voltage transients, preventing them from harming our digital circuitry. Voltage regulators placed throughout the digital system power distribution network can assist in maintaining constant voltage levels. Quality DC power supplies can reject noise as well. Unfortunately, transients are only one form of external noise with which the digital designer has to contend.

EMP and RFI

The same equipment producing the dreaded line transients can produce other noise as well. Whenever a switch, relay, or contact opens or closes, particularly in an inductive circuit, sparks are generated. Sparks contain a wide variety of frequencies that radiate through space as if broadcast by an antenna. This radiation is described as **EMP—electromagnetic pulses** or **RFI—radio-frequency interference**. Radiation of this nature can induce noise signals into a digital system in a manner similar to tuning in a radio station. The noise can show up not only on the power supply lines, but also directly on the signal lines. Shielding and special wiring techniques can help minimize this problem.

ESD

Another external source of trouble is **electrostatic discharge (ESD)**. ESD is a spark produced by static electricity discharge. On a dry day a person can accumulate a charge of over 35,000 V simply by walking across a rug. Touching an IC lead can transfer all the static charge from your body to the chip. Your ICs will not be thrilled to be zapped with a charge of this magnitude. The discharge of static electricity into the metal cases housing your circuitry can also be harmful even though you may not be directly contacting a chip. The designer should take precautions to prevent static sensitive chips from being directly connected to external equipment (keyboards, CRTs) that may be subjected to ESD. In addition, special precautions should be taken when handling any static sensitive IC. (See Appendix B for further details.)

Internally Generated Noise

Internally generated noise refers to electrical noise created within a digital circuit, and in this regard the components and technologies we use are often our own worse enemies. Logic devices are like excited children requiring our supervision and control to prevent them from running amuck. High frequency signals—and the positive and negative going edges of digital signals constitute high frequency signals—play by rules that are vastly different from those of low frequency signals: Wires are not wires at high frequencies, they are transmission lines. Signal lines that are not physically connected may actually be electrically connected because of stray capacitance. The list of potential problems goes on and on. We begin discussing sources of internally generated noise by looking at a prime culprit, the logic chip.

Switching Noise

Most digital devices require greater amounts of electrical current while switching than when "resting" in a fixed state. Because many of the gate's internal transistors are momentarily on during switching, **low impedance** paths exist between the supply voltage and ground yielding the increased current flow.

The source of current flow is the system power supply that must be properly designed to supply adequate current to all system components. However, the numerous chips comprising a digital system do not necessarily receive current in a consistent manner. It is more likely that the chips will demand "bursts" of current as they switch, causing the system voltage levels to fluctuate. Naturally, variations in power supply levels will alter logic levels. In addition, the changes in system power levels are very brief in duration, a characteristic that produces many high frequency components contributing to EMP and RFI problems. The noise signals can be coupled into adjacent signal lines, producing false logic responses.

To minimize the adverse consequences of switching noise, **decoupling capacitors** are used. Decoupling capacitors are placed between the "voltage supply rail" and the "ground rail" of each IC package. This is shown in Figure 12.13. The leads of the decoupling capacitor must be kept as short as possible to be effective. Basically, the capacitors act as minicurrent sources for their adjacent chips. When a gate within the IC package switches, the instantaneous current demand required for switching is supplied by the capacitor's stored charge. After switching occurs the capacitor charge is replenished by the system power supply. The net effect is that the voltage disturbances on the system power supply lines are minimized.

The size of the decoupling capacitor can be calculated using the capacitor current relationship formula:

$$I = C \frac{\Delta V}{\Delta t}$$

Rearranging, $C = I\Delta t/\Delta V$ where I is the current demand during an output change of state (voltage change high to low or low to high divided by the load resistance), ΔV is the change in V_{CC} level due to the change in state, and Δt is the length of time it takes the output to switch.

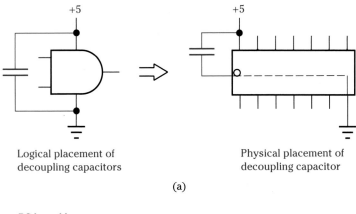

Logical placement of
decoupling capacitors

Physical placement of
decoupling capacitor

(a)

PC board layout

0.1 μfd decoupling capacitor

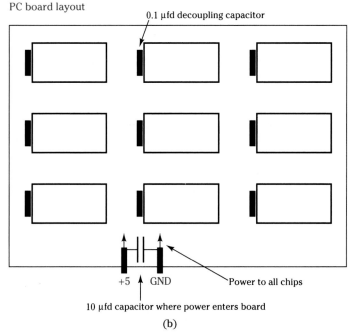

+5   ▲   GND   ↘Power to all chips

10 μfd capacitor where power enters board

(b)

**Figure 12.13
Decoupling Capacitor
Placement**

Compute the decoupling capacitor size required for a chip containing four gates. When switching each gate demands 50 mA of current for 15 ns. The V_{CC} supply voltage varies by 0.1 V during this time.

Solution   The capacitance required is calculated using $C = I\Delta t / \Delta V$. Multiply this value by 4 to determine the effective capacitance value required for the worst case situation where all four gates switch simultaneously.

$$C = \frac{50 \text{ mA} \times 15 \text{ ns}}{0.1} = \frac{7.5 \times 10^{-10}}{0.1} = 0.007 \text{ } \mu\text{fd}$$

$$0.007 \text{ } \mu\text{fd} \times 4 = 0.03 \text{ } \mu\text{fd} \text{ for the chip}$$

*DESIGN
EXAMPLE 12.6*

Generally, specific calculations are not required to determine decoupling capacitor sizes. Some rules of thumb for the major logic families include:

TTL—0.01 to 0.1 μfd for every two to five chips
CMOS—0.01 to 0.1 μfd for every two to five chips
        1 μfd tantalum capacitor every ten chips
ECL—0.01 μfd for every four to six chips

Decoupling capacitors generally are of mica, ceramic, or polystyrene construction. Since many digital systems require several printed circuit boards, a 10 to 100 μfd capacitor is also recommended between the power supply V_{CC} and ground lines as they enter the individual boards for the CMOS and TTL families. For ECL a 1 μfd capacitor in parallel with a 100 pfd capacitor is recommended. Figure 12.13 also shows typical decoupling capacitor placement on a circuit board.

In addition to the switching current demands, the power supply must also provide current to charge any capacitance associated with each gate output. This output capacitance occurs naturally and is called **stray capacitance**. Any two conductors separated by a dielectric (insulation), such as lengths of wire, solder connections, and printed circuit vias, have some capacitance that is just as real as capacitance contributed by an actual capacitor component. The power supply provides current to charge these capacitances during switching, placing added demands and noise on the power supply distribution lines. The designer can help him- or herself by carefully laying out the printed circuit board housing the digital logic to minimize lengths. The added output capacitance also contributes to slow rise and fall times.

Crosstalk

A basic tenet of electricity states that current flow through a wire results in the establishment of a magnetic field about the wire. If the magnetic field intercepts another wire, and if the magnetic field is changing (as is the case with digital signals), then current is induced to flow within the second wire. Transformers operate on this principle; however, this phenomenon is not desirable in digital systems. Separate signal wires are used in a digital system specifically to carry separate and distinct signals. When the signal from one line is inadvertently induced into another, the problem is referred to as **crosstalk**.

Crosstalk as just described is due to **inductive coupling**, but **capacitive coupling** can lead to crosstalk as well. As mentioned earlier, throughout any electronic system **stray capacitance** exists and is formed anytime two conductors are separated by a dielectric. Thus, two adjacent wires have some capacitance between them, which can couple an AC signal from one to the other. These two forms of crosstalk are illustrated in Figure 12.14.

Crosstalk can be minimized by preventing long parallel runs of wire. This reduces inductive and capacitive coupling between the wires. In addition, special wiring techniques and shielding can help. These solutions are discussed in later sections.

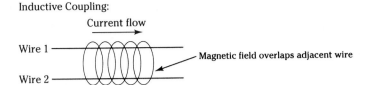

Inductive Coupling:

Capacitive Coupling:

Figure 12.14
Forms of Crosstalk

Ground Bounce

We mentioned earlier the noise problems associated with logic devices when switching states. This leads to voltage variations on power supply +V and ground lines. Variations on the ground line are particularly troublesome since most logic levels are referenced to ground. If the ground line voltage varies from 0 V to 0.5 V because of a disturbance, output signal levels will also vary by the same amount. (See Figure 12.15.) Naturally, a sizable deviation from a logic zero ground level causes false states. Voltage variations on the ground line caused by circuit switching noise are know as **ground bounce**. Again, proper attention to wiring and power system layout can minimize this problem.

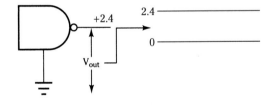

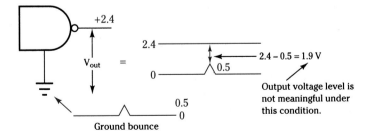

Figure 12.15
A Digital Circuit's
Reaction to Ground
Bounce

Simultaneous Switching

Simultaneous switching means that many logic device outputs are changing state at the same time or within a very narrow "window" of time. The combined effect of the switching noise associated with each gate can produce significant voltage variations on ground and signal lines. Simultaneous switching is a fact of life in many bus systems. Special care and attention must be paid to the physical layout of bus drivers on printed circuit boards or silicon wafers. In some cases, for example, drivers may need to be spaced around the periphery of the board/chip to minimize the effects of switching noise. Several of the wiring techniques discussed subsequently are also helpful. In addition, careful attention to the design of the power supply distribution system can help eliminate ground bounce.

Ground Loops

Critical to the performance of any digital system is a good power supply ground. This requires substantially more than just dangling a wire over the circuit and making connections to it. A poorly constructed ground distribution system will exhibit high impedance, resulting in voltage drops along its length. A voltage potential along the ground path (which supposedly is always 0 V) can seriously affect device switching threshold levels and can radiate noise to other portions of the circuitry. A ground path constructed with high impedance is known as a **ground loop** and is detrimental to circuit performance.

All conductors have some inductance and capacitance associated with them for these are fundamental electrical properties. However, in high speed and high frequency digital networks, even small amounts of inductance and capacitance can create problems, particularly when in the ground circuit. As current flows through these elements, voltage drops are produced causing portions of the ground line to have a different potential from others. For example, when a logic gate switches, the inductance associated with the ground lines develops a momentary voltage drop in response to the change in current flow. This produces a "spike" of noise on the ground line (ground bounce), which also shows up on signal lines. Naturally, we want to minimize this disturbance or our signals will become worthless. Since inductance is the primary reason for the noise, the solution begins by designing a low inductance power distribution system.

Inductance is related to the ground loop area, so minimizing the area will minimize the inductance. This is accomplished in several ways. Keeping both power and ground lines as short as possible reduces the wiring length and reduces inductance. Decoupling capacitors should be installed with very short leads to minimize the inductance associated with the capacitor lead wires. In fact, some newer ICs are designed with the power and ground terminals centered on the chip for this very reason. Older chips use a diagonal pin arrangement for power and ground, but this arrangement places the pins farthest from the chip and adds wiring inductance to the lines.

Since inductance is also related to the magnetic field established by current flow, minimizing the magnetic field also provides beneficial effects. Shielding against inductive (or capacitive) coupling with metal is helpful. Special printed circuit board construction, coaxial cable, or twisted pair cable helps as well since this kind of wiring can cancel or reduce magnetic fields. For example, power and ground lines should be run alongside each other on the printed circuit board (usually, one on top and one on the bottom of the board) so that the magnetic fields generated by the opposing current flow cancel each other. Many printed circuit boards employ a **ground plane** where one copper layer of the printed circuit board is all ground. With this technique all power lines are directly above the ground plane and magnetic effects are readily canceled. (See Figure 12.16.) In effect, this minimizes the ground loop area. Alternately, a grid structure of ground lines can also effectively battle magnetic fields. (See Figure 12.17.) One added benefit from minimizing ground loops stems from the fact that reducing inductance also reduces susceptibility to EMP and RFI problems.

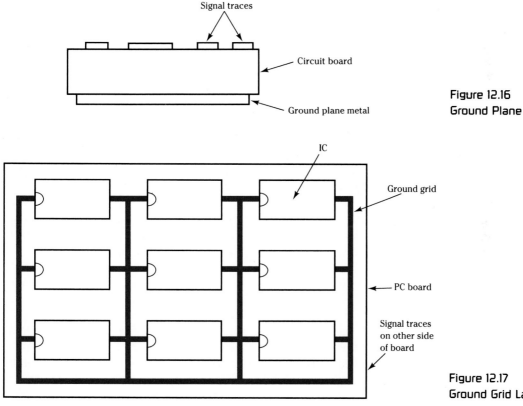

Figure 12.16 Ground Plane

Figure 12.17 Ground Grid Layout

Although you may mentally picture the design considerations discussed as applying only to logic family/printed circuit board design, they are also real-life concerns for designers using and designing on silicon. Even though all circuitry may be confined to only a few dense ICs, the problems relating to speed, power, fan-out, noise margin, and signal transmission do not go away. They occur at all levels of digital design.

12.10 ■ DIGITAL SIGNAL TRANSMISSION LINES

We know now that for a variety of reasons, such as slow rise and fall times or crosstalk, digital signal quality degrades. Fast switching digital signals "see" wire as a **transmission line** rather than just a length of wire and is another reason for digital signals falling short of ideal characteristics. Transmission line theory is used extensively in the electronics communication field and is also utilized to account for signal behavior in the digital world.

A transmission line is a complicated "beast" that is usually modeled as a resistor, inductor, capacitor network (RLC) as shown in Figure 12.18. As such, digital signals traveling over a transmission line are subjected to the charging and discharging effects of the inductors and capacitor. This leads to some unusual effects since the "wire" is actually an electronic circuit. Problems occur when there is an impedance mismatch among the signal source, the transmission line, and/or the receiver. Impedance mismatches lead to signal **reflections** where signal energy bounces back and forth along the transmission line between the source and the receiver. In certain cases the initial signal, the **incident wave**, combines with a reflection, the **reflected wave**. The combined signal may pull the input level of the receiving device above or below a normal logic level. In addition, multiple reflections can create pulses that may bother edge sensitive devices. Generally, reflections produce signal delays, ringing and overshoot on pulse edges, and reduce circuit noise margins.

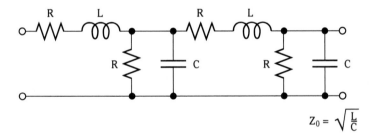

**Figure 12.18
Transmission Line
Model**

$$Z_0 = \sqrt{\frac{L}{C}}$$

Since the wire carrying a signal resembles a typical wire, how do we know when it is going to act like a transmission line? The answer is: when the propagation delay time of the transmission line is equal to or greater than the rise/fall time of the signal. This implies that transmission lines slow down a signal just as logic does and that the faster the signal changes the more likely that transmission line effects will occur. Of course, the length of the line is as important as the signal rate of change since long runs of wire also invite transmission line problems. Therefore, transmission line effects ultimately depend on the logic technology (rate of change) in use as well as the length of line. Typical transmission line delays are in the order of several nanoseconds per foot. Rise/fall times vary from a few nanoseconds to tens of nanoseconds.

For high speed logic families wire only 6 inches or 7 inches in length is sufficiently long to constitute a transmission line.

When it is clear that certain signal lines meet the criteria to be classified as transmission lines, then additional design measures need to be examined. Some transmission lines can be left as is if the signals they are carrying do not require instantaneous or real clean transitions between states. Some computer system memory address and data lines may fall into this category since these signals typically have plenty of time to settle down. On the other hand, clock and strobe signals are required to have fast, clean edges. Transmission line problems on these lines may lead to poor timing and false switching.

Many of the suggestions for good circuit board layout also apply to maintaining good transmission line performance. Some of the printed circuit board construction techniques used as well as specific kinds of transmission cable can keep transmission lines uniform and signal behavior predictable. Every transmission line has a **characteristic impedance** that is independent of wire length. The characteristic impedance is equal to $\sqrt{L/C}$. Since the physical attributes of the wire determine the values of L and C (inductance and capacitance), the wire's physical construction also determines the characteristic impedance.

In practical design the characteristic impedance is calculated using physical parameters such as length, spacing, dielectric constants, and other elements related to the construction of the transmission cable. The ideal transmission system consists of a source, transmission line, and receiver all having the same impedance. This assures that all signal energy emanating from the source travels along the transmission line and is totally absorbed by the receiving load. When impedances are not matched, such as when the load impedance is greater than the transmission line impedance, some energy is absorbed by the load while the remainder reflects back toward the source. Reflections can occur in or out of phase depending on the impedances of the receiving load and the transmission line. Regardless, it is apparent that any impedance discontinuity results in reflections. The problem is compounded when multiple loads are placed onto the transmission line since each has the potential to cause an impedance mismatch. Furthermore, typical digital circuits have low output impedances and high input impedances, so mismatches are present right from the beginning. Two steps are taken to control the mismatch— selecting the proper transmission line medium and **line terminations**.

Transmission Cable

A controlled transmission line cannot be created by point-to-point wiring because, as mentioned before, the characteristic impedance of the line is determined by the line's physical characteristics. Therefore, special transmission wire and printed circuit techniques are used to maintain constant characteristic impedance (Figure 12.19). Generally, lower characteristic impedance implies less propagation delay. Some commonly used transmission lines include:

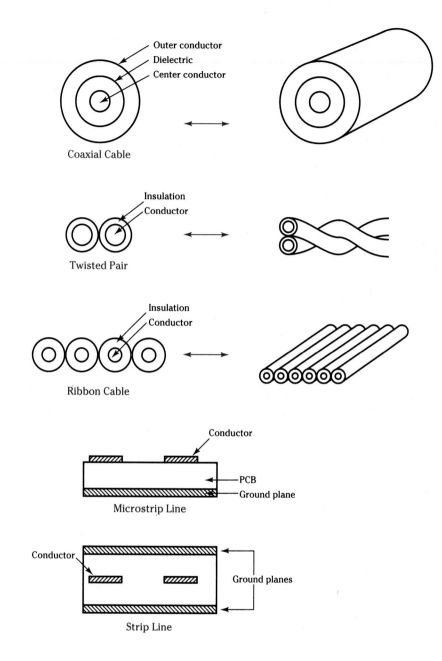

Figure 12.19
Conductor Outlines

- **Coaxial cable**   This cable consists of a center conductor surrounded by a dielectric material that is itself surrounded by a shielded ground wire. The center conductor carries the digital signal while the ground conductor is typically attached to the system ground at both ends of the cable. Coax (short for coaxial) cable minimizes magnetic fields because the two conductors carry the same current but in opposite directions, resulting in magnetic field cancellation. This significantly minimizes crosstalk. Coax also maintains a uniform characteristic impedance provided the wire is not bent or pinched. Typical coax characteristic impedances are 50 ohms, 75 ohms, 93 ohms, and 150 ohms.
- **Twisted pair**   Twisted pair cable is formed from two wires twisted together (roughly 30 turns per foot). Similar to telephone wire, twisted pair is much cheaper than coax, but still provides good magnetic field cancellation. A typical characteristic impedance is 110 ohms. Twisted pair is used extensively for long wiring runs in digital systems.
- **Ribbon cable**   Ribbon cable is made from multiple wires laid out side by side. Typically, every other wire carries a signal while the intermediate wires are grounded to reduce crosstalk.
- **Microstrip lines**   A microstrip is a printed circuit board trace (wire) of specific thickness and width that, in part, determines the characteristic impedance. The microstrip runs along one side of the board while a ground plane covers the other side. Microstrips are inexpensive because they are a normal part of printed circuit board construction.
- **Strip line**   A strip line is similar to a microstrip except that the strip is surrounded by a ground plane on the top and on the bottom. This construction can greatly improve the signal line's immunity to crosstalk.

Terminators

Impedance mismatches occur in transmission line systems even when using uniform transmission cable. Line termination is used to minimize mismatches in critical cases. A terminator is nothing more than an impedance placed onto the transmission line near the last receiver (the logic gate farthest from the source). Typical terminators are constructed from resistors, resistor networks, or resistor-capacitor networks and are used to balance the transmission line to receiver impedance. Tristate buses sometimes require termination to prevent noise pickup if all attached devices become inactive. Figure 12.20 illustrates several termination networks.

Series termination is used to eliminate a discontinuity between the source and the transmission line. To be effective the value of R plus the output impedance of the driving gate should equal the characteristic impedance of the transmission line.

Voltage divider termination is formed by a pull-up and pull-down resistor combination whose parallel value equals that of the line's characteristic impedance. This terminating network is placed close to the load. However, a voltage

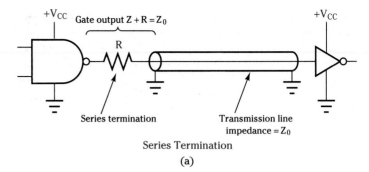

Series Termination

(a)

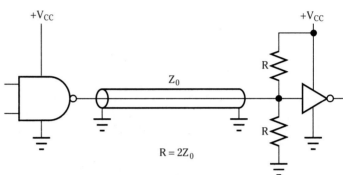

Voltage Divider Termination

(b)

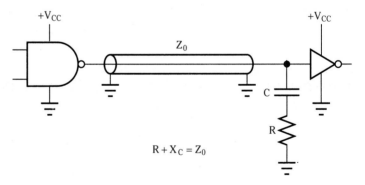

AC Termination

(c)

Figure 12.20
Termination Networks

divider draws current from the power supply and consequently dissipates power, making this network a power hungry solution.

AC **termination** is recommended for basic terminating requirements. The insertion of a capacitor blocks DC current flow and eliminates the power dissipation considerations found with voltage divider termination. The value of capacitance is such that the capacitive reactance is very low (only several ohms) at the frequency of interest. Then R equals the line characteristic impedance.

Sometimes diodes are also used as terminators. Some logic devices have the diodes built into the ICs for this and other noise reduction purposes. In all cases the lead length of the terminating components must be as short as possible to minimize lead inductance.

Line Drivers and Line Receivers

Sending signals down transmission lines and bus structures often places current demands on the driving source that standard logic gates cannot meet. The large number of receiving devices placed onto the bus along with the high values of capacitance associated with the devices contribute to the high current demand. Special logic devices are often used to provide the necessary current, and sometimes voltage levels, for bus usage. **Line drivers** are used as the driving devices under these conditions and **line receivers** are used at the receiving end. As an example, a 74ALS244A-1 line driver can safely supply up to 48 mA of current for line driving purposes. This is in contrast to a basic inverter's current limitation of 8 mA.

The IBM PC has room for six 8K $\times$ 8 ROM chips. Thirteen address lines are required to meet the addressing needs of the chips. Since the microprocessor chip supplying address information cannot drive the ROM chips directly, show how 74244s can be used in this line driving application.

Solution The IBM PC uses two 74244 chips to buffer address lines A_0–A_{12}. An enabling signal called AEN (address enable) takes the 74244 out of the tristate mode when low.

*DESIGN
EXAMPLE 12.7*

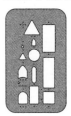

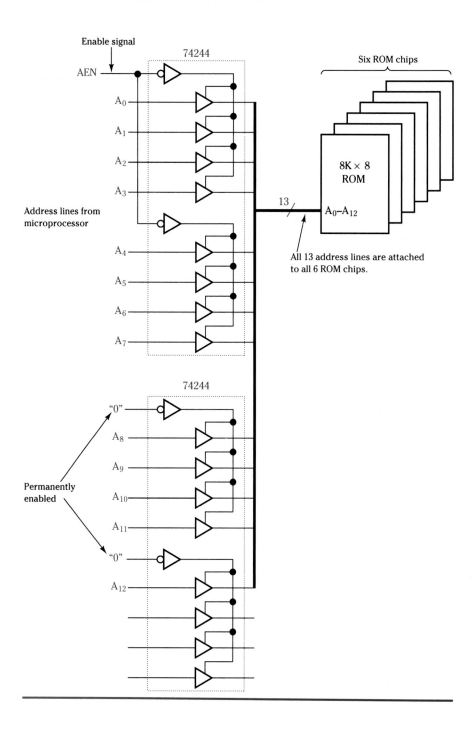

Enable signal

AEN

74244

Six ROM chips

A_0

A_1

A_2

A_3

Address lines from microprocessor

8K × 8 ROM

A_0–A_{12}

13

All 13 address lines are attached to all 6 ROM chips.

A_4

A_5

A_6

A_7

74244

"0"

A_8

A_9

Permanently enabled

A_{10}

A_{11}

"0"

A_{12}

- Many logic families are widely used for logic design. Popular families include TTL, CMOS, and ECL.

- Each logic family is characterized by its operating specifications. These include parameters such as speed, power, fan-out, and input and output voltage levels.

- Many logic families include several subfamilies offering speed and power advantages to the logic designer.

- Many new logic devices are fabricated using existing TTL, CMOS, or ECL technologies, but they are not part of any distinct logic family. These newer parts are very dense and can offer the designer significant advantages over existing logic families.

- Digital circuit performance is affected by noise signals. The source of noise can be internal to the device or induced externally.

- Noise problems in digital circuits can be minimized by careful attention to circuit layout, use of decoupling capacitors, and proper selection of transmission media.

SUMMARY

Sections 12.1–12.6

1. What is the state of a TTL output device if the output voltage is:
   *(a) 0.01 V   (b) 0.23 V   (c) 2.5 V   (d) 5.1 V   (e) 3.2 V   (f) 1.3 V
2. What are the high and low logic level voltages for a CMOS part if V_{CC} is:
   *(a) 3 V   (b) 5 V   (c) 8 V   (d) 10 V
3. What is the state of the output of 10K ECL devices if their output voltages are:
   (a) −1.05 V   *(b) −1.72 V   (c) −1.21 V   (d) −0.8 V
4. Name one advantage and one disadvantage of ECL over TTL and CMOS technologies.
5. Use manufacturers' data books and determine for the following part numbers, output current in the high state, output current in the low state (or supply current), propagation delay time from the high to low level, and propagation delay time from the low to high level:
   (a) 7486   (b) 74LS86   (c) 74AS86   (d) 74HCT86   (e) 10107-ECL

Section 12.7

6. What is the speed-power product for the following gate specifications:
   *(a) t_{pd} = 6 ns, power dissipation = 5 mW
   (b) t_{pd} = 11 ns, power dissipation = 2.2 mW

PROBLEMS

---

* See Appendix F: Answers to Selected Problems.

Section 12.8

*7. The following NAND-NAND network is created using a 7410. Show how the unused inputs should be tied to prevent unwanted noise problems.

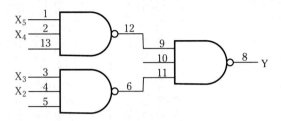

8. Redraw the circuit given in problem 7 as a NOR-NOR network using a 7427. Show how unused inputs should be treated to allow the function $Y = \overline{(\overline{X_5 + X_4}) + (\overline{X_3 + X_2})}$ to be created.

9. Using data sheets for the 7400 and 74LS32, determine the maximum fan-out for:
   (a) the 7400 driving 74LS32s
   (b) the 74LS32 driving 7400s

10. Using a manufacturer's data book, determine the sink current for a 7400 driving ten:
    *(a) 7400s   (b) 74LS00s   (c) 74ALS00s

*11. If a data sheet specifies $I_{OL} = 16$ mA and $I_{IL} = 1.6$ mA, the suggested fan-out is ten. If the measured sink current is actually 0.9 mA, how conservative are the manufacturer's specifications?

12. Determine the noise margin of a logic gate with the following specifications:
    (a) $V_{IL(max)} = 0.6$ $V_{IH(min)} = 2.3$ $V_{OH(min)} = 4.2$ $V_{OL(max)} = 0.01$
    (b) $V_{IL(max)} = 0.53$ $V_{IH(min)} = 1.9$ $V_{OH(min)} = 4.9$ $V_{OL(max)} = 0.2$

Section 12.9

13. Determine the likely cause of noise, using the following chip waveforms:

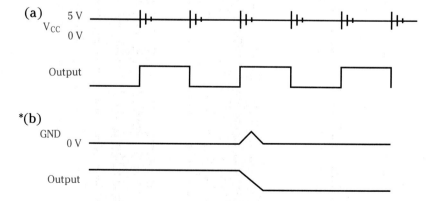

---

* See Appendix F: Answers to Selected Problems.

(c)

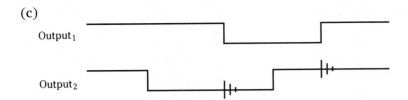

Output$_1$

Output$_2$

14. A CMOS gate output voltage is 3.5 volts (V). What happens to the level of this voltage if there is a +0.35 V fluctuation in ground potential?

*15. Determine the decoupling capacitor size required to decouple a quadruple NAND gate. Each gate requires 45 mA for 8 ns during simultaneous switching. The supply voltage fluctuates 0.15 V during this time.

16. Draw the physical layout of a 3-bit ripple carry adder comprised of 7408, 7432, and 7486 TTL chips. Do not show the signal connections, but show the power and ground plane printed circuit board layout. Using recommended decoupling capacitor sizes, show a reasonable placement and connection for the decoupling capacitors on the layout.

*17. Sketch the wire assignments for a length of 44 strand ribbon cable transmitting 16 data bits and 6 check bits between memory and ECC logic so that crosstalk is minimized.

Section 12.10

18. A 75-ohm transmission cable is connected from a bus driver to a logic gate input. Reflections are occurring on the line, so a voltage divider terminator is chosen to alleviate the problem. Sketch the circuit layout and calculate the resistor values used in the termination network.

---

* See Appendix F: Answers to Selected Problems.

CHAPTER 13

STATE MACHINE DESIGN

OBJECTIVES

To explain how state machines are used.

To differentiate between the Mealy and Moore state machine models.

To design state machines using traditional design techniques.

To illustrate how programmable logic hardware and software are used to design state machines.

To explore alternative state machine implementations, such as multiplexer, PROM, and microcode based designs.

PREVIEW

State machines are sequential circuits designed to progress through a predetermined number of states. As the state machine passes through each state, the output signals created in conjunction with the design are then used for control purposes and often to determine the next state of the circuit. State machine design ranges in complexity from simple counters to complex microprocessor chips. For example, the sequence of events occurring within a microprocessor chip as it executes a program are controlled by a state machine—the microprocessor's control logic. (Examples of these events include moving data, changing register contents, and reading information from memory.) This chapter introduces state machines, some of the techniques used in their design, and several technology based implementations.

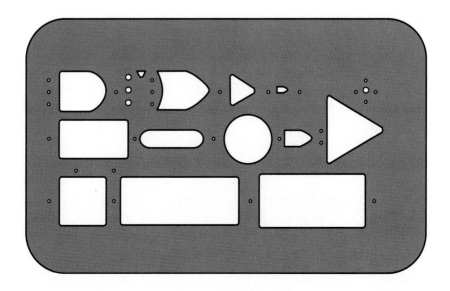

13.1 ■ STATE MACHINES

There are many ways to implement a state machine and a host of methods to design and analyze the circuitry. A typical state machine consists of a number of flip-flops (the exact number depends on the number of desired states) controlled by combinatorial circuits. The flip-flops exist in a **present state** and under the direction of combinatorial circuitry and a clock pulse they will change to the **next state**. Both the present state and the next state are determined by the designer's application for the system. In addition, input and output lines are often included as part of the state machine circuitry. This implies that the flip-flip outputs may not necessarily be the state machine outputs. Both synchronous and asynchronous state machine designs are used, although due to their glitch-free nature, synchronous designs predominate. We will confine our discussion to synchronous designs.

You can mentally picture a state machine by considering a basic binary synchronous counter. The counter starts in a present state of zero and progresses on each clock pulse to its next designated state. Any state machine accomplishes the same purpose—moving from state to state when specified conditions exist. We will design a state machine counter later in this chapter to understand these points further.

Figure 13.1 illustrates the basic elements of a state machine. External inputs feed a combinatorial logic network. The combinatorial logic controls the sequencing of the storage device states and assists in generating external outputs. An important characteristic of a state machine is the storage element feeding its signals back into the combinatorial logic. This means that the external outputs and subsequent states are determined by current states. As you might imagine, this can become rather complicated.

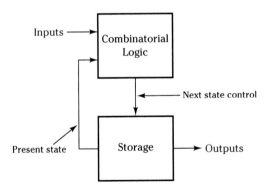

Figure 13.1
Basic State Machine
Arrangement

13.2 ■ MEALY AND MOORE STATE MACHINE MODELS

Most state machine designs can be classified as one of two design models. The model for the **Moore** machine is shown in Figure 13.2, whereas the model for the **Mealy** machine is shown in Figure 13.3. In both models input signals

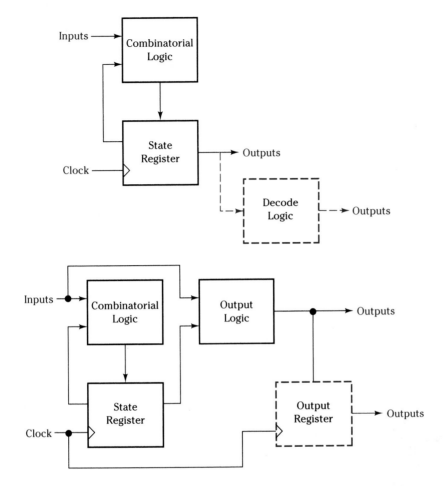

Figure 13.2
Moore State Machine
Model

Figure 13.3
Mealy State Machine
Model

feed a combinatorial logic network designed to control the sequencing of the
present state register. This register is a group of flip-flops that performs a
counting sequence as the machine passes through its various state assignments. The present state register outputs also feed, either directly or indirectly,
the machine output lines. In the Moore machine model outputs may be obtained
directly from the present state register if the machine state corresponds to the
desired output state. Often the machine output does not directly reflect the
register state and decoding logic (shown in dotted lines) may be added to
create the necessary output signals.

The Mealy machine produces output signals combined from input signals
and register state. Since the outputs are partially attained from nonregistered
input signals that can change, additional output registers may be required to
maintain output levels. The addition of an output register delays the state
machine output signal by one clock period. Incidentally, even though not
shown, an assumption is made that the present state registers in both models
are controlled by a synchronous clock. The clock signal is not considered to
be a state machine input, but rather, a separate control signal.

13.3 ■ STATE DIAGRAMS

State diagrams or flow diagrams are pictorial representations of the sequential activity in a state machine and are often a first step in conceptualizing design ideas. Figure 13.4 shows a very basic state diagram for a MOD 8 counter to illustrate the important features of the diagram.

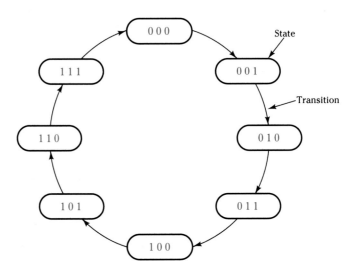

Figure 13.4
MOD 8 Counter State Diagram

Each oval represents one of the counter states. In practical diagrams the actual state values can be written in the oval as it is in this example, or some symbolic letters can be used to represent the state. The lines with arrows connecting the ovals represent transitions between states. Transitions occur synchronized with a clock pulse, and in this example a transition takes place for each clock pulse. The diagram clearly shows that the counter cycles through a normal counting sequence. In more complex state machines transitions still occur with clock pulses, but they are conditioned by the present state value and the input signal levels. Therefore, a transition may not always take place simply because there is a clock pulse. On a state diagram any conditions dictating a transition are drawn next to the transition line indicating exactly when the transition is to occur. Figure 13.5 shows specifically how Moore and Mealy state diagrams are constructed.

The main difference between the traditional Mealy and Moore diagrams is in representing output states. Since Moore machine outputs are dependent on the present state of the machine, they are indicated in the oval along with the state. Mealy outputs depend on both present state and input values and are represented on the diagram along the transition line causing them to occur. In other words, the input variable levels causing the transition to occur and the output levels that result once the transition is complete are listed along the transition line. Admittedly, these are fine points but they are useful when converting ideas represented by a state diagram into a functional circuit.

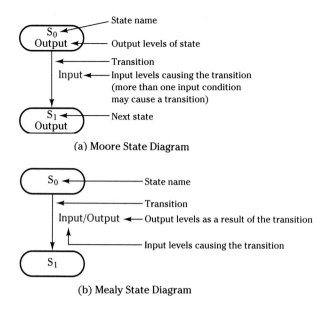

(a) Moore State Diagram

(b) Mealy State Diagram

Figure 13.5
State Diagram
Formats

Draw the state diagram for a decade counter. Assume a count of 0000 represents state S_0.

DESIGN EXAMPLE 13.1

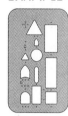

Solution The state diagram represents the states through which the state machine will pass. Since a decade counter passes through ten states and then repeats the sequence, the state diagram will indicate a progression from S_0 through S_9 with state S_9 continuing back to S_0.

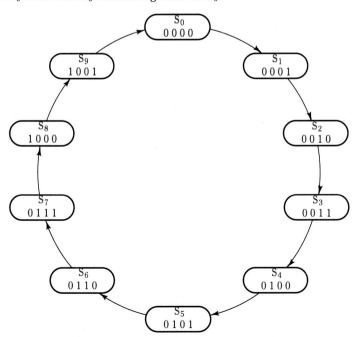

Show the modifications necessary in the decade counter state diagram in Design Example 13.1 if the counter is constructed to halt at state S_9 and not restart counting until a synchronous reset occurs.

Solution   States that are maintained (hold states) are indicated by a looping transition line around the oval representing the state. External input signals causing a transition are indicated along the transition line they control.

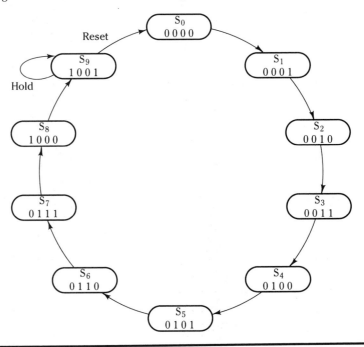

13.4 ■ ALTERNATE METHODS USED TO REPRESENT STATE MACHINES

A state diagram is only one of several ways to represent the behavior of a state machine. In fact, typical state machine design methods depend on many tables and graphs to synthesize a circuit completely. A **transition table** is the state machine's truth table. As shown in Figure 13.6, the table contains information on the present and next state of the machine and pertinent information on input and output levels, as well as the levels on the flip-flop D inputs. The state diagram accompanying Figure 13.6 shows how similar information is conveyed on the transition table.

For example, when the machine is in state S_0, it can move to state S_1 when input X is low (indicated by $\overline{X}$ on the state diagram). As shown on the transition table, the flip-flop D inputs must be set to proper levels ($D_2 = 0$, $D_1 = 1$) for the transition to take place. Outputs Y and Z are high and low, respectively, when this occurs (indicated by $Y\overline{Z}$ on the state diagram). When in state S_0, the machine can move to state S_2 when X is high (indicated by X on the state

State Diagram:

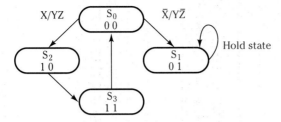

Transition Table:

PRESENT STATE	INPUTS	NEXT STATE	FLIP-FLOP	OUTPUTS
Q_2 Q_1	X	Q_2 Q_1	D_2 D_1	Y Z
0 0 (S0)	0	0 1	0 1	1 0
0 0 (S0)	1	1 0	1 0	1 1
1 0 (S2)	d	1 1	1 1	1 1
1 1 (S3)	d	0 0	0 0	1 1
0 1 (S1)	d	0 1	0 1	1 0

Figure 13.6
How State Diagrams
and Transition Tables
are Related

diagram). Flip-flop inputs are indicated as being set to $D_2 = 1$ and $D_1 = 0$ for the transition to occur. Outputs Y and Z both go high as a result of this transition (indicated by YZ on the state diagram). Information is extracted from the transition table during the design process to create the state machine control logic. Often both state diagrams and transition tables are used together to solidify the details of a state machine design.

Figure 13.7 depicts how a **flowchart** represents state machine behavior. This technique is similar to the flowcharts used in software programming. In

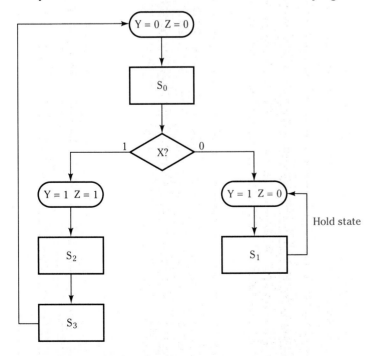

Figure 13.7
Flowchart

this case rectangles show the state of the machine, whereas diamond shaped symbols represent decisions made on the basis of input and state conditions. Oval shaped symbols hold the levels of output pins. This particular flowchart matches the state diagram of Figure 13.6. You may wish to compare the two and select the method that appeals to you. However, for the remainder of this section we will design using the state diagram and the transition table. (*Note:* It is also possible to describe state machine behavior using **state equations**. This approach is used in computer assisted design work. For specific information regarding this approach, refer to Appendix D.)

DESIGN EXAMPLE 13.3

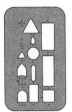

Create the transition table depicted by the following state diagram. Assume that D flip-flops are used in the design.

Solution   From the state diagram extract the present state and the input conditions necessary to cause a transition from that present state to another state. On the transition table, between the present state and next state, indicate the levels of the input signals causing a specific transition to occur. If an input has no bearing on the transition, list the input as a don't care (d) value. The flip-flop D inputs are listed with the levels required to cause a transition to the listed next state condition. Output values occurring after the transition are listed as well.

State Diagram:

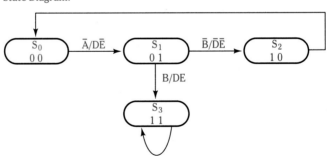

Transition Table:

PRESENT STATE		INPUTS		NEXT STATE		FLIP-FLOPS		OUTPUTS	
Q_2	Q_1	A	B	Q_2	Q_1	D_2	D_1	D	E
0	0	0	d	0	1	0	1	1	0
0	1	d	0	1	0	1	0	0	0
0	1	d	1	1	1	1	1	1	1
1	0	d	d	0	0	0	0	0	0
1	1	d	d	1	1	1	1	1	1

Output levels after transition

Input levels causing transition. These levels force the transition; the transition actually occurs with a clock pulse.

That is, Two paths from state 01

Next state after transition

D flip-flop input values to put flip-flops into the next state

13.5 ■ FOUR-BIT SYNCHRONOUS COUNTER DESIGN USING STATE MACHINE TECHNIQUES

Up to this point, we have described state machines in general terms and discussed various representations of state machine operation. These discussions are necessary because the complexities of state machines forces us to adopt some systematic methods so that we can proceed with their design. However, actual design examples convert generalities into specifics and help us make sense of all the options. You may recall the 4-bit J-K flip-flop synchronous counter design from Chapter 8 (Figure 8.14). This design was obtained intuitively by noting the times that individual flip-flops change level during a normal counting sequence. AND gates were used to provide proper levels to the J-K inputs prior to a clock pulse. Since we have a solid understanding of counter operation we can reexamine the counter from a state machine point of view. We will redesign the counter using state machine methods to illustrate a general procedure for subsequent designs.

Figure 13.8 shows the transition table representing a synchronous 4-bit counter. The individual flip-flop outputs comprising the present state of the counter are listed in a normal counting sequence. The next state columns list the same flip-flop outputs after the occurrence of a clock pulse. Notice that no input conditions are listed since the counter changes state simply by receiving a clock pulse. In a pure binary counter such as this it is not surprising that the next state is equal to the present state plus 1. The transition table at this point of development only tells us the progression of states. Additional information is required in order to design the combinatorial network controlling the sequence of state transitions. The completed transition table containing this information is shown in Figure 13.9.

PRESENT STATE				NEXT STATE			
Q_4	Q_3	Q_2	Q_1	Q_4	Q_3	Q_2	Q_1
0	0	0	0	0	0	0	1
0	0	0	1	0	0	1	0
0	0	1	0	0	0	1	1
0	0	1	1	0	1	0	0
0	1	0	0	0	1	0	1
0	1	0	1	0	1	1	0
0	1	1	0	0	1	1	1
0	1	1	1	1	0	0	0
1	0	0	0	1	0	0	1
1	0	0	1	1	0	1	0
1	0	1	0	1	0	1	1
1	0	1	1	1	1	0	0
1	1	0	0	1	1	0	1
1	1	0	1	1	1	1	0
1	1	1	0	1	1	1	1
1	1	1	1	0	0	0	0

Figure 13.8
4-Bit Counter Transition Table

PRESENT STATE				NEXT STATE				JKs			
Q_4	Q_3	Q_2	Q_1	Q_4	Q_3	Q_2	Q_1	$J_4 K_4$	$J_3 K_3$	$J_2 K_2$	$J_1 K_1$
0	0	0	0	0	0	0	1	0 d	0 d	0 d	1 d
0	0	0	1	0	0	1	0	0 d	0 d	1 d	d 1
0	0	1	0	0	0	1	1	0 d	0 d	d 0	1 d
0	0	1	1	0	1	0	0	0 d	1 d	d 1	d 1
0	1	0	0	0	1	0	1	0 d	d 0	0 d	1 d
0	1	0	1	0	1	1	0	0 d	d 0	1 d	d 1
0	1	1	0	0	1	1	1	0 d	d 0	d 0	1 d
0	1	1	1	1	0	0	0	1 d	d 1	d 1	d 1
1	0	0	0	1	0	0	1	d 0	0 d	0 d	1 d
1	0	0	1	1	0	1	0	d 0	0 d	1 d	d 1
1	0	1	0	1	0	1	1	d 0	0 d	d 0	1 d
1	0	1	1	1	1	0	0	d 0	1 d	d 1	d 1
1	1	0	0	1	1	0	1	d 0	d 0	0 d	1 d
1	1	0	1	1	1	1	0	d 0	d 0	1 d	d 1
1	1	1	0	1	1	1	1	d 0	d 0	d 0	1 d
1	1	1	1	0	0	0	0	d 1	d 1	d 1	d 1

Figure 13.9
Complete Transition Table for a 4-Bit Synchronous Counter

The controlling inputs for the counter are the J-K lines of the individual flip-flops. Since there are four flip-flops, there are eight controlling inputs that must be accounted for on the transition table. We must determine the proper levels that should be applied to each input prior to a clock pulse to ensure that the next state transition occurs. In other words, before the flip-flops can go on to the next state, we have to set up the correct signals at each J and K input to make the transition happen. This is the bulk of the work required for this circuit and for most state machines. Incidentally, no inputs or outputs are listed on this transition table because a counter is a special kind of state machine. The counter does not require inputs to make the transition to the next state (the transition occurs naturally) and the outputs are taken directly from the flip-flops (no additional logic is required).

How do we determine the correct J-K levels? Starting at present state = 0000 and comparing this to the anticipated next state (0001), we clearly see that only flip-flop output Q_1 should change. Q_1 is controlled by inputs J_1 and K_1. To ensure that Q_1 goes high on the clock pulse, we must establish conditions on J_1 and K_1 that set the flip-flop. Recalling J-K flip-flop operation, the set operation occurs when $J = 1$ and $K = 0$. These values could be listed under J_1 and K_1 on the transition table, but another approach can help simplify the circuitry. Since Q_1 is low prior to the clock pulse, it can be set into the next state by either a direct set ($J = 1$, $K = 0$) or a toggle condition ($J = 1$, $K = 1$). In either condition J must be high for a set to transpire; the status of K is unimportant. Therefore, we list J as a 1 and label K as a don't care (d) value as the conditions for the transition from a present state = 0000 to the next state = 0001. This process is carried for each state transition.

As you glance at the transition table, you will notice the many different J-K combinations listed for each possible transition. Although it may appear that entering all these J-K levels is time-consuming, the process is significantly speeded up by constructing a separate **flip-flop excitation table**, which simply lists how the flip-flop changes state. (The flip-flop input levels or stimuli are often referred to as the excitation levels.) This is shown in Figure 13.10. Since one of the four J-K combinations is always used, filling in the transition table is simplified by referencing the flip-flop excitation table. All J-K combinations include one input don't care condition following reasoning similar to that discussed previously.

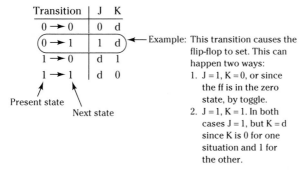

Figure 13.10
J-K Excitation Table

Once the transition table is complete, logic design begins using basic logic development techniques. Figure 13.11 shows how K-maps (the typical technique) are constructed for each J and K input utilizing the present state flip-flop Q values as K-map variables. This should make sense since the present state of the flip-flops ultimately determines the flip-flops' next state. The resulting equations determine the characteristics of the combinatorial logic network. The circuitry obtained from the logic equations is shown in Figure 13.12 and matches the circuitry discused in Chapter 8.

Figure 13.13 shows how the 4-bit counter fits the Moore machine description discussed earlier. Except for the lack of inputs, the present state register

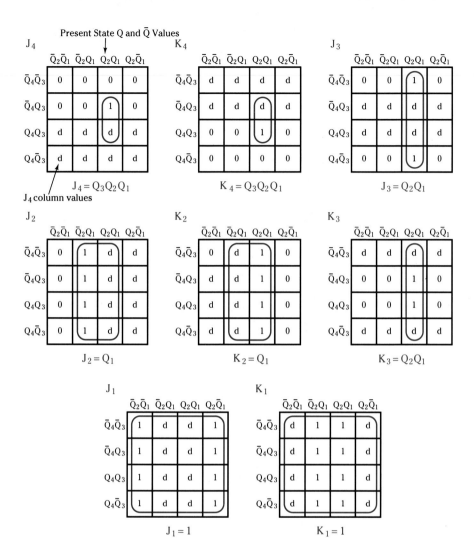

Figure 13.11
K-Maps Derived from
the Transition Table

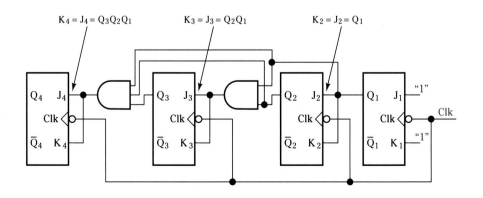

Figure 13.12
4-Bit Synchronous
Counter

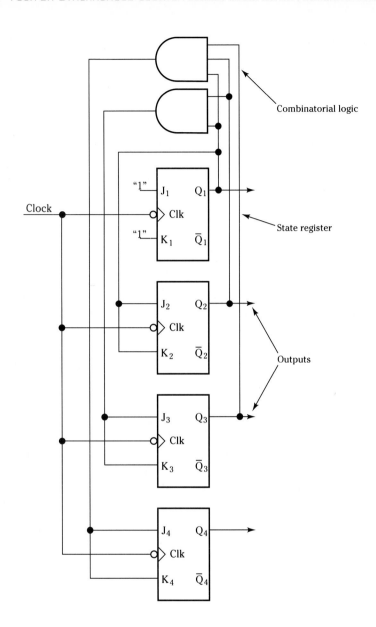

Figure 13.13
4-Bit Synchronous
Counter Modeled as a
State Machine

and combinatorial logic network duplicate the essential elements shown in the Moore machine model.

Summarizing the state machine design procedure:

1. Conceptualize the design.
2. Create state diagrams and transition tables to describe circuit operation.
3. Determine the combinatorial logic equations representing the circuitry that moves the state sequence from each present state to each next state. Use logic reduction techniques to assist in this effort.

DESIGN EXAMPLE 13.4

Using state machine design techniques, design a truncated count synchronous MOD 3 counter. Draw the circuit and timing diagrams to verify operation.

Solution   Construct a transition table showing present and next states. Also indicate the J-K levels required to cause the next state transition. Use a J-K excitation table for convenience. Since two flip-flops are required and only three of the four states are necessary, state 11 is treated as a don't care state.

State 11 = Don't Care:

PRESENT STATE		NEXT STATE		FLIP-FLOPS			
Q_2	Q_1	Q_2	Q_1	J_2	K_2	J_1	K_1
0	0	0	1	0	d	1	d
0	1	1	0	1	d	d	1
1	0	0	0	d	1	0	d

J-K Excitation Table:

TRANSITION	J	K
0 $\longrightarrow$ 0	0	d
0 $\longrightarrow$ 1	1	d
1 $\longrightarrow$ 0	d	1
1 $\longrightarrow$ 1	d	0

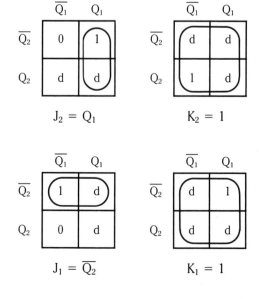

$J_2 = Q_1$

$K_2 = 1$

$J_1 = \overline{Q_2}$

$K_1 = 1$

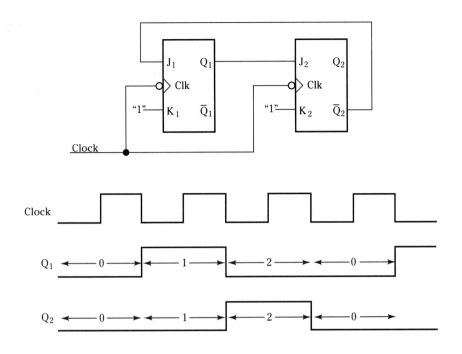

Design a synchronous counter for the following count sequence: 0, 2, 3, 5, 7.

Solution The atypical binary sequence contains five distinct states and requires three flip-flops to attain these states. The state diagram and J-K excitation
table are drawn for convenience.

*DESIGN
EXAMPLE 13.5*

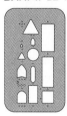

State Diagram:

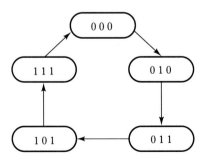

Excitation Table:

TRANSITION	J	K
0 ⟶ 0	0	d
0 ⟶ 1	1	d
1 ⟶ 0	d	1
1 ⟶ 1	d	0

Transition Table:

PRESENT STATE			NEXT STATE			FLIP-FLOPS					
Q_3	Q_2	Q_1	Q_3	Q_2	Q_1	J_3	K_3	J_2	K_2	J_1	K_1
0	0	0	0	1	0	0	d	1	d	0	d
0	1	0	0	1	1	0	d	d	0	1	d
0	1	1	1	0	1	1	d	d	1	d	0
1	0	1	1	1	1	d	0	1	d	d	0
1	1	1	0	0	0	d	1	d	1	d	1
0	0	1	d	d	d	d	d	d	d	d	d
1	0	0	d	d	d	d	d	d	d	d	d
1	1	0	d	d	d	d	d	d	d	d	d

K-Maps:

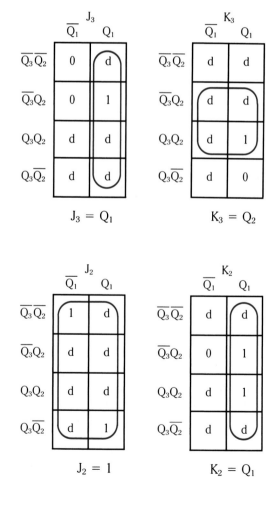

$J_3 = Q_1$ $K_3 = Q_2$

$J_2 = 1$ $K_2 = Q_1$

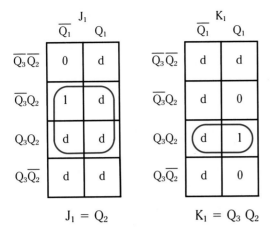

$$J_1 = Q_2 \qquad K_1 = Q_3\,Q_2$$

Circuit:

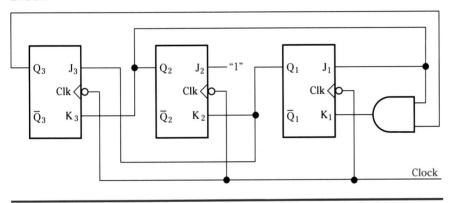

13.6 ■ STATE MACHINE DESIGNS CONTAINING INPUTS AND OUTPUTS

State machine designs that are more complex than counters have next state sequences determined by both input signals and the present state. In addition, outputs not necessarily equal to present state values may be required. The transition table must reflect the effects of input signal levels on both next state and output values. In other words, once a transition to the next state is listed on the transition table, the output levels created by this change are also listed. Once the transition table is set up to account for input level changes, K-map or other reduction techniques are utilized to create the corresponding register control and output circuit logic. In designs like this the K-map variables are the present state values and the input signal values. Design Example 13.6 shows how.

DESIGN EXAMPLE 13.6

Design a control system to process bottles on a conveyor line. The system must provide control signals to accomplish the following:

ACTIVITY	OUTPUT	INPUT
1. Place the bottle on the conveyor	$Y_1 = 1$	$X = 0$
2. Start the conveyor	$Y_2 = 1$	$X = 0$
3. Stop at the label/cap station	$Y_2 = 0$	$X = 1$
4. Paste label	$Y_3 = 1$	
5. Cap	$Y_4 = 1$	

The output signals are low except for the activities noted above. Input X is the sensor that senses the position of the bottle on the conveyor. X is low when the bottle is moving; high when the bottle reaches the paste/cap station. Assume that the appropriate equipment is in place to respond to the control signals. Use D flip-flops.

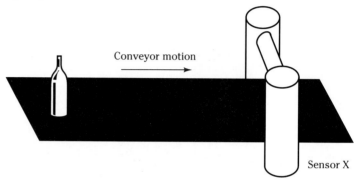

Conveyor motion

Sensor X

Solution　Define states, draw the state diagram, complete the transition table, and design the circuit.

Define States:

S_0	000	Place bottle
S_1	001	Start conveyor
S_2	010	Stop conveyor
S_3	011	Paste label
S_4	100	Cap

State Diagram:

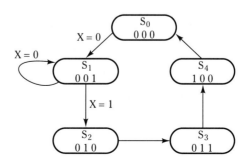

Transition Table:

PRESENT STATE			INPUT	NEXT STATE			FLIP-FLOPS			OUTPUTS			
Q_3	Q_2	Q_1	X	Q_3	Q_2	Q_1	D_3	D_2	D_1	Y_4	Y_3	Y_2	Y_1
0	0	0	0	0	0	0	0	0	0	0	0	0	1
0	0	0	1	d	d	d	d	d	d	d	d	d	d
0	0	1	0	0	0	1	0	0	1	0	0	1	0
0	0	1	1	0	1	0	0	1	0	0	0	0	0
0	1	0	0	d	d	d	d	d	d	d	d	d	d
0	1	0	1	0	1	1	0	1	1	0	1	0	0
0	1	1	0	d	d	d	d	d	d	d	d	d	d
0	1	1	1	1	0	0	1	0	0	1	0	0	0
1	0	0	0	d	d	d	d	d	d	d	d	d	d
1	0	0	1	0	0	0	0	0	0	0	0	0	0
1	0	1	0	d	d	d	d	d	d	d	d	d	d
1	0	1	1	d	d	d	d	d	d	d	d	d	d
1	1	0	0	d	d	d	d	d	d	d	d	d	d
1	1	0	1	d	d	d	d	d	d	d	d	d	d
1	1	1	0	d	d	d	d	d	d	d	d	d	d
1	1	1	1	d	d	d	d	d	d	d	d	d	d

D_3

	$\overline{Q_1}\,\overline{X}$	$\overline{Q_1}\,X$	$Q_1\,X$	$Q_1\,\overline{X}$
$\overline{Q_3}\,\overline{Q_2}$	0	d	0	0
$\overline{Q_3}\,Q_2$	d	0	1	d
$Q_3\,Q_2$	d	d	d	d
$Q_3\,\overline{Q_2}$	d	0	d	d

$$D_3 = Q_1 Q_2$$

D_2

	$\overline{Q_1}\,\overline{X}$	$\overline{Q_1}\,X$	$Q_1\,X$	$Q_1\,\overline{X}$
$\overline{Q_3}\,\overline{Q_2}$	0	d	1	0
$\overline{Q_3}\,Q_2$	d	1	0	d
$Q_3\,Q_2$	d	d	d	d
$Q_3\,\overline{Q_2}$	d	0	d	d

$$D_2 = \overline{Q_3}\,\overline{Q_2}\,X + Q_2\,\overline{Q_1}$$

D_1

	$\overline{Q_1}\,\overline{X}$	$\overline{Q_1}\,X$	$Q_1\,X$	$Q_1\,\overline{X}$
$\overline{Q_3}\,\overline{Q_2}$	1	d	0	1
$\overline{Q_3}\,Q_2$	d	1	0	d
$Q_3\,Q_2$	d	d	d	d
$Q_3\,\overline{Q_2}$	d	0	d	d

$$D_1 = \overline{X} + \overline{Q_3}\,\overline{Q_1}$$

Y_4

	$\overline{Q_1}\,\overline{X}$	$\overline{Q_1}\,X$	$Q_1\,X$	$Q_1\,\overline{X}$
$\overline{Q_3}\,\overline{Q_2}$	0	d	0	0
$\overline{Q_3}\,Q_2$	d	0	1	d
$Q_3\,Q_2$	d	d	d	d
$Q_3\,\overline{Q_2}$	d	0	d	d

$$Y_4 = Q_2 Q_1$$

Y_3

	$\overline{Q_1}\,\overline{X}$	$\overline{Q_1}\,X$	$Q_1\,X$	$Q_1\,\overline{X}$
$\overline{Q_3}\,\overline{Q_2}$	0	d	0	0
$\overline{Q_3}\,Q_2$	d	1	0	d
$Q_3\,Q_2$	d	d	d	d
$Q_3\,\overline{Q_2}$	d	0	d	d

$$Y_3 = Q_2\,\overline{Q_1}$$

Y_2

	$\overline{Q_1}\,\overline{X}$	$\overline{Q_1}\,X$	$Q_1\,X$	$Q_1\,\overline{X}$
$\overline{Q_3}\,\overline{Q_2}$	0	d	0	1
$\overline{Q_3}\,Q_2$	d	0	0	d
$Q_3\,Q_2$	d	d	d	d
$Q_3\,\overline{Q_2}$	d	0	d	d

$$Y_2 = Q_1\,\overline{X}$$

Y_1

	$\overline{Q_1}\,\overline{X}$	$\overline{Q_1}\,X$	$Q_1\,X$	$Q_1\,\overline{X}$
$\overline{Q_3}\,\overline{Q_2}$	1	d	0	0
$\overline{Q_3}\,Q_2$	d	0	0	d
$Q_3\,Q_2$	d	d	d	d
$Q_3\,\overline{Q_2}$	d	0	d	d

$$Y_1 = \overline{Q_1}\,\overline{X}$$

Circuit:

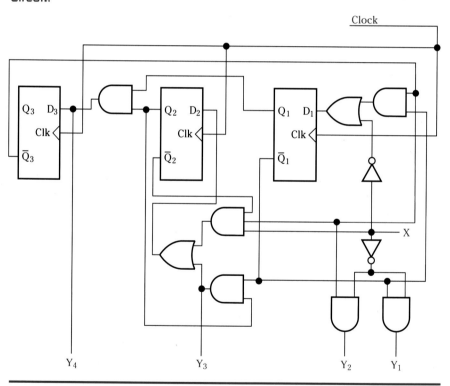

13.7 ■ MULTIPLEXER BASED STATE MACHINES

An alternative design approach to the state machine configurations we have discussed is to use multiplexers for the combinatorial control logic. The advantage to this approach is that every design consists of the same number of standard building blocks arranged in a regular structure. This makes state machine design easy to understand for the nondesigner as well as make the design process itself relatively straightforward. The disadvantage of the multiplexer technique is that it does not minimize the number of circuit components required. Figure 13.14 shows a very simple circuit arrangement involving only a counter and multiplexer. This design is little more than a data selector logic circuit (as was discussed in Chapter 6) and not really a state machine, but it is useful to illustrate how counters and multiplexers react. The circuit also has practical application when the state register is simply a counter (a predictable sequence) and output values are easily determined from the present state. A circuit such as this can be used for waveform generation as shown in the figure.

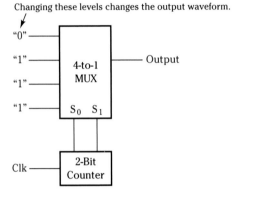

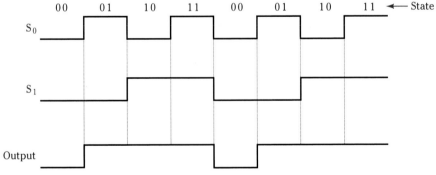

Figure 13.14
MUX Based Waveform
Generator

Design a two-phase clock using multiplexers and a counter corresponding to the following output waveforms:

DESIGN
EXAMPLE 13.7

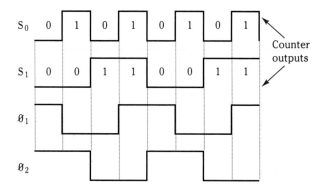

Solution A two-phase clock waveform contains two waveforms of the same frequency and duty cycle that are phase shifted with respect to each other. Use a counter to drive two multiplexers. Each multiplexer is set to function as a waveform generator, one for $\emptyset_1$ and one for $\emptyset_2$.

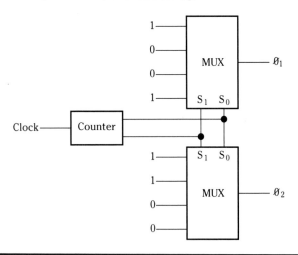

Figure 13.15 illustrates the typical configuration for a complete multiplexer based state machine. Each D flip-flop comprising the state register has an associated multiplexer to provide the correct logic levels for the D inputs. The multiplexer size is determined by the number of states in the system and the number of input signals. The state register outputs drive the multiplexer select inputs, causing the multiplexer output to reflect the level of one of the multiplexer inputs. The multiplexer inputs contain signals derived from the state register or from dedicated circuit inputs. Connected this way, the multiplexer output influences the next state of the state register.

Figure 13.16 shows the state diagram for a circuit we will design with multiplexers. The accompanying transition table shows the present state, input, and next state values assigned according to the state diagram. Present state conditions are listed several times, once for each transition possibility that can

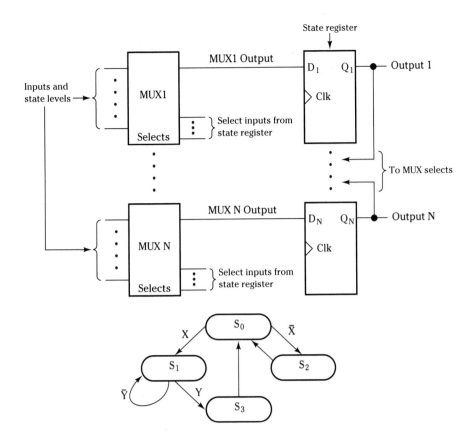

State register

MUX1 Output

Inputs and state levels →

MUX1

Selects

Select inputs from state register

D_1 Q_1 — Output 1

Clk

To MUX selects

MUX N Output

MUX N

Selects

Select inputs from state register

D_N Q_N — Output N

Clk

Figure 13.15
Block Diagram of a Multiplexer Based State Machine

PRESENT STATE	INPUTS	NEXT STATE	FLIP-FLOPS	MULTIPLEXERS	
Q_2 Q_1	X Y	Q_2 Q_1	D_2 D_1	MUX2	MUX1
0 0	1 d	0 1	0 1	$\bar{X}$	X
0 0	0 d	1 0	1 0		
0 1	d 0	0 1	0 1	Y	1
0 1	d 1	1 1	1 1		
1 0	none	0 0	0 0	0	0
1 1	none	0 0	0 0	0	0

Figure 13.16
State Diagram and Transition Table for a Multiplexer Based Design

occur as dictated by the input levels. For instance, there are two possible outcomes from state S_0 that are ultimately determined by the level of input X. Since the multiplexers provide the D flip-flops with the signals that determine the next state, we examine the next state levels and compare them to the input levels. For the transition from state S_0 to either S_1 or S_2 input X is the determining signal. Notice that Q_1's level in the next state column is identical to the level for input X, whether the next state is 01 or 10. Therefore, the multiplexer (MUX1) controlling flip-flop Q_1 must present the X value to the D_1 input prior to a clock pulse for the state transition to take place. The next state Q_2 column shows that the multiplexer (MUX2) controlling the Q_2 flip-flop requires the complement of X for the correct transition to take place. Therefore, $\overline{X}$ is assigned to an input of MUX2.

The comparisons between next states and inputs continue until the transition table is complete. When unconditional transitions occur (not dependent on input levels), the multiplexer assignments are simply that of the next state.

Figure 13.17 shows the complete multiplexer state machine. The state register flip-flop outputs feed the multiplexer select inputs and the multiplexer outputs feed the flip-flop D inputs. Multiplexer input assignments are taken from the transition table as described. For instance, when the state register is in state S_0, flip-flop input D_1 should be connected to input X prior to the next clock pulse. Since the flip-flops are presently selecting the first multiplexer input, which is connected to X, this condition is met. On the next clock pulse flip-flop Q_1 will store the value of X. In a similar fashion, the second multiplexer provides the complement of X to flip-flop input D_2.

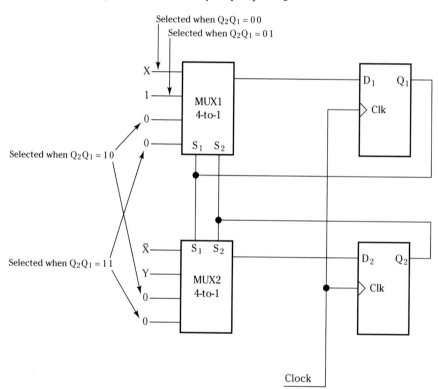

Figure 13.17
Multiplexer Based
State Machine Circuit

In this example the circuit outputs are taken directly from the state register. If additional outputs need to be generated, they would be added to the transition table and determined by standard reduction techniques. Furthermore, this particular design fits nicely into 4-to-1 multiplexers. Since a perfect fit is not always assured, some multiplexer inputs may not be used in other designs. Although this procedure is not an efficient use of circuitry, the design simplicity trade-off is often worthwhile.

Design a MUX based state machine from the following state diagram:

DESIGN
EXAMPLE 13.8

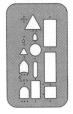

State Diagram:

S_0 $0\,0$ $\xrightarrow{\bar{A}/D\bar{E}}$ S_1 $0\,1$ $\xrightarrow{\bar{B}/D\bar{E}}$ S_2 $1\,0$

$S_1 \xrightarrow{B/DE} S_3$ $1\,1$

Solution   Using the state diagram, create a transition table specifying next states, flip-flop input conditions, and multiplexer input conditions. State machine outputs should also be defined and designed from the table using a K-map. K-map variables should include all inputs and next state outputs.

PRESENT STATE		INPUTS		NEXT STATE		FLIP-FLOPS		MUX		OUTPUTS	
Q_2	Q_1	A	B	Q_2	Q_1	D_2	D_1	M_2	M_1	D	E
0	0	1	d	0	0	0	0	0	$\bar{A}$	d	d
0	0	0	d	0	1	0	1			1	0
0	1	d	0	1	0	1	0	1	B	0	0
0	1	d	1	1	1	1	1			1	1
1	0	d	d	0	0	0	0	0	0	d	d
1	1	d	d	1	1	1	1	1	1	1	1

D

	$\bar{A}\bar{B}$	$\bar{A}B$	AB	$A\bar{B}$
$\bar{Q_2}\bar{Q_1}$	d	d	d	d
$\bar{Q_2}Q_1$	1	1	d	d
Q_2Q_1	1	1	1	1
$Q_2\bar{Q_1}$	0	d	d	0

$$D = Q_1$$

E

	$\bar{A}\bar{B}$	$\bar{A}B$	AB	$A\bar{B}$
$\bar{Q_2}\bar{Q_1}$	d	d	d	d
$\bar{Q_2}Q_1$	0	0	d	d
Q_2Q_1	1	d	d	1
$Q_2\bar{Q_1}$	0	d	d	0

$$E = Q_2Q_1$$

D and E output K-maps. Unused combinations are treated as don't cares.

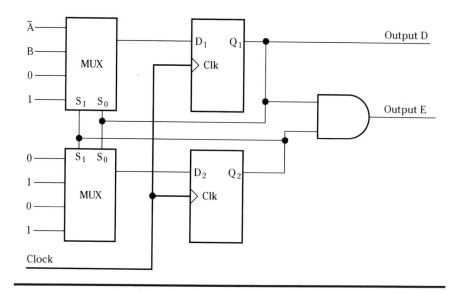

13.8 ■ PROM BASED STATE MACHINES

Another state machine design alternative is using a PROM for both the next state determination and the output logic. Figure 13.18 shows a block diagram arrangement of the components for this state machine approach. The outputs from the state register feed a portion of the PROM address lines, while the remainder of the address lines are controlled by state machine input signals. The PROM output lines have two functions. One is to provide the output signals for the state machine while the other is to provide the necessary signals to move the state register to the next state. Typically, some PROM output pins are dedicated strictly for output use, whereas others are used only for next state generation. The data burned into the PROM determines the necessary output and next state generation signals.

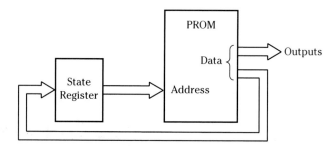

Figure 13.18
PROM Based State
Machine

Figure 13.19 is essentially the same transition table used for the MUX based design in Figures 13.16 and 13.17 except for the addition of several outputs (O_1, O_2, and O_3). The table in Figure 13.19 shows how next state signals are determined. Since the state register is constructed with D flip-flops, the D inputs

PRESENT STATE	INPUTS	NEXT STATE	FLIP-FLOPS	PROM OUTPUTS		
Q_2 Q_1	X Y	Q_2 Q_1	D_2 D_1	Q_5 Q_4	Q_3 Q_2 Q_1	
0 0	1 d	0 1	0 1	0 1	1 0 1	
0 0	0 d	1 0	1 0	1 0	1 1 0	
0 1	d 0	0 1	0 1	0 1	1 1 1	
0 1	d 1	1 1	1 1	1 1	0 1 1	
1 0	none	0 0	0 0	0 0	0 0 0	
1 1	none	0 0	0 0	0 0	0 0 0	

That is, to move from state 0 0 to state 0 1,
PROM bits $Q_5 Q_4 = 0\,1$. This will set flip-flop
inputs $D_2 D_1$ to the proper levels
for the transition.

Next state   State machine
signals   outputs
              (present state)

Figure 13.19
Transition Table for a
PROM Based State
Machine

merely need to have the levels called for by the next state before the clocking signal occurs. Therefore, the table shows several identical columns—the next state outputs, the D inputs, and bits O_5 and Q_4 of the PROM output. Burning the appropriate PROM locations with this information provides all the logical decision making for next state generation.

Figure 13.20 shows the circuitry requirements for this design. Inputs X and Y, as well as Q_1 and Q_2 from the state register, feed the PROM's addressing lines. PROM outputs O_1, O_2, and O_3 are the state machine output signals, whereas O_4 and O_5 feed back to the state register D inputs. Assuming that some power on reset circuitry is in place for the state register, PROM addresses 0, 1, 2, or 3 will be selected depending on the level of input X. (The choice of

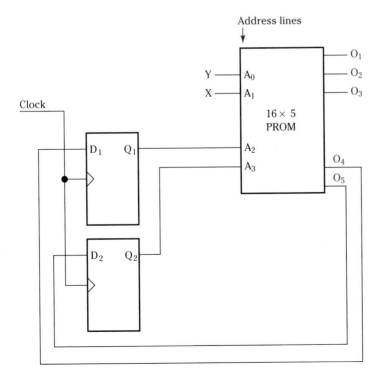

Figure 13.20
PROM Based State
Machine Circuit

An Alternate PROM Based State Machine Configuration

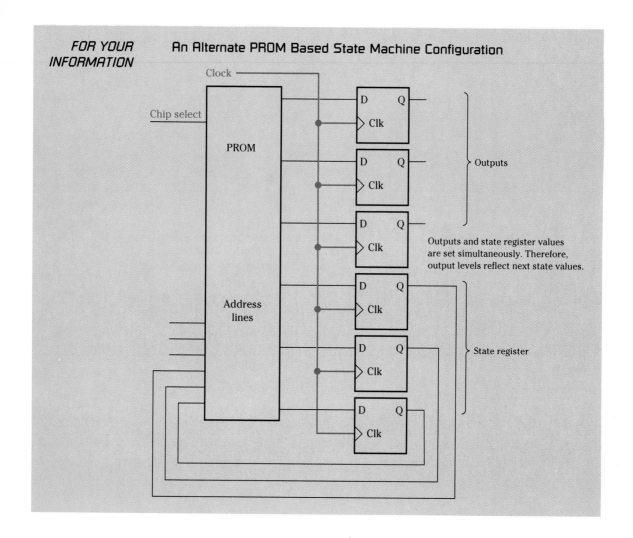

four initial addresses stems from the don't care condition for the Y input when in present state 00.) The address selected provides output signals and next state information for the state register. From this point on outputs and next state signals are generated as the PROM is cycled through its addresses.

Figure 13.21 shows how the present state and input information from the transition table is translated into PROM data. For instance, if the state register is in state 00, then PROM address lines A_3, A_2 equal 00. One of the first four PROM addresses will be selected as determined by the levels on A_1 (X variable) and A_0 (Y variable). Since the Y variable is considered a don't care parameter for state 00, then the level on X determines which address is selected. If X is low, then O_5, O_4, O_3, O_2, O_1 = 10110. The two most significant bits feed the state register flip-flops so that on the following clock pulse the register moves to state 10. The three least significant bits are the designated output signals

Q_2 A_3	Q_1 A_2	X A_1	Y A_0	O_5	O_4	O_3	O_2	O_1
0	0.	0	0	1	0	1	1	0
0	0	0	1	1	0	1	1	0
0	0	1	0	0	1	1	0	1
0	0	1	1	0	1	1	0	1
0	1	0	0	0	1	1	1	1
0	1	0	1	0	1	1	1	1
0	1	1	0	1	1	0	1	1
0	1	1	1	1	1	0	1	1
1	0	0	0	0	0	0	0	0
1	0	0	1	0	0	0	0	0
1	0	1	0	0	0	0	0	0
1	0	1	1	0	0	0	0	0
1	1	0	0	0	0	0	0	0
1	1	0	1	0	0	0	0	0
1	1	1	0	0	0	0	0	0
1	1	1	1	0	0	0	0	0

Next state signals   Output signals

Figure 13.21
Translating State Information into PROM Data

for the current state. If X is high, then a different PROM address is read (address 2 or 3) so that O_5, O_4, O_3, O_2, O_1 = 01101. The state register moves to state 01 on the next clock pulse as dictated by PROM outputs O_5 and O_4. The state machine outputs are set by bits O_3, O_2, and O_1.

Some of the address data are repeated when states are not determined by specific input levels (such as states 10 and 11). The number of addresses and data bits for this state machine dictates a 16 × 5 PROM, an unlikely size. Naturally, the designer would specify a PROM as close as possible to the required size. In addition, provisions would be necessary to ensure that the PROM's chip select and enable circuitry are functional.

The primary advantage of PROM based designs is that changes in output or next state signals can be easily modified without redesigning the circuitry. Only new PROM or EPROM data is required. One disadvantage is the lack of system speed. Since the delay from change of state to new output signals is determined by the state register delay and the relatively slow access time of the PROM, a PROM based design will not perform with the same speed as a complete logic based circuit. Again, the designer faces a trade-off between speed and simplicity. It should also be noted that some of the programmable sequencer chips available (the PROSE device, for example) are created around a state register-PROM arrangement. PALASM2 can be used to design with these parts.

Design a PROM based state machine to control railroad crossing gates.

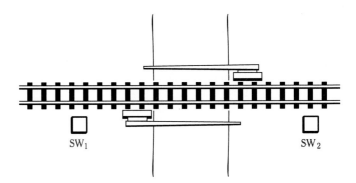

Two switches (SW$_1$ and SW$_2$) provide a high level output when a train crosses them. The gates are maintained in an open position by a low logic level; a high level closes the gates. The gates remain open in state 00 when both switches are low. Once a train crosses a switch (from either direction) the gates close. Then the state machine passes to state 01. The state machine moves to state 10 once the switch initially indicating the presence of a train reverts to the low level. The state machine goes back to state 00 when all switches are low at which time the gates reopen.

Solution   Create a transition table describing present state, inputs, next state, and output operation. From this create the necessary PROM data. Use the alternate PROM configuration since the gate output signal depends on the next state.

| PRESENT STATE | | INPUTS | | NEXT STATE | | OUTPUT |
Q_2	Q_1	SW$_1$	SW$_2$	Q_2	Q_1	GATE
0	0	0	0	0	0	0
0	0	1	0	0	1	1
0	0	0	1	0	1	1
0	1	1	0	0	1	1
0	1	0	1	0	1	1
0	1	1	1	1	0	1
1	0	1	1	1	0	1
1	0	0	1	1	0	1
1	0	1	0	1	0	1
1	0	0	0	0	0	0

PROM Data:

ADDRESS				OUTPUTS		
A_3	A_2	A_1	A_0	O_2	O_1	O_0
0	0	0	0	0	0	0
0	0	0	1	0	1	1
0	0	1	0	0	1	1
0	0	1	1	0	0	0
0	1	0	0	0	0	0
0	1	0	1	0	1	1
0	1	1	0	0	1	1
0	1	1	1	1	0	1
1	0	0	0	0	0	0
1	0	0	1	1	0	1
1	0	1	0	1	0	1
1	0	1	1	1	0	1
1	1	0	0	0	0	0
1	1	0	1	0	0	0
1	1	1	0	0	0	0
1	1	1	1	0	0	0

These addresses should never be accessed in this design.

Present state SW_1 SW_2 Next state values   Gate output signal

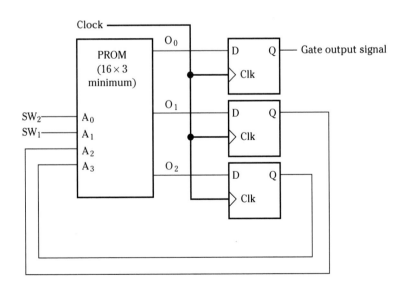

13.9 ■ MICROCODE BASED STATE MACHINES

Probably the most high level approach to state machine design is a **microcoded** system. Microcode or microinstruction state machines are similar to PROM based systems except that the pattern of PROM outputs is predefined into **control fields.** Each control field is responsible for a specific task in the state machine design. For instance, one field may control the selection of the next sequential address while another field may determine if a jump to another portion of the memory is necessary. Thus, microcode is a predefined arrangement of bits that cause a specific hardware response to occur. In this arrangement designing the state machine becomes more a matter of software programming. The designer is more concerned with the **microcode instruction set** rather than the hardware because the hardware based on the microcode is so uniform and well defined. The microcode program determines the actual characteristics of the system. Rather than redesigning a state machine from scratch, new microcode instructions are written to run on the existing hardware. In fact, most computer systems, from basic microprocessors to mainframes, are controlled with microcode. As you can imagine, this topic is well beyond the scope of this text since it forms the hardware transition point between logic design and computer system design. But it is important to realize that this aspect of design exists, particularly since microcode based sequencer chips are available for general design use.

SUMMARY

- State machines are sequential circuits designed to create control signals for the sequencing of events in other circuits.

- State machines are well-defined circuits generally consisting of a state register and combinatorial logic.

- State machine designs are classified using two general models—Mealy and Moore.

- For the purposes of design state machine operation is described using state diagrams, flowcharts, transition tables, or state equations.

- State machines may be designed and constructed in a variety of ways. Typically, discrete logic, multiplexers, PROMs, or microcode based designs are used.

PROBLEMS

Section 13.3

*1. Draw the state diagram for a MOD 12 counter.
*2. Modify the state diagram for the MOD 12 counter to include an indication that counting will not begin until an input START signal is given.
 3. Draw the state diagram for a circuit passing through seven states. The circuit has one input A and two outputs X and Y. The outputs are both

---

* See Appendix F: Answers to Selected Problems.

low when A is low and the state machine is in the 000 starting state. When A goes high, the machine cycles through the remaining states as long as A remains high. If A goes low while in states 010, 100, or 110, the machine remains in those states until A is raised high again. If A goes low in any other state, the machine advances to the nearest state where it will remain until A is raised high again. The machine outputs are both low in the 000 state. In all subsequent even states $X = 0$ and $Y = 1$. In all odd states $X = 1$ and $Y = 0$.

Section 13.4

4. Draw the transition table for the MOD 12 counter state diagram shown in problem 1.
*5. Modify the transition table from problem 4 to include a START input signal that prevents counting from beginning until START is high.
6. Draw the transition table for the state machine described by a state diagram in problem 3. Use D flip-flops in the design.
7. Redraw the transition table from problem 6 using J-K flip-flops in the design.
8. Draw the transition table for a synchronous MOD 8 counter using D flip-flops. Redraw using J-K flip-flops.
9. Draw the transition table for the MOD 16 counter designed in Figure 13.9 using D flip-flops.

Section 13.5

10. Using K-maps and the transition table derived in problem 9, draw the circuitry for the D flip-flop based MOD 16 counter.
11. Redesign the MOD 3 counter from Design Example 13.4 using D flip-flops.
*12. Design a J-K flip-flop based state machine to function as a MOD 9 counter with the following counting sequence: 2, 3, 5, 6, 7, 8, 10, 12, 15
13. Redesign the counter in problem 12 so that counting cannot continue unless the COUNT control line is high.
14. Design a MOD 5 synchronous counter with a normal counting sequence 0–4. In order to prevent problems that would occur if the counter powered up into states 5, 6, or 7, design into the counter the means to move automatically to the 000 state if the counter accidentally finds itself in one of these illegal states.
15. A company's name appears on a lighted billboard. The company name has five letters in it and each letter is backlighted by an individual light bulb. The company president wants to put some glitter in the company's advertising. Design the digital controls to a lighting system that will allow the letters to be lit in the following sequence: All letters off, then the first and last are on, then the first two and last two are on, then all are on. The sequence repeats after this. Draw the state diagram and transition table using J-K flip-flops.

---

* See Appendix F: Answers to Selected Problems.

16. Using state diagrams and J-K flip-flop/transition tables, design the circuit to control the pneumatic tube deposit system for a drive-up bank teller. The specifications are as follows:

The system is ready for use when the container door is closed and the vacuum pressure is off. A customer pushes a button when service is desired. This opens the door so the customer can withdraw the deposit canister. When the customer replaces the canister, the button is repushed. This closes the door and activates the vacuum to draw the canister into the bank. The system then returns to a ready state. Input is the service button; outputs are door open (high is open, low is closed) and vacuum pressure (high is pressure on, low is pressure off).

Section 13.7

*17. Draw the output waveform for the following waveform generator:

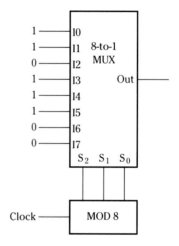

*18. Design a MOD 8 synchronous counter as a state machine using a multiplexer based design. Draw the transition table and circuit.

19 Redesign the state machine depicted by the state diagram in Figure 13.6 using multiplexers. Include the design for the state machine and for the state machine outputs.

20. Using multiplexers, design a state machine for a circuit passing through seven states. The circuit has one input A and two outputs X and Y. The outputs are both low when A is low and the state machine is in the 000 starting state. When A goes high, the machine cycles through the remaining states as long as A remains high. If A goes low while in states 010, 100, or 110 the machine remains in those states until A is raised high again. If A goes low in any other state, the machine advances to the nearest state where it will remain until A is raised high again. The machine outputs are

---

* See Appendix F: Answers to Selected Problems.

both low in the 000 state. In all subsequent even states X = 0 and Y = 1. In all odd states X = 1 and Y = 0. The state diagram for this circuit was developed in problem 3.

Section 13.8

*21. Use a PROM based state machine approach to design a synchronous MOD 8 counter. How does this design compare to the other MOD 8 state machine designs tried in this problem section in terms of circuit complexity and design complexity?

22. Modify the MOD 8 design (include more PROM outputs) to provide an active low decoder output for each counter state.

23. Use a PROM based state machine to implement the circuit defined by the following state diagram:

State Diagram:

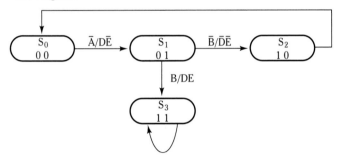

MANUFACTURERS' DATA SHEETS

2114A
1024 x 4 BIT STATIC RAM

	2114AL-1	2114AL-2	2114AL-3	2114AL-4	2114A-4	2114A-5
Max. Access Time (ns)	100	120	150	200	200	250
Max. Current (mA)	40	40	40	40	70	70

- **HMOS II Technology**
- **Low Power, High Speed**
- **Identical Cycle and Access Times**
- **Single +5V Supply ±10%**
- **High Density 18 Pin Package**
- **Completely Static Memory—No Clock or Timing Strobe Required**

- **Directly TTL Compatible: All Inputs and Outputs**
- **Common Data Input and Output Using Three-State Outputs**
- **High Reliability Plastic or Cerdip**
- **Available in EXPRESS**
  — Standard Temperature Range
  — Extended Temperature Range

The Intel 2114A is a 4096-bit static Random Access Memory organized as 1024 words by 4-bits using HMOS II, a high performance MOS technology. It uses fully DC stable (static) circuitry throughout, in both the array and the decoding, therefore it requires no clocks or refreshing to operate. Data access is particularly simple since address setup times are not required. The data is read out nondestructively and has the same polarity as the input data. Common input/output pins are provided.

The 2114A is designed for memory applications where the high performance and high reliability of HMOS II, low cost, large bit storage, and simple interfacing are important design objectives. The 2114A is assembled in an 18-pin package for the highest possible density.

It is directly TTL compatible in all respects: inputs, outputs, and a single +5V supply. A separate Chip Select ($\overline{CS}$) lead allows easy selection of an individual package when outputs are or-tied.

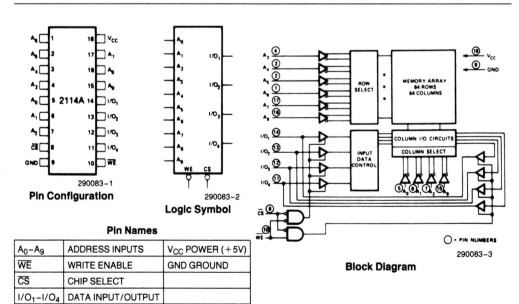

Pin Configuration

290083-1

Logic Symbol

290083-2

Block Diagram

290083-3

Pin Names

A$_0$–A$_9$	ADDRESS INPUTS	V$_{CC}$ POWER (+5V)
$\overline{WE}$	WRITE ENABLE	GND GROUND
$\overline{CS}$	CHIP SELECT	
I/O$_1$–I/O$_4$	DATA INPUT/OUTPUT	

ABSOLUTE MAXIMUM RATINGS*

Temperature Under Bias .... −10°C to +80°C

Storage Temperature .... −65°C to +150°C

Voltage on any Pin
  With Respect to Ground .... −3.5V to +7V

Power Dissipation........................1.0W

D.C. Output Current ....5 mA

Notice: Stresses above those listed under "Absolute Maximum Ratings" may cause permanent damage to the device. This is a stress rating only and functional operation of the device at these or any other conditions above those indicated in the operational sections of this specification is not implied. Exposure to absolute maximum rating conditions for extended periods may affect device reliability.

D.C. AND OPERATING CHARACTERISTICS

T_A = 0°C to 70°C, V_{CC} = 5V ±10%, unless otherwise noted

Symbol	Parameter	2114AL-1/L-2/L-3/L-4			2114A-4/-5			Unit	Conditions
		Min	Typ(1)	Max	Min	Typ(1)	Max		
$|I_{LI}|$	Input Load Current (All Input Pins)		0.01	1			1	μA	V_{IN} = 0V to 5.5V
$|I_{LO}|$	I/O Leakage Current		0.1	10			10	μA	$\overline{CS}$ = V_{IH} $V_{I/O}$ = 0V to 5.5V
I_{CC}	Power Supply Current		25	40		50	70	mA	V_{CC} = max, $I_{I/O}$ = 0 mA, T_A = 0°C
V_{IL}	Input Low Voltage	−3.0		0.8	−3.0		0.8	V	
V_{IH}	Input High Voltage	2.0		6.0	2.0		6.0	V	
I_{OL}	Output Low Current	4.0	9.0		4.0	9.0		mA	V_{OL} = 0.4V
I_{OH}	Output High Current	−2.0	−2.5		−2.0	−2.5		mA	V_{OH} = 2.4V
I_{OS}(2)	Output Short Circuit Current			40			40	mA	V_{OUT} = GND

NOTES:
1. Typical values are for T_A = 25°C and V_{CC} = 5.0V.
2. Duration not to exceed 1 second.

CAPACITANCE T_A = 25°C, f = 1.0 MHz

Symbol	Test	Max	Unit	Conditions
$C_{I/O}$	Input/Output Capacitance	5	pF	$V_{I/O}$ = 0V
C_{IN}	Input Capacitance	5	pF	V_{IN} = 0V

NOTE:
This parameter is periodically sampled and not 100% tested.

A.C. CONDITIONS OF TEST

Input Pulse Levels ....0.8V to 2.0V

Input Rise and Fall Times ....10 ns

Input and Output Timing Levels........0.8V to 2.0V

Output Load ....1 TTL Gate and C_L = 100 pF

LOAD FOR T_{OTD} AND T_{OTW}

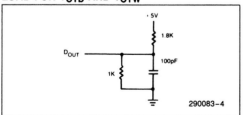

Figure 1

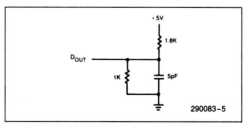

Figure 2

2114 Datasheet. Reprinted by permission of Intel Corporation, Copyright/Intel Corporation 1989.

A.C. CHARACTERISTICS T_A = 0°C to 70°C, V_{CC} = 5V ±10%, unless otherwise noted.

READ CYCLE[1]

Symbol	Parameter	2114AL-1		2114AL-2		2114AL-3		2114A-4/L-4		2114A-5		Unit
		Min	Max	Min	Max	Min	Max	Min	Max	Min	Max	
t_{RC}	Read Cycle Time	100		120		150		200		250		ns
t_A	Access Time		100		120		150		200		250	ns
t_{CO}	Chip Selection to Output Valid		70		70		70		70		85	ns
t_{CX}[3]	Chip Selection to Output Active	10		10		10		10		10		ns
t_{OTD}[3]	Output 3-state from Deselection		30		35		40		50		60	ns
t_{OHA}	Output Hold from Address Change	15		15		15		15		15		ns

WRITE CYCLE[2]

Symbol	Parameter	2114AL-1		2114AL-2		2114AL-3		2114A-4/L-4		2114A-5		Unit
		Min	Max	Min	Max	Min	Max	Min	Max	Min	Max	
t_{WC}	Write Cycle Time	100		120		150		200		250		ns
t_W	Write Time	75		75		90		120		135		ns
t_{WR}	Write Release Time	0		0		0		0		0		ns
t_{OTW}[3]	Output 3-state from Write		30		35		40		50		60	ns
t_{DW}	Data to Write Time Overlap	70		70		90		120		135		ns
t_{DH}	Data Hold from Write Time	0		0		0		0		0		ns

NOTES:
1. A Read occurs during the overlap of a low $\overline{CS}$ and a high $\overline{WE}$.
2. A Write occurs during the overlap of a low $\overline{CS}$ and a low $\overline{WE}$. t_W is measured from the latter of $\overline{CS}$ or $\overline{WE}$ going low to the earlier of $\overline{CS}$ or $\overline{WE}$ going high.
3. Measured at ±500 mV with 1 TTL Gate and C_L = 5.00 pF.

WAVEFORMS

READ CYCLE[3]

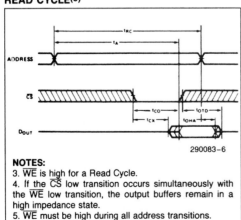

290083-6

NOTES:
3. $\overline{WE}$ is high for a Read Cycle.
4. If the $\overline{CS}$ low transition occurs simultaneously with the $\overline{WE}$ low transition, the output buffers remain in a high impedance state.
5. $\overline{WE}$ must be high during all address transitions.

WRITE CYCLE

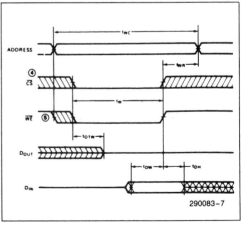

290083-7

2114 Datasheet. Reprinted by permission of Intel Corporation, Copyright/Intel Corporation 1989.

TYPICAL D.C. AND A.C. CHARACTERISTICS

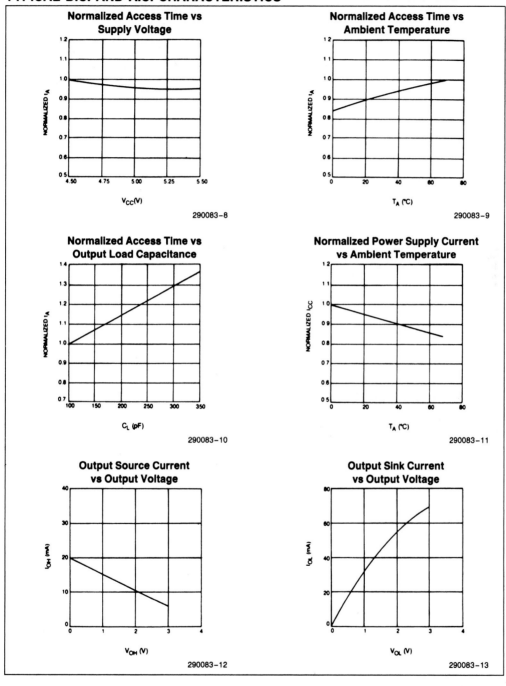

2114 Datasheet. Reprinted by permission of Intel Corporation, Copyright/Intel Corporation 1989.

TEXAS INSTRUMENTS

TMS4256, TMS4257
262,144-BIT DYNAMIC RANDOM-ACCESS MEMORIES

MAY 1983 — REVISED JANUARY 1988

- **262,144 × 1 Organization**
- **Single 5-V Power Supply**
  - **5% Tolerance Required for TMS4256-8**
  - **10% Tolerance Required for TMS4256-10, -12, -15, and TMS4257-10, -12, -15**
- **JEDEC Standardized Pinouts**
- **Performance Ranges:**

DEVICE	ACCESS TIME ROW ADDRESS (MAX)	ACCESS TIME COLUMN ADDRESS (MAX)	READ OR WRITE CYCLE (MIN)	V$_{DD}$ TOLERANCE
'4256-8	80 ns	40 ns	160 ns	± 5%
'4256-10 '4257-10	100 ns	50 ns	200 ns	± 10%
'4256-12 '4257-12	120 ns	60 ns	220 ns	± 10%
'4256-15 '4257-15	150 ns	75 ns	260 ns	± 10%

- **Long Refresh Period . . . 4 ms (Max)**
- **Operations of the TMS4256/TMS4257 Can Be Controlled by TI's SN74ALS2967, SN74ALS2968, and THCT4502 Dynamic RAM Controllers**
- **All Inputs, Outputs, and Clocks Fully TTL Compatible**
- **3-State Unlatched Outputs**
- **Common I/O Capability with "Early Write" Feature**
- **Page Mode ('4256) or Nibble-Mode ('4257)**
- **Low Power Dissipation**
- **$\overline{RAS}$-Only Refresh Mode**
- **Hidden Refresh Mode**
- **$\overline{CAS}$-Before-$\overline{RAS}$ Refresh Mode**
- **Available with MIL-STD-883B Processing and L(0 °C to 70 °C), E(−40 °C to 85 °C), or S(−55 °C to 100 °C) Temperature Ranges (SMJ4256, with 10% Power Supply)**

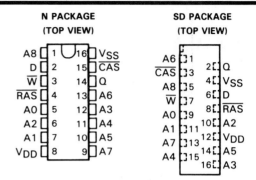

N PACKAGE (TOP VIEW)

A8	1		16	V$_{SS}$
D	2		15	$\overline{CAS}$
$\overline{W}$	3		14	Q
$\overline{RAS}$	4		13	A6
A0	5		12	A3
A2	6		11	A4
A1	7		10	A5
V$_{DD}$	8		9	A7

SD PACKAGE (TOP VIEW)

A6	1		2	Q
$\overline{CAS}$	3		4	V$_{SS}$
A8	5		6	D
$\overline{W}$	7		8	$\overline{RAS}$
A0	9		10	A2
A1	11		12	V$_{DD}$
A7	13		14	A5
A4	15		16	A3

FM PACKAGE (TOP VIEW)

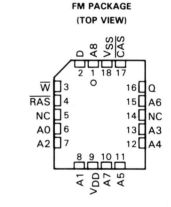

PIN NOMENCLATURE	
A0-A8	Address Inputs
$\overline{CAS}$	Column-Address Strobe
D	Data In
NC	No Connection
Q	Data Out
$\overline{RAS}$	Row-Address Strobe
V$_{DD}$	5-V Power Supply
V$_{SS}$	Ground
$\overline{W}$	Write Enable

description

The TMS4256 and TMS4257 are high-speed, 262,144-bit dynamic random-access memories, organized as 262,144 words of one bit each. They employ state-of-the-art SMOS (scaled MOS) N-channel double-level polysilicon/polycide gate technology for very high performance combined with low cost and improved reliability.

The '4256-8 with a 5% voltage tolerance has a maximum $\overline{RAS}$ access time of 80 ns. The '4256/'4257-10, -12, and -15 with 10% voltage tolerances have maximum $\overline{RAS}$ access times of 100 ns, 120 ns, and 150 ns, respectively.

New SMOS technology permits operation from a single 5-V supply, reducing system power supply and decoupling requirements, and easing board layout. I_{DD} peaks are 125 mA typical, and a -1 V input voltage undershoot can be tolerated, minimizing system noise considerations.

All inputs and outputs, including clocks, are compatible with Series 74 TTL. All address and data-in lines are latched on-chip to simplify system design. Data out is unlatched to allow greater system flexibility.

The '4256 and '4257 are offered in 16-pin plastic dual-in-line, 16-pin plastic zig-zag in-line (ZIP), and 18-lead plastic chip carrier packages. They are guaranteed for operation from 0 °C to 70 °C. The dual-in-line package is designed for insertion in mounting-hole rows on 7,62-mm (300-mil) centers.

operation

address (A0 through A8)

Eighteen address bits are required to decode 1 of 262,144 storage cell locations. Nine row-address bits are set up on pins A0 through A8 and latched onto the chip by the row-address strobe ($\overline{RAS}$). Then the nine column-address bits are set up on pins A0 through A8 and latched onto the chip by the column-address strobe ($\overline{CAS}$). All addresses must be stable on or before the falling edges of $\overline{RAS}$ and $\overline{CAS}$. $\overline{RAS}$ is similar to a chip enable in that it activates the sense amplifiers as well as the row decoder. $\overline{CAS}$ is used as a chip select, activating the column decoder and the input and output buffers.

write enable ($\overline{W}$)

The read or write mode is selected through the write-enable ($\overline{W}$) input. A logic high on the $\overline{W}$ input selects the read mode and a logic low selects the write mode. The write-enable terminal can be driven from standard TTL circuits without a pull-up resistor. The data input is disabled when the read mode is selected. When $\overline{W}$ goes low prior to $\overline{CAS}$, data out will remain in the high-impedance state for the entire cycle, permitting common I/O operation.

data in (D)

Data is written during a write or read-modify-write cycle. Depending on the mode of operation, the falling edge of $\overline{CAS}$ or $\overline{W}$ strobes data into the on-chip data latch. This latch can be driven from standard TTL circuits without a pull-up resistor. In an early write cycle, $\overline{W}$ is brought low prior to $\overline{CAS}$ and the data is strobed in by $\overline{CAS}$ with setup and hold times referenced to this signal. In a delayed-write or read-modify-write cycle, $\overline{CAS}$ will already be low, thus the data will be strobed in by $\overline{W}$ with setup and hold times referenced to this signal.

data out (Q)

The three-state output buffer provides direct TTL compatibility (no pull-up resistor required) with a fanout of two Series 74 TTL loads. Data out is the same polarity as data in. The output is in the high-impedance (floating) state until $\overline{CAS}$ is brought low. In a read cycle the output goes active after the access time interval $t_{a(C)}$ that begins with the negative transition of $\overline{CAS}$ as long as $t_{a(R)}$ is satisfied. The output becomes valid after the access time has elapsed and remains valid while $\overline{CAS}$ is low; $\overline{CAS}$ going high returns it to a high-impedance state. In a read-modify-write cycle, the output will follow the sequence for the read cycle.

refresh

A refresh operation must be performed at least once every four milliseconds to retain data. This can be achieved by strobing each of the 256 rows (A0-A7). A normal read or write cycle will refresh all bits in each row that is selected. A $\overline{RAS}$-only operation can be used by holding $\overline{CAS}$ at the high (inactive) level, thus conserving power as the output buffer remains in the high-impedance state.

$\overline{CAS}$-before-$\overline{RAS}$ refresh

The $\overline{CAS}$-before-$\overline{RAS}$ refresh is utilized by bringing $\overline{CAS}$ low earlier than $\overline{RAS}$ (see parameter t_{CLRL}) and holding it low after $\overline{RAS}$ falls (see parameter t_{RLCHR}). For successive $\overline{CAS}$-before-$\overline{RAS}$ refresh cycles, $\overline{CAS}$ can remain low while cycling $\overline{RAS}$. The external address is ignored and the refresh address is generated internally.

hidden refresh

Hidden refresh may be performed while maintaining valid data at the output pin. This is accomplished by holding $\overline{CAS}$ at V_{IL} after a read operation and cycling $\overline{RAS}$ after a specified precharge period, similar to a $\overline{CAS}$-before-$\overline{RAS}$ refresh cycle. The external address is also ignored during the hidden refresh cycles. The data at the output pin remains valid up to the maximum $\overline{CAS}$ low pulse duration, $t_{w(CL)}$.

page mode (TMS4256)

Page-mode operation allows effectively faster memory access by keeping the same row address and strobing random column addresses onto the chip. Thus, the time required to set up and strobe sequential row addresses for the same page is eliminated. The maximum number of columns that can be addressed is determined by $t_{w(RL)}$, the maximum $\overline{RAS}$ low pulse duration.

nibble mode (TMS4257)

Nibble-mode operation allows high-speed serial read, write, or read-modify-write access of 1 to 4 bits of data. The first bit is accessed in the normal manner with read data coming out at $t_{a(C)}$ time. The next sequential nibble bits can be read or written by cycling $\overline{CAS}$ while $\overline{RAS}$ remains low. The first bit is determined by the row and column addresses, which need to be supplied only for the first access. Column A8 and row A8 (CA8, RA8) provide the two binary bits for initial selection of the nibble addresses. Thereafter, the falling edge of $\overline{CAS}$ will access the next bit of the circular 4-bit nibble in the following sequence:

In nibble-mode, all normal memory operations (read, write, or read-modify-write) may be performed in any desired combination.

power-up

To achieve proper device operation, an initial pause of 200 μs is required after power up, followed by a minimum of eight initialization cycles.

logic symbol[†]

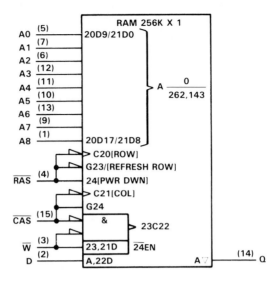

[†]This symbol is in accordance with ANSI/IEEE Std. 91-1084 and IEC Publication 617-12.
The pin numbers shown are for the 16-pin dual-in-line package.

functional block diagram

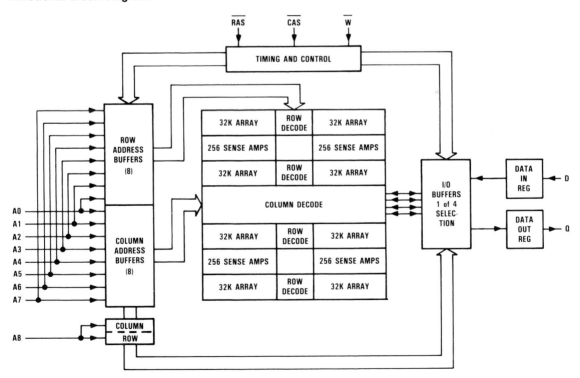

TMS4256 Datasheet. Printed by permission of the copyright holder, Texas Instruments Incorporated.

absolute maximum ratings over operating free-air temperature range (unless otherwise noted)†

Voltage range for any pin, including V_{DD} supply (see Note 1) . . . . −1 V to 7 V
Short circuit output current . . . . .50 mA
Power dissipation . . . . .1 W
Operating free-air temperature range . . . . .0 °C to 70 °C
Storage temperature range . . . . −65 °C to 150 °C

†Stresses beyond those listed under ''Absolute Maximum Ratings'' may cause permanent damage to the device. This is a stress rating only and functional operation of the device at these or any other conditions beyond those indicated in the ''Recommended Operating Conditions'' section of this specification is not implied. Exposure to absolute-maximum-rated conditions for extended periods may affect device reliability.

NOTE 1: All voltage values in this data sheet are with respect to V_{SS}.

recommended operating conditions

		MIN	NOM	MAX	UNIT
V_{DD}	Supply voltage ('4256/'4257-10, -12, -15)	4.5	5	5.5	V
V_{DD}	Supply voltage ('4256-8)	4.75	5	5.25	V
V_{SS}	Supply voltage		0		V
V_{IH}	High-level input voltage	2.4		6.5	V
V_{IL}	Low-level input voltage (see Note 2)	−1		0.8	V
T_A	Operating free-air temperature	0		70	°C

NOTE 2: The algebraic convention, where the more negative (less positive) limit is designated as maximum, is used in this data sheet for logic voltage levels only.

TMS4256 Datasheet. Printed by permission of the copyright holder, Texas Instruments Incorporated.

electrical characteristics over full ranges of recommended operating conditions (unless otherwise noted)

PARAMETER		TEST CONDITIONS	TMS4256-8		TMS4256-10 TMS4257-10		UNIT
			MIN	MAX	MIN	MAX	
V_{OH}	High-level output voltage	$I_{OH} = -5$ mA	2.4		2.4		V
V_{OL}	Low-level output voltage	$I_{OL} = 4.2$ mA		0.4		0.4	V
I_I	Input current (leakage)	$V_I = 0$ V to 6.5 V, $V_{DD} = 5$ V, All other pins = 0 V to 6.5 V		± 10		± 10	μA
I_O	Output current (leakage)	$V_O = 0$ V to 5.5 V, $V_{DD} = 5$ V, $\overline{CAS}$ high		± 10		± 10	μA
I_{DD1}	Average operating current during read or write cycle	t_c = minimum cycle, Output open		70		70	mA
I_{DD2}	Standby current	After 1 memory cycle, $\overline{RAS}$ and $\overline{CAS}$ high, Output open		4.5		4.5	mA
I_{DD3}	Average refresh current	t_c = minimum cycle, $\overline{RAS}$ cycling, $\overline{CAS}$ high, Output open		70		58	mA
I_{DD4}	Average page-mode current	$t_{c(P)}$ = minimum cycle, $\overline{RAS}$ low, $\overline{CAS}$ cycling, Output open		60		50	mA
I_{DD5}	Average nibble-mode current	$t_{c(N)}$ = minimum cycle, $\overline{RAS}$ low, $\overline{CAS}$ cycling, Output open				45	mA

PARAMETER		TEST CONDITIONS	TMS4256-12 TMS4257-12		TMS4256-15 TMS4257-15		UNIT
			MIN	MAX	MIN	MAX	
V_{OH}	High-level output voltage	$I_{OH} = -5$ mA	2.4		2.4		V
V_{OL}	Low-level output voltage	$I_{OL} = 4.2$ mA		0.4		0.4	V
I_I	Input current (leakage)	$V_I = 0$ V to 6.5 V, $V_{DD} = 5$ V, All other pins = 0 V to 6.5 V		± 10		± 10	μA
I_O	Output current (leakage)	$V_O = 0$ V to 5.5 V, $V_{DD} = 5$ V, $\overline{CAS}$ high		± 10		± 10	μA
I_{DD1}	Average operating current during read or write cycle	t_c = minimum cycle, Output open		65		60	mA
I_{DD2}	Standby current	After 1 memory cycle, $\overline{RAS}$ and $\overline{CAS}$ high, Output open		4.5		4.5	mA
I_{DD3}	Average refresh current	t_c = minimum cycle, $\overline{RAS}$ cycling, $\overline{CAS}$ high, Output open		53		48	mA
I_{DD4}	Average page-mode current	$t_{c(P)}$ = minimum cycle, $\overline{RAS}$ low, $\overline{CAS}$ cycling, Output open		45		40	mA
I_{DD5}	Average nibble-mode current	$t_{c(N)}$ = minimum cycle, $\overline{RAS}$ low, $\overline{CAS}$ cycling, Output open		40		35	mA

capacitance over recommended supply voltage range and operating free-air temperature range, f = 1 MHz

PARAMETER		MAX	UNIT
$C_{i(A)}$	Input capacitance, address inputs	5	pF
$C_{i(D)}$	Input capacitance, data input	5	pF
$C_{i(RC)}$	Input capacitance strobe inputs	5	pF
$C_{i(W)}$	Input capacitance, write enable input	7	pF
C_O	Output capacitance	7	pF

switching characteristics over recommended supply voltage range and operating free-air temperature range

PARAMETER		TEST CONDITIONS	ALT. SYMBOL	TMS4256-8		TMS4256-10 TMS4257-10		UNIT
				MIN	MAX	MIN	MAX	
$t_{a(C)}$	Access time from $\overline{CAS}$	$t_{RLCL} \geq$ MAX, C_L = 100 pF, Load = 2 Series 74 TTL gates	t_{CAC}		40		50	ns
$t_{a(R)}$	Access time from $\overline{RAS}$	t_{RLCL} = MAX, C_L = 100 pF, Load = 2 Series 74 TTL gates	t_{RAC}		80		100	ns
$t_{dis(CH)}$	Output disable time after $\overline{CAS}$ high	C_L = 100 pF, Load = 2 Series 74 TTL gates	t_{OFF}	0	20	0	30	ns

PARAMETER		TEST CONDITIONS	ALT. SYMBOL	TMS4256-12 TMS4257-12		TMS4256-15 TMS4257-15		UNIT
				MIN	MAX	MIN	MAX	
$t_{a(C)}$	Access time from $\overline{CAS}$	$t_{RLCL} \geq$ MAX, C_L = 100 pF, Load = 2 Series 74 TTL gates	t_{CAC}		60		75	ns
$t_{a(R)}$	Access time from $\overline{RAS}$	t_{RLCL} = MAX, C_L = 100 pF, Load = 2 Series 74 TTL gates	t_{RAC}		120		150	ns
$t_{dis(CH)}$	Output disable time after $\overline{CAS}$ high	C_L = 100 pF, Load = 2 Series 74 TTL gates	t_{OFF}	0	30	0	30	ns

TMS4256 Datasheet. Printed by permission of the copyright holder, Texas Instruments Incorporated.

timing requirements over recommended supply voltage range and operating free-air temperature range

PARAMETER		ALT. SYMBOL	TMS4256-8		TMS4256-10 TMS4257-10		UNIT
			MIN	MAX	MIN	MAX	
$t_{c(P)}$	Page-mode cycle time (read or write cycle)	t_{PC}	70		100		ns
$t_{c(PM)}$	Page-mode cycle time (read-modify-write cycle)	t_{PCM}	95		135		ns
$t_{c(rd)}$	Read cycle time[†]	t_{RC}	160		200		ns
$t_{c(W)}$	Write cycle time	t_{WC}	160		200		ns
$t_{c(rdW)}$	Read-write/read-modify-write cycle time	t_{RWC}	185		235		ns
$t_{w(CH)P}$	Pulse duration, $\overline{CAS}$ high (page mode)	t_{CP}	20		40		ns
$t_{w(CH)}$	Pulse duration, $\overline{CAS}$ high (non-page mode)	t_{CPN}	25		25		ns
$t_{w(CL)}$	Pulse duration, $\overline{CAS}$ low[‡]	t_{CAS}	40	10,000	50	10,000	ns
$t_{w(RH)}$	Pulse duration, $\overline{RAS}$ high	t_{RP}	70		90		ns
$t_{w(RL)}$	Pulse duration, $\overline{RAS}$ low[§]	t_{RAS}	80	10,000	100	10,000	ns
$t_{w(W)}$	Write pulse duration	t_{WP}	20		30		ns
t_t	Transition times (rise and fall) for $\overline{RAS}$ and $\overline{CAS}$	t_T	3	50	3	50	ns
$t_{su(CA)}$	Column-address setup time	t_{ASC}	0		0		ns
$t_{su(RA)}$	Row-address setup time	t_{ASR}	0		0		ns
$t_{su(D)}$	Data setup time	t_{DS}	0		0		ns
$t_{su(rd)}$	Read-command setup time	t_{RCS}	0		0		ns
$t_{su(WCL)}$	Early write-command setup time before $\overline{CAS}$ low	t_{WCS}	0		0		ns
$t_{su(WCH)}$	Write-command setup time before $\overline{CAS}$ high	t_{CWL}	20		30		ns
$t_{su(WRH)}$	Write-command setup time before $\overline{RAS}$ high	t_{RWL}	20		30		ns
$t_{h(CLCA)}$	Column-address hold time after $\overline{CAS}$ low	t_{CAH}	15		15		ns
$t_{h(RA)}$	Row-address hold time	t_{RAH}	15		15		ns
$t_{h(RLCA)}$	Column-address hold time after $\overline{RAS}$ low	t_{AR}	55		65		ns
$t_{h(CLD)}$	Data hold time after $\overline{CAS}$ low	t_{DH}	20		30		ns
$t_{h(RLD)}$	Data hold time after $\overline{RAS}$ low	t_{DHR}	60		80		ns
$t_{h(WLD)}$	Data hold time after $\overline{W}$ low	t_{DH}	20		30		ns
$t_{h(CHrd)}$	Read-command hold time after $\overline{CAS}$ high	t_{RCH}	0		0		ns
$t_{h(RHrd)}$	Read-command hold time after $\overline{RAS}$ high	t_{RRH}	10		10		ns
$t_{h(CLW)}$	Write-command hold time after $\overline{CAS}$ low	t_{WCH}	20		30		ns
$t_{h(RLW)}$	Write-command hold time after $\overline{RAS}$ low	t_{WCR}	65		80		ns

Continued next page.

NOTE 3: Timing measurements are referenced to V_{IL} max and V_{IH} min.

[†]All cycle times assume t_t = 5 ns.

[‡]In a read-modify-write cycle, t_{CLWL} and $t_{su(WCH)}$ must be observed. Depending on the user's transition times, this may require additional $\overline{CAS}$ low time ($t_{w(CL)}$). This applies to page-mode read-modify-write also.

[§]In a read-modify-write cycle, t_{RLWL} and $t_{su(WRH)}$ must be observed. Depending on the user's transition times, this may require additional $\overline{RAS}$ low time ($t_{w(RL)}$).

TMS4256 Datasheet. Printed by permission of e copyright holder, Texas Instruments Incorporated.

timing requirements over recommended supply voltage range and operating free-air temperature range (continued)

PARAMETER		ALT. SYMBOL	TMS4256-8		TMS4256-10 TMS4257-10		UNIT
			MIN	MAX	MIN	MAX	
t_{RLCH}	Delay time, $\overline{RAS}$ low to $\overline{CAS}$ high	t_{CSH}	80		100		ns
t_{CHRL}	Delay time, $\overline{CAS}$ high to $\overline{RAS}$ low	t_{CRP}	0		0		ns
t_{CLRH}	Delay time, $\overline{CAS}$ low to $\overline{RAS}$ high	t_{RSH}	40		50		ns
t_{RLCHR}	Delay time, $\overline{RAS}$ low to $\overline{CAS}$ high¶	t_{CHR}	20		20		ns
t_{CLRL}	Delay time, $\overline{CAS}$ low to $\overline{RAS}$ low¶	t_{CSR}	10		10		ns
t_{RHCL}	Delay time, $\overline{RAS}$ high to $\overline{CAS}$ low¶	t_{RPC}	0		0		ns
t_{CLWL}	Delay time, $\overline{CAS}$ low to $\overline{W}$ low (read-modify-write cycle only)	t_{CWD}	40		50		ns
t_{RLCL}	Delay time, $\overline{RAS}$ low to $\overline{CAS}$ low (maximum value specified only to guarantee access time)	t_{RCD}	25	40	25	50	ns
t_{RLWL}	Delay time, $\overline{RAS}$ low to $\overline{W}$ low (read-modify-write cycle only)	t_{RWD}	80		100		ns
t_{rf}	Refresh time interval	t_{REF}		4		4	ms

Continued next page.

NOTE 3: Timing measurements are referenced to V_{IL} max and V_{IH} min.
¶ $\overline{CAS}$-before-$\overline{RAS}$ refresh only.

TMS4256 Datasheet. Printed by permission of the copyright holder, Texas Instruments Incorporated.

timing requirements over recommended supply voltage range and operating free-air temperature range (continued)

PARAMETER		ALT. SYMBOL	TMS4256-12 TMS4257-12		TMS4256-15 TMS4257-15		UNIT
			MIN	MAX	MIN	MAX	
$t_{c(P)}$	Page-mode cycle time (read or write cycle)	t_{PC}	120		145		ns
$t_{c(PM)}$	Page-mode cycle time (read-modify-write cycle)	t_{PCM}	160		190		ns
$t_{c(rd)}$	Read cycle time[†]	t_{RC}	220		260		ns
$t_{c(W)}$	Write cycle time	t_{WC}	220		260		ns
$t_{c(rdW)}$	Read-write/read-modify-write cycle time	t_{RWC}	260		305		ns
$t_{w(CH)P}$	Pulse duration, $\overline{CAS}$ high (page mode)	t_{CP}	50		60		ns
$t_{w(CH)}$	Pulse duration, $\overline{CAS}$ high (non-page mode)	t_{CPN}	25		25		ns
$t_{w(CL)}$	Pulse duration, $\overline{CAS}$ low[‡]	t_{CAS}	60	10,000	75	10,000	ns
$t_{w(RH)}$	Pulse duration, $\overline{RAS}$ high	t_{RP}	90		100		ns
$t_{w(RL)}$	Pulse duration, $\overline{RAS}$ low[§]	t_{RAS}	120	10,000	150	10,000	ns
$t_{w(W)}$	Write pulse duration	t_{WP}	30		45		ns
t_t	Transition times (rise and fall) for $\overline{RAS}$ and $\overline{CAS}$	t_T	3	50	3	50	ns
$t_{su(CA)}$	Column-address setup time	t_{ASC}	0		0		ns
$t_{su(RA)}$	Row-address setup time	t_{ASR}	0		0		ns
$t_{su(D)}$	Data setup time	t_{DS}	0		0		ns
$t_{su(rd)}$	Read-command setup time	t_{RCS}	0		0		ns
$t_{su(WCL)}$	Early write-command setup time before $\overline{CAS}$ low	t_{WCS}	0		0		ns
$t_{su(WCH)}$	Write-command setup time before $\overline{CAS}$ high	t_{CWL}	35		45		ns
$t_{su(WRH)}$	Write-command setup time before $\overline{RAS}$ high	t_{RWL}	35		45		ns
$t_{h(CLCA)}$	Column-address hold time after $\overline{CAS}$ low	t_{CAH}	20		25		ns
$t_{h(RA)}$	Row-address hold time	t_{RAH}	15		15		ns
$t_{h(RLCA)}$	Column-address hold time after $\overline{RAS}$ low	t_{AR}	80		100		ns
$t_{h(CLD)}$	Data hold time after $\overline{CAS}$ low	t_{DH}	30		45		ns
$t_{h(RLD)}$	Data hold time after $\overline{RAS}$ low	t_{DHR}	90		120		ns
$t_{h(WLD)}$	Data hold time after $\overline{W}$ low	t_{DH}	30		45		ns
$t_{h(CHrd)}$	Read-command hold time after $\overline{CAS}$ high	t_{RCH}	0		0		ns
$t_{h(RHrd)}$	Read-command hold time after $\overline{RAS}$ high	t_{RRH}	10		10		ns
$t_{h(CLW)}$	Write-command hold time after $\overline{CAS}$ low	t_{WCH}	30		45		ns
$t_{h(RLW)}$	Write-command hold time after $\overline{RAS}$ low	t_{WCR}	90		120		ns

Continued next page.

NOTE 3: Timing measurements are referenced to V_{IL} max and V_{IH} min.

[†]All cycle times assume $t_t = 5$ ns.

[‡]In a read-modify-write cycle, t_{CLWL} and $t_{su(WCH)}$ must be observed. Depending on the user's transition times, this may require additional $\overline{CAS}$ low time $(t_{w(CL)})$. This applies to page-mode read-modify-write also.

[§]In a read-modify-write cycle, t_{RLWL} and $t_{su(WRH)}$ must be observed. Depending on the user's transition times, this may require additional $\overline{RAS}$ low time $(t_{w(RL)})$.

timing requirements over recommended supply voltage range and operating free-air temperature range (concluded)

PARAMETER		ALT. SYMBOL	TMS4256-12 TMS4257-12		TMS4256-15 TMS4257-15		UNIT
			MIN	MAX	MIN	MAX	
t_{RLCH}	Delay time, $\overline{RAS}$ low to $\overline{CAS}$ high	t_{CSH}	120		150		ns
t_{CHRL}	Delay time, $\overline{CAS}$ high to $\overline{RAS}$ low	t_{CRP}	0		0		ns
t_{CLRH}	Delay time, $\overline{CAS}$ low to $\overline{RAS}$ high	t_{RSH}	60		75		ns
t_{RLCHR}	Delay time, $\overline{RAS}$ low to $\overline{CAS}$ high¶	t_{CHR}	25		30		ns
t_{CLRL}	Delay time, $\overline{CAS}$ low to $\overline{RAS}$ low¶	t_{CSR}	10		20		ns
t_{RHCL}	Delay time, $\overline{RAS}$ high to $\overline{CAS}$ low¶	t_{RPC}	0		0		ns
t_{CLWL}	Delay time, $\overline{CAS}$ low to $\overline{W}$ low (read-modify-write cycle only)	t_{CWD}	60		70		ns
t_{RLCL}	Delay time, $\overline{RAS}$ low to $\overline{CAS}$ low (maximum value specified only to guarantee access time)	t_{RCD}	25	60	25	75	ns
t_{RLWL}	Delay time, $\overline{RAS}$ low to $\overline{W}$ low (read-modify-write cycle only)	t_{RWD}	120		145		ns
t_{rf}	Refresh time interval	t_{REF}		4		4	ms

NOTE 3: Timing measurements are referenced to V_{IL} max and V_{IH} min.
¶$\overline{CAS}$-before-$\overline{RAS}$ refresh only.

NIBBLE-MODE CYCLE

switching characteristics over recommended supply voltage range and operating free-air temperature range

PARAMETER		ALT. SYMBOL	TMS4257-10		TMS4257-12		TMS4257-15		UNIT
			MIN	MAX	MIN	MAX	MIN	MAX	
$t_{a(CN)}$	Nibble-mode access from $\overline{CAS}$	t_{NCAC}		25		30		40	ns

timing requirements over recommended supply voltage range and operating free-air temperature range

PARAMETER		ALT. SYMBOL	TMS4257-10		TMS4257-12		TMS4257-15		UNIT
			MIN	MAX	MIN	MAX	MIN	MAX	
$t_{c(N)}$	Nibble-mode cycle time	t_{NC}	50		60		75		ns
$t_{c(rdWN)}$	Nibble-mode read-modify-write cycle time	t_{NRMW}	70		85		105		
t_{CLRHN}	Nibble-mode delay time, $\overline{CAS}$ low to $\overline{RAS}$ high	t_{NRSH}	25		30		40		
t_{CLWLN}	Nibble-mode delay time, $\overline{CAS}$ to $\overline{W}$ delay	t_{NCWD}	20		25		30		ns
$t_{w(CLN)}$	Nibble-mode pulse duration, $\overline{CAS}$ low	t_{NCAS}	25		30		40		
$t_{w(CHN)}$	Nibble-mode pulse duration, $\overline{CAS}$ high	t_{NCP}	15		20		25		
$t_{su(WCHN)}$	Nibble-mode write command setup before $\overline{CAS}$ high	t_{NCWL}	20		25		35		

NOTE 3: Timing measurements are referenced to V_{IL} max and V_{IH} min.

read cycle timing

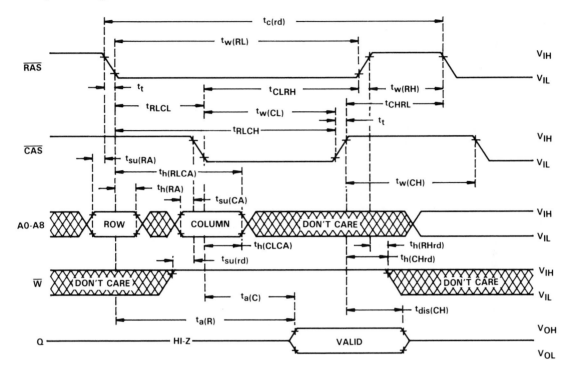

TMS4256 Datasheet. Printed by permission of the copyright holder, Texas Instruments Incorporated.

early write cycle timing

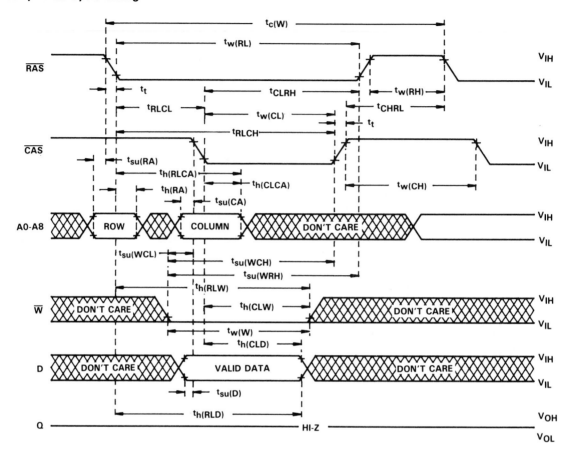

TMS4256 Datasheet. Printed by permission of the copyright holder, Texas Instruments Incorporated.

write cycle timing

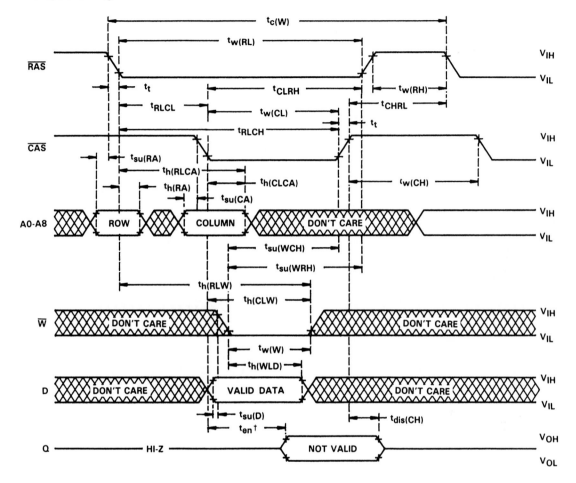

†The enable time (t_{en}) for a write cycle is equal in duration to the access time from $\overline{CAS}$ ($t_{a(C)}$) in a read cycle; but the active levels at the output are invalid.

TMS4256 Datasheet. Printed by permission of the copyright holder, Texas Instruments Incorporated.

read-write/read-modify-write cycle timing

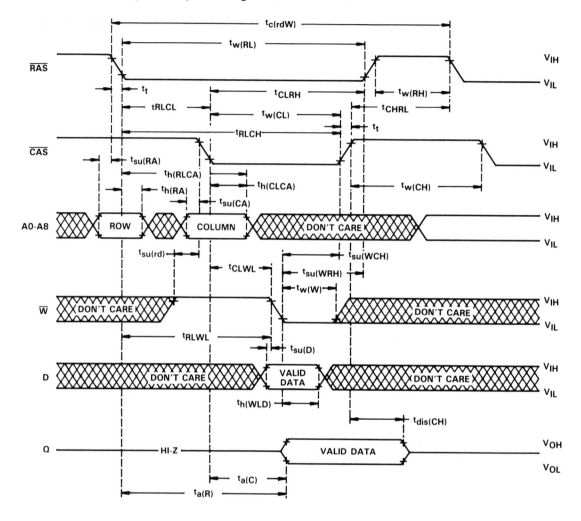

TMS4256 Datasheet. Printed by permission of the copyright holder, Texas Instruments Incorporated.

page-mode read cycle timing

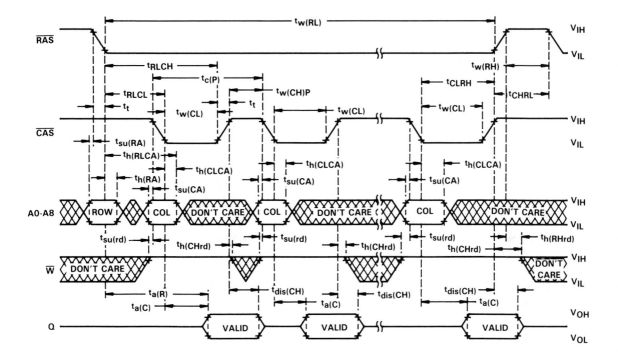

NOTE 4: A write cycle or a read-modify-write cycle can be intermixed with read cycles as long as the write and read-modify-write timing specifications are not violated.

TMS4256 Datasheet. Printed by permission of the copyright holder, Texas Instruments Incorporated.

page-mode write cycle timing

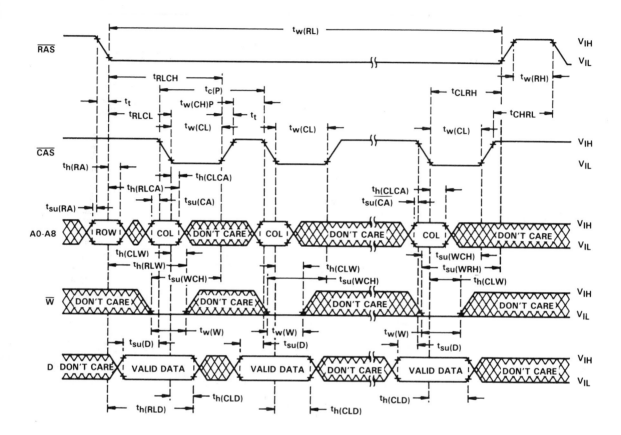

NOTE 5: A read cycle or a read-modify-write cycle can be intermixed with write cycles as long as read and read-modify-write timing specifications are not violated.

TMS4256 Datasheet. Printed by permission of the copyright holder, Texas Instruments Incorporated.

page-mode read-modify-write cycle timing

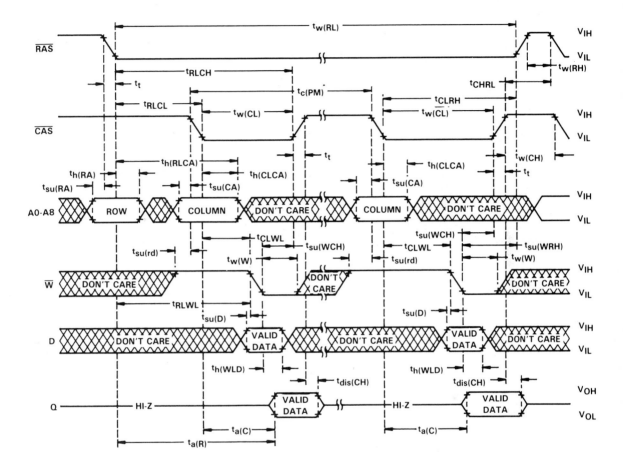

NOTE 6: A read or a write cycle can be intermixed with read-modify-write cycles as long as the read and write timing specifications are not violated.

TMS4256 Datasheet. Printed by permission of the copyright holder, Texas Instruments Incorporated.

nibble-mode read cycle timing

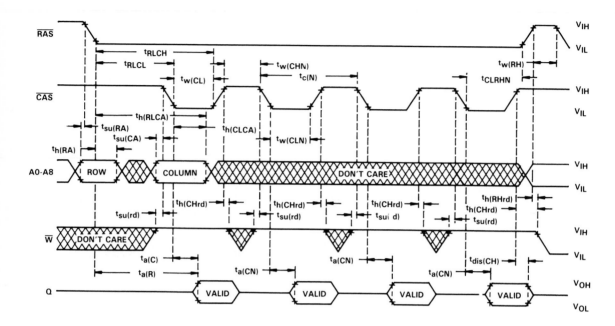

TMS4256 Datasheet. Printed by permission of the copyright holder, Texas Instruments Incorporated.

nibble-mode write cycle timing

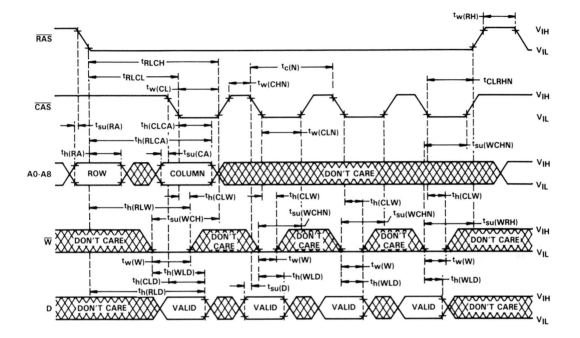

TMS4256 Datasheet. Printed by permission of the copyright holder, Texas Instruments Incorporated.

nibble-mode read-modify-write-cycle timing

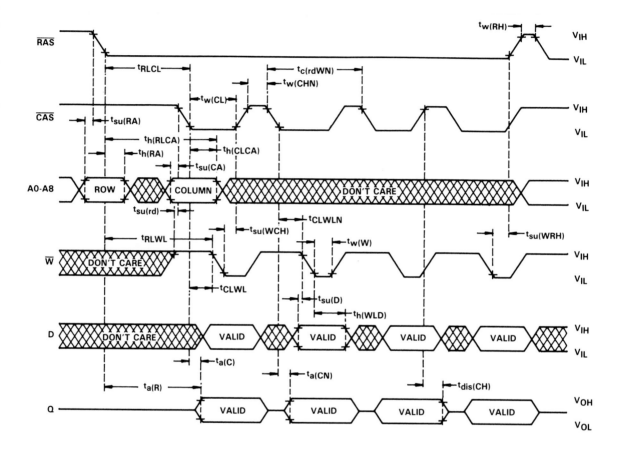

TMS4256 Datasheet. Printed by permission of the copyright holder, Texas Instruments Incorporated.

$\overline{\text{RAS}}$-only refresh cycle timing

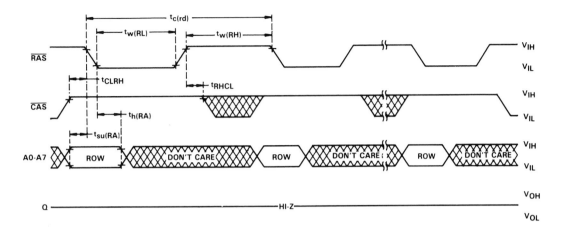

hidden refresh cycle timing

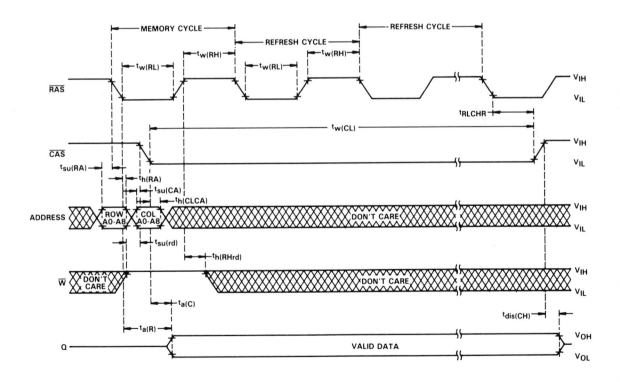

TMS4256 Datasheet. Printed by permission of the copyright holder, Texas Instruments Incorporated.

automatic ($\overline{CAS}$-before-$\overline{RAS}$) refresh cycle timing

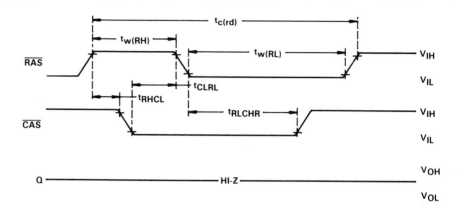

MECHANICAL DATA

16-pin plastic zig-zag in-line package (SD suffix)

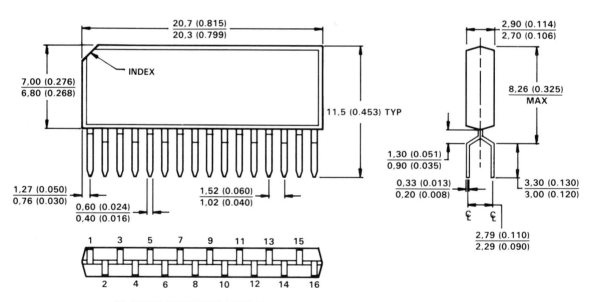

ALL LINEAR DIMENSIONS ARE IN MILLIMETERS AND PARENTHETICALLY IN INCHES

TMS4256 Datasheet. Printed by permission of the copyright holder, Texas Instruments Incorporated.

TMS2732A
32,768 BIT UV ERASABLE PROGRAMMABLE READ-ONLY MEMORY

AUGUST 1983—REVISED FEBRUARY 1988

- Organization . . . 4096 × 8

- Single 5-V Power Supply

- All Inputs/Outputs Fully TTL Compatible

- Max Access/Min Cycle Times
TMS2732A-17	170 ns
TMS2732A-20	200 ns
TMS2732A-25	250 ns
TMS2732A-45	450 ns

- Low Standby Power Dissipation . . .
  158 mW (Maximum)

- JEDEC Approved Pinout . . . Industry
  Standard

- 21-V Power Supply Required for
  Programming

- N-Channel Silicon-Gate Technology

- PEP4 Version Available with 168 Hour
  Burn-In, and Extended Guaranteed Operating
  Temperature Range from −10 °C to 85 °C
  (TMS2732A-_ _JP4)

J PACKAGE
(TOP VIEW)

A7	1	24 V_{CC}
A6	2	23 A8
A5	3	22 A9
A4	4	21 A11
A3	5	20 $\overline{G}/V_{PP}$
A2	6	19 A10
A1	7	18 $\overline{E}$
A0	8	17 Q8
Q1	9	16 Q7
Q2	10	15 Q6
Q3	11	14 Q5
GND	12	13 Q4

PIN NOMENCLATURE	
A0-A11	Address Inputs
$\overline{E}$	Chip Enable
$\overline{G}/V_{PP}$	Output Enable/21 V
GND	Ground
Q1-Q8	Outputs
V_{CC}	5-V Power Supply

description

The TMS2732A is an ultraviolet light-erasable, electrically programmable read-only memory. It has 32,768 bits organized as 4,096 words of 8-bit length. The TMS2732A only requires a single 5-volt power supply with a tolerance of ±5%.

The TMS2732A provides two output control lines: Output Enable ($\overline{G}/V_{PP}$) and Chip Enable ($\overline{E}$). This feature allows the $\overline{G}/V_{PP}$ control line to eliminate bus contention in multibus microprocessor systems. The TMS2732A has a power-down mode that reduces maximum power dissipation from 657 mW to 158 mW when the device is placed on standby.

This EPROM is supplied in a 24-pin dual-in-line ceramic package and is designed for operation from 0 °C to 70 °C. The TMS2732A is also offered in the PEP4 version with an extended guaranteed operating temperature range of −10 °C to 85 °C and 168 hour burn-in (TMS2732A-_ _JP4).

operation

The six modes of operation for the TMS2732A are listed in the following table.

FUNCTION (PINS)	MODE					
	Read	Output Disable	Power Down (Standby)	Program	Program Verification	Inhibit Programming
$\overline{E}$ (18)	V_{IL}	$X^{\dagger}$	V_{IH}	V_{IL}	V_{IL}	V_{IH}
$\overline{G}/V_{PP}$ (20)	V_{IL}	V_{IH}	$X^{\dagger}$	21 V	V_{IL}	21 V
V_{CC} (24)	5 V	5 V	5 V	5 V	5 V	5 V
Q1-Q8 (9 to 11, 13 to 17)	Q	HI-Z	HI-Z	D	Q	HI-Z

$^{\dagger}X = V_{IH}$ or V_{IL}

read/output disable

The two control pins ($\overline{E}$ and $\overline{G}/V_{PP}$) must have low-level TTL signals in order to provide data at the outputs. Chip enable ($\overline{E}$) should be used for device selection. Output enable ($\overline{G}/V_{PP}$) should be used to gate data to the output pins.

power down

The power-down mode reduces the maximum power dissipation from 657 mW to 158 mW. A TTL high-level signal applied to $\overline{E}$ selects the power-down mode. In this mode, the outputs assume a high-impedance state, independent of $\overline{G}/V_{PP}$.

erasure

The TMS2732A is erased by exposing the chip to shortwave ultraviolet light that has a wavelength of 253.7 nanometers (2537 angstroms). The recommended minimum exposure dose (UV intensity × exposure time) is fifteen watt-seconds per square centimeter. The lamp should be located about 2.5 centimeters (1 inch) above the chip during erasure. After erasure, all bits are at a high level. It should be noted that normal ambient light contains the correct wavelength for erasure. Therefore, when using the TMS2732A, the window should be covered with an opaque label.

programming

Note that the application of a voltage in excess of 22 V to $\overline{G}/V_{PP}$ may damage the TMS2732A.

After erasure (all bits in logic 1 state), logic 0s are programmed into the desired locations. A logic 0 can only be erased by ultraviolet light. In the program mode, $\overline{G}/V_{PP}$ is taken from a TTL low level to 21 V and data to be programmed are applied in parallel to output pins Q1-Q8. The location to be programmed is addressed. Once data and addresses are stable, a 10-millisecond TTL low-level pulse is applied to $\overline{E}$. The maximum width of this pulse is 11 milliseconds. The programming pulse must be applied at each location that is to be programmed. Locations may be programmed in any order.

Several TMS2732As can be programmed simultaneously by connecting them in parallel and following the programming sequence previously described.

program inhibit

The program inhibit is useful when programming multiple TMS2732As connected in parallel with different data. Program inhibit can be implemented by applying a high-level signal to $\overline{E}$ of the device that is not to be programmed

program verify

After the EPROM has been programmed, the programmed bits should be verified. To verify bit states, $\overline{G}/V_{PP}$ and $\overline{E}$ are set to V_{IL}.

logic symbol†

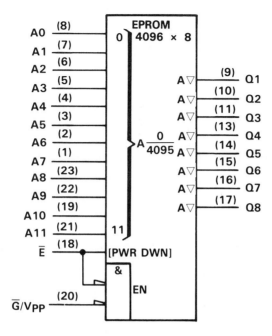

†This symbol is in accordance with ANSI/IEEE Std 91-1984 and IEC Publication 617-12.

absolute maximum ratings over operating free-air temperature range (unless otherwise noted)‡

Supply voltage range, V_{CC} . . . . −0.3 V to 7 V
Supply voltage range, V_{PP} . . . . −0.3 V to 22 V
Input voltage range (except program) . . . . −0.3 to 7 V
Output voltage range . . . . −0.3 V to 7 V
Operating free-air temperature range . . . . 0°C to 70°C
Storage temperature range . . . . −65°C to 150°C

‡Stresses beyond those listed under ''Absolute Maximum Ratings'' may cause permanent damage to the device. This is a stress rating only, and functional operation of the device at these or any other conditions beyond those indicated in the ''Recommended Operating Conditions'' section of this specification is not implied. Exposure to absolute-maximum-rated conditions for extended periods may affect device reliability.

TMS2732A Datasheet. Printed by permission of the copyright holder, Texas Instruments Incorporated.

recommended operating conditions

PARAMETER		MIN	NOM	MAX	UNIT
V_{CC}	Supply voltage (see Note 1)	4.75	5	5.25	V
V_{PP}	Supply voltage (see Note 2)		V_{CC}		V
V_{IH}	High-level input voltage	2		$V_{CC}+1$	V
V_{IL}	Low-level input voltage	−0.1		0.8	V
T_A	Operating free-air temperature	0		70	°C

NOTES: 1. V_{CC} must be applied before or at the same time as V_{PP} and removed after or at the same time as V_{PP}. The device must not be inserted into or removed from the board when V_{PP} or V_{CC} is applied.

2. V_{PP} can be connected to V_{CC} directly (except in the program mode). V_{CC} supply current in this case would be $I_{CC} + I_{PP}$. During programming, V_{PP} must be maintained at 21 V (±0.5 V).

electrical characteristics over full ranges of recommended operating conditions

PARAMETER		TEST CONDITIONS	MIN	MAX	UNIT
V_{OH}	High-level output voltage	$I_{OH} = -400\ \mu A$	2.4		V
V_{OL}	Low-level output voltage	$I_{OL} = 2.1$ mA		0.45	V
I_I	Input current (leakage)	$V_I = 0$ V to 5.25 V		±10	μA
I_O	Output current (leakage)	$V_O = 0.4$ V to 5.25 V		±10	μA
I_{CC1}	V_{CC} supply current (standby)	$\overline{E}$ at V_{IH}, $\overline{G}/V_{PP}$ at V_{IL}		30	mA
I_{CC2}	V_{CC} supply current (active)	$\overline{E}$ and $\overline{G}/V_{PP}$ at V_{IL}		125	mA

capacitance over recommended supply voltage range and operating free-air temperature range, f = 1 MHz[†]

PARAMETER		TEST CONDITIONS	TYP[‡]	MAX	UNIT
C_i Input capacitance	All except $\overline{G}/V_{PP}$	$V_I = 0$ V	6	9	pF
	$\overline{G}/V_{PP}$			20	
C_o Output capacitance		$V_O = 0$ V	8	12	pF

[†]These parameters are tested on sample basis only.
[‡]Typical values are at $T_A = 25$ °C and nominal voltages.

switching characteristics over recommended supply voltage range and operating free-air temperature range

PARAMETER		TEST CONDITIONS	TMS2732A-17 MIN	TMS2732A-17 MAX	TMS2732A-20 MIN	TMS2732A-20 MAX	TMS2732A-25 MIN	TMS2732A-25 MAX	TMS2732A-45 MIN	TMS2732A-45 MAX	UNIT
$t_{a(A)}$	Access time from address	$C_L = 100$ pF, 1 Series 74 TTL load, $t_r \leq 20$ ns, $t_f \leq 20$ ns, See Figure 1 and Note 3		170		200		250		450	ns
$t_{a(E)}$	Access time from $\overline{E}$			170		200		250		450	ns
$t_{en(G)}$	Output enable time from $\overline{G}/V_{PP}$			65		70		100		150	ns
t_{dis}[†]	Output disable time from $\overline{E}$ or $\overline{G}$, whichever occurs first		0	60	0	60	0	85	0	130	ns
$t_v(A)$	Output data valid time after change of address, $\overline{E}$, or $\overline{G}/V_{PP}$, whichever occurs first		0		0		0		0		ns

NOTE 3: For all switching characteristics and timing measurements, input pulse levels are 0.40 V and 2.4 V. Input and output reference levels are 0.8 V and 2.0 V.
[†]Value calculated from 0.5 V delta to measured output level. This parameter is only sampled, not 100% tested.

TMS2732A Datasheet. Printed by permission of the copyright holder, Texas Instruments Incorporated.

recommended conditions for programming, T_A = 25 °C (see Note 4)

		MIN	NOM	MAX	UNIT
V_{CC}	Supply voltage	4.75	5	5.25	V
V_{PP}	Supply voltage	20.5	21	21.5	V
V_{IH}	High-level input voltage	2		$V_{CC}+1$	V
V_{IL}	Low-level input voltage	−0.1		0.8	V
$t_{w(E)}$	$\overline{E}$ pulse duration	9	10	11	ms
$t_{su(A)}$	Address setup time	2			µs
$t_{su(D)}$	Data setup time	2			µs
$t_{su(VPP)}$	$\overline{G}$/Vpp setup time	2			µs
$t_{h(A)}$	Address hold time	0			µs
$t_{h(D)}$	Data hold time	2			µs
$t_{h(VPP)}$	$\overline{G}$/Vpp hold time	2			µs
$t_{rec(PG)}$	$\overline{G}$/Vpp recovery time	2			µs
$t_{r(PG)G}$	$\overline{G}$/Vpp rise time during programming	50			ns
t_{EHD}	Delay time, data valid after $\overline{E}$ low			1	µs

NOTE 4:   When programming the TMS2732A, connect a 0.1 µF capacitor between $\overline{G}$/Vpp and GND to suppress spurious voltage transients which may damage the device.

programming characteristics, T_A = 25 °C

PARAMETER		TEST CONDITIONS	MIN	TYP	MAX	UNIT
V_{IH}	High-level input voltage		2		$V_{CC}+1$	V
V_{IL}	Low-level input voltage		−0.1		0.8	V
V_{OH}	High-level output voltage (verify)	I_{OH} = −400 µA	2.4			V
V_{OL}	Low-level output voltage (verify)	I_{OL} = 2.1 mA			0.45	V
I_I	Input current (all inputs)	V_I = V_{IL} or V_{IH}			10	µA
I_{PP}	Supply current	$\overline{E}$ = V_{IL}, $\overline{G}$ = Vpp			50	mA
I_{CC}	Supply current				125	mA
$t_{dis(PR)}$	Output disable time		0		130	ns

TMS2732A Datasheet. Printed by permission of the copyright holder, Texas Instruments Incorporated.

PARAMETER MEASUREMENT INFORMATION

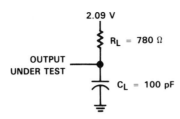

FIGURE 1. TYPICAL OUTPUT LOAD CIRCUIT

AC testing input/output wave forms

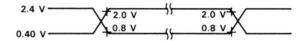

A.C. testing inputs are driven at 2.4 V for logic 1 and 0.4 V for logic 0. Timing measurements are made at 2.0 V for logic 1 and 0.8 V for logic 0 for both inputs and outputs.

read cycle timing

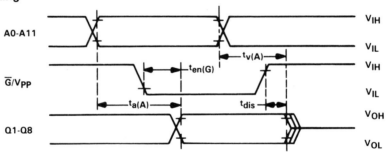

standby mode

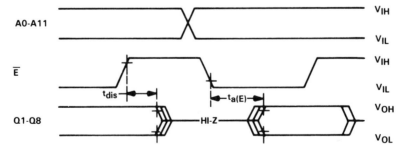

NOTE 3:   For all switching characteristics and timing measurements, input pulse levels are 0.40 V and 2.4 V. Input and output timing reference levels are 0.8 V and 2.0 V.

TMS2732A Datasheet. Printed by permission of the copyright holder, Texas Instruments Incorporated.

program cycle timing

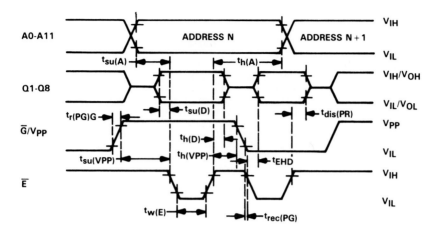

NOTE 3: For all switching characteristics and timing measurements, input pulse levels are 0.40 V and 2.4 V. Input and output timing reference levels are 0.8 V and 2.0 V.

TMS2732A Datasheet. Printed by permission of the copyright holder, Texas Instruments Incorporated.

CY7C164
CY7C166

16,384 x 4 Static R/W RAM

Features

- **Automatic power-down when deselected**
- **Output Enable ($\overline{OE}$) Feature (7C166)**
- **CMOS for optimum speed/ power**
- **High speed**
  — 20 ns t_{AA}
- **Low active power**
  — 440 mW
- **Low standby power**
  — 110 mW
- **TTL compatible inputs and outputs**
- **2V data retention (L version)**
- **Capable of withstanding greater than 2001V electrostatic discharge**

Functional Description

The CY7C164 and CY7C166 are high performance CMOS static RAMs organized as 16,384 x 4 bits. Easy memory expansion is provided by an active LOW chip enable ($\overline{CE}$) and three-state drivers. The CY7C166 has an active low output enable ($\overline{OE}$) feature. Both devices have an automatic power-down feature, reducing the power consumption by 60% when deselected.

Writing to the device is accomplished when the chip enable ($\overline{CE}$) and write enable ($\overline{WE}$) inputs are both LOW (and the output enable ($\overline{OE}$) is LOW for the 7C166). Data on the four input/output pins (I/O$_0$ through I/O$_3$) is written into the memory location specified on the address pins (A$_0$ through A$_{13}$).

Reading the device is accomplished by taking chip enable ($\overline{CE}$) LOW (and $\overline{OE}$ LOW for 7C166), while write enable ($\overline{WE}$) remains HIGH. Under these conditions the contents of the memory location specified on the address pins will appear on the four data I/O pins.

The I/O pins stay in high impedance state when chip enable ($\overline{CE}$) is HIGH, or write enable ($\overline{WE}$) is LOW (or output enable ($\overline{OE}$) is HIGH for 7C166). A die coat is used to insure alpha immunity.

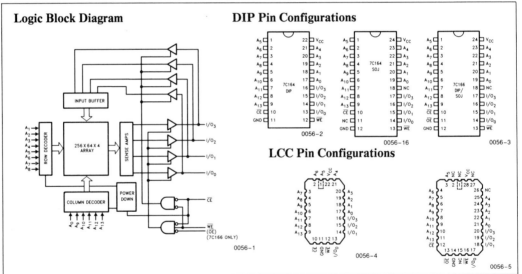

Selection Guide

		7C164/164L-20 7C166/166L-20	7C164/164L-25 7C166/166L-25	7C164/164L-35 7C166/166L-35	7C164/164L-45 7C166/166L-45
Maximum Access Time (ns)		20	25	35	45
Maximum Operating Current (mA)	Commercial	80	70	70	50
	Military		80	70	70
Maximum Standby Current (mA)	Commercial	40/20	20/20	20/20	20/20
	Military		40/20	20/20	20/20

CY7C164 Datasheet. Datasheet courtesy of Cypress Semiconductor.

Maximum Ratings
(Above which the useful life may be impaired. For user guidelines, not tested.)

Storage Temperature .... $-65°C$ to $+150°C$

Ambient Temperature with
Power Applied .... $-55°C$ to $+125°C$

Supply Voltage to Ground Potential .... $-0.5V$ to $+7.0V$

DC Voltage Applied to Outputs
in High Z State....................... $-0.5V$ to $+7.0V$

DC Input Voltage .... $-3.0V$ to $+7.0V$

Output Current into Outputs (Low) .... 20 mA

Static Discharge Voltage .... $>2001V$
(Per MIL-STD-883 Method 3015)

Latch-up Current........................... >200 mA

Operating Range

Range	Ambient Temperature	V_{CC}
Commercial	0°C to $+70°C$	5V $\pm10\%$
Military[3]	$-55°C$ to $+125°C$	5V $\pm10\%$

Electrical Characteristics Over Operating Range[4]

Parameters	Description	Test Conditions		7C164/164L-20 7C166/166L-20 Min.	7C164/164L-20 7C166/166L-20 Max.	7C164/164L-25, 35 7C166/166L-25, 35 Min.	7C164/164L-25, 35 7C166/166L-25, 35 Max.	7C164/164L-45 7C166/166L-45 Min.	7C164/164L-45 7C166/166L-45 Max.	Units
V_{OH}	Output HIGH Voltage	V_{CC} = Min., I_{OH} = -4.0 mA		2.4		2.4		2.4		V
V_{OL}	Output LOW Voltage	V_{CC} = Min., I_{OL} = 8.0 mA			0.4		0.4		0.4	V
V_{IH}	Input HIGH Voltage			2.2	V_{CC}	2.2	V_{CC}	2.2	V_{CC}	V
V_{IL}	Input LOW Voltage			-3.0	0.8	-3.0	0.8	-3.0	0.8	V
I_{IX}	Input Load Current	GND $\leq V_I \leq V_{CC}$		-10	$+10$	-10	$+10$	-10	$+10$	μA
I_{OZ}	Output Leakage Current	GND $\leq V_O \leq V_{CC}$, Output Disabled		-10	$+10$	-10	$+10$	-10	$+10$	μA
I_{OS}	Output Short Circuit Current[1]	V_{CC} = Max., V_{OUT} = GND			-350		-350		-350	mA
I_{CC}	V_{CC} Operating Supply Current	V_{CC} = Max. I_{OUT} = 0 mA	Coml.		80		70		50	mA
			Mil. 25				80		70	
			35				70			
I_{SB_1}	Automatic $\overline{CE}$[2] Power Down Current	Max. V_{CC}, $\overline{CE} \geq V_{IH}$ Min. Duty Cycle = 100%	Coml.		40		20		20	mA
			Mil. 25				40		20	
			35				20			
I_{SB_2}	Automatic $\overline{CE}$[2] Power Down Current	Max. V_{CC}, $\overline{CE} \geq V_{CC} - 0.3V$ $V_{IN} \geq V_{CC} - 0.3V$ or $V_{IN} \leq 0.3V$	Coml.		20		20		20	mA
			Mil.				20		20	

Capacitance[5]

Parameters	Description	Test Conditions	Max.	Units
C_{IN}	Input Capacitance	T_A = 25°C, f = 1 MHz, V_{CC} = 5.0V	5	pF
C_{OUT}	Output Capacitance		7	

Notes:
1. Not more than 1 output should be shorted at one time. Duration of the short circuit should not exceed 30 seconds.
2. A pull-up resistor to V_{CC} on the $\overline{CE}$ input is required to keep the device deselected during V_{CC} power-up, otherwise I_{SB} will exceed values given.
3. T_A is the "instant on" case temperature.
4. See the last page of this specification for Group A subgroup testing information.
5. Tested initially and after any design or process changes that may affect these parameters.

AC Test Loads and Waveforms

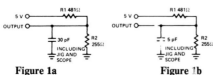

Figure 1a   Figure 1b

Equivalent to: THÉVENIN EQUIVALENT

OUTPUT ○—–ᴡᴡ—○ 1.73 V   0056–8
            167Ω

Figure 2

Switching Characteristics Over Operating Range[4, 6]

Parameters	Description		7C164/164L-20 7C166/166L-20 Min.	Max.	7C164/164L-25 7C166/166L-25 Min.	Max.	7C164/164L-35 7C166/166L-35 Min.	Max.	7C164/164L-45 7C166/166L-45 Min.	Max.	Units
READ CYCLE											
t_{RC}	Read Cycle Time		20		25		35		45		ns
t_{AA}	Address to Data Valid			20		25		35		45	ns
t_{OHA}	Output Hold from Address Change		3		3		3		3		ns
t_{ACE}	$\overline{CE}$ LOW to Data Valid			20		25		35		45	ns
t_{DOE}	$\overline{OE}$ LOW to Data Valid	7C166		12		15		25		30	ns
t_{LZOE}	$\overline{OE}$ LOW to LOW Z	7C166	3		3		3		3		ns
t_{HZOE}	$\overline{OE}$ HIGH to HIGH Z	7C166		8		15		15		15	ns
t_{LZCE}	$\overline{CE}$ LOW to Low Z[8]		5		5		5		5		ns
t_{HZCE}	$\overline{CE}$ HIGH to High Z[7, 8]			10		10		15		15	ns
t_{PU}	$\overline{CE}$ LOW to Power Up		0		0		0		0		ns
t_{PD}	$\overline{CE}$ HIGH to Power Down			20		25		35		45	ns
WRITE CYCLE[9]											
t_{WC}	Write Cycle Time		20		20		30		40		ns
t_{SCE}	$\overline{CE}$ LOW to Write End		15		20		25		35		ns
t_{AW}	Address Set-up to Write End		15		20		25		35		ns
t_{HA}	Address Hold from Write End		0		0		0		0		ns
t_{SA}	Address Set-up to Write Start		0		0		0		0		ns
t_{PWE}	$\overline{WE}$ Pulse Width		15		20		25		35		ns
t_{SD}	Data Set-up to Write End		11		13		15		20		ns
t_{HD}	Data Hold from Write End		0		0		0		5		ns
t_{LZWE}	$\overline{WE}$ HIGH to Low Z[8]		3		3		3		3		ns
t_{HZWE}	$\overline{WE}$ LOW to High Z[7, 8]		0	0	0	7	0	10	0	15	ns

Notes:

6. Test conditions assume signal transition times of 5 ns or less, timing reference levels of 1.5V, input pulse levels of 0 to 3.0V and output loading of the specified I_{OL}/I_{OH} and 30 pF load capacitance.

7. t_{HZCE} and t_{HZWE} are specified with $C_L = 5$ pF as in *Figure 1b*. Transition is measured ± 500 mV from steady state voltage.

8. At any given temperature and voltage condition, t_{HZCE} is less than t_{LZCE} for any given device. These parameters are guaranteed and not 100% tested.

9. The internal write time of the memory is defined by the overlap of $\overline{CE}$ LOW and $\overline{WE}$ LOW. Both signals must be LOW to initiate a write and either signal can terminate a write by going HIGH. The data input setup and hold timing should be referenced to the rising edge of the signal that terminates the write.

10. $\overline{WE}$ is HIGH for read cycle.

11. Device is continuously selected, $\overline{CE} = V_{IL}$. (7C166: $\overline{OE} = V_{IL}$ also.)

12. Address valid prior to or coincident with $\overline{CE}$ transition low.

13. 7C166 only: Data I/O will be high impedance if $\overline{OE} = V_{IH}$.

CY7C164 Datasheet. Datasheet courtesy of Cypress Semiconductor.

Data Retention Characteristics (L Version Only)

Parameter	Description	Test Conditions	CY7C164/CY7C166		Units
			Min.	Max.	
V_{DR}	V_{CC} For Retention of Data	$V_{CC} = 2.0V$, $\overline{CE} \geq V_{CC} - 0.2V$ $V_{IN} \geq V_{CC} - 0.2V$ or $V_{IN} \leq 0.2V$	2.0		V
I_{CCDR}	Data Retention Current			1000	μA
t_{CDR}	Chip Deselect to Data Retention Time		0		ns
t_R	Operation Recovery Time		t_{RC}[14]		ns
I_{LI}	Input Leakage Current			2	μA

Note:
14. t_{RC} = read cycle time.

Data Retention Waveform

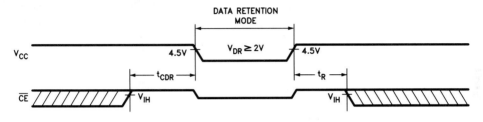

0056-13

Switching Waveforms

Read Cycle No. 1 (Notes 10, 11)

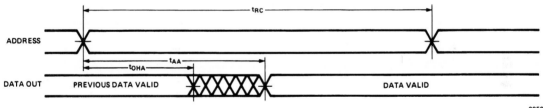

0056-9

CY7C164 Datasheet. Datasheet courtesy of Cypress Semiconductor.

Switching Waveforms (Continued)

Read Cycle No. 2 (Notes 10, 12)

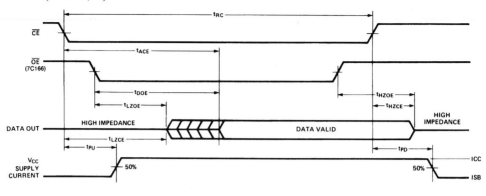

0056–10

Write Cycle No. 1 ($\overline{WE}$ Controlled) (Notes 9, 13)

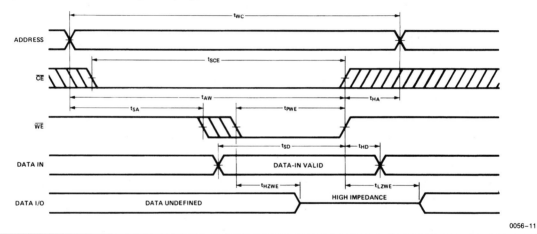

0056–11

Write Cycle No. 2 ($\overline{CE}$ Controlled) (Notes 9, 13)

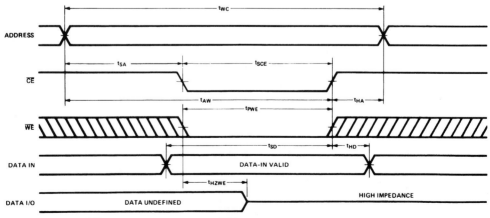

0056–12

Note: If $\overline{CE}$ goes HIGH simultaneously with $\overline{WE}$ HIGH, the output remains in a high impedance state.

CY7C164 Datasheet. Datasheet courtesy of Cypress Semiconductor.

Typical DC and AC Characteristics

NORMALIZED SUPPLY CURRENT vs. SUPPLY VOLTAGE

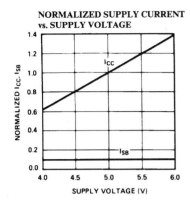

NORMALIZED SUPPLY CURRENT vs. AMBIENT TEMPERATURE

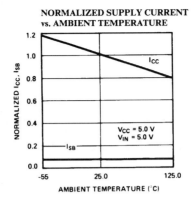

OUTPUT SOURCE CURRENT vs. OUTPUT VOLTAGE

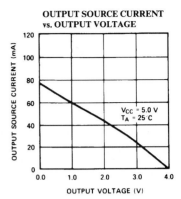

NORMALIZED ACCESS TIME vs. SUPPLY VOLTAGE

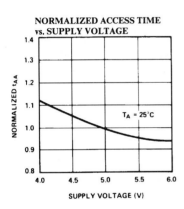

NORMALIZED ACCESS TIME vs. AMBIENT TEMPERATURE

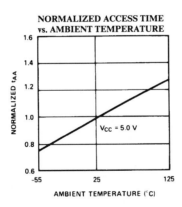

OUTPUT SINK CURRENT vs. OUTPUT VOLTAGE

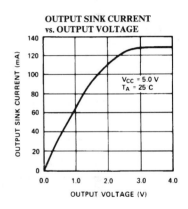

TYPICAL POWER-ON CURRENT vs. SUPPLY VOLTAGE

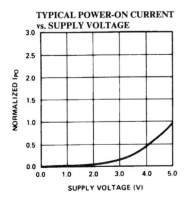

TYPICAL ACCESS TIME CHANGE vs. OUTPUT LOADING

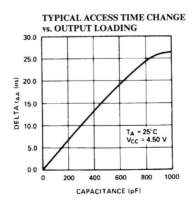

NORMALIZED I_{CC} vs. CYCLE TIME

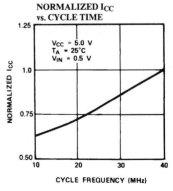

0056–14

CY7C164 Datasheet. Datasheet courtesy of Cypress Semiconductor.

7C164 Truth Table

CE	WE	Input/Outputs	Mode
H	X	High Z	Deselect Power Down
L	H	Data Out	Read
L	L	Data In	Write

7C166 Truth Table

CE	WE	OE	Inputs/Outputs	Mode
H	X	X	High Z	Deselect Power Down
L	H	L	Data Out	Read
L	L	X	Data In	Write
L	H	H	High Z	Deselect

Ordering Information

Speed (ns)	Ordering Code	Package Type	Operating Range
20	CY7C164-20PC	P9	Commercial
	CY7C164L-20PC	P9	
	CY7C164-20VC	V13	
	CY7C164L-20VC	V13	
	CY7C164-20DC	D10	
	CY7C164L-20DC	D10	
	CY7C164-20LC	L52	
	CY7C164L-20LC	L52	
25	CY7C164-25PC	P9	Commercial
	CY7C164L-25PC	P9	
	CY7C164-25VC	V13	
	CY7C164L-25VC	V13	
	CY7C164-25DC	D10	
	CY7C164L-25DC	D10	
	CY7C164-25LC	L52	
	CY7C164L-25LC	L52	
	CY7C164-25DMB	D10	Military
	CY7C164L-25DMB	D10	
	CY7C164-25LMB	L52	
	CY7C164L-25LMB	L52	
35	CY7C164-35PC	P9	Commercial
	CY7C164L-35PC	P9	
	CY7C164-35VC	V13	
	CY7C164L-35VC	V13	
	CY7C164-35DC	D10	
	CY7C164L-35DC	D10	
	CY7C164-35LC	L52	
	CY7C164L-35LC	L52	
	CY7C164-35DMB	D10	Military
	CY7C164L-35DMB	D10	
	CY7C164-35LMB	L52	
	CY7C164L-35LMB	L52	

Speed (ns)	Ordering Code	Package Type	Operating Range
45	CY7C164-45PC	P9	Commercial
	CY7C164L-45PC	P9	
	CY7C164-45VC	V13	
	CY7C164L-45VC	V13	
	CY7C164-45DC	D10	
	CY7C164L-45DC	D10	
	CY7C164-45LC	L52	
	CY7C164L-45LC	L52	
	CY7C164-45DMB	D10	Military
	CY7C164L-45DMB	D10	
	CY7C164-45LMB	L52	
	CY7C164L-45LMB	L52	

Speed (ns)	Ordering Code	Package Type	Operating Range
20	CY7C166-20PC	P13	Commercial
	CY7C166L-20PC	P13	
	CY7C166-20VC	V13	
	CY7C166L-20VC	V13	
	CY7C166-20DC	D14	
	CY7C166L-20DC	D14	
	CY7C166-20LC	L54	
	CY7C166L-20LC	L54	
25	CY7C166-25PC	P13	Commercial
	CY7C166L-25PC	P13	
	CY7C166-25VC	V13	
	CY7C166L-25VC	V13	
	CY7C166-25DC	D14	
	CY7C166L-25DC	D14	
	CY7C166-25LC	L54	
	CY7C166L-25LC	L54	
	CY7C166-25DMB	D14	Military
	CY7C166L-25DMB	D14	
	CY7C166-25LMB	L54	
	CY7C166L-25LMB	L54	

CY7C164 Datasheet. Datasheet courtesy of Cypress Semiconductor.

Ordering Information (Continued)

Speed (ns)	Ordering Code	Package Type	Operating Range
35	CY7C166-35PC	P13	Commercial
	CY7C166L-35PC	P13	
	CY7C166-35VC	V13	
	CY7C166L-35VC	V13	
	CY7C166-35DC	D14	
	CY7C166L-35DC	D14	
	CY7C166-35LC	L54	
	CY7C166L-35LC	L54	
	CY7C166-35DMB	D14	Military
	CY7C166L-35DMB	D14	
	CY7C166-35LMB	L54	
	CY7C166L-35LMB	L54	
45	CY7C166-45PC	P13	Commercial
	CY7C166L-45PC	P13	
	CY7C166-45VC	V13	
	CY7C166L-45VC	V13	
	CY7C166-45DC	D14	
	CY7C166L-45DC	D14	
	CY7C166-45LC	L54	
	CY7C166L-45LC	L54	
	CY7C166-45DMB	D14	Military
	CY7C166L-45DMB	D14	
	CY7C166-45LMB	L54	
	CY7C166L-45LMB	L54	

Bit Map

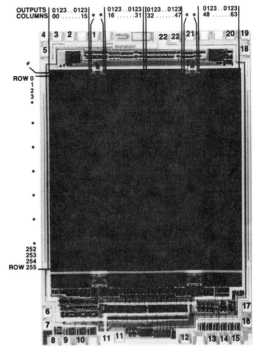

0056–15

Address Designators

Address Name	Address Function	Pin Number
A5	X3	1
A6	X4	2
A7	X5	3
A8	X6	4
A9	X7	5
A10	Y5	6
A11	Y4	7
A12	Y0	8
A13	Y1	9
A0	Y2	17
A1	Y3	18
A2	X0	19
A3	X1	20
A4	X2	21

CY7C164 Datasheet. Datasheet courtesy of Cypress Semiconductor.

MILITARY SPECIFICATIONS
Group A Subgroup Testing

DC Characteristics

Parameters	Subgroups
V_{OH}	1,2,3
V_{OL}	1,2,3
V_{IH}	1,2,3
V_{IL}	1,2,3
I_{IX}	1,2,3
I_{OZ}	1,2,3
I_{OS}	1,2,3
I_{CC}	1,2,3
I_{SB1}	1,2,3
I_{SB2}	1,2,3

Switching Characteristics

Parameters	Subgroups
READ CYCLE	
t_{RC}	7,8,9,10,11
t_{AA}	7,8,9,10,11
t_{OHA}	7,8,9,10,11
t_{ACE}	7,8,9,10,11
t_{DOE}[1]	7,8,9,10,11
WRITE CYCLE	
t_{WC}	7,8,9,10,11
t_{SCE}	7,8,9,10,11
t_{AW}	7,8,9,10,11
t_{HA}	7,8,9,10,11
t_{SA}	7,8,9,10,11
t_{PWE}	7,8,9,10,11
t_{SD}	7,8,9,10,11
t_{HD}	7,8,9,10,11

Data Retention Characteristics (L Version Only)

Parameters	Subgroups
V_{DR}	1,2,3
I_{CCDR}	1,2,3

Note:
1. 7C166 only.

Document #: 38-00032-C

Package Diagrams

22 Lead (300 MIL) Molded DIP P9

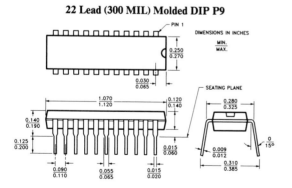

22 Lead (300 MIL) Cerdip D10

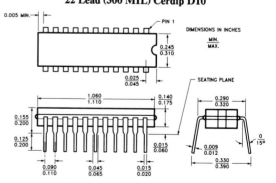

CY7C164 Datasheet. Datasheet courtesy of Cypress Semiconductor.

24 Lead (300 MIL) Molded DIP P13

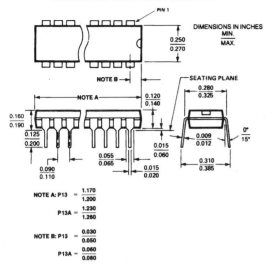

NOTE A: P13 = 1.170 / 1.200

P13A = 1.230 / 1.260

NOTE B: P13 = 0.030 / 0.050

P13A = 0.060 / 0.080

24 Lead (300 MIL) Cerdip D14

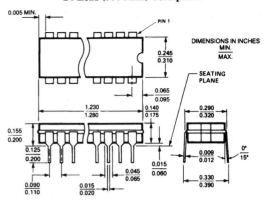

22 Pin Rectangular Leadless Chip Carrier L52

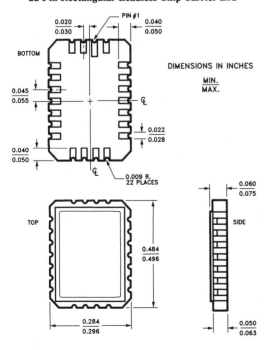

28 Pin Rectangular Leadless Chip Carrier L54

(MIL-M-38510 C-11A)

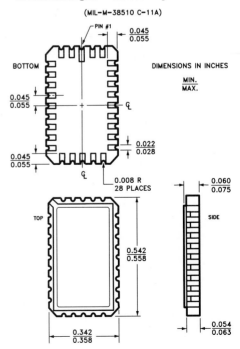

CY7C164 Datasheet. Datasheet courtesy of Cypress Semiconductor.

24 Lead Molded SOJ V13

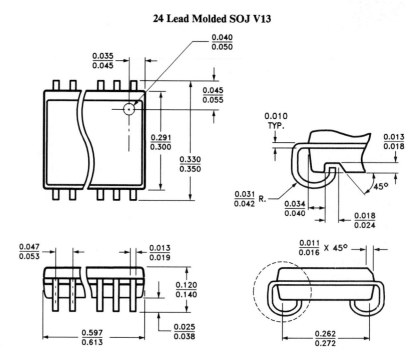

CY7C164 Datasheet. Datasheet courtesy of Cypress Semiconductor.

1K	Commercial	X2201A	1024 x 1 Bit

Nonvolatile Static RAM

FEATURES
- **Single 5V Supply**
- **Fully TTL Compatible**
- **Infinite E²PROM Array Recall, RAM Read and Write Cycles**
- **Access Time of 300 ns Max.**
- **Nonvolatile Store Inhibit: V_{CC} = 3V Typical**
- **100 Year Data Retention**

DESCRIPTION
The Xicor X2201A is a 1024 x 1 NOVRAM featuring a high-speed static RAM overlaid bit-for-bit with a nonvolatile E²PROM. The X2201A is fabricated with the same reliable N-channel floating gate MOS technology used in all Xicor 5V nonvolatile memories.

The NOVRAM design allows data to be easily transferred from RAM to E²PROM (store) and from E²PROM to RAM (recall). The store operation is completed in 10 ms or less and the recall is typically completed in 1 μs.

Xicor NOVRAMs are designed for unlimited write operations to RAM, either from the host or recalls from E²PROM. The E²PROM array is designed for a minimum 10,000 store cycles. Data retention is specified to be greater than 100 years.

PIN CONFIGURATION

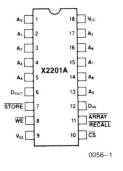

0056–1

PIN NAMES

A_0–A_9	Address Inputs
D_{IN}	Data Input
D_{OUT}	Data Out
$\overline{WE}$	Write Enable
$\overline{CS}$	Chip Select
$\overline{ARRAY\ RECALL}$	Array Recall
$\overline{STORE}$	Store
V_{CC}	+5V
V_{SS}	Ground

FUNCTIONAL DIAGRAM

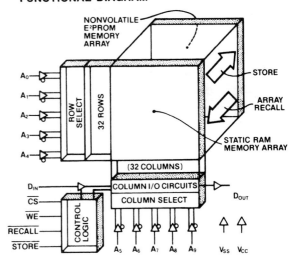

X2201A Datasheet. Courtesy of Xicor, Inc.

ABSOLUTE MAXIMUM RATINGS*

Temperature Under Bias . . . . −10°C to +85°C
Storage Temperature . . . . −65°C to +150°C
Voltage on any Pin with
   Respect to Ground . . . . −1.0V to +7V
D.C. Output Current . . . . 5 mA
Lead Temperature
   (Soldering, 10 Seconds) . . . . 300°C

*COMMENT

Stresses above those listed under "Absolute Maximum Ratings" may cause permanent damage to the device. This is a stress rating only and the functional operation of the device at these or any other conditions above those indicated in the operational sections of this specification is not implied. Exposure to absolute maximum rating conditions for extended periods may affect device reliability.

D.C. OPERATING CHARACTERISTICS

T_A = 0°C to +70°C, V_{CC} = +5V ±10%, unless otherwise specified.

Symbol	Parameter	Limits		Units	Test Conditions
		Min.	Max.		
I_{CC}	Power Supply Current		60	mA	All Inputs = V_{CC} $I_{I/O}$ = 0 mA
I_{LI}	Input Load Current		10	μA	V_{IN} = GND to V_{CC}
I_{LO}	Output Leakage Current		10	μA	V_{OUT} = GND to V_{CC}
V_{IL}	Input Low Voltage	−1.0	0.8	V	
V_{IH}	Input High Voltage	2.0	V_{CC} + 1.0	V	
V_{OL}	Output Low Voltage		0.4	V	I_{OL} = 4.2 mA
V_{OH}	Output High Voltage	2.4		V	I_{OH} = −2 mA

CAPACITANCE T_A = 25°C, f = 1.0 MHz, V_{CC} = 5V

Symbol	Test	Max.	Units	Conditions
$C_{I/O}$[1]	Input/Output Capacitance	8	pF	$V_{I/O}$ = 0V
C_{IN}[1]	Input Capacitance	6	pF	V_{IN} = 0V

A.C. CONDITIONS OF TEST

Input Pulse Levels	0V to 3.0V
Input Rise and Fall Times	10 ns
Input and Output Timing Levels	1.5V
Output Load	1 TTL Gate and C_L = 100 pF

MODE SELECTION

Inputs				Input Output I/O	Mode
CS	WE	ARRAY RECALL	STORE		
H	X	H	H	Output High Z	Not Selected[2]
L	H	H	H	Output Data	Read RAM
L	L	H	H	Input Data High	Write "1" RAM
L	L	H	H	Input Data Low	Write "0" RAM
X	H	L	H	Output High Z	Array Recall
H	X	L	H	Output High Z	Array Recall
X	H	H	L	Output High Z	Nonvolatile Storing[3]
H	X	H	L	Output High Z	Nonvolatile Storing[3]

Notes: (1) This parameter is periodically sampled and not 100% tested.

(2) Chip is deselected but may be automatically completing a store cycle.

(3) $\overline{STORE}$ = L is required only to initiate the store cycle, after which the store cycle will be automatically completed ($\overline{STORE}$ = X).

X2201A Datasheet. Courtesy of Xicor, Inc.

A.C. CHARACTERISTICS

$T_A = 0°C$ to $+70°C$, $V_{CC} = +5V \pm 10\%$, unless otherwise specified.

Read Cycle Limits

Symbol	Parameter	Min.	Max.	Units
t_{RC}	Read Cycle Time	300		ns
t_A	Access Time		300	ns
t_{CO}	Chip Select to Output Valid		200	ns
t_{OH}	Output Hold from Address Change	50		ns
t_{LZ}	Chip Select to Output in Low Z	10		ns
t_{HZ}	Chip Deselect to Output in High Z	10	100	ns

Read Cycle

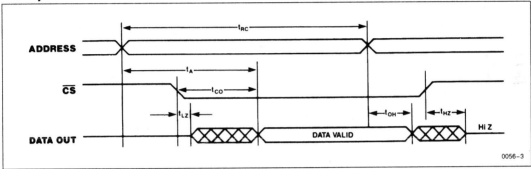

X2201A Datasheet. Courtesy of Xicor, Inc.

Write Cycle Limits

Symbol	Parameter	Min.	Max.	Units
t_{WC}	Write Cycle Time	300		ns
t_{CW}	Chip Select to End of Write	150		ns
t_{AS}	Address Setup Time	50		ns
t_{WP}	Write Pulse Width	150		ns
t_{WR}	Write Recovery Time	25		ns
t_{DW}	Data Valid to End of Write	100		ns
t_{DH}	Data Hold Time	0		ns
t_{WZ}	Write Enable to Output in High Z	10	100	ns
t_{OW}	Output Active from End of Write	10		ns

Write Cycle

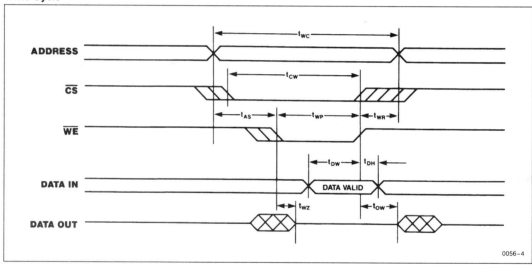

X2201A Datasheet. Courtesy of Xicor, Inc.

Early Write Cycle

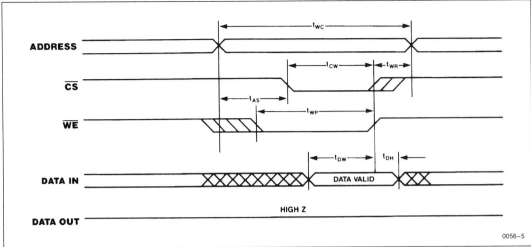

Store Cycle Limits

Symbol	Parameter	Min.	Max.	Units
t_{STC}	Store Cycle Time		10	ms
t_{STP}	Store Pulse Width	100		ns
t_{STZ}	Store to Output in High Z		500	ns
t_{OST}	Output Active from End of Store	10		ns

Store Cycle

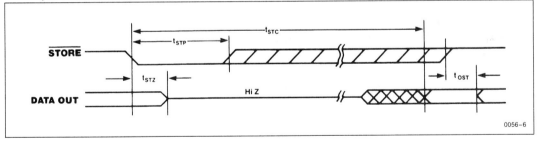

X2201A Datasheet. Courtesy of Xicor, Inc.

Array Recall Cycle Limits

Symbol	Parameter	Min.	Max.	Units
t_{RCC}	Array Recall Cycle Time	1200		ns
t_{RCP}	Recall Pulse Width[4]	450		ns
t_{RCZ}	Recall to Output in High Z		150	ns
t_{ORC}	Output Active from End of Recall	10		ns
t_{ARC}	Recalled Data Access Time from End of Recall		750	ns

Array Recall Cycle

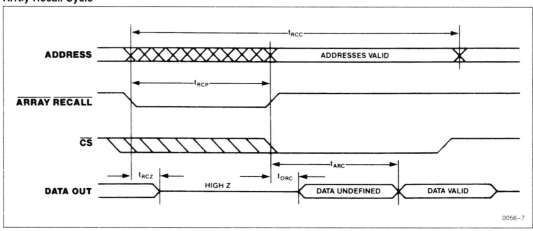

Note: (4) Array Recall rise time must be less than 1 μs.

X2201A Datasheet. Courtesy of Xicor, Inc.

PIN DESCRIPTIONS AND DEVICE OPERATION

Addresses (A₀–A₉)

The address inputs select a memory location during a read or write operation.

Chip Select ($\overline{CS}$)

The Chip Select input must be LOW to enable read/write operations with the RAM array. $\overline{CS}$ HIGH will place the D_{OUT} in the high impedance state.

Write Enable ($\overline{WE}$)

The Write Enable input controls the D_{OUT} buffer, determining whether a RAM read or write operation is enabled. $\overline{WE}$ HIGH enables a read and $\overline{WE}$ LOW enables a write.

Data In (D_{IN})

Data is written into the device via the D_{IN} input.

Data Out (D_{OUT})

Data from a selected address is output on the D_{OUT} output. This pin is in the high impedance state when either $\overline{CS}$ is HIGH or when $\overline{WE}$ is LOW.

$\overline{STORE}$

The $\overline{STORE}$ input, when LOW, will initiate the transfer of the entire contents of the RAM array to the E²PROM array. The $\overline{WE}$ and $\overline{ARRAY\ RECALL}$ inputs are inhibited during the store cycle. The store operation will be completed in 10 ms or less.

A store operation has priority over RAM read/write operations. If $\overline{STORE}$ is asserted during a read operation, the read will be discontinued. If $\overline{STORE}$ is asserted during a RAM write operation, the write will be immediately terminated and the store performed. The data at the RAM address that was being written will be unknown in both the RAM and E²PROM.

$\overline{ARRAY\ RECALL}$

The $\overline{ARRAY\ RECALL}$ input, when LOW, will initiate the transfer of the entire contents of the E²PROM array to the RAM array. The transfer of data will typically be completed in 1 μs or less.

An array recall has priority over RAM read/write operations and will terminate both operations when $\overline{ARRAY\ RECALL}$ is asserted. $\overline{ARRAY\ RECALL}$ LOW will also inhibit the $\overline{STORE}$ input.

WRITE PROTECTION

The X2201A has three write protect features that are employed to protect the contents of the nonvolatile memory.

- V_{CC} Sense—All functions are inhibited when V_{CC} is ≤3V, typically.

- Write Inhibit—Holding either $\overline{STORE}$ HIGH or $\overline{ARRAY\ RECALL}$ LOW during power-up or power-down will prevent an inadvertent store operation and E²PROM data integrity will be maintained.

- Noise Protection—A $\overline{STORE}$ pulse of less than 20 ns will *not* initiate a store cycle.

ENDURANCE

The endurance specification of a device is characterized by the predicted *first* bit failure to occur in the entire memory (device or system) array rather than the average or typical value for the array. Since endurance is limited by the number of electrons trapped in the oxide during data changes, Xicor NOVRAMs are designed to minimize the number of changes an E²PROM bit cell undergoes during store operations. Only those bits in the E²PROM that are different from their corresponding location in the RAM will be "cycled" during a nonvolatile store. This characteristic reduces unnecessary cycling of any of the rest of the bits in the array, thereby increasing the potential endurance of each bit and increasing the potential endurance of the entire array. Reliability data documented in RR504, the *Xicor Reliability Report on Endurance,* and additional reports are available from Xicor.

Part Number	Store Cycles	Data Changes Per Bit
X2201A	10,000	1,000

SYMBOL TABLE

WAVEFORM	INPUTS	OUTPUTS
	Must be steady	Will be steady
	May change from Low to High	Will change from Low to High
	May change from High to Low	Will change from High to Low
	Don't Care: Changes Allowed	Changing: State Not Known
	N/A	Center Line is High Impedance

X2201A Datasheet. Courtesy of Xicor, Inc.

**Normalized Active Supply Current
vs. Ambient Temperature**

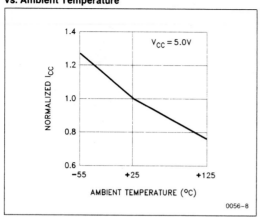

**Normalized Access Time
vs. Ambient Temperature**

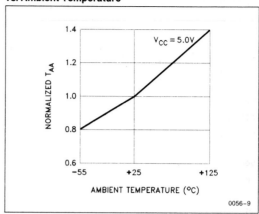

X2201A Datasheet. Courtesy of Xicor, Inc.

PAL16R8 Family

16L8, 16R8
16R6, 16R4

Features/Benefits

- Standard 20-pin architectures
- TTL and CMOS versions
- High speed, as fast as 10 ns tPD for PAL16R8D Series
- Low power, as low as zero standby for PALC16R8Z Series
- Security fuse/cell on all devices

Description

The PAL16R8 Series offers the four most popular PAL device architectures. It also provides the fastest PAL devices in the industry.

The PAL16R8 Series consists of four devices, each with sixteen array inputs and eight outputs. The devices have either 0, 4, 6, or 8 registered outputs, with the remaining being combinatorial.

The PAL device transfer function is the familiar Boolean sum of products. The PAL device consists of a programmable AND array driving a fixed OR array. Product terms with all bits programmed (disconnected) assume the logical high state, and product terms with both true and complement of any signal connected assume the logical low state.

Variable Input/Output Pin Ratio

The registered devices in the series have eight dedicated input lines, and each combinatorial output is an I/O pin. The combinatorial device has ten dedicated input lines, and only six of the eight combinatorial outputs are I/O pins. Buffers for device inputs have complementary outputs to provide user-programmable input signal polarity. Unused input pins should be tied directly to VCC or GND.

Programmable Three-State Outputs

Each output has a three-state output buffer with programmable three-state control. On combinatorial outputs, a product term controls the buffer, allowing enable and disable to be a function of any combination of device inputs or output feedback. The output provides a bidirectional I/O pin in the combinatorial configuration, and may be configured as a dedicated input if the buffer is always disabled.

Registers with Feedback

Registered outputs are provided for data storage and synchronization. Registers are composed of D-type flip-flops which are loaded on the low-to-high transition of the clock input.

Ordering Information — Newer Products

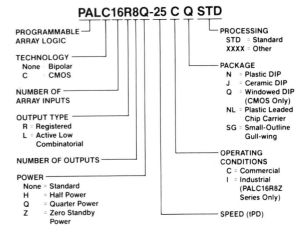

Ordering Information — Older Products

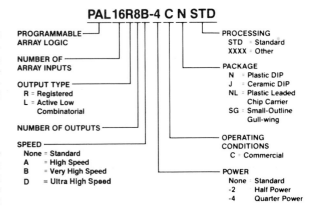

Packages

The commercial PAL16R8 Series is available in the plastic DIP (N), ceramic DIP (J), plastic leaded chip carrier (NL), and small outline (SG) packages. The CMOS versions are also available in windowed (Q) packages.

Polarity

All outputs are active low.

Performance

Several speed/power versions are available (see table). The D Series offers the fastest TTL programmable logic devices in the industry at 10 ns tPD.

Preload

The CMOS Series offers register preload for device testability. The register can be preloaded from outputs by using super-voltages in order to simplify functional testing.

	DEDICATED INPUTS	OUTPUTS	
		COMBINATORIAL	REGISTERED
PAL16L8	10	8 (6 I/O)	0
PAL16R8	8	0	8
PAL16R6	8	2 I/O	6
PAL16R4	8	4 I/O	4

SUFFIX	t_{PD} (ns)	I_{CC} (mA)
A	25	180
A-2	35	90
A-4	55	50
B	15	180
B-2	25	90
B-4	35	55
(C)Q-25	25	45
D	10	180

DIP/SO Pinouts

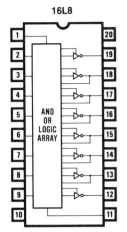

16L8

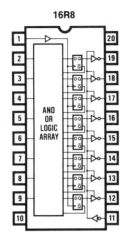

16R8

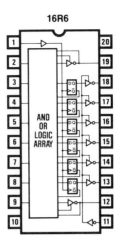

16R6

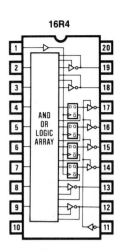

16R4

PLCC Pinouts

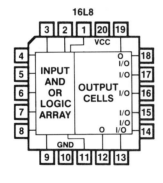

16L8

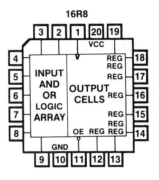

16R8

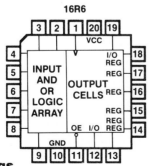

16R6

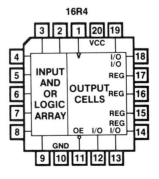

16R4

Package Drawings

LOGIC DIAGRAM

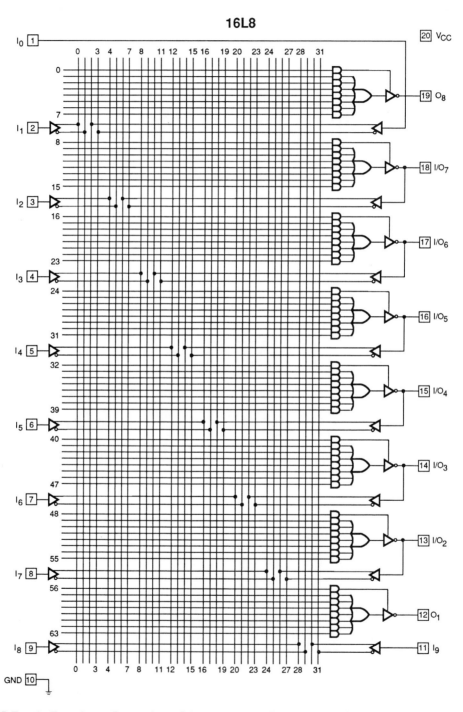

16L8

LOGIC DIAGRAM

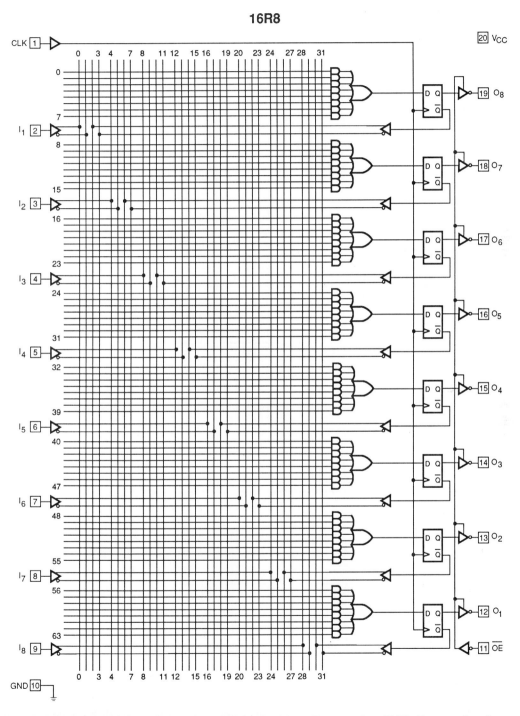

16R8

LOGIC DIAGRAM

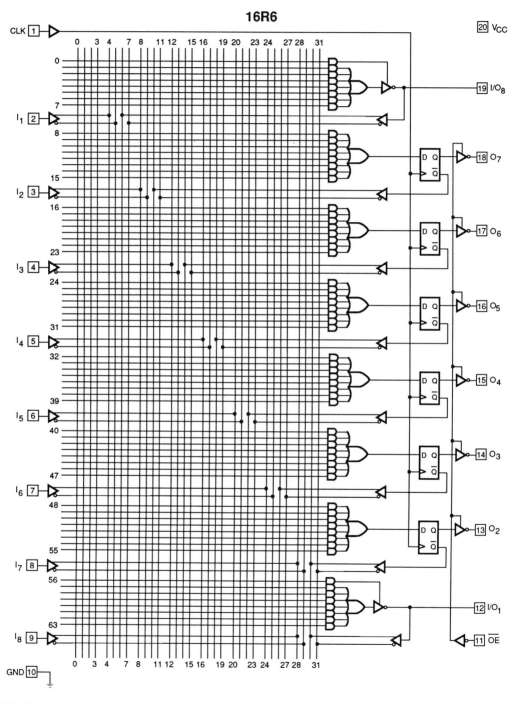

16R6

LOGIC DIAGRAM

16R4

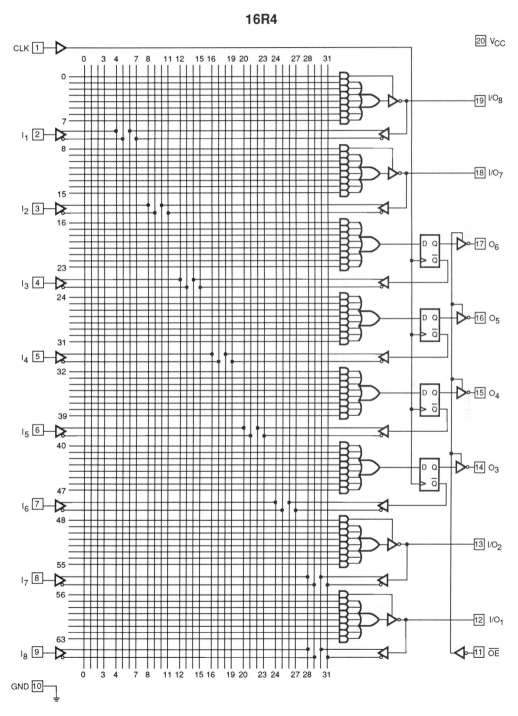

8-Bit Monolithic
Multipling D/A Converter
AD1408/AD1508*

PRELIMINARY TECHNICAL DATA

FEATURES
Improved Replacement for Industry Standard 1408/1508
Improved Settling Time: 250ns typ
Improved Linearity: ±0.1% Accuracy Guaranteed Over
   Temperature Range (–9 Grade)
High Output Voltage Compliance: +0.5V to –5.0V
Low Power Consumption: 157mW typ
High Speed 2-Quadrant Multiplying Input: 4.0mA/μs
   Slew Rate
Single Chip Monolithic Construction
Hermetic 16-Pin Ceramic DIP

AD1408/AD1508 FUNCTIONAL BLOCK DIAGRAM

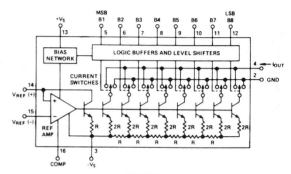

TO-116

PRODUCT DESCRIPTION

The AD1408 and AD1508 are low cost monolithic integrated
circuit 8-bit multiplying digital-to-analog converters, consisting
of matched bipolar switches, a precision resistor network and a
control amplifier. The single chip is mounted in a hermetically
sealed ceramic 16 lead dual-in-line package.

Advanced circuit design and precision processing techniques
result in significant performance advantages over older industry standard 1408/1508 devices. The maximum linearity error
over the specified operating temperature range is guaranteed
to be less than ±¼LSB (–9 grade) while settling time to ±½LSB
is reduced to 250ns typ. The temperature coefficient of gain
is typically 20ppm/°C and monotonicity is guaranteed over
the entire operating temperature range.

The AD1408/AD1508 is recommended for all low-cost 8-bit
DAC requirements; it is also suitable for upgrading overall
performance where older, less accurate and slower 1408/1508.
devices have been designed in. The AD1408 series is specified
for operation over the 0 to +75°C temperature range, the
AD1508 series for operation over the extended temperature
range of –55°C to +125°C.

PRODUCT HIGHLIGHTS

1. Monolithic IC construction makes the AD1408/AD1508 an
   optimum choice for applications where low cost is a major
   consideration.

2. The AD1408/AD1508 directly replaces other devices of
   this type.

3. Versatile design configuration allows voltage or current outputs, variable or fixed reference inputs, CMOS or TTL
   logic compatibility and a wide choice of accuracy and temperature range specifications.

4. Accuracies within ±¼LSB allow performance improvement
   of older applications without redesign.

5. Faster settling time (250ns typ) permits use in higher speed
   applications.

6. Low power consumption improves stability and reduces
   warm-up time.

7. The AD1408/AD1508 multiplies in two quadrants when a
   varying reference voltage is applied. When multiplication is
   not required, a fixed reference is used.

*Covered by Patent Numbers 3,961,326; 4,141,004.

SPECIFICATIONS (typical @ +25°C and V_{CC} = +5.0V dc, V_{EE} = -15V dc unless otherwise noted)

MAXIMUM RATINGS

RATING	SYMBOL	VALUE	UNIT
POWER SUPPLY VOLTAGE	V_{CC}	+5.5	V dc
	V_{EE}	-16.5	V dc
DIGITAL INPUT VOLTAGE	V_5 thru V_{12}	+5.5, 0	V dc
APPLIED OUTPUT VOLTAGE	V_0	+0.5, -5.2	V dc
REFERENCE CURRENT	I_{14}	5.0	mA
REFERENCE AMPLIFIER INPUTS	V_{14}, V_{15}	V_{CC}, V_{EE}	V dc
POWER DISSIPATION			
(Package Limitation)		1000	mW
Derate above T_A = +25°C	P_D	6.7	mW/°C
OPERATING TEMPERATURE RANGE			
AD1408 Series	T_A	0 to +75	°C
AD1508 Series	T_A	-55 to +125	°C
STORAGE TEMPERATURE RANGE	T_{STG}	-65 to +150	°C

ELECTRICAL CHARACTERISTICS

(V_{CC} = +5.0V dc, V_{EE} = -15V dc, $\frac{V_{REF}}{R14}$ = 2.0mA, **AD1508 Series**: T_A = -55°C to +125°C
AD1408 Series: T_A = 0 to +75°C unless otherwise noted. All digital inputs at high logic level.)

CHARACTERISTIC	SYMBOL	MIN	TYP	MAX	UNIT
RELATIVE ACCURACY					
(Error Relative to Full Scale I_O)					
AD1508-9, AD1408-9	E_r	—	—	±0.10	%
AD1508-8, AD1408-8	E_r	—	—	±0.19	%
AD1408-7	E_r	—	—	±0.39	%
SETTLING TIME					
to Within 1/2LSB [Includes t_{PLH}]					
(T_A = +25°C)	t_S	—	250	—	ns
PROPAGATION DELAY TIME					
T_A = +25°C	t_{PLH}, t_{PHL}	—	30	100	ns
OUTPUT FULL SCALE CURRENT DRIFT	TCI_O	—	-20	—	ppm/°C
DIGITAL INPUT LOGIC LEVELS (MSB)					
High Level, Logic "1"	V_{IH}	2.0	—	—	V dc
Low Level, Logic "0"	V_{IL}	—	—	0.8	V dc
DIGITAL INPUT CURRENT (MSB)					
High Level, V_{IN} = 5.0V	I_{IH}	—	0	0.04	mA
Low Level, V_{IL} = 0.8V	I_{IL}	—	-0.4	-0.8	mA
REFERENCE INPUT BIAS CURRENT					
(Pin 15)	I_{15}	—	-1.0	-3.0	μA
OUTPUT CURRENT RANGE					
V_{EE} = -5.0V	I_{OR}	0	2.0	2.1	mA
V_{EE} = -6.0V to -15V	I_{OR}	0	2.0	4.2	mA
OUTPUT CURRENT					
V_{REF} = 2.000V, R14 = 1000Ω	I_O	1.9	1.99	2.1	mA
OUTPUT CURRENT					
(All Bits Low)	I_O (min)	—	0	4.0	μA
OUTPUT VOLTAGE COMPLIANCE					
(E_1 ≤ 0.19% at T_A = +25°C)					
V_{EE} = -5V	V_O	—	—	-0.6, +0.5	V dc
V_{EE} below -10V	V_O	—	—	-5.0, +0.5	V dc
REFERENCE CURRENT SLEW RATE	SRI_{REF}	—	4.0	—	mA/μs
OUTPUT CURRENT POWER SUPPLY					
SENSITIVITY	$PSSI_O$	—	0.5	2.7	μA/V
POWER SUPPLY CURRENT					
(All Bits Low)	I_{CC}	—	+9	+14	mA
	I_{EE}	—	-7.5	-13	mA
POWER SUPPLY VOLTAGE RANGE					
(T_A = +25°C)	V_{CCR}	+4.5	+5.0	+5.5	V dc
	V_{EER}	-4.5	-15	-16.5	V dc
POWER DISSIPATION					
All Bits Low					
V_{EE} = -5.0V dc	P_D	—	82	135	mW
V_{EE} = -15V dc	P_D	—	157	265	mW
All Bits High					
V_{EE} = -5.0V dc	P_D	—	70	—	mW
V_{EE} = -15V dc	P_D	—	132	—	mW

Specifications subject to change without notice.

AD1408 Datasheet. Courtesy of Analog Devices.

Applying the AD1408/AD1508

APPLYING THE AD1408/1508

Reference Amplifier Drive and Compensation

Figures 2a and 2b are the connection diagrams for using the AD1408/AD1508 in basic voltage output modes. In Figure 2a, a positive reference voltage, V_{REF}, is converted to a current by resistor R14. This reference current determines the scale factor for the output current such that the full scale output is 1LSB (1/256) less than the reference current. R15 provides bias current compensation to the reference control amplifier to minimize temperature drift; it is nominally equal to R14 although it needn't be a stable precision resistor. This configuration develops a negative output voltage across R_L and requires a positive V_{REF}.

If a negative V_{REF} is to be used, connections to the reference control amplifier must be reversed as shown in Figure 2b. This circuit also delivers a negative output voltage, but presents a high impedance to the reference source. The negative V_{REF} must be at least 4 volts above the V_{EE} supply.

Two quadrant multiplication may be performed by applying a bipolar ac signal as the reference as long as pin 14 is positive relative to pin 15 (reference current must flow into pin 14). If the ac reference is applied to pin 14 through R14, a negative voltage equal to the negative peak of the ac reference must be applied through R15 to pin 15; if the ac reference is applied to pin 15 through R15, a positive voltage equal to the positive peak of the ac reference must be applied through R14 to pin 14.

When a dc reference is used, capacitive bypass from reference to ground will improve noise rejection.

The compensation capacitor, C, provides proper phase margin for the reference control amplifier. As R14 is increased, the closed-loop gain of the amplifier is decreased, therefore C must be increased. For R14 = 1.0kΩ, 2.5kΩ and 5.0kΩ, minimum values of capacitance are 15pF, 37pF and 75pF respectively. C may be tied to either V_{EE} or ground, but tying it to V_{EE} increases negative supply noise rejection. If the reference is driven by a high-impedance current source, heavy compensation of the amplifier is required; this causes a reduction in overall bandwidth.

Output Current Range

The nominal value for output current range is 0 to 1.992mA as determined by a 2mA reference current. If V_{EE} is more negative than –7.0 volts, this range may be increased to a maximum of 0 to 4.2mA. An increase in speed may be realized at increased output current levels, but power consumption will increase, possibly causing small shifts in linearity.

Pin 1, range control, may be grounded or unconnected. Although older devices of this type require different terminations for various applications, the AD1408/AD1508 compensates automatically. This pin is not connected internally, therefore any previously installed connections will be tolerated.

Output Voltage Range

The voltage on pin 4 is restricted to a +0.5 to –0.6 volt range when V_{EE} = –5V. When V_{EE} is more negative than –10 volts, this range is extended to +0.5 to –5.0 volts. If the current into pin 14 is 2mA (full-scale output current = 1.992mA), a 2.5kΩ resistor between the output, pin 4, and ground will provide a 0 to –4.980 volt full-scale. If R_L exceeds 500Ω however, the settling time of the device is increased.

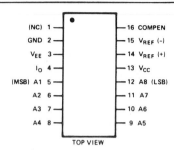

Figure 1. Pin Connections

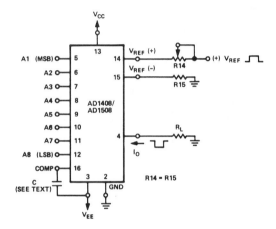

a. Connections for Use with Positive Reference

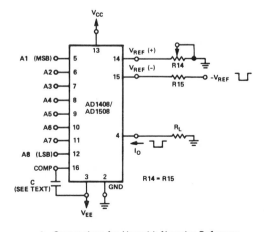

b. Connections for Use with Negative Reference

Figure 2. Basic Connections

Voltage Output

A low impedance voltage output may be derived from the output current of the AD1408/AD1508 by using an output amplifier as shown in Figure 3. The output current I_O flows in R_O to create a positive-going voltage range at the output of amplifier A1. R_O may be chosen for the desired range of output voltage; the complete circuit transfer function is given in Figure 3.

If a bipolar output voltage range is desired, R_{BP}, shown dotted, must be installed. Its purpose is to provide an offset equal to one-half of full-scale at the output of A1. The procedure for calibrating the circuit of Figure 3 is as follows:

Calibration for Unipolar Outputs (No R_{BP})

1. With all bits "OFF", adjust the A1 null-pot, R1, for V_{OUT} = 0.00V.

2. With all bits "ON", adjust R_{REF} for V_{OUT} = (Nominal Full Scale) − 1LSB = +9.961 volts

Calibration for Bipolar Outputs (R_{BP} installed, R1 not required)

1. With all bits "OFF", adjust R_{BP} for V_{OUT} = −F.S. = −5.000 volts

2. With Bit 1 (MSB) "ON", and all other Bits "OFF", adjust R_{REF} for V_{OUT} = 0.000V.

3. With all bits "ON", verify that E_{OUT} = +5.000V − 1LSB = 4.961V.

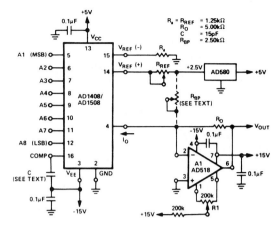

$$V_{OUT} = \frac{V_{REF}}{R_{REF}}(R_O)\left[\frac{A1}{2} + \frac{A2}{4} + \frac{A3}{8} + \frac{A4}{16} + \frac{A5}{32} + \frac{A6}{64} + \frac{A7}{128} + \frac{A8}{256}\right]$$

ADJUST V_{REF}, R_{REF} OR R_O SO THAT WITH ALL DIGITAL INPUTS
AT LOGIC "1", V_{OUT} = 9.961 VOLTS:

$$V_{OUT} = \frac{2.5}{1.25k\Omega}(5k\Omega)\left[\frac{1}{2} + \frac{1}{4} + \frac{1}{8} + \frac{1}{16} + \frac{1}{32} + \frac{1}{64} + \frac{1}{128} + \frac{1}{256}\right] = 9.961 \text{ VOLTS}$$

Figure 3. Typical Connection Diagram, AD1408/AD1508, Voltage Output, Fixed Reference

---

AD1408/AD1508 ORDERING GUIDE

Model	Accuracy (±% F.S.)	Temperature Range (°C)	Package Style[1]
AD1408-7D	0.39	0 to +75	Q16A
AD1408-8D	0.19	0 to +75	Q16A
AD1408-9D	0.10	0 to +75	Q16A
AD1508-8D	0.19	−55 to +125	Q16A
AD1508-9D	0.10	−55 to +125	Q16A

[1] See Section 19 for package outline information.

AD1408 Datasheet. Courtesy of Analog Devices.

PRECAUTIONS FOR THE HANDLING OF MOS DEVICES

Any MOS circuit can be catastrophically damaged by excessive electrostatic discharge or transient voltages. The following procedures are recommended to avoid accidental circuit damage.

I. **Testing MOS Circuits:**

    1. All units should be handled directly from the conductive or antistatic plastic tube in which they were shipped if possible. This action minimizes touching of individual leads.

    2. If units are to be tested without using the tube carrier, the following precautions should be taken:

        a. Table surfaces which potentially will come in contact with the devices either directly or indirectly (such as through shipping tubes) must be metal or of another conductive material and should be electrically connected to the test equipment and to the test operator (a grounding bracelet is recommended).

        b. The units should be transported in bundled antistatic tubes or metal trays, both of which will assume a common potential when placed on a conductive table top.

        c. Do not band tubes together with adhesive tape or rubber bands without first wrapping them in a conductive layer.

II. **Test Equipment (Including Environmental Equipment):**

    1. All equipment must be properly returned to the same reference potential (ground) as the devices, the operator, and the container for the devices.

    2. Devices to be tested should be protected from high voltage surges developed by:

        a. Turning electrical equipment on or off.

        b. Relay switching.

        c. Transients from voltage sources (AC line or power supplies).

III. **Assembling MOS Devices Onto PC Boards:**

    1. The MOS circuits should be mounted on the PC board last.

2. Similar precautions should be taken as in Item I above, at the assembly work station.

3. Soldering irons or solder baths should be at the same reference (ground) potential as the devices.

4. Plastic materials which are not antistatic treated should be kept away from devices as they develop and maintain high levels of static charge.

IV. Device Handling:

1. Handling of devices should be kept to a minimum. If handling is required, avoid touching the leads directly.

V. General:

1. The handler should take every precaution that the device will see the same reference potential when moved.

2. Anyone handling individual devices should develop a habit of first touching the container in which the units are stored before touching the units.

3. Before placing the units into a PC board, the handler should touch the PC board first.

4. Personnel should not wear clothing which will build up static charge. They should wear smocks and clothing made of 100% cotton rather than wool or synthetic fibers.

5. Be careful of electrostatic build up through the movement of air over plastic material. This is especially true of acid sinks.

6. Personnel or operators should always wear grounded wrist straps when working with MOS devices.

7. A 1 meg ohm resistance ground strap is recommended and will protect people up to 5,000 volts AC RMS or DC by limiting current to 5 milliamperes.

8. Antistatic ionized air equipment is very effective and useful in preventing electrostatic damage.

9. Low humidity maximizes potential static problems. Maintaining humidity levels above 45% is one of the most effective ways to guard against static handling problems.

Courtesy of Xicor, Inc.

Appendix C

PALASM2— PROGRAMMABLE LOGIC DEVELOPMENT SOFTWARE

Notice:

PALASM2 software is available **free** from Advanced Micro Devices/Monolithic Memories for **qualified users**. Requests for software can be made by calling 800-222-9323. This offer is subject to the company policies of AMD/MMI.

PALASM2

PALASM2 (PAL Assembler) software is created by Advanced Micro Devices/ Monolithic Memories (AMD/MMI) for PAL and programmable logic sequencer design. We will use this computer aided design package to show the basic software procedures involved in PAL design and to create several PAL based circuits. PALASM2 is a good choice for this study since it is similar to other commonly used software packages, runs on inexpensive personal computers, and is readily available.

If you have access to other PAL development software, you will still find the information and examples in this appendix useful.

Running PALASM2 software is similar to running any other computer software package in that it must be installed on the computer system first. Usually this involves copying the software from floppy disks to a hard disk drive or making backup floppy disk copies. The manufacturer's documentation explains the process. Assuming that the software is installed properly, you as the software user can expect to see the following from PALASM2.

Main Menu

When PALASM2 is invoked, a main menu screen is displayed on the computer monitor as shown in Figure C.1. Access to all the PALASM2 procedures are

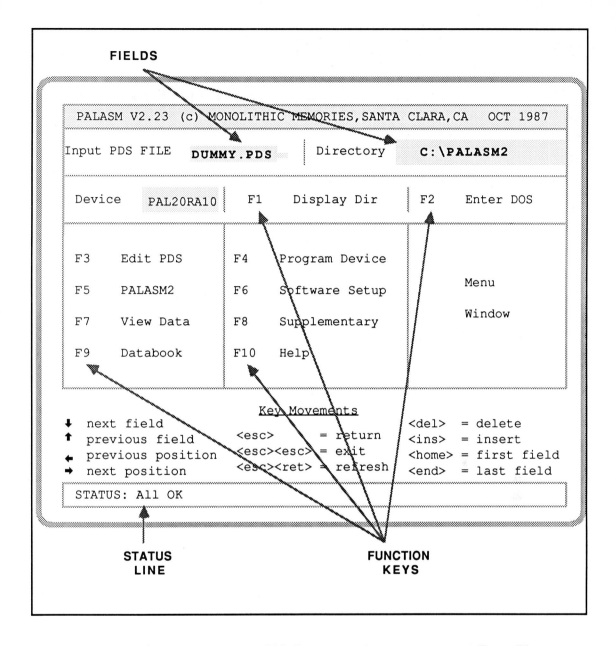

FIELDS

PALASM V2.23 (c) MONOLITHIC MEMORIES,SANTA CLARA,CA OCT 1987

Input PDS FILE **DUMMY.PDS** │ Directory **C:\PALASM2**

Device PAL20RA10 │ F1 Display Dir │ F2 Enter DOS

F3   Edit PDS   F4   Program Device

F5   PALASM2   F6   Software Setup   Menu

F7   View Data   F8   Supplementary   Window

F9   Databook   F10   Help

Key Movements

↓ next field
↑ previous field
← previous position   <esc>   = return
→ next position   <esc><esc> = exit
                            <esc><ret> = refresh

<del> = delete
<ins> = insert
<home> = first field
<end> = last field

STATUS: All OK

STATUS
LINE

FUNCTION
KEYS

Figure C.1
The Main Menu

through the main menu screen. The cursor initially rests on an "input PDS field," which identifies the data input file containing the PAL logic equations and simulation information. A "directory field" also appears to assist the software in locating the input file. The user can change either field simply by typing over the existing information. If the file already exists from previous design

Popular Programmable Logic (PAL) Development Software Packages

SOFTWARE	COMPANY
PALASM2	Advanced Micro Devices/Monolithic Memories
PLPL	Advanced Micro Devices/Monolithic Memories
CUPL	Personal CAD Systems
ABEL	Data I/O Corp.
LOG/iC	ISDATA

work, the PAL part number is displayed in the "device field." When first beginning a design, the "status line" at the bottom of the menu indicates that the file cannot be found. This is normal since the input file has not been created. Some familiarity with the computer operating system (DOS) is a useful aid toward identifying some of the files used and created by PALASM2.

"Function keys" F1 through F10 select the various options offered by PALASM2. F5, for instance, begins the PAL development process; F3 calls up the word processor program used to create the input file. The following list identifies the purpose of the various function keys:

F1—displays the files in the PALASM2 directory
F2—allows entry into DOS without quitting the PALASM2 program
F3—invokes the word processor (editor) used for generating the input file
F4—this option invokes the communications software used to program physically a PAL chip
F5—runs the PALASM2 development programs (syntax check, expansion, minimization, fuse plot assembly, and simulation)
F6—this option is used to configure PALASM2 to point automatically to the word processing program and programming files, as well as provide other support information
F7—is "view data" and provides output timing displays for the various PALASM2 processes
F8—provides access to useful supplementary programs such as a binary to hex converter
F9—lists data regarding the various programmable devices supported by PALASM2
F10—is a help option

CREATING THE INPUT FILE

With some preliminary PALASM2 information in hand, we will now concentrate on the primary task facing the designer—creating an input file. The input file, through Boolean equations or state equations (discussed in Appendix D), describes how a particular PAL device should be configured. The file is created using a word processing program of the designer's choice; function key F3 transfers control from PALASM2 to the word processor software. Since almost any word processor can be used, most people tend to use a program with which they are familiar. Only two restrictions apply: The input file name cannot be more than eight characters long (for MS-DOS) and the file must be a pure ASCII file. ASCII file creation is a typical function of most word processors and should not be a hindrance. In addition, the recommended extension for input file names is ".PDS" (i.e., TEST1.PDS).

The Declarations Section

The input file consists of several key sections, each covering specific aspects of the PAL design. The first section is the **Declarations,** a list of basic design information including part number and pin assignments. Figure C.2 shows the format for the Declarations section. The "keywords" such as TITLE and CHIP inform the PALASM2 software of the particulars concerning your design. (Only the CHIP keyword is actually essential in the Declarations section since it identifies the part pin definition.)

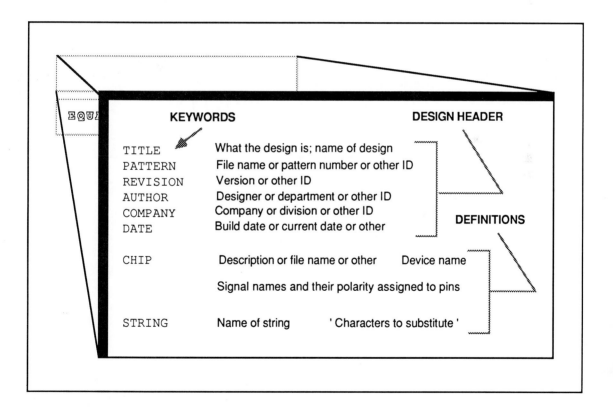

Figure C.2
Declaration's Format

To illustrate the Declarations section and the other sections to follow, we will create a simple combinatorial circuit using a 16L8 PAL. Figure C.3 shows the Declarations section and the truth table for the circuit. Every keyword is used except the STRING keyword. (STRING is used to combine several pins into a single identifying name for convenience.)

Keywords

TITLE	Sample_equation
PATTERN	Active_high_output_TT
REVISION	01
AUTHOR	Prestopnik
COMPANY	FMCC
DATE	8/20/88
CHIP	Simple_eq PAL16L8

→;PINS

1	2	3	4	5	6	7	8	9	10	←These are listed only for convenience.
NC	A	B	C	D	NC	NC	NC	NC	GND	

→;PINS

11	12	13	14	15	16	17	18	19	20	Pin names
NC	NC	NC	NC	NC	NC	NC	/Y	NC	VCC	PALASM2 uses these pin names while preparing the part.

Comments

Truth table used for this part:

A	B	C	D	Y
0	0	0	0	0
0	0	0	1	0
0	0	1	0	0
0	0	1	1	0
0	1	0	0	0
0	1	0	1	0
0	1	1	0	0
0	1	1	1	1
1	0	0	0	1
1	0	0	1	1
1	0	1	0	1
1	0	1	1	1
1	1	0	0	0
1	1	0	1	1
1	1	1	0	0
1	1	1	1	1

Figure C.3
Printout Showing the
PALASM2 Declaration
Section

Figures C.2 and C.3 explain the purpose of the keywords in this example. However, the CHIP keyword deserves special attention, which is provided by Figure C.4 and the following paragraph.

The CHIP keyword is immediately followed by an identifying part name and a specific PAL part number—PAL16L8 in this example. PALASM2 requires the part number in preparation for the creation of a fuse map. Following the

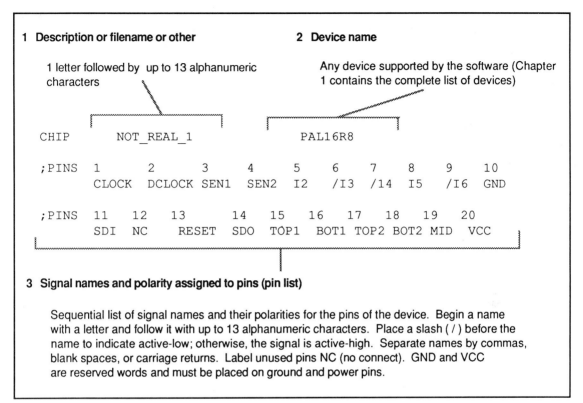

1 Description or filename or other

1 letter followed by up to 13 alphanumeric characters

2 Device name

Any device supported by the software (Chapter 1 contains the complete list of devices)

```
CHIP        NOT_REAL_1              PAL16R8

;PINS   1       2       3       4    5    6     7     8     9    10
        CLOCK   DCLOCK  SEN1    SEN2 I2   /I3   /I4   I5    /I6  GND

;PINS   11      12      13      14   15   16    17    18    19   20
        SDI     NC      RESET   SDO  TOP1 BOT1  TOP2  BOT2  MID  VCC
```

3 Signal names and polarity assigned to pins (pin list)

Sequential list of signal names and their polarities for the pins of the device. Begin a name with a letter and follow it with up to 13 alphanumeric characters. Place a slash (/) before the name to indicate active-low; otherwise, the signal is active-high. Separate names by commas, blank spaces, or carriage returns. Label unused pins NC (no connect). GND and VCC are reserved words and must be placed on ground and power pins.

Figure C.4
CHIP Syntax and Pin List

part number, pin assignments are given. The descriptive names the designer assigns to each input and output are listed here. (The actual pin numbers are only listed for visual convenience so that correspondence between pin numbers and pin names is noted. PALASM2 ignores any text preceded by a semicolon (;). Therefore, the semicolon is used to add comments to the input file, such as the pin numbers. It is in your best interest to sprinkle comments liberally

PALASM2 Character Symbols

FOR YOUR INFORMATION

CHARACTER	FUNCTION
,	Pin list separator
;	Precedes comments
/	Not (inversion)
*	AND
+	OR
:+:	Exclusive-OR
=	Combinatorial equation operator
:=	Registered equation operator
*=	Latched equation operator

throughout the input file so that you or other designers can understand how you developed the part.

PALASM2 reads the pin names that you assign to your part, beginning with the first as pin 1. PALASM2 then proceeds sequentially, assigning pin descriptions to physical pins. Pin names are separated by spaces, commas, or carriage returns. When pins are unused, assign the name NC (no connection) to them. GND and VCC are used for the power pins. For the example given in Figure C.3, pins 2, 3, 4, and 5 are assigned the input variable names A, B, C, and D, whereas pin 18 is the output /Y. Since the PAL16L8 is an active-low device (chosen because the example truth table implements $\overline{Y}$), the output pin name is listed as /Y. The slash is the PALASM2 symbol for negation.

The Equations Section

The Equations section follows the Declarations section in the input file. (See Figure C.5.) In this example the truth table fundamental products are entered directly as a standard sum of products (SOP) expression following the equation keyword, except for the inversion explained here. There are two items of note. First, PALASM2 logic symbols for negation, AND, and OR are used when writing the logic equations. For instance, ANDing two variables, such as D and E, is written D*E. Every logic operation must be noted with the correct symbols for the equation to be correct. Second, the $\overline{Y}$ output is listed as /Y because the device output is active-low (see Equation Polarity section).

DESIGN EXAMPLE C.1

Write the following expressions using PALASM2 notation:
(a) $Y = A\overline{B}C$ (b) $Y = \overline{A}B\overline{C} + A\overline{B}\,\overline{C}$

Solution   AND operations are defined with the * symbol; ORs with the + symbol, and inversion with the / symbol.

(a) Y=A*/B*C
(b) Y=/A*B*/C + A*/B*/C

Equation Polarity

PALASM2 places several restrictions on equation and output pin definitions. When using active-high output PAL devices, both the pin name and the equation name must have matching polarities. For example, an active-high device output represented as Halt in the Declarations section must be represented as Halt = "expression" in the Equations section. If the pin was defined as /Halt in the Declarations, then the equation should read /Halt = "expression." Active-low outputs are defined differently. Either the pin definition or the equation will list the output name in complement form, but not both. This tends to be confusing because it appears that two versions of the output name and an equation have to be manipulated to represent the circuit correctly. The following procedure can help:

TITLE   Sample_equation
PATTERN   Active_high_output_TT
REVISION   01
AUTHOR   Prestopnik
COMPANY   FMCC
DATE   8/20/88
CHIP   Simple_eq PAL16L8

;PINS	1	2	3	4	5	6	7	8	9	10
	NC	A	B	C	D	NC	NC	NC	NC	GND

;PINS	11	12	13	14	15	16	17	18	19	20
	NC	NC	NC	NC	NC	NC	NC	/Y	NC	VCC

EQUATIONS

Y=/(/A*B*C*D + A*/B*/C*/D + A*/B*/C*D + A*/B*C*/D + A*/B*C*D
+ A*B*/C*D + A*B*C*D)

A	B	C	D	Y
0	0	0	0	0
0	0	0	1	0
0	0	1	0	0
0	0	1	1	0
0	1	0	0	0
0	1	0	1	0
0	1	1	0	0
0	1	1	1	1
1	0	0	0	1
1	0	0	1	1
1	0	1	0	1
1	0	1	1	1
1	1	0	0	0
1	1	0	1	1
1	1	1	0	0
1	1	1	1	1

Truth table for PALASM2 expression

Figure C.5
PALASM2 Printout
Showing the
Equations Section

For active-high devices: Write equations to match the expected output function. Then match the equation polarity to that of the pin list. For instance, if an output is listed as X and the equation is the AND of A with B, then the equation reads, X=A*B. If the pin definition is /X, then the equation reads, /X=A*B. Equations are defined in a normal representation for both cases.

For active-low devices: Write the expression for the output in complement form, making sure that the polarity of the pin definition and the equation are opposite. For example, if an output is defined as X and is created by ANDing A with B, the complement form of the expression would state that the output is the complement of the AND function. Stated algebraically, X=/(A*B). This makes sense since the only difference between active-high and active-low notation is a single level of inversion. But since PALASM2 requires opposite notation on active-low outputs, X must be listed in the equation as /X=/(A*B). If X was defined in the pin list as /X, then the expression would read X=/(A*B). In either case the output expression is the complement of the "normal" functional expression.

For the preceding reasons the expression for Y in Figure C.5 is equal to the complement of the SOP expression. Also note that this is not nearly so difficult or confusing using PAL devices with programmable output polarity since the designer can program active-high or active-low outputs. However, many of the most common PAL devices are active-low, so the polarity concept should be understood.

DESIGN EXAMPLE C.2

Show how the expression $Y = A\overline{B}C$ is written for PALASM2 use if the programmable device outputs are active-high and then again if the outputs are active-low. Indicate the pin definition as well.

Solution   Active-high outputs must have matching polarities in the pin definition and Equations section. Active-low polarities must differ.

	PIN DEFINITION	EQUATION
Active-high	Y	Y=A*/B*C
	/Y	/Y=A*/B*C
Active-low	Y	/Y=/(A*/B*C)
	/Y	Y=/(A*/B*C)

Alternate Polarity Representation

Understanding active-low outputs tends to be difficult because we generally view everything as active-high. We would rather write out the SOP expression directly from truth tables, as we have done in the past, and be finished with the expression. Naturally, we could use an active-high output PAL and follow past practice, but this is not always feasible. We have to consider cost and functional constraints. A PAL device having active-high outputs does not mean it has the right number of inputs, outputs, product terms, power/speed specifications, or price tag for every application. You get the picture. So, if we must use an active-low part, how can we make design life easier?

Borrowing from a design technique used in the chapter on basic combinatorial design, we will create PAL circuitry utilizing the complement design method whenever we have active-low parts. The complement truth table for the example used earlier (Figure C.5) is shown in Figure C.6 along with the Equations section implementing it. The SOP expression is written for all one levels on the complement truth table. However, there is a subtle difference between the complement method discussed in earlier chapters and what we will do now. Typically, we would take the complement equation, reduce it using Boolean algebra or K-maps, and then recomplement the equation, resulting in an expression matching the original uncomplemented truth table output. Manual reduction is unnecessary with PALASM2 (see next section), and in the case of an active-low PAL device recomplementing the expression also becomes unnecessary because the PAL chip's active-low output characteristic does it for us. The AND-OR network represented by the complement expression will be created by the software, but the inverter between the OR gate and the output in an active-low device complements the entire function. Hence, the chip output will represent the original uncomplemented truth table.

A	B	C	D	Y	$\overline{Y}$	
0	0	0	0	0	1	
0	0	0	1	0	1	Complement truth
0	0	1	0	0	1	table output
0	0	1	1	0	1	
0	1	0	0	0	1	
0	1	0	1	0	1	
0	1	1	0	0	1	
0	1	1	1	1	0	
1	0	0	0	1	0	
1	0	0	1	1	0	
1	0	1	0	1	0	
1	0	1	1	1	0	
1	1	0	0	0	1	
1	1	0	1	1	0	
1	1	1	0	0	1	
1	1	1	1	1	0	

```
TITLE       Sample_equation
PATTERN     Comptest
REVISION    01
AUTHOR      Prestopnik
COMPANY     FMCC
DATE        8/20/88
CHIP        Simple_eq PAL16L8
;PINS       1    2    3    4    5    6    7    8    9    10
            NC   A    B    C    D    NC   NC   NC   NC   GND

;PINS       11   12   13   14   15   16   17   18   19   20
            NC   NC   NC   NC   NC   NC   NC   /Y   NC   VCC
```

Expressions

EQUATIONS

Y=/A*/B*/C*/D + /A*/B*/C*D + /A*/B*C*/D + /A*/B*C*D + /A*B*/C*/D
+ /A*B*/C*D + /A*B*C*/D + A*B*/C*/D + A*B*C*/D

Figure C.6
PALASM2 Equation
Section Using a
Complement Equation

To summarize:

1. Complement all truth table output entries.
2. Write the SOP expression for the complement truth table in the input file Equations section.
3. Run PALASM2 to process the part.

Implement the following truth table using the complement procedure and PALASM2.

XYZ	F	$\overline{F}$
000	0	1
001	1	0
010	1	0
011	0	1
100	1	0
101	0	1
110	1	0
111	1	0

Solution Complement the F output column to $\overline{F}$ as shown. Write the PALASM2 expression for $\overline{F}$. However, be sure to change the polarity for $\overline{F}$ in the expression since active-low output devices are represented with opposite polarities in the pin definition and expression.

$$F = /X*/Y*/Z + /X*Y*Z + X*/Y*Z$$

Equation Simplification

You may have noticed that all equations written in the input file were taken directly from the truth tables without doing any of the traditional minimization steps. This is a drastic and significant departure from our previous design methods considering all the benefits reduction provides. We are not giving up on reduction; rather, with programmable logic design we simply let the computer do the minimization work. One of the PALASM2 processing steps, and this is true for other programmable logic software packages as well, is to minimize logic equations. This is done primarily to reduce the number of product terms required for complex equations. Even though the typical AND array consists of 32-input AND gates, there may only be six or seven AND gate outputs assigned to any OR gate. Therefore, SOP expressions of more than six or seven product terms are not possible. It is more important to reduce the number of AND gates rather than the number of AND gate inputs. This is often a critical design consideration that affects the selection of a specific PAL device.

RUNNING PALASM2

Once the input file is complete (the input file Simulation section discussion is being deferred until later), PALASM2 is used to process the design. Function key F5 initiates this procedure, producing a submenu containing PALASM2's five processing functions: **syntax check, logic expansion, minimization, assembly,** and **simulation.**

The syntax check verifies that all keywords and associated information used in the input file have been entered correctly. Warning messages appear at the bottom of the menu screen if the input file contains errors. Errors can be studied by selecting function key F7, which invokes the view submenu. View options are used extensively in the design process. For instance, the "runtime" option under F7 allows you to view the syntax checker output.

When the input file is error-free, the logic expander program is executed. This run prepares the logic equations for the next step—minimization. The expander takes functions such as Exclusive-OR and expands them into their AND/OR equivalents. The minimization step reduces equations down to their simplest form. If any errors occur during these runs, the F7 view option can be utilized for an explanation.

Finally, the fuse plot can be created. The assembly portion of PALASM2 accomplishes this and creates several output files in the process. These are shown in Figures C.7 and C.8 for the complement example discussed in Figure C.6.

PALASM XPLOT, V2.23 - MARKET RELEASE (2-1-88)
(C) - COPYRIGHT MONOLITHIC MEMORIES INC, 1988

Title	:	Sample_equation	Author	:	Prestopnik
Pattern	:	Active_high_output_TT	Company	:	FMCC
Revision	:	01	Date	:	8/20/88

PAL16L8
SIMPLE_EQ

```
                  11   1111  1111  2222  2222  2233
      0123  4567  8901  2345  6789  0123  4567  8901

 0 XXXX XXXX XXXX XXXX XXXX XXXX XXXX XXXX
 1 XXXX XXXX XXXX XXXX XXXX XXXX XXXX XXXX
 2 XXXX XXXX XXXX XXXX XXXX XXXX XXXX XXXX
 3 XXXX XXXX XXXX XXXX XXXX XXXX XXXX XXXX
 4 XXXX XXXX XXXX XXXX XXXX XXXX XXXX XXXX
 5 XXXX XXXX XXXX XXXX XXXX XXXX XXXX XXXX
 6 XXXX XXXX XXXX XXXX XXXX XXXX XXXX XXXX
 7 XXXX XXXX XXXX XXXX XXXX XXXX XXXX XXXX ←Unprogrammed fuse

 8 ---- ---- ---- ---- ---- ---- ---- ----  /Programmed fuse
 9 ---- X--- ---- -X-- ---- ---- ---- ----
10 -X-- ---- -X-- ---- ---- ---- ---- ----
11 -X-- -X-- ---- ---- ---- ---- ---- ----
12 XXXX XXXX XXXX XXXX XXXX XXXX XXXX XXXX
13 XXXX XXXX XXXX XXXX XXXX XXXX XXXX XXXX
14 XXXX XXXX XXXX XXXX XXXX XXXX XXXX XXXX
15 XXXX XXXX XXXX XXXX XXXX XXXX XXXX XXXX

16 XXXX XXXX XXXX XXXX XXXX XXXX XXXX XXXX
17 XXXX XXXX XXXX XXXX XXXX XXXX XXXX XXXX
18 XXXX XXXX XXXX XXXX XXXX XXXX XXXX XXXX
19 XXXX XXXX XXXX XXXX XXXX XXXX XXXX XXXX
20 XXXX XXXX XXXX XXXX XXXX XXXX XXXX XXXX
21 XXXX XXXX XXXX XXXX XXXX XXXX XXXX XXXX
22 XXXX XXXX XXXX XXXX XXXX XXXX XXXX XXXX
23 XXXX XXXX XXXX XXXX XXXX XXXX XXXX XXXX
```

Figure C.7
PALASM2 Fuse Map

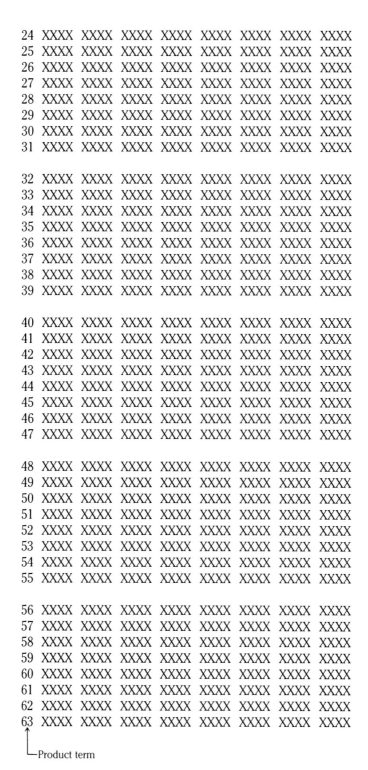

```
24 XXXX XXXX XXXX XXXX XXXX XXXX XXXX XXXX
25 XXXX XXXX XXXX XXXX XXXX XXXX XXXX XXXX
26 XXXX XXXX XXXX XXXX XXXX XXXX XXXX XXXX
27 XXXX XXXX XXXX XXXX XXXX XXXX XXXX XXXX
28 XXXX XXXX XXXX XXXX XXXX XXXX XXXX XXXX
29 XXXX XXXX XXXX XXXX XXXX XXXX XXXX XXXX
30 XXXX XXXX XXXX XXXX XXXX XXXX XXXX XXXX
31 XXXX XXXX XXXX XXXX XXXX XXXX XXXX XXXX

32 XXXX XXXX XXXX XXXX XXXX XXXX XXXX XXXX
33 XXXX XXXX XXXX XXXX XXXX XXXX XXXX XXXX
34 XXXX XXXX XXXX XXXX XXXX XXXX XXXX XXXX
35 XXXX XXXX XXXX XXXX XXXX XXXX XXXX XXXX
36 XXXX XXXX XXXX XXXX XXXX XXXX XXXX XXXX
37 XXXX XXXX XXXX XXXX XXXX XXXX XXXX XXXX
38 XXXX XXXX XXXX XXXX XXXX XXXX XXXX XXXX
39 XXXX XXXX XXXX XXXX XXXX XXXX XXXX XXXX

40 XXXX XXXX XXXX XXXX XXXX XXXX XXXX XXXX
41 XXXX XXXX XXXX XXXX XXXX XXXX XXXX XXXX
42 XXXX XXXX XXXX XXXX XXXX XXXX XXXX XXXX
43 XXXX XXXX XXXX XXXX XXXX XXXX XXXX XXXX
44 XXXX XXXX XXXX XXXX XXXX XXXX XXXX XXXX
45 XXXX XXXX XXXX XXXX XXXX XXXX XXXX XXXX
46 XXXX XXXX XXXX XXXX XXXX XXXX XXXX XXXX
47 XXXX XXXX XXXX XXXX XXXX XXXX XXXX XXXX

48 XXXX XXXX XXXX XXXX XXXX XXXX XXXX XXXX
49 XXXX XXXX XXXX XXXX XXXX XXXX XXXX XXXX
50 XXXX XXXX XXXX XXXX XXXX XXXX XXXX XXXX
51 XXXX XXXX XXXX XXXX XXXX XXXX XXXX XXXX
52 XXXX XXXX XXXX XXXX XXXX XXXX XXXX XXXX
53 XXXX XXXX XXXX XXXX XXXX XXXX XXXX XXXX
54 XXXX XXXX XXXX XXXX XXXX XXXX XXXX XXXX
55 XXXX XXXX XXXX XXXX XXXX XXXX XXXX XXXX

56 XXXX XXXX XXXX XXXX XXXX XXXX XXXX XXXX
57 XXXX XXXX XXXX XXXX XXXX XXXX XXXX XXXX
58 XXXX XXXX XXXX XXXX XXXX XXXX XXXX XXXX
59 XXXX XXXX XXXX XXXX XXXX XXXX XXXX XXXX
60 XXXX XXXX XXXX XXXX XXXX XXXX XXXX XXXX
61 XXXX XXXX XXXX XXXX XXXX XXXX XXXX XXXX
62 XXXX XXXX XXXX XXXX XXXX XXXX XXXX XXXX
63 XXXX XXXX XXXX XXXX XXXX XXXX XXXX XXXX
```

└─Product term

Figure C.7
PALASM2 Fuse Map
(continued)

TOTAL FUSES BLOWN: 122

PALASM XPLOT, V2.23 - MARKET RELEASE (2-1-88)
(C) - COPYRIGHT MONOLITHIC MEMORIES INC, 1988

Title	:	Sample_equation	Author	:	Prestopnik
Pattern	:	Active_high_output_TT	Company	:	FMCC
Revision	:	01	Date	:	8/20/88

PAL16L8
SIMPLE_EQ*
QP20* ⟵———— Total pins
QF2048* ⟵———————— Total fuses on device
G0*F0*
L0256 11111111111111111111111111111111*
L0288 11110111111110111111111111111111*
L0320 10111111101111111111111111111111*
L0352 10111011111111111111111111111111*
C0F9A*
2A0D

                ⌐ Unprogrammed fuse (0)   ⌐Programmed fuse (1)

**Figure C.8
JEDEC Fuse Map**

Figure C.7 is a standard fuse plot that is generated not only by PALASM2, but also by all PAL development software. Its organization shows how the PAL AND array is programmed. Each X on the map indicates an intact fuse; each dash (–) shows a blown or programmed fuse. The numbers listed vertically on the left represent AND gate inputs while the numbers listed along the top represent the PAL input lines. These numbers correspond to those listed on the manufacturers' data sheets. Using these numbers and a PAL data sheet, you can verify the logical circuit created by the PAL development software. Notice how the fuse plot indicates that the majority of the fuses are intact. Since our design is for a small AND-OR network, this makes sense—most of the PAL circuitry is unused.

The JEDEC fuse file is eventually transferred to a PAL programmer when the PAL device is physically programmed. This file is a condensed version of the fuse plot and is not typically used by the designer for verification.

Determine from the fuse plot listing in Figure C.7:
(a) the logic expression
(b) the logic circuit
(c) if the logic expression matches the truth table from Figure C.6

Solution The fuse plot indicates how fuses were blown based on the design input file logic equation. Remember that PALASM2 minimizes the equation so that a simplified circuit is likely. Use a 16L8 data sheet along with the fuse plot to determine the actual circuit.

**DESIGN
EXAMPLE C.4**

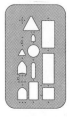

(a) Reproduced below is a portion of the fuse plot:

These inputs assigned in input design file, all others are unused

$\overline{AA}$ $\overline{BB}$ $\overline{CC}$ $\overline{DD}$

	0123	4567	11 8901	1111 2345	1111 6789	2222 0123	2222 4567	2233 8901	
8	----	----	----	----	----	----	----	----	
9	----	X---	----	-X--	----	----	----	----	$= \overline{BD}$
10	-X--	----	-X--	----	----	----	----	----	$= \overline{AC}$
11	-X--	-X--	----	----	----	----	----	----	$= \overline{AB}$

AND gate inputs

Input lines

This AND enables the tristate output buffer.

Since these product terms feed an OR gate, the resulting equation is $\overline{Y} = \overline{BD} + \overline{AC} + \overline{AB}$. This entire equation is inverted after the OR gate to produce Y.

(b) Circuit:

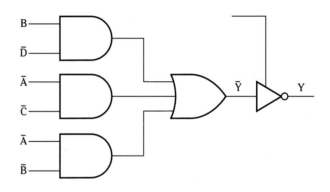

(c) From the truth table of Figure C.6:

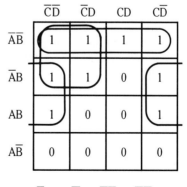

	$\overline{CD}$	$\overline{C}D$	CD	$C\overline{D}$
$\overline{AB}$	1	1	1	1
$\overline{A}B$	1	1	0	1
AB	1	0	0	1
$A\overline{B}$	0	0	0	0

$$\overline{Y} = \overline{BD} + \overline{AC} + \overline{AB}$$

PALASM2 automatically reduced the equation from the design file.

SIMULATION

It is relatively quick and easy to generate a fuse plot using PALASM2, but this does not ensure by any stretch of the imagination that the design is correct. Simulation tests should be run on every design to verify that the circuit functions as expected. Simulation commands are entered in the **Simulation section** of the input file immediately following the circuit logic equations. The keyword "SIMULATION" identifies the simulation commands in the design input file. After running the simulation software, you can analyze the output timing diagram and timing history created from the simulation run.

Figure C.9 shows the PALASM2 simulation commands. The commands allow the designer to initialize input and flip-flop values, change input levels, check for output responses, and do conditional testing. A sequence of simulation commands are referred to as **simulation vectors.** The following examples of some key simulation commands are presented to illustrate their usefulness.

Description of Simulation Commands

Command	Description
PRLDF	Initializes register outputs on preloadable devices
SETF	Specifies new input values
CLOCKF	Generates a clock signal on the dedicated clock pin
CHECK	Verifies that the expected values and the simulated values are the same
TRACE_ON	Defines specific signals to record in a special output file
TRACE_OFF	Turns off the TRACE_ON command.
FOR . . . TO . . . DO loop	Iterates a set of commands a fixed number of times
WHILE . . . DO loop	Iterates a set of commands until a condition is satisfied
IF . . . THEN . . . ELSE	Conditional branching

Figure C.9
Simulation Commands

- PRLDF (preload)—This command is used to set registered device outputs to a known logic level. For instance, to initialize an output pin named STOP to a high level and an output pin named GO to a low logic level, the PRLDF simulation command is written as

PRLDF STOP /GO

  An uncomplemented output name is initialized high, whereas a complemented output name is initialized low. In the input file Simulation section the slash has nothing to do with the actual signal name; it merely indicates that a low logic level is desired for the designated pin. This convention is used with many of the simulation commands.

- SETF—SETF applies logic levels to input pins. To apply a high level to an input listed as A and a low level to an input listed as B, SETF is written as

SETF A /B

- CLOCKF—This command defines the action of flip-flop clock lines. Each CLOCKF command produces a simulation clock pulse that undergoes a low to high and then a high to low transition (positive going edge followed by a negative going edge). The pin name of the clock line follows the clock command. In addition, before using CLOCKF for the first time, PALASM2 requires that the clock signal be initialized low with the SETF command. For example, if clock signal CLK2 is to be pulsed for the first time in the simulation run, the following simulation commands would accomplish the task:

SETF /CLK2
CLOCKF CLK2

- TRACE_ON and TRACE_OFF—PALASM2 automatically generates a full set of timing diagrams based on the simulation results. The inputs and outputs are listed on the timing diagram in the order that they appear on the input file pin list. Using typical PAL chips (20 pins), the timing results will occupy several display screens. In contrast, the TRACE_ON and TRACE_OFF commands allow the designer to select specific signals for viewing. This, of course, is much more convenient than viewing everything at one time. The two commands can be used anywhere throughout the simulation run as desired. To view selectively input and output lines A, B, and Y from a 20-pin PAL simulation run, we enter the commands as follows:

TRACE_ON A B Y
other commands
TRACE_OFF

If the input or output name is preceded by a slash, then the complement waveform is displayed. The waveforms can be observed using function key F7. Several viewing options are available. Simulation history and history waveform displays give complete simulation results. Simulation trace and trace waveform presents only the TRACE_ON/TRACE_OFF information. Both history waveform and trace waveform are typical timing diagram displays, whereas simulation history and trace history displays contain the same waveform information, but are listed as a pattern of high (H), low (L), Hi-Z (Z), and undefined (X) symbols. Simulation history and trace history files can be printed out for analysis.

Armed with an understanding of some of the simulation commands, we will now examine the simulation run for the Figure C.5 example. The complete input file, including simulation commands, is shown in Figure C.10. Since the circuit is a combinatorial network, all truth table inputs are specified using the SETF command. The simulation history output is shown in Figure C.11. (The waveform display conveys similar information.)

TITLE	Sample_equation								
PATTERN	Active_high_output_TT								
REVISION	01								
AUTHOR	Prestopnik								
COMPANY	FMCC								
DATE	8/20/88								
CHIP	Simple_eq PAL16L8								
;PINS	1	2	3	4	5	6	7	8	9	10
	NC	A	B	C	D	NC	NC	NC	NC	GND
;PINS	11	12	13	14	15	16	17	18	19	20
	NC	NC	NC	NC	NC	NC	NC	/Y	NC	VCC

EQUATIONS

$Y = /(/A*B*C*D + A*/B*/C*/D + A*/B*/C*D + A*/B*C*/D + A*/B*C*D + A*B*/C*D + A*B*C*D)$

SIMULATION

SETF /A /B /C /D ←——— Tests ABCD for input combination 0000
SETF /A /B /C D
SETF /A /B C /D
SETF /A /B C D
SETF /A B /C /D
SETF /A B /C D
SETF /A B C /D
SETF /A B C D
SETF A /B /C /D ←——— Tests ABCD for input combination 1000
SETF A /B /C D
SETF A /B C /D
SETF A /B C D
SETF A B /C /D
SETF A B /C D
SETF A B C /D
SETF A B C D

Figure C.10
PALASM2 Printout
with Simulation
Commands

PALASM SIMULATION, V2.23 - MARKET RELEASE (2-1-88)
(C) - COPYRIGHT MONOLITHIC MEMORIES INC, 1988
PALASM SIMULATION HISTORY LISTING

Title	:	Sample_equation	Author	:	Prestopnik
Pattern	:	Active_high_output_TT	Company	:	FMCC
Revision	:	01	Date	:	8/20/88

PAL16L8
SIMPLE_EQ ⟶ Input combination ABCD = 0000
                  That is, SETF /A /B /C /D
Page : 1

```
         g g g g g g g g g g   g g g g g g  ←— Simulation commands where g = SETF
A       (L) L L L L L L L HH   H H H H H H
B        L L L L HHHHL L       L L HHHH
C        L L HHL L HHL L       HHL L HH
D       (L) HL HL HL HL H      L HL HL H
GND      L L L L L L L L L L   L L L L L L
/Y      (L) L L L L L L HHH    HHL HL H ←— Matches truth table from
VCC     (HHHHHHHHHH)          HHHHHH       Figure C.6 → OK to program PAL
```

Pin names

Output = low

Figure C.11
PALASM2 History
Listing

The simulation commands listed in Figure C.11 indicate that the circuit is tested for all input combinations. Inputs A, B, C, and D are first set to all zeros (SETF /A /B /C /D). The simulation history listing (Figure C.11) shows that the output is low for this input combination. Since this is the expected output, the circuit is correct for this input combination. Further analysis of the simulation history shows that the circuit functions correctly for every input combination. The PAL can now be physically programmed.

**DESIGN
EXAMPLE C.5**

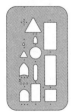

Write the simulation commands for the PAL circuit of Figure C.6 so that inputs A, B, C, and D are tested for input combinations 1101, 1110, 1111. Also supply trace-on commands to compare input B with output /Y.

Solution   Use SETF commands to set input logic levels. Use TRACE_ON and TRACE_OFF to compare signals on specific pins. Thus

    TRACE_ON B, /Y
    SETF A B /C D
    SETF A B C /D
    SETF A B C D
    TRACE_OFF

PROGRAMMING THE DEVICE

Programming is accomplished using the PALASM2 F4 option. This option switches control from PALASM2's main menu to a communications software package of your choice. Typically, you will use the programming software supplied with the PAL programmer attached to your computer. Programming is straightforward. The PAL chip is placed into the PAL programmer socket, and, using the communications software, the JEDEC fuse information is transferred from the computer to the programmer. The communications software for your PAL programmer may offer several options, such as blank part testing, that may be run during this portion of the process. The chip is then burned by the PAL programmer, and if simulation was run, a functional test also takes place. The customized PAL chip is now ready for use—welcome to the world of ASIC!

REGISTERED OUTPUT DEVICES

Registered PAL chips have flip-flops or latches feeding the output pins and they are designated in the input file using special symbols. For instance, the equation $Y = ABC + DEF$ implies a simple combinatorial circuit with output Y. If the equation read $Y := ABC + DEF$, the output Y is actually the output of a D flip-flop. (A latch would be specified as $Y * = ABC + DEF$.) The SOP expression feeds the D input of the flip-flop while Y comes from the flip-flop Q output. These two examples are summarized in Figure C.12.

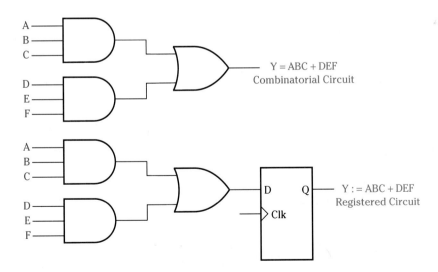

Figure C.12
PAL Output Notation

To gain experience using registered devices, we will complete a PAL circuit design from input file creation through simulation using the following design example.

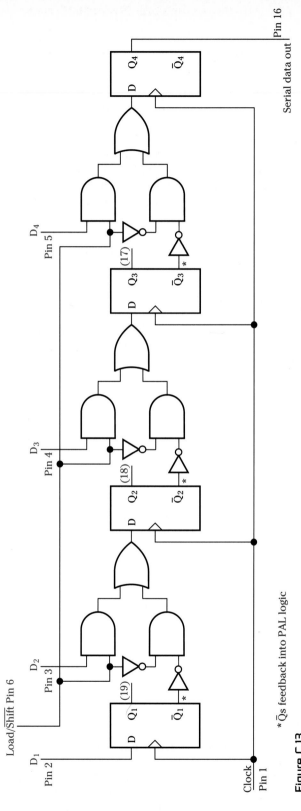

Figure C.13
4-Bit PISO Using a
16R8 PAL

*$\bar{Q}$s feedback into PAL logic

Design a 4-bit parallel-in, serial-out (PISO) shift register using a 16R8 PAL chip.

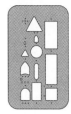

Solution In order to begin, we need a PISO design in mind. Figure C.13 shows a design that fulfills our need. The circuit contains five inputs—four data inputs and a load/shift control signal. When load/shift is high, data is loaded into the four D flip-flops on an active clock pulse. Shifting occurs with each clock pulse when the load/shift line is low. The last flip-flop in the register contains the serial data output line. However, since every flip-flop output in the 16R8 PAL device feeds an output pin, the shift register will also function as a PIPO (parallel-in, parallel-out). Two functions for the price of one.

Figure C.14 shows the Declarations and Equations sections of the input file for this shift register design. Since the 16R8 has registered outputs, the notation (:=) is used in the Equations section. The equations are also listed using the notation for an active-low device. Each equation specifies the conditions under which data is transferred into the flip-flops, either from the parallel inputs or via shifting. For instance, flip-flop Q2 (identified as DT2) receives information at its D input from parallel input D2 when load/shift is high (D2*LDSH) or from flip-flop Q1 when load/shift is low (DT1*/LDSH).

TITLE	Four_bit_shift_reg
PATTERN	01
REVISION	01
AUTHOR	Prestopnik
COMPANY	FMCC
DATE	8/10/88

CHIP	Shiftreg	PAL16R8

;PINS	1	2	3	4	5	6	7	8	9	10
	CLK	D1	D2	D3	D4	LDSH	NC	NC	NC	GND

;PINS	11	12	13	14	15	16	17	18	19	20
	OE	NC	NC	NC	NC	OUT	DT3	DT2	DT1	VCC

EQUATIONS

```
/DT1   :=  /D1
/DT2   :=  /(D2*LDSH + DT1*/LDSH)
/DT3   :=  /(D3*LDSH + DT2*/LDSH)
/OUT   :=  /(D4*LDSH + DT3*/LDSH)
```

Figure C.14
Shift Register
PALASM2 Printout

Inspection of the 16R8 data sheet shows that the $\overline{Q}$ outputs of the 16R8 are wired back into the AND array. These paths will be utilized to connect the individual flip-flops into a shift register string. However, we normally would route the Q signal from one flip-flop to the D input of the next in the shift register string. Since we can only use the $\overline{Q}$ output, we have a difference in level to rectify. The 16R8 takes care of the level difference easily because the flip-flop $\overline{Q}$ output passes through an inverter/buffer before connecting with the AND array. PALASM2 makes the appropriate inverted connection between the $\overline{Q}$ output and the AND gate automatically, converting $\overline{Q}$ to Q. Figure C.15 shows a portion of the 16R8 data sheet and the PALASM2 fuse plot to illustrate this point. Basically, this means that equations can be defined with outputs taken from either Q or $\overline{Q}$ without regards to the inversion.

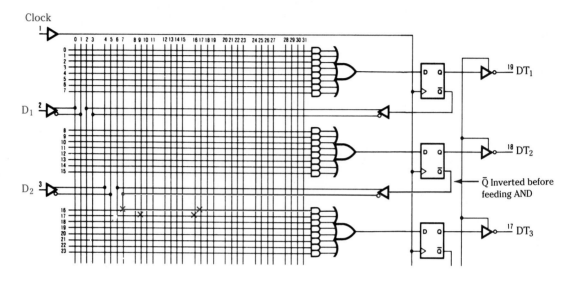

PALASM XPLOT, V2.23 - MARKET RELEASE (2-1-88)
(C) - COPYRIGHT MONOLITHIC MEMORIES INC, 1988

Title	:	Four_bit_shift_reg	Author	:	Prestopnik
Pattern	:	01	Company	:	FMCC
Revision	:	01	Date	:	8/10/88

PAL16R8
SHIFTREG

			11	1111	1111	2222	2222	2233
	0123	4567	8901	2345	6789	0123	4567	8901
0	–X––	––––	––––	––––	––––	––––	––––	––––
1	XXXX	XXXX	XXXX	XXXX	XXXX	XXXX	XXXX	XXXX
2	XXXX	XXXX	XXXX	XXXX	XXXX	XXXX	XXXX	XXXX
3	XXXX	XXXX	XXXX	XXXX	XXXX	XXXX	XXXX	XXXX
4	XXXX	XXXX	XXXX	XXXX	XXXX	XXXX	XXXX	XXXX
5	XXXX	XXXX	XXXX	XXXX	XXXX	XXXX	XXXX	XXXX
6	XXXX	XXXX	XXXX	XXXX	XXXX	XXXX	XXXX	XXXX
7	XXXX	XXXX	XXXX	XXXX	XXXX	XXXX	XXXX	XXXX

Figure C.15
Fuse Map—PAL
Comparison

```
 8  —X—  ————  ————  ————  —X—   ————  ————  ————
 9  ————  —X—   ————  ————  X—    ————  ————  ————
10  XXXX  XXXX  XXXX  XXXX  XXXX  XXXX  XXXX  XXXX
11  XXXX  XXXX  XXXX  XXXX  XXXX  XXXX  XXXX  XXXX
12  XXXX  XXXX  XXXX  XXXX  XXXX  XXXX  XXXX  XXXX
13  XXXX  XXXX  XXXX  XXXX  XXXX  XXXX  XXXX  XXXX
14  XXXX  XXXX  XXXX  XXXX  XXXX  XXXX  XXXX  XXXX
15  XXXX  XXXX  XXXX  XXXX  XXXX  XXXX  XXXX  XXXX

16  ————  —X—   ————  ————  —X—   ————  ————  ————
17  ————  ————  —X—   ————  X—    ————  ————  ————
18  XXXX  XXXX  XXXX  XXXX  XXXX  XXXX  XXXX  XXXX
19  XXXX  XXXX  XXXX  XXXX  XXXX  XXXX  XXXX  XXXX
20  XXXX  XXXX  XXXX  XXXX  XXXX  XXXX  XXXX  XXXX
```

This fuse connects $\overline{Q_2}$ through an inverter to the AND-OR logic feeding the D of Q_3.

Figure C.15
Fuse Map—PAL
Comparison
(continued)

The simulation run for the shift register is shown in Figure C.16 and the trace history diagram confirming proper operation is shown in Figure C.17.

```
SIMULATION
TRACE_ON CLK LDSH /DT1 /DT2 /DT3 /OUT
SETF /OE  ←———————— Enables tristate outputs
SETF /CLK  ←———————— Initial setting of clock
SETF D1 /D2 /D3 /D4 LDSH  ←———————— Set parallel inputs = 1000 and LDSH to load
CLOCKF CLK  ←———————— Load 1000 into register
SETF /LDSH /D1  ←———————— Set LDSH = shift and D₁ to 0
FOR I := 1 TO 10 DO
        BEGIN
            CLOCKF CLK       10 clock pulses
        END
TRACE_OFF
```

Figure C.16
Shift Register
Simulation Commands

The simulation commands load a 1000 pattern into the shift register using SETF. Input LDSH (load/shift) is high during this time. A single clock pulse actually accomplishes the load. Then LDSH is brought low to enable shifting. Data input D1 is also brought low at this time so that zeros fill the shift register as data is shifted out. Shifting takes place on each positive going clock edge. Notice also how a FOR..TO..DO loop (one of the simulation commands) easily creates ten clock pulses—much simpler than using ten separate CLOCKF statements.

The trace history diagram shows how the data pattern is shifted through the shift register. A thorough simulation sequence would test shifting and loading with a variety of test patterns to verify fully all aspects of shift register operation. Note that the history display shows complement output levels since

PALASM SIMULATION, V2.23 - MARKET RELEASE (2-1-88)
(C) - COPYRIGHT MONOLITHIC MEMORIES INC, 1988
PALASM SIMULATION SELECTIVE TRACE LISTING

Title	: Four_bit_shift_reg	Author	: Prestopnik
Pattern	: 01	Company	: FMCC
Revision	: 01	Date	: 8/10/88

PAL16R8
SHIFTREG
Page : 1

```
         g g g   c g   c     c     c     c     c c c c   c  ←——Simulation commands  g = SETF
  CLK    X L L HH L L HH L   HH L HH L HH L H   L HL HL HL HL H  L                   c = CLOCK
  LDSH   X X HH HH L L L L   L L L L L L L L L   L L L L L L L L L  L                p = PRLDF
 /DT1    X X X X L L L L HH  HH HH HH HH HH HH   HH HH HH HH HH HH  H
 /DT2    X X X X HH HH L L   L HH HH HH HH HH HH  HH HH HH HH HH HH  H
 /DT3    X X X X HH HH HH HH HL L L HH HH HH HH  HH HH HH HH HH HH  H
 /OUT    X X X X HH HH HH HH HH HH HH L L HH HH   HH HH HH HH HH HH  H
```

Pin names

Shifting

Load register with 1000
(Complement values seen here)

Figure C.17
Shift Register History
Listing

the 16R8 is an active-low output device. Using trace commands, you can obtain a complement display to make analysis easier.

A final note illustrating the compactness of PALs: Building this shift register with basic TTL parts would require the following: 2-7474, 2-7408, 1-7404, 1-7432. The logical function of six TTL chips is reduced to one PAL chip. (And the PAL chip is not even completely utilized.)

As you can see, programmable logic developed with computer aided design (CAD) software greatly simplifies and speeds up the design process. Although we have only discussed programmable logic design using basic PAL parts, PALASM2 supports the design of much more complex PALs and logic sequencers. This appendix also provides insight into the future of logic design. Computer assisted design is a crucial element in future logic circuit development. Software will constantly change to meet the needs of newer and more dense technologies—a cost-efficient way to meet new design challenges. The software's use as a design tool requires that all designers become proficient with computers and CAD systems as well as with the fundamentals of logic design.

1. Create the PALASM2 Declarations "CHIPS" section by assigning input variables SW_1, SW_2, SW_3, SW_4, SW_5 to input pins 2 through 6 of a PAL16L8. Assign inputs Output Enable, Reset, and Test1 to pins 7, 8, and 9. Assign inputs ACK, INT, and HLT to pins 11, 13, and 14. Outputs X_0 through X_4 are assigned to pins 15 through 19.
2. Write the following SOP expressions using PALASM2 symbols for AND, OR, and inversion:

   *(a) $\overline{Q}\overline{R}\overline{S}T + \overline{Q}R\overline{S}T + Q\overline{R}\overline{S}T + QRST$

   (b) $\overline{START}$ STOP $\overline{GO}$ + $\overline{START}$ STOP GO + START STOP $\overline{GO}$
3. Show how the following expressions are written in PALASM2 for an active-high output PAL device when the pin definition for the output is TIMER_ENABLE:

   (a) $AB\overline{C}$ (b) $\overline{A}B\overline{C} + A\overline{B}C$ (c) $WX\overline{Y}Z + \overline{W}Z$
4. Show how the following expressions are written in PALASM2 for an active-high output PAL device when the pin definition for the output is $\overline{TIMER\_ENABLE}$:

   *(a) $AB\overline{C}$ (b) $\overline{A}B\overline{C} + A\overline{B}C$ (c) $WX\overline{Y}Z + \overline{W}Z$
5. Show how the following expressions are written in PALASM2 for an active-low output PAL device when the pin definition for the output is TIMER_ENABLE:

   *(a) $AB\overline{C}$ (b) $\overline{A}B\overline{C} + A\overline{B}C$ (c) $WX\overline{Y}Z + \overline{W}Z$
6. Show how the following expressions are written in PALASM2 for an active-low output PAL device when the pin definition for the output is $\overline{TIMER\_ENABLE}$:

   (a) $AB\overline{C}$ (b) $\overline{A}B\overline{C} + A\overline{B}C$ (c) $WX\overline{Y}Z + \overline{W}Z$
7. Take advantage of the complement method and write the PALASM2 expressions for the following truth table functions so that they can be implemented on an active-low output PAL device.

*(a)

A	B	C	Y
0	0	0	0
0	0	1	1
0	1	0	1
0	1	1	0
1	0	0	0
1	0	1	1
1	1	0	0
1	1	1	1

(b)

W	X	Y	T
0	0	0	0
0	0	1	0
0	1	0	0
0	1	1	1
1	0	0	0
1	0	1	1
1	1	0	0
1	1	1	0

(c)

D	C	B	A	Z
0	0	0	0	1
0	0	0	1	1
0	0	1	0	1
0	0	1	1	0
0	1	0	0	0
0	1	0	1	1
0	1	1	0	0
0	1	1	1	1
1	0	0	0	0
1	0	0	1	1
1	0	1	0	1
1	0	1	1	0
1	1	0	0	0
1	1	0	1	0
1	1	1	0	1
1	1	1	1	1

---

* See Appendix F: Answers to Selected Problems.

*8. What logic function is created according to the following 16L8 fuse plot? Assume that input pins 2, 3, and 4 are assigned variables X, Y, and Z, respectively. The output OUT is assigned to pin 16.

	0	4	8	12	16	20	24	28
	⋮	⋮	⋮	⋮	⋮	⋮	⋮	⋮
23	XXXX	XXXX	XXXX	XXXX	XXXX	XXXX	XXXX	XXXX
24	-----	-----	-----	-----	-----	-----	-----	-----
25	-X--	X---	X---	-----	-----	-----	-----	-----
26	X---	-X--	X---	-----	-----	-----	-----	-----
27	X---	X---	X---	-----	-----	-----	-----	-----
28	XXXX	XXXX	XXXX	XXXX	XXXX	XXXX	XXXX	XXXX
29	XXXX	XXXX	XXXX	XXXX	XXXX	XXXX	XXXX	XXXX
30	XXXX	XXXX	XXXX	XXXX	XXXX	XXXX	XXXX	XXXX
31	XXXX	XXXX	XXXX	XXXX	XXXX	XXXX	XXXX	XXXX
32	XXXX	XXXX	XXXX	XXXX	XXXX	XXXX	XXXX	XXXX

9. Show how the SETF simulation command could be used to set variables W, X, Y, and Z to the following levels:

*(a) 0010   (b) 0101   (c) 1101   (d) 1111

10. Write the simulation commands to show how variables TEST1, TEST2, and ACK can be tested for input combinations 000, 010, 100, and 110 during simulation. Show how these specific tests can be seen in the trace history file.

11. Write the following SOP expressions if the PAL output X is taken from a D flip-flop:

*(a) $\overline{Q}\overline{R}\overline{S}\overline{T} + \overline{Q}R\overline{S}T + Q\overline{R}\overline{S}\overline{T} + QRST$

(b) $\overline{START}\ STOP\ \overline{GO} + \overline{START}\ STOP\ GO + START\ STOP\ \overline{GO}$

12. Write the following SOP expressions if the PAL output X is taken from a D latch:

(a) $\overline{Q}\overline{R}\overline{S}\overline{T} + \overline{Q}R\overline{S}T + Q\overline{R}\overline{S}\overline{T} + QRST$

(b) $\overline{START}\ STOP\ \overline{GO} + \overline{START}\ STOP\ GO + START\ STOP\ \overline{GO}$

13. The simulation run in Figure C.16 for the 4-bit shift register does not fully test the circuit. Add simulation vectors to test the following conditions:

(a) Loading is inhibited when the LDSH line is low.

(b) The bit pattern 1101 successfully shifts through the circuit.

(c) As data is shifted out of the register, either a high or low level shifts in to replace the data.

14. Using PALASM2 or similar PAL programming software, as well as a suitable word processing package, create the logic input file to implement:

$$Y = \overline{A}\overline{B}\overline{C}\overline{D}EFG + \overline{A}BC\overline{D}EFG + \overline{A}BC\overline{D}EFG + A\overline{B}\overline{C}D\overline{E}FG + AB\overline{C}DEF\overline{G}$$

15. Run PALASM2 to process the input file. Use a 16L8 PAL. Check the fuse map output and verify that proper fuses have been blown.

---

* See Appendix F: Answers to Selected Problems.

16. Write the simulation vectors to test that the circuit developed in problem 14 produces a high output for every product term in the equation.
17. Implement the following truth table using PALASM2 and a PAL16L8. Simulate completely.

W	X	Y	Z	A
0	0	0	0	0
0	0	0	1	1
0	0	1	0	0
0	0	1	1	0
0	1	0	0	0
0	1	0	1	1
0	1	1	0	1
0	1	1	1	0
1	0	0	0	0
1	0	0	1	1
1	0	1	0	0
1	0	1	1	1
1	1	0	0	0
1	1	0	1	0
1	1	1	0	1
1	1	1	1	1

18. Use a programmable logic device to create a circuit that will detect the presence of an illegal BCD value. When an illegal value is detected, a flip-flop should be set. A separate flip-flop reset should also be provided to clear out the system. Use a 16R8 and assume a clocking signal is available.
19. Create a MOD 8 synchronous counter using a 16R8. Simulate the design to verify proper operation.

STATE MACHINE DESIGN USING PALASM2

USING PROGRAMMABLE LOGIC HARDWARE AND SOFTWARE TO CREATE STATE MACHINE DESIGNS

Having worked through several state machine designs in Chapter 13, we know that one point is clear—it is a lot of work! Programmable logic, and more importantly, the software supporting the logic can make a significant difference in state machine design. Using programmable logic as a state machine design approach can significantly reduce the design effort. We will confine our programmable logic designs to PAL devices in this appendix, however; later sections will describe how advances in programmable logic technology are dramatically extending state machine design capabilities.

PALASM2 software can greatly assist in state machine development. It is quite obvious that we can take equations developed from previous state machine design techniques and enter them as Boolean equations using PALASM2. The software would process these statements and create the desired circuit. However, this approach does not reduce the design work nor shorten the time spent creating transition tables and K-maps. PALASM2 as well as most programmable logic software packages offer an alternative. Using the software, **state equations** describing the desired hardware function are entered in the design input file. The software creates reduced logic equations, which are verified with subsequent simulation runs. When the design objectives are satisfied, the PAL chip is programmed.

PALASM2 and State Machine Design

Using PALASM2 to create a state machine design is similar to producing any other PAL design. First, an input file, starting with the familiar Declarations section (discussed in Appendix C), is required to describe the logic. Following the declarations, a **State section** identifies the machine states, state equations, and output equations. Next, a Conditions section lists the input conditions causing transitions to occur.

PALASM2 State Section

The State section of the PALASM2 input file begins with the keyword STATE. This section contains three components that spell out the desired state machine behavior:

1. **Global defaults**   Defaults are simply statements in the machine description that ease design work. In a typical design a state that the state machine should enter when all else fails is designated. This guards against the possibility that the system will "lock up" or fail to function if a glitch accidentally forces the machine into an unassigned state. Rather than specifying this "recovery" state in each state equation, a single default statement accomplishes the same thing. Other typical defaults include the machine type (Mealy or Moore) and default output levels.

2. **State assignments**   State assignments associate output pin levels with state names. For instance, a state identified as S_0 exists when outputs Y_1, Y_2, and $Y_3 = 010$. A PALASM2 state assignment for this condition is

$$S_0 = /Y1*Y2*/Y3$$

3. **State and output equations**   State equations define the conditions causing a transition to occur and are written in the following form:

State_Name := Condition1 → Next_State + . . .

+ ConditionN → Next_State +→ Default_State

(The state name is defined previously under state assignments; each state name has a corresponding state equation. The conditions defining transitions are determined from input levels and/or other states.) The state equation previously defined can be verbally described to read as follows: From the current state proceed to the next state if condition1 is met, or proceed to a different next state if conditionN is met; otherwise go to the default state. The number of conditions and states is unlimited, so a state equation carries enormous design power. (The default state condition can be eliminated from the state equation if the condition was previously specified as a default state in the Global Defaults section.)

Output equations define how the state machine outputs react. If machine states and output levels are the same, such as in a counter, then output equations are unnecessary. When state and output levels differ, the output equations define the relationship. Output equations have several different forms depending on the design:

a. Registered Mealy outputs:

State_Name.OUTF := Condition1 → Outputs . . .

+ ConditionN → Outputs +→Default

b. Combinatorial Mealy outputs:

State_Name.OUTF = Condition1 → Outputs ...
$$+ \text{ConditionN} \rightarrow \text{Outputs} +\rightarrow \text{Default}$$

c. Registered Moore outputs:

$$\text{State\_Name.OUTF} := \text{Outputs}$$

d. Combinatorial Moore outputs:

$$\text{State\_Name.OUTF} = \text{Outputs}$$

PALASM2 Conditions Section

The information in the Conditions section uniquely identifies an input or groups of inputs. Condition names used in this section were previously identified in the State section as part of the state and output equations. The condition names usually identify the specific input combinations that cause a transition from one state to another. Condition statement syntax is

$$\text{Condition\_Name} = \text{Input1} * \text{Inputx} \ldots + \text{Inputy} * \text{Inputn}.$$

That is

$$\text{Flag\_Up} = \text{Sun} * \text{Morning} + \text{Special\_Occasion}$$

CREATING A PALASM2 STATE MACHINE DESIGN

Figure D.1 shows the input file describing the synchronous 4-bit counter designed in Chapter 13. However, the counter design is now based on state equations rather than on Boolean equations. This allows us to avoid much of the state machine logical design work and gives us the opportunity to concentrate solely on the state machine behavioral design. PALASM2 does the actual circuit design.

TITLE	Four_bit_synchronous_counter									
PATTERN	Fourcntr.pds									
REVISION	01									
AUTHOR	Prestopnik									
COMPANY	FMCC									
DATE	8/27/88									
CHIP	Counter PAL16R8									
;PINS	1	2	3	4	5	6	7	8	9	10
	CLK	NC	NC	NC	NC	NC	NC	NC	NC	GND
;PINS	11	12	13	14	15	16	17	18	19	20
	OE	NC	NC	NC	NC	/Q1	/Q2	/Q3	/Q4	VCC

Declarations section

Figure D.1
Input File for a
Synchronous Counter
Using State Equations

STATE
MOORE_MACHINE
DEFAULT_BRANCH S0

;STATE ASSIGNMENTS   STATES ARE THE SAME AS THE OUTPUTS

S0=/Q4*/Q3*/Q2*/Q1 ⎫
S1=/Q4*/Q3*/Q2*Q1 ⎪
S2=/Q4*/Q3*Q2*/Q1 ⎪
S3=/Q4*/Q3*Q2*Q1 ⎪
S4=/Q4*Q3*/Q2*/Q1 ⎪
S5=/Q4*Q3*/Q2*Q1 ⎪
S6=/Q4*Q3*Q2*/Q1 ⎪
S7=/Q4*Q3*Q2*Q1 ⎬ State assignments
S8=Q4*/Q3*/Q2*/Q1 ⎪
S9=Q4*/Q3*/Q2*Q1 ⎪
S10=Q4*/Q3*Q2*/Q1 ⎪
S11=Q4*/Q3*Q2*Q1 ⎪
S12=Q4*Q3*/Q2*/Q1 ⎪
S13=Q4*Q3*/Q2*Q1 ⎪
S14=Q4*Q3*Q2*/Q1 ⎪
S15=Q4*Q3*Q2*Q1 ⎭

State
section

S0 := VCC→S1 ⎫
S1 := VCC→S2 ⎪
S2 := VCC→S3 ⎪
S3 := VCC→S4 ⎪
S4 := VCC→S5 ⎪
S5 := VCC→S6 ⎪
S6 := VCC→S7 ⎪
S7 := VCC→S8 ⎬ State equations
S8 := VCC→S9 ⎪
S9 := VCC→S10 ⎪
S10 := VCC→S11 ⎪
S11 := VCC→S12 ⎪
S12 := VCC→S13 ⎪
S13 := VCC→S14 ⎪
S14 := VCC→S15 ⎪
S15 := VCC→S0 ⎭

←———No output equations required since outputs and states are the same.
←———No conditions section since the only condition is the predefined V_{cc}.

SIMULATION

PRLDF /Q4 /Q3 /Q2 /Q1   ←—— All flip-flops initialized to zero
SETF /OE ←—— Enable outputs
SETF /CLK ←—— Initialize clock
FOR J := 1 TO 25 DO
    BEGIN
        CLOCKF CLK ⎬ 25 clock pulses
    END

Figure D.1
Input File for a
Synchronous Counter
Using State Equations
(continued)

The input file Declarations section contains basic design information and the chip pin assignments. A PAL16R8 is used in this synchronous counter example, although it contains more flip-flops than are actually needed. Since the 4-bit counter contains 16 states, a device with only four flip-flops would suffice. Generally, the selection of the appropriate programmable logic device begins by estimating device capabilities such as the number of flip-flops, output polarity, I/O pin requirements, and other fundamental logic concerns. Many programmable logic devices are available specifically to assist with state machine design and may be preferred over a basic PAL device. Even so, simple state machine design is easily carried out with inexpensive PAL chips.

The State section identifies that this design is a Moore machine. This designation is selected since outputs are taken directly from the present state register, the counter in this case, and not created with additional logic and input signals. Recall that we identified the counter states as $S_0, S_1, \ldots, S_{15}$. The first state, S_0, is assigned as the default state. If the circuit accidentally entered an incorrect state or had input signal problems (it can't in this design since every state is used and there are no inputs), the machine would be forced automatically to the S_0 state. Using the $S_0, S_1, \ldots, S_{15}$ names, state assignments are made for each possible counter state. For instance, state S_2 is identified as the state that occurs when register outputs $Q_4, Q_3, Q_2, Q_1 = 0010$. This state assignment is written in PALASM2 as S2=/Q4*/Q3*Q2*/Q1.

Following the state assignments are the state equations. The state equations define how and when transitions between states take place. For instance, when the circuit is in state S_5, the state equation S5 := VCC→S6 means that upon condition VCC the machine should go to state S_6. VCC is a PALASM2 unconditional transition statement. A state equation containing VCC implies that the next state occurs immediately upon receipt of the next clock pulse. This makes sense in a counter since each clock pulse should move the counter to the next state. That is why every state equation in this example has the VCC condition. When a circuit design changes state on some other input condition, a meaningful name for the condition is given in the state equation. Furthermore, the condition name assigned is fully defined in the input file Conditions section following the state equations. In this example no Conditions section is required since the only transition condition given is the predefined VCC condition.

There is also no need for output equations in this design because the output levels and the state levels are the same. If output equations were necessary, they would be included with the state equations. Rounding out the input file is the Simulation section, which is used to verify design operation. Once the design input file is complete, PALASM2 is put through its paces. The simulation history file is shown in Figure D.2.

From this design it is evident that a working circuit can be designed very easily from just a behavioral description of the state machine. This is much less complex and tedious than the more traditional state machine design approach and has the added benefit that design work is completed quickly.

As a final point, the initial state of the PAL flip-flops (or any flip-flops for that matter) are undetermined at power-up. Some flip-flops turn on high and some turn on low, but the actual starting state is random. If the state machine

PALASM SIMULATION, V2.23 - MARKET RELEASE (2-1-88)
(C) - COPYRIGHT MONOLITHIC MEMORIES INC, 1988
PALASM SIMULATION HISTORY LISTING

Title   : Four_bit_synchronous_couAuthor   : Prestopnik
Pattern   : Fourcntr.pds   Company   : FMCC
Revision   : 01   Date   : 8/27/88

PAL16R8
COUNTER
Page : 1

```
       p g g   c   c      c   c   c      c   c   c   c      c   c   c
CLK    XXL HHL HHL H   HL HHL HHL HH   L HHL HHL HHL   HHL HHL HHL H
GND    LLLLLLLLLL     LLLLLLLLLL      LLLLLLLLLL       LLLLLLLLLL
OE     HLLLLLLLLL     LLLLLLLLLL      LLLLLLLLLL       LLLLLLLLLL
/Q1    HHL L HHHLLL   HHHLLL HHHL     LL HHHHL LL HH   HL LL HHHHL LL
/Q2    HHL L HHHHHH   LLLLLL HHHH     HHL LLLLL LL HH  HHHHL LLLLL
/Q3    HHL L HHHHHH   HHHHHHHL LLL    LLLLLLLL L HH    HHHHHHHHHH
/Q4    HHL L HHHHHH   HHHHHHHHHH      HHHHHHHHL L       LLLLLLLLLL
VCC    HHHHHHHHHH     HHHHHHHHHH      HHHHHHHHHH        HHHHHHHHHH
```

Active-low
output   1111   1110   1101

PAL16R8
COUNTER
Page : 2

```
       c   c   c      c   c   c   c      c   c   c      c   c   c
CLK    HL HHL HHL HH   L HHL HHL HHL   HHL HHL HHL H   HL HHL HHL
GND    LLLLLLLLLL     LLLLLLLLLL      LLLLLLLLLL       LLLLLLLL
OE     LLLLLLLLLL     LLLLLLLLLL      LLLLLLLLLL       LLLLLLLL
/Q1    HHHL LL HHHL   LL HHHL LL HH   HL LL HHHL LL    HHHL LL HH
/Q2    HHHHHHL LLL    LL HHHHHHL L    LLLL HHHHHH      LLLLLL HH
/Q3    LLLLLLLLLL     LL HHHHHHHH     HHHHL LLLLL      LLLLLL HH
/Q4    LLLLLLLLLL     LL HHHHHHHH     HHHHHHHHHH       HHHHHHL L
VCC    HHHHHHHHHH     HHHHHHHHHH      HHHHHHHHHH       HHHHHHHH
```

                    0000 1111   ——— Counter recycles

Figure D.2
Synchronous Counter
History Listing

must have a specific starting state (we were not particular in the counter design), then provisions must be made to ensure that the starting state is attained regardless of the power-on state. In designs that do not utilize every possible flip-flop state, this can be particularly troublesome if the circuit powers-up into one of the unused states. Unless a "path" exists from the unused state to a real state, the state machine will be permanently stuck in the illegal state. Generally, separate clear or reset lines are used to ensure a definite starting state. One of the advantages of more specialized programmable logic chips is the ability to specify a power-on state. PALASM2 has this capability when used with special state machine chips (called PLS and PROSE devices).

Design Example D.1 shows how a state machine can be designed using PALASM2.

Your company just received a government contract to design a rocket launch control sequencer. The circuit must provide the following capabilities and be designed using a PAL16R8:

1. Do nothing—defined as state S_0 (000)
2. Turn on a launch sequence light—defined as state S_1 (001)
3. Retract gantry arm—defined as state S_2 (010)
4. Blow explosive bolts—defined as state S_3 (011)
5. Ignition—defined as state S_4 (100)

Various inputs determined next states:
input X_1—when high, initiates the launch sequence
input X_2—when high, indicates that the gantry arm is completely retracted
input X_3—when high, indicates that the explosive bolts are successfully detonated

Several outputs are required:
output Y_1—when high, turns on launch sequence light
output Y_2—when high, provides gantry arm motion
output Y_3—when high, detonates explosive bolts
output Y_4—when high, provides ignition signal

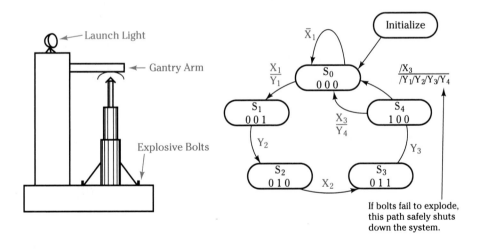

Solution Create an input file describing the system. Use PALASM2 to process the design.

```
TITLE       Rocket_Launch_Sequencer
PATTERN     Rocket.pds
REVISION    01
AUTHOR      Prestopnik
COMPANY     FMCC
DATE        8/28/88
CHIP        Rocket PAL16R8
```

;PINS	1	2	3	4	5	6	7	8	9	10
	CLK	X1	X2	X3	NC	NC	NC	NC	NC	GND

;PINS	11	12	13	14	15	16	17	18	19	20
	OE	/Q1	/Q2	/Q3	NC	Y1	Y2	Y3	Y4	VCC

```
STATE
MEALY_MACHINE
DEFAULT_BRANCH S0
DEFAULT_OUTPUT /Y1, /Y2, /Y3, /Y4
```

$$
\left.\begin{array}{l}
S0 = /Q3*/Q2*/Q1 \\
S1 = /Q3*/Q2*Q1 \\
S2 = /Q3*Q2*/Q1 \\
S3 = /Q3*Q2*Q1 \\
S4 = Q3*/Q2*/Q1
\end{array}\right\} \text{Define states}
$$

$$
\left.\begin{array}{l}
S0 := SEQST{\rightarrow}S1 \\
S1 := VCC{\rightarrow}S2 \\
S2 := RETRACT{\rightarrow}S3 \\
S3 := VCC{\rightarrow}S4 \\
S4 := BLOWN{\rightarrow}S0 \\
\quad + NOBLOW{\rightarrow}S0
\end{array}\right\} \text{State assignments}
$$

$$
\left.\begin{array}{l}
S0.OUTF := SEQST{\rightarrow}Y1*/Y2*/Y3*/Y4 \\
S1.OUTF := VCC{\rightarrow}Y1*Y2*/Y3*/Y4 \\
S2.OUTF := RETRACT{\rightarrow}Y1*Y2*/Y3*/Y4 \\
S3.OUTF := VCC{\rightarrow}Y1*Y2*Y3*/Y4 \\
S4.OUTF := BLOWN{\rightarrow}Y1*Y2*Y3*Y4 \\
\quad + NOBLOW{\rightarrow}/Y1*/Y2*/Y3*/Y4
\end{array}\right\} \text{Output equations (SEQST = sequence start)}
$$

$$
\left.\begin{array}{l}
CONDITIONS \\
SEQST = X1 \\
RETRACT = X2 \\
BLOWN = X3 \\
NOBLOW = /X3
\end{array}\right\} \text{Conditions governing output and states defined above}
$$

SIMULATION

PRLDF /Q1 /Q2 /Q3 /Y1 /Y2 /Y3 /Y4   ← Initializes flip-flops and outputs
SETF /OE   ← Enable outputs
SETF /X1 /X2 /X3   ← All inputs low
SETF /CLK   ← Initialize clock
CLOCKF CLK   ← Clock pulse
SETF X1   ← $X_1 = 1$
CLOCKF CLK ⎫
CLOCKF CLK ⎭ 2 clock pulses
SETF X2   ← $X_2 = 1$
CLOCKF CLK ⎫
CLOCKF CLK ⎭ 2 clock pulses
SETF X3   ← $X_3 = 1$
CLOCKF CLK ⎫
CLOCKF CLK ⎭ 2 clock pulses
;NEXT TEST MISFIRE
SETF /X1 /X2 /X3
SETF /CLK
CLOCKF CLK
SETF X1
CLOCKF CLK
CLOCKF CLK
SETF X2
CLOCKF CLK
CLOCKF CLK
CLOCKF CLK
CLOCKF CLK
;NEXT TEST GANTRY ARM PROBLEM

SETF /X1 /X2 /X3
SETF /CLK
CLOCKF CLK
SETF X1
CLOCKF CLK
CLOCKF CLK
SETF /X2
CLOCKF CLK
CLOCKF CLK
SETF X3
CLOCKF CLK
CLOCKF CLK

PALASM SIMULATION, V2.23 - MARKET RELEASE (2-1-88)
(C) - COPYRIGHT MONOLITHIC MEMORIES INC, 1988
PALASM SIMULATION HISTORY LISTING

Title	:	Rocket_Launch_Sequencer	Author	:	Prestopnik
Pattern	:	Rocket.pds	Company	:	FMCC
Revision	:	01	Date	:	8/28/88

PAL16R8
ROCKET
Page : 1

┌─ Last clock for first test sequence
                                                              ↓

```
        pggg  cg      c   cg   c     cg   c   cg     cg   c    cg ←c = clock;
CLK  XXXLHHLLHH    LHHLLHHLHH    LLHHLHHLLH    HLLHHLHHLL      g = SETF
X1   XXLLLLLLHHH   HHHHHHHHHH    HHHHHHHHLLL   LLHHHHHHHH
X2   XXLLLLLLLLL   LLLLHHHHHH    HHHHHHHHLLL   LLLLLLLLLH
X3   XXLLLLLLLLL   LLLLLLLLLLL   LHHHHHHHLLL   LLLLLLLLLL
GND  LLLLLLLLLL    LLLLLLLLLL    LLLLLLLLLL    LLLLLLLLLL
OE   HLLLLLLLLL    LLLLLLLLLL    LLLLLLLLLL    LLLLLLLLLL
/Q1  HHLLLHHHHH    HHLLLLHHHL    LLLHHHHHHH    HHHHLLLHHH
/Q2  HHLLLHHHHH    HHHHHHHLLLL   LLLHHHHHHH    HHHHHHHLLL
/Q3  HHLLLLLLLH    HHHHHHHHHH    HHHLLLHHHH    HHHHHHHHHH
Y1   LLLLLLLLLL    LLHHHHHHHH    HHHHHHHHHH    LLLLHHHHHH
Y2   LLLLLLLLLL    LLLLLLLHHHH   HHHHHHHHHH    LLLLLLLHHH
Y3   LLLLLLLLLL    LLLLLLLLLL    LLLHHHHHHH    LLLLLLLLLL
Y4   LLLLLLLLLL    LLLLLLLLLL    LLLLLLHHHH    LLLLLLLLLL
VCC  HHHHHHHHHH    HHHHHHHHHH    HHHHHHHHHH    HHHHHHHHHH
```

Complement values { /Q1 /Q2 /Q3

┌─ Outputs low
   Q3Q2Q1 = 000

┌─ Q1 and Y1 go high on clock when X1 = 1

┌─ Q2 and Y2 go high on clock when X2 = 1

┌─ Output values reset

┌─ Last clock = misfire test
    ↓
  ┌─ Gantry arm test

```
 c   c    c      cg   cg   c     c    c    cg     c    c
HHLHHLHHLH    HLLHHLLHHL    HHLHHLHHLL    HHLHHL
HHHHHHHHHH    HHLLLLHHHH    HHHHHHHHHH    HHHHHH
HHHHHHHHHH    HHLLLLLLLL    LLLLLLLLLL    LLLLLL
LLLLLLLLLL    LLLLLLLLLL    LLLLLLLLLH    HHHHHH
LLLLLLLLLL    LLLLLLLLLL    LLLLLLLLLL    LLLLLL
LLLLLLLLLL    LLLLLLLLLL    LLLLLLLLLL    LLLLLL
HLLLHHHHHH    LLLLHHHHHH    HLLLHHHHHH    HLLLHH
LLLLHHHHHH    HHHHLLLLHH    HHHHLLLHHH    HHHHLL
HHHHLLLHHH    HHHHHHHHHH    HHHHHHHHHH    HHHHHH
HHHHHHHLLL    HHHHHHHHHLL   LHHHHHHLLL    LHHHHH
HHHHHHHLLL    LLLLHHHHLL    LLLLHHHLLL    LLLLHH
LLLLHHHLLL    LLLLLLLLLL    LLLLLLLLLL    LLLLLL
LLLLLLLLLL    LLLLLLLLLL    LLLLLLLLLL    LLLLLL
HHHHHHHHHH    HHHHHHHHHH    HHHHHHHHHH    HHHHHH
```

┌─ System moves back to state so outputs reset due to misfire (no X3)

IEEE/ANSI LOGIC SYMBOLS

In recent years a series of standardized logic symbols has been developed with the express purpose of conveying the function of logic devices, ranging from the very simple to the very complex, in a consistent descriptive format. Known formally as the ANSI/IEEE Std 91-1984 and informally as the "new logic symbols," the symbols are fabricated using a simple rectangular box known as an **element outline** around which inputs, outputs, and descriptive information are placed. Figure E.1 shows how common logic gate functions are represented using this standard.

The AND and OR symbols are drawn to show that any number of inputs is possible. The mnemonic symbol within each rectangle shown in the figure denotes the function of the block. These symbols are called **qualifying symbols** and are used to represent the logic capabilities of the device in a manner similar to the way the shape of the older symbols conveyed the logic function of a device. For instance, an & symbol identifies the AND operation. It should be noted that the standard does not eliminate the traditional AND, OR, EX-OR, and inverter symbols; they may be used if desired. However, the newer symbols are preferred.

At this point it is worth explaining why these symbols are useful since there is no meaningful difference between the old and new symbols. The newer symbols may be combined in an arrangement that allows a logic diagram to be simple in format but to provide extensive information regarding the function of complex logic circuitry. For example, there are no symbols using older techniques to represent adequately all the operations in a presettable synchronous counter. The newer symbols allow a descriptive symbol for any complex circuit to be developed. There are more practical reasons as well for using the newer symbols. Many manufacturers have adopted these symbols for their products and included the newer symbols in their data books. In addition, companies working with the federal government are required to use the symbols.

Combining Symbols

Figure E.2 shows how several logic symbols can be grouped into a single symbol. Three 2-input AND gates may be "stacked" together to condense the drawing and form an array. The fact that three individual AND gates are included is made known by maintaining three distinct rectangles in the array. Since all three functions are the same, a single qualifying symbol can be used as shown in E.2c. Only the first AND gate has the qualifying symbol shown; the AND function is implied in the remaining two blocks.

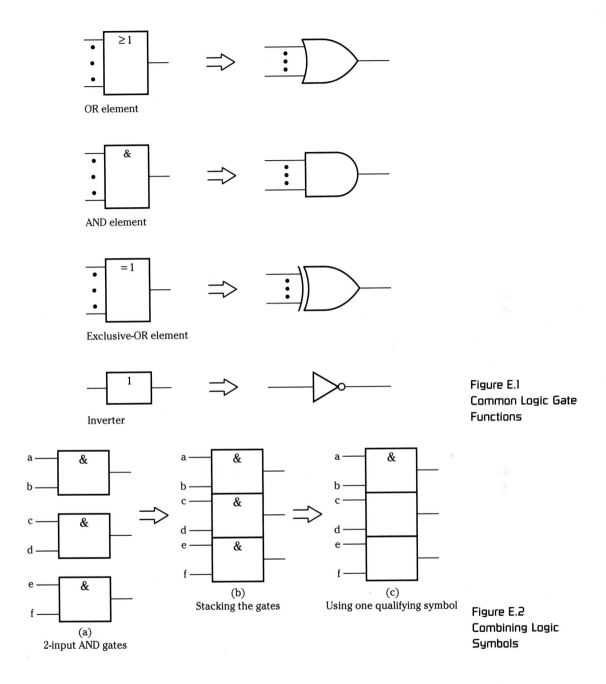

Figure E.1
Common Logic Gate
Functions

Figure E.2
Combining Logic
Symbols

Control Block

Figure E.3 illustrates how signals common to more than one element in the symbol are indicated. A **common control block** is drawn on the top or bottom of the symbol and has as its inputs the signals common to all other elements in the structure.

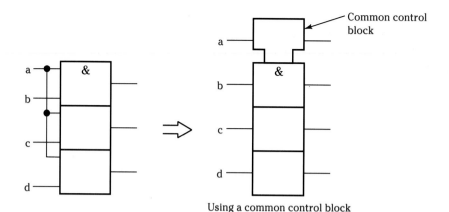

Figure E.3
Use of the Common
Control Block

Using a common control block

The figure shows that input "a" is common to all three AND gates in the array. By placing the "a" input on the control block rather than on the individual AND gate blocks, this common connection is indicated. This is a convenient notation for more complex functions where a single input controls the operation of several subfunctions of a complex circuit. By examining the symbol, you can determine the exact nature of each control line.

Dynamic Inputs

Figure E.4 shows how dynamic input signals, such as a clock edge, are indicated on the logic symbol. The wedge symbol we have used for flip-flops is retained from previous usage and indicates that a change in flip-flop state is synchronous with either a positive or negative going clock transition. Bubbles or wedges are used to indicate negative level sensitive inputs or negative edge triggered clock inputs. The two symbols shown in the figure are for a D flip-flop and a J-K flip-flop. Except for the specific pin information and the fact that all inputs are drawn on the left and all outputs are drawn on the right, the symbol is unchanged from the traditional notation.

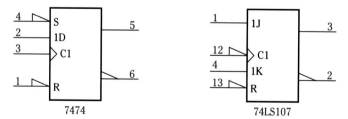

Figure E.4
How Dynamic Input
Signals Are Indicated
on the Logic Symbol

Dependency Notation

Dependency notation utilizes letters within the logic symbol to indicate the relationship between input and output symbols. This notation is used in complex logic symbols and is the primary technique by which the new logic symbols convey information about the capabilities of a logic function. Eleven dependencies are included in the standard; these are shown in Figure E.5. For example, the letters EN are used to identify an enabling input and the inputs and outputs controlled by it. The letter M indicates an input and the inputs and outputs controlled by the selection of a particular mode of operation.

SUMMARY OF DEPENDENCY NOTATION

Type of Dependency	Letter*	Effect on internal logic state of, or action of, the affected input or output:	
		Affecting input at its 1-state	Affecting input at its 0-state
ADDRESS	A	Permits action (address selected)	Prevents action (address not selected)
CONTROL	C	Permits action	Prevents action
ENABLE	EN	Permits action	(1) Prevents action of affected inputs (2) Imposes external high-impedance state on open-circuit and 3-state outputs (internal state of 3-state output is unaffected) (3) Imposes high-impedance L-level on passive-pulldown outputs and high-impedance H-level on passive-pullup outputs (4) Imposes 0-state on other outputs
AND	G	Does not alter state (permits action)	Imposes 0-state
MODE	M	Permits action (mode selected)	Prevents action (mode not selected)
NEGATE	N	Complements state	Does not alter state (no effect)
RESET	R	Affected output reacts as it would to S=0, R=1	No effect
SET	S	Affected output reacts as it would to S=1, R=0	No effect
OR	V	Imposes 1-state	Does not alter state (permits action)
TRANSMISSION	X	Transmission path established	No transmission path established
INTER-CONNECTION	Z	Imposes 1-state	Imposes 0-state

* These letters appear at the affecting input (or output) and are followed by a number represented in the general cases by the letter M. Each input or output affected by that input (or output) is labeled with that same number.

Figure E.5
Summary of
Dependency Notation

Dependency notation is put into action by:

a. Labeling the controlling input with the appropriate dependency letter and an identifying number
b. Labeling each input and output affected by the controlling input with the same identifying number

Figure E.6 shows some simple examples of dependency notation. For instance, the AND dependency example shows that input "b" is identified with the G1 dependency notation. The G stands for the AND operation, while the 1 is simply chosen for convenience. This means that any other input listed with the number 1 is ANDed with the "b" input signal. Therefore, input "a" and "b" are ANDed together. Input "c" is ANDed with the complement of "b" as indicated by the bar over the 1. If any outputs had a 1 listed, then the output would be controlled by the AND function.

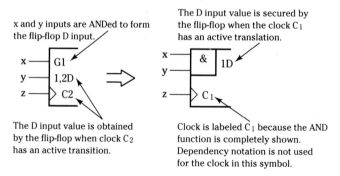

Figure E.6
Examples of
Dependency Notation

A complete understanding of the new logic symbols, particularly the details of dependency notation, require studying the ANSI/IEEE Std 91-1984 for logic function graphic symbols and ANSI/IEEE Std 991-1986 standard for logic circuit diagrams. A complete list of graphic symbols is shown in Figure E.7, whereas several common logic functions utilizing these symbols are shown in Figure E.8.

1	2	3	4	5	6
LOGICAL INVERSION					
ELECTRICAL INVERSION					
DYNAMIC INPUT					
NON LOGIC CONNECTION					
ANALOG CONNECTION					
REVERSE SIGNAL FLOW					
BIDIRECTIONAL SIGNAL FLOW					
INPUT WITH HYSTERESIS					
OPEN CIRCUIT OUTPUTS					
PASSIVE - PULLUP OUTPUT PASSIVE - PULLDOWN OUTPUT					
3-STATE OUTPUT					
SPECIALLY AMPLIFIED OUTPUT					
VIRTUAL INPUT					
INPUT CONTROLLED DELAY AT AN OUTPUT					
MULTIPLE LINES PROVIDING A SINGLE LOGIC INPUT					
GROUPING OF LINES REPRESENTING A NUMERIC VALUE					
GROUPING OF LINES WITH SIMILAR NAMES					

Courtesy of IEEE Standards Board.

Figure E.7
Graphic Symbols for
the ANSI/IEEE
Standard

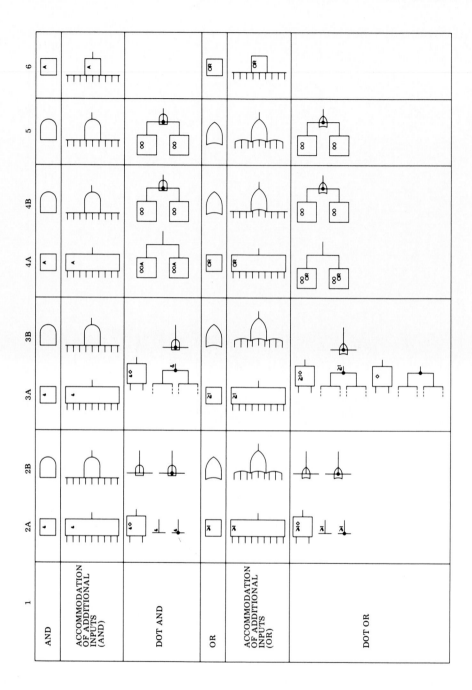

Figure E.7
Graphic Symbols for
the ANSI/IEEE
Standard (continued)   Courtesy of IEEE Standards Board.

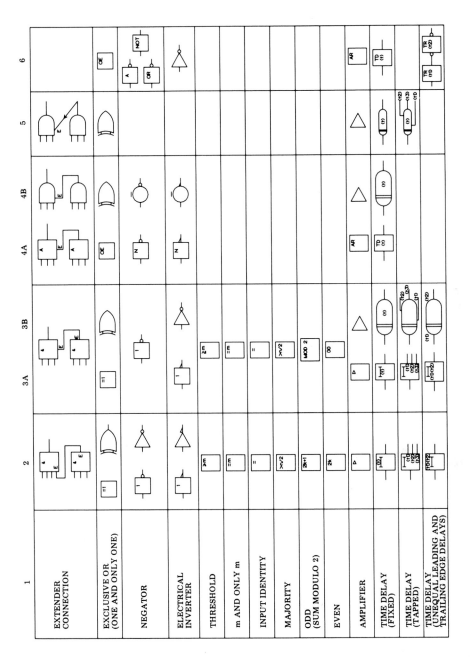

Figure E.7
Graphic Symbols for
the ANSI/IEEE
Standard (continued)

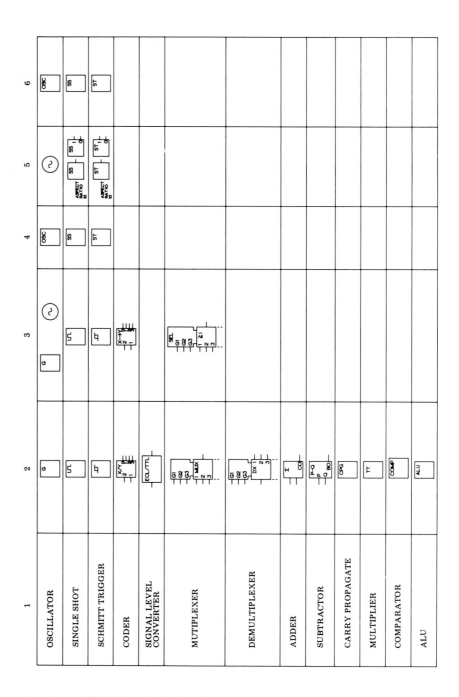

Figure E.7
Graphic Symbols for
the ANSI/IEEE
Standard (continued)

Courtesy of IEEE Standards Board.

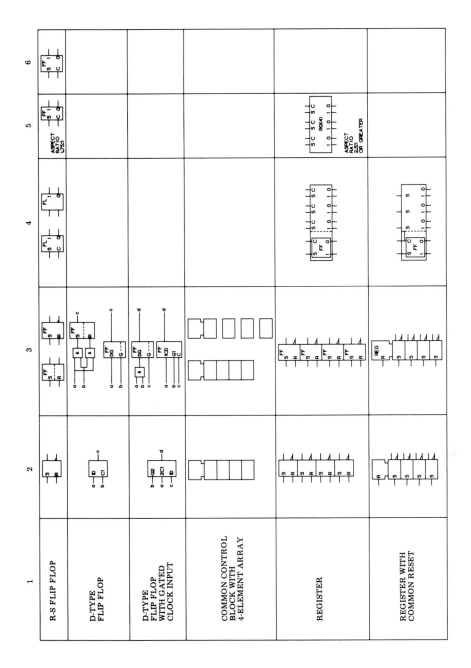

Figure E.7
Graphic Symbols for
the ANSI/IEEE
Standard (continued)

Courtesy of IEEE Standards Board.

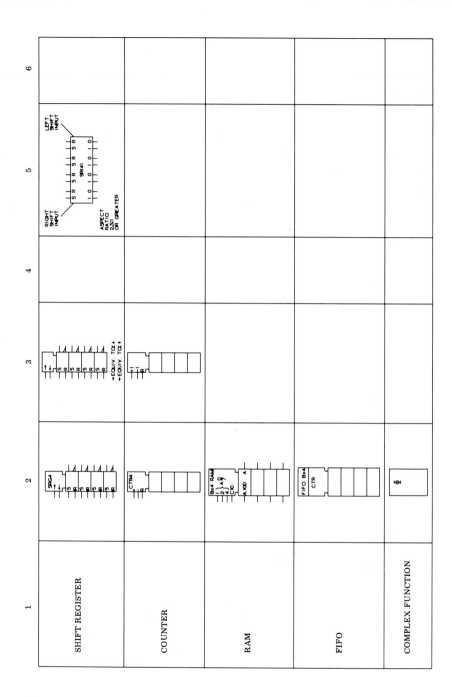

Figure E.7
Graphic Symbols for
the ANSI/IEEE
Standard (continued)   Courtesy of IEEE Standards Board.

STANDARD LINE LABELS:	2	3	4	5	6
1					
ENABLE	EN				
SET WHEN 1; RESET WHEN 0	D	D			
SET (TOGGLE WHEN BOTH) / RESET	J K	J K	S C		
SET / RESET	S R	S R		S C	S C
TOGGLE (COMPLEMENT)	T	T	T	T	T
SHIFT m POSITIONS	→m ←m	→m ←m			
COUNT UP BY m	+m	+m			
COUNT DOWN BY m	−m	−m			
QUERY INPUT OF CONTENT-ADDRESSABLE MEMORY	?				
MATCH OUTPUT OF CONTENT-ADDRESSABLE MEMORY	!				
SET CONTENT EQUAL m	CT=m				
CONTENT OUTPUT	CT=m CT<m CT>m ●m				
INPUT MUST BE A "1"	1→				
OUTPUT ALWAYS A "1"	1→				

Figure E.7
Graphic Symbols for
the ANSI/IEEE
Standard (continued)

Courtesy of IEEE Standards Board.

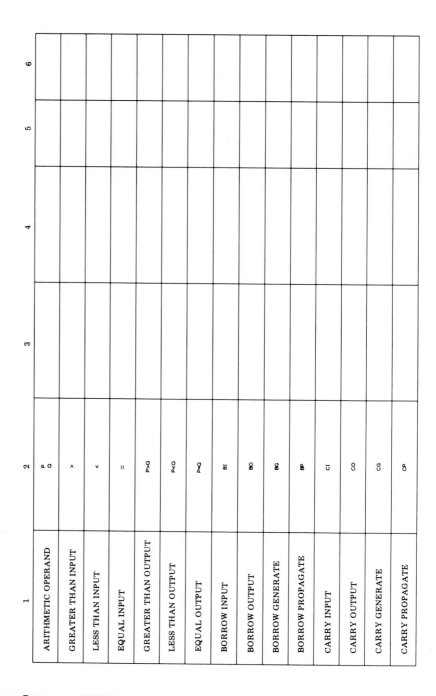

1	2	3	4	5	6
ARITHMETIC OPERAND	P Q				
GREATER THAN INPUT	>				
LESS THAN INPUT	<				
EQUAL INPUT	=				
GREATER THAN OUTPUT	P>Q				
LESS THAN OUTPUT	P<Q				
EQUAL OUTPUT	P=Q				
BORROW INPUT	BI				
BORROW OUTPUT	BO				
BORROW GENERATE	BG				
BORROW PROPAGATE	BP				
CARRY INPUT	CI				
CARRY OUTPUT	CO				
CARRY GENERATE	CG				
CARRY PROPAGATE	CP				

Figure E.7
Graphic Symbols for
the ANSI/IEEE
Standard (continued)

Courtesy of IEEE Standards Board.

DEPENDENCY NOTATION LINE LABELS: (note "m" is replaced by a number)

	2	3
ADDRESS	Am	
CONTROL	Cm	C
HOLD	Cm	H0 HOLD H-STATE / H1 HOLD L-STATE
ENABLE	ENm	
AND (GATE)	Gm	Gm / G
MODE	Mm	
NEGATE (XOR)	Nm	
RESET	Rm	
SET	Sm	
OR	Vm	
TRANSMISSION	Xm	
INTERCONNECTION	Zm	
D INPUT AFFECTED BY A CONTROL INPUT	Ci / D	C / D　C / D
R INPUT AFFECTED BY AN AND INPUT	Gi / R	Gi / R　Gi / R G / R　G / R

Courtesy of IEEE Standards Board.

Figure E.7
Graphic Symbols for
the ANSI/IEEE
Standard (continued)

74164 8-bit shift register with parallel outputs:

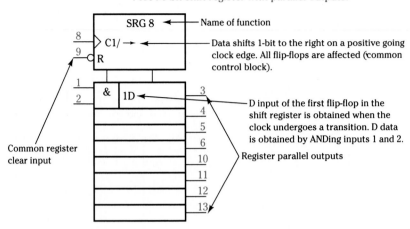

SRG 8 ← ———— Name of function

Data shifts 1-bit to the right on a positive going clock edge. All flip-flops are affected (common control block).

D input of the first flip-flop in the shift register is obtained when the clock undergoes a transition. D data is obtained by ANDing inputs 1 and 2.

Register parallel outputs

Common register clear input

74192 4-bit synchronous up/down counter:

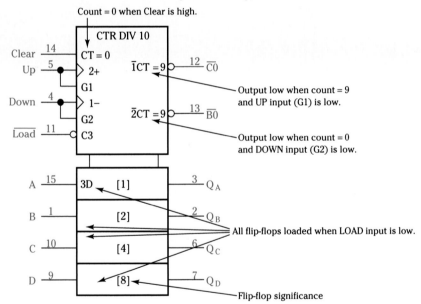

Count = 0 when Clear is high.

Output low when count = 9 and UP input (G1) is low.

Output low when count = 0 and DOWN input (G2) is low.

All flip-flops loaded when LOAD input is low.

Flip-flop significance

Figure E.8
Examples of the ANSI/
IEEE Standard Using
Common TTL Parts

Appendix F

ANSWERS TO SELECTED PROBLEMS

Chapter 1

1. (a) Analog   (c) Digital   (f) Music is analog in nature, but stored as digital information on a compact disc.
2. (a) 38   (e) 85.625
3. (c) 1100.1   (f) 11010.001

4. (c) 10000.001
5. (b) 0111
7. (d) 1024   (g) 1048576
8. (a) 3   (e) 4095
13. (a) 1,200,000,000,000 nanoseconds

Chapter 2

1. $Z12 = X1W5T9$
   Unique combination = 111.
   Nonunique combinations = 000 through 110.

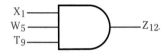

$$X_1 \\ W_5 \\ T_9 \quad \rightarrow Z_{12}.$$

3. Truth table outputs = 1 for input combinations. 00001 through 11111 (31 combinations).
   Output = 0 for input combination = 00000.

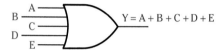

$$A \\ B \\ C \\ D \\ E \quad Y = A + B + C + D + E$$

6. Truth table outputs = 1 for input combinations. 00000 through 11110 (31 combinations).
   Output = 0 for input combination 11111.

$$A \\ B \\ C \\ D \\ E \quad Y = \overline{ABCDE}$$

8. $2^{10} = 1024$ One combination produces a high output 1111111111; all 1023 remaining combinations produce a low output.
11. $T = 0$ for $WXY = 001$; $T = 0$ for $WXY = 101$; $T = 1$ for $WXY = 010$.
13. (a) $(\overline{W + X})Y\overline{Z} = Q$ (b) $Q = 0$ for $WXYZ = 1010$;   (c) $Q = 1$ for $WXYZ = 0010$
16. $\overline{\overline{WXYZ}}$
19. $T = \overline{\overline{X}Y} + Z$
20. (b) $ENABLE = A_0\overline{A_1}A_2\overline{A_3}A_4A_5\overline{A_6}\overline{A_7}$

21.
   X

   $\overline{X}$

22.

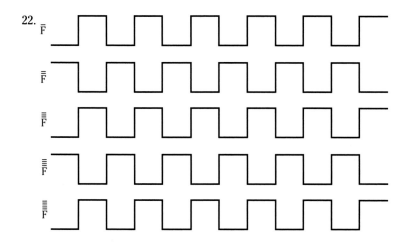

24.

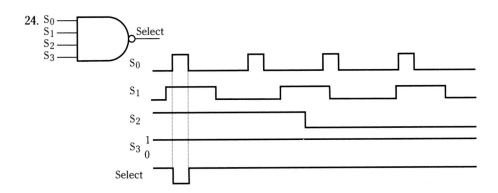

26.

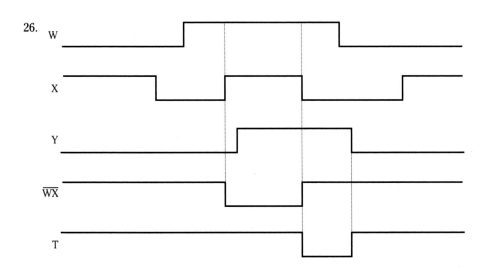

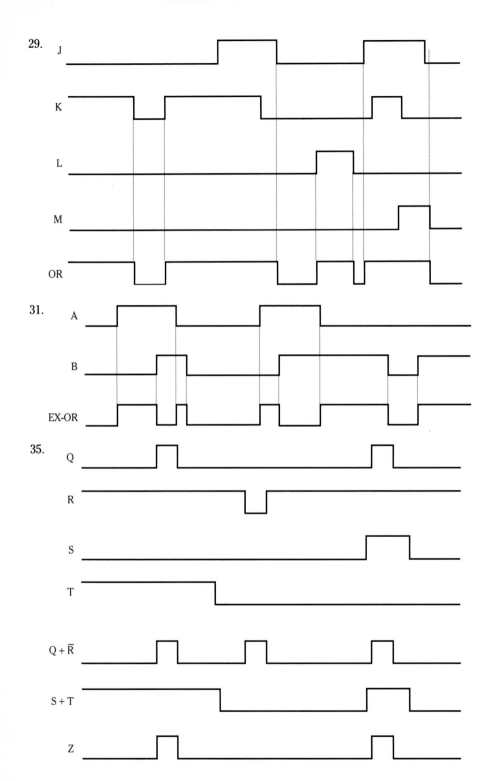

29.

31.

35.

37.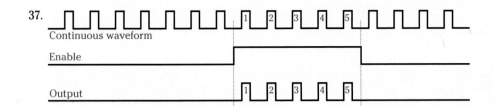

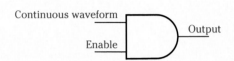

Chapter 3

1. (a) $T = X + YZ$
2. (a) $X = AC + B\overline{C}$
3. (a) $Q = Y + \overline{X}$
4. (b) $X = \overline{A} + B + \overline{C} + \overline{D}$
5. (a) $T = \overline{(\overline{UV} + V\overline{X})} + \overline{(\overline{UV} + U\overline{X})}$
6. (a)

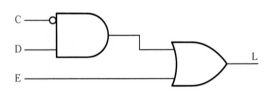

8. (a) C is high when A AND B are both low.
   (d) K is high when L is low OR M is high OR N is low.
10. (a)

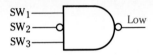

11. (a) $F = (A + B + C)(A + B + \overline{C})(\overline{A} + B + \overline{C})(\overline{A} + \overline{B} + C)$
13. (a) $X = \overline{C}\overline{B}A + \overline{C}B\overline{A} + \overline{C}BA + C\overline{B}A + CB\overline{A}$
16. Sensors $= S_4, S_3, S_2, S_1$; Output $= \overline{S}_4 S_3 S_2 S_1 + S_4\overline{S}_3 S_2 S_1 + S_4 S_3\overline{S}_2 S_1 + S_4 S_3 S_2\overline{S}_1$

Chapter 4

1. (a) $Y = A\overline{C} + AB$
2. (a) $Y = \overline{A}C$
3. (a) $AB\overline{C}\overline{D} + AB\overline{C}D + \overline{A}BCD + \overline{A}\overline{B}C\overline{D}$ (i) $B + \overline{D}$
5. (a)

	$\overline{Y}\overline{Z}$	$\overline{Y}Z$	YZ	$Y\overline{Z}$
$\overline{W}\overline{X}$	1	1	1	0
$\overline{W}X$	0	0	0	0
WX	0	1	0	0
$W\overline{X}$	0	0	1	1

$G = \overline{W}\overline{X}\overline{Y} + \overline{W}\overline{X}Y + W\overline{X}Y + WX\overline{Y}Z$

7. (a) Prime Implicant Table:

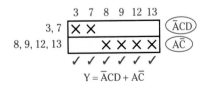

$Y = \overline{A}CD + A\overline{C}$

Chapter 5

1. (a) 101101   (d) 1111111
2. (a) 01111
3. (a) 1101001   (b) 1010110
7. (a) 00000001   (c) 1111100
8. (a) 11101
9. (a) 0011010

10. (a) 11110
11. (a) 1001
12. (a) 1001
15. (b) 00100101   (c) 11100111
16. (a) 00100101   (b) 11101000
17. (a) 11111010

19.

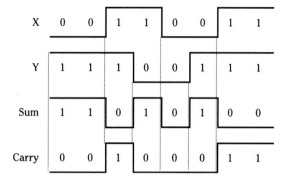

20.

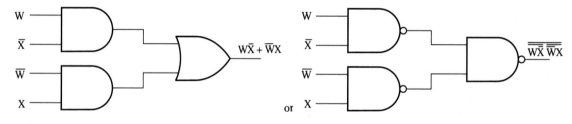

24.

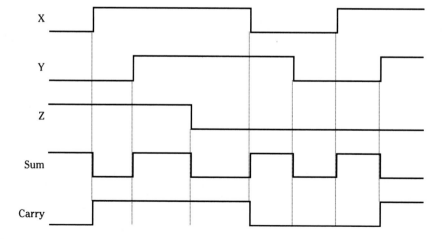

27.

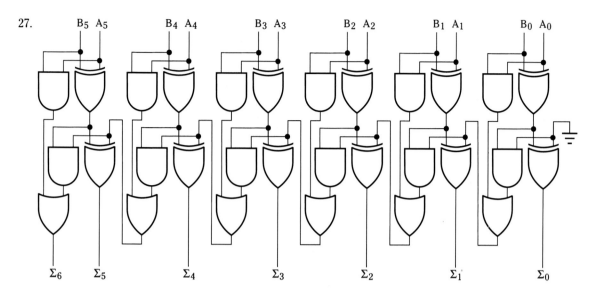

$B_5\ A_5$ $B_4\ A_4$ $B_3\ A_3$ $B_2\ A_2$ $B_1\ A_1$ $B_0\ A_0$

Σ_6 Σ_5 Σ_4 Σ_3 Σ_2 Σ_1 Σ_0

29. $C_4 = C_0P_0P_1P_2P_3 + G_0P_1P_2P_3 + G_1P_2P_3 + G_2P_3 + G_3$
$C_5 = C_0P_0P_1P_2P_3P_4 + G_0P_1P_2P_3P_4 + G_1P_2P_3P_4 +$
$G_2P_3P_4 + G_3P_4 + G_4$

31.

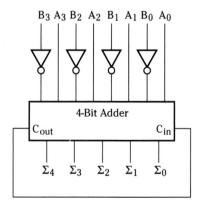

$B_3\ A_3\ B_2\ A_2\ B_1\ A_1\ B_0\ A_0$

4-Bit Adder

C_{out} C_{in}

Σ_4 Σ_3 Σ_2 Σ_1 Σ_0

33. (b) Mode bit = 1
$S_3S_2S_1S_0 = 0110$

Chapter 6

1. (a) 0110 0010 0111
2. (a) 7950
4. (a) parity bit = 0
5. (a) parity bit = 0
7. (a) 111001110

8. (a) 01001101
9. (a) 1011, 1111   (b) 0010, 1010
10. The output is high whenever X and Z are high.
    The decoder output is high regardless of the level
    of Y. The decoder decodes XYZ = 101 and 111.

13. Inputs = ABCD.

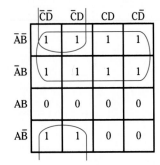

$Y = \bar{A} + \bar{B}\bar{C}$ = circuit equation

14. (a) 7

18.

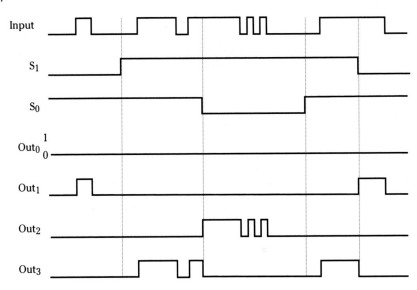

21.

BCD				EXCESS 3			
A	B	C	D	EA	EB	EC	ED
0	0	0	0	0	0	1	1
0	0	0	1	0	1	0	0
0	0	1	0	0	1	0	1
0	0	1	1	0	1	1	0
0	1	0	0	0	1	1	1
0	1	0	1	1	0	0	0
0	1	1	0	1	0	0	1
0	1	1	1	1	0	1	0
1	0	0	0	1	0	1	1
1	0	0	1	1	1	0	0

$EA = A + BD + BC$

$EB = B\bar{C}\bar{D} + \bar{B}D + \bar{B}C$

$EC = \bar{C}\bar{D} + CD$

$ED = \bar{D}$

Reduced equations

Chapter 7

1. (a) $Q = \overline{Q} = ?$ (b) $Q = 1, \overline{Q} = 0$

3.

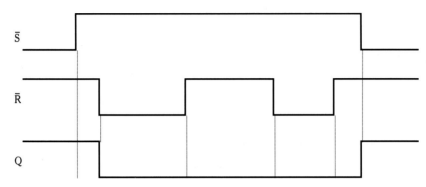

5.

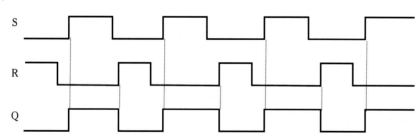

8.

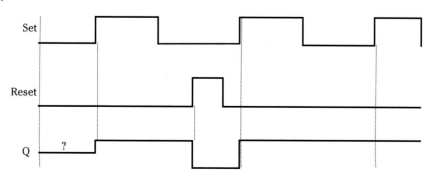

11.

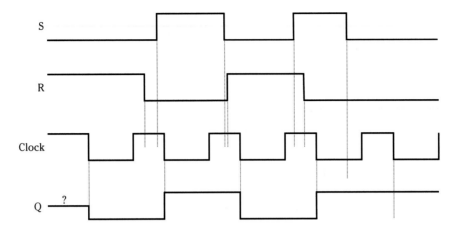

12. $Q_1 = 1$, $Q_2 = 0$, $Q_3 = 1$

15.

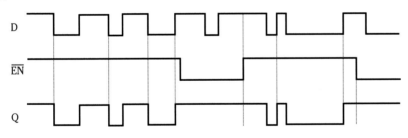

16.

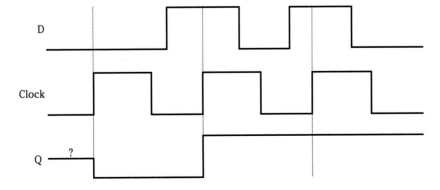

19. (a)

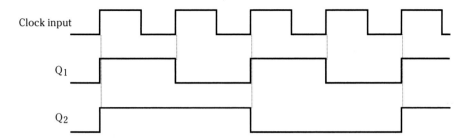

21.

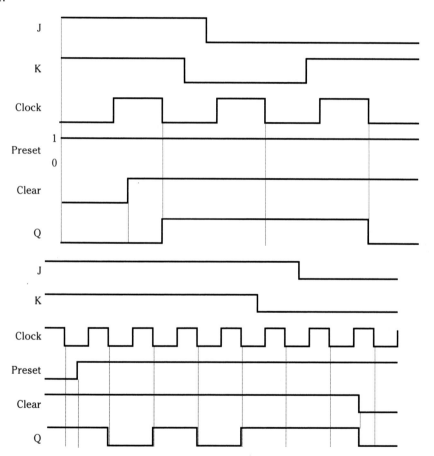

26. (a) Dynamic hazard   (d) No hazard

27.

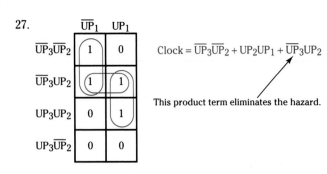

$$Clock = \overline{UP}_3\overline{UP}_2 + UP_2UP_1 + \overline{UP}_3UP_2$$

This product term eliminates the hazard.

29. Race condition between D and CLK signals

30. (a) 142.875 kohms

Chapter 8

2.

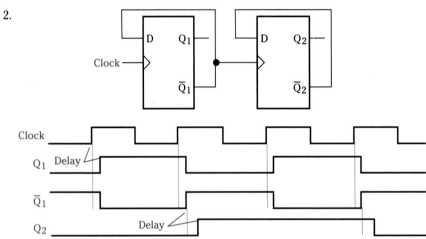

3. Nine

6.

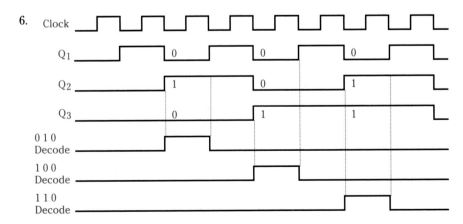

7. 1024

9. (b) $\div 10$, $\div 60$, $\div 120$, $\div 600$, $\div 1200$

11.

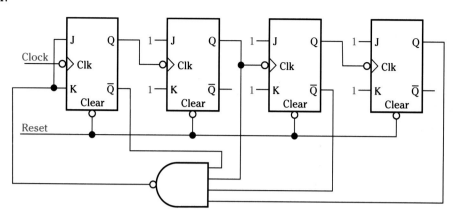

14.

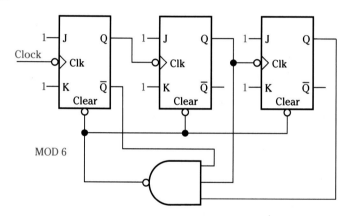

16.

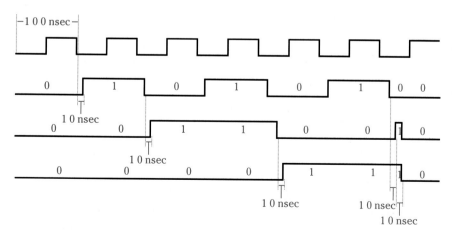

18. (a) 12.5 MHz

21.

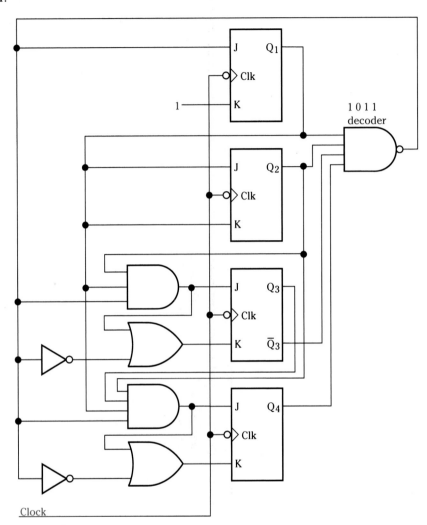

23.

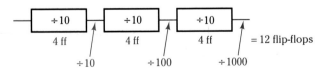

Direct design = 10 flip-flops

25. 25.6 μsec; 25.6 μsec

31.

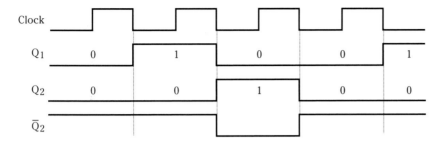

33. (a)

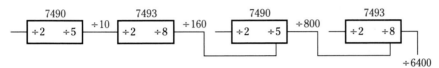

37. (a) Load = 8 pulses; Unload = 7 pulses

39.

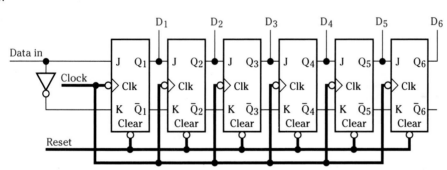

41.

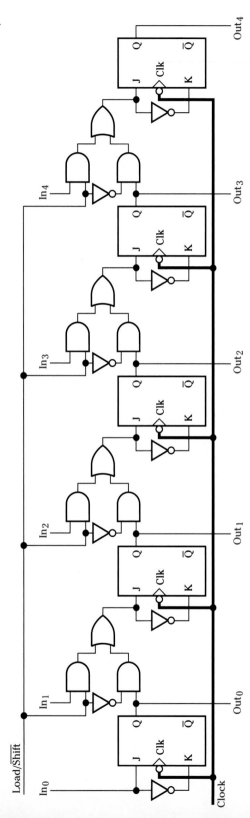

43.

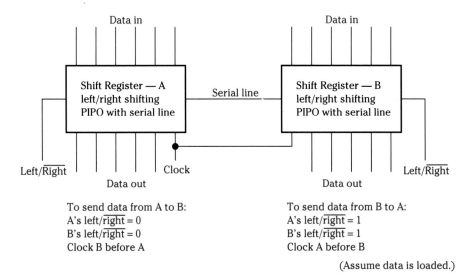

To send data from A to B:
A's left/$\overline{\text{right}}$ = 0
B's left/$\overline{\text{right}}$ = 0
Clock B before A

To send data from B to A:
A's left/$\overline{\text{right}}$ = 1
B's left/$\overline{\text{right}}$ = 1
Clock A before B

(Assume data is loaded.)

45. Ring counters: 10, 12, and 18 states; Johnson counters: 20, 24, and 36 states.

46. (b)

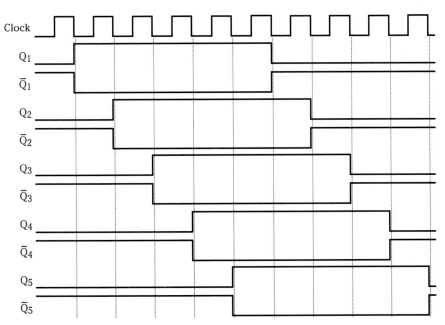

Chapter 9

3. (a) 128K × 4 = 4-bit word size; 128K address space; 17 address lines
4. (a) 524,288 cells
5. (c) 1010 written to address 0011010011
8. 256K × 1 requires 18 address lines; 3F2D5 =
   row   column
   11111100 1:011010101.

16.

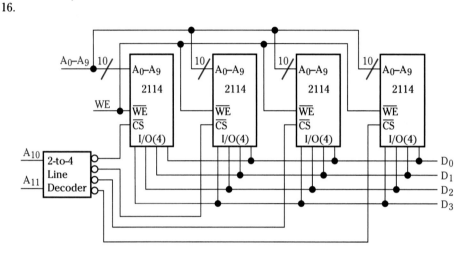

19.

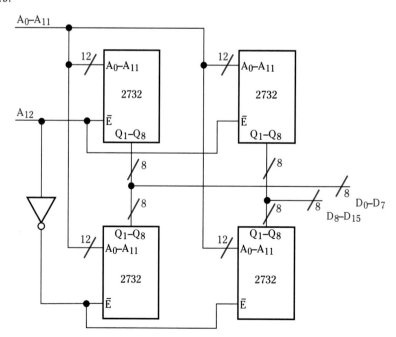

10. 2732A-17 and 2732A-20 access time is too long for the other versions.
12. (b) 256K × 16 = 4,194,304 cells, needs 1024 2114s
13. (a) 2 chips
14. (a) 16 chips

22. (a) Third address, second bit is stuck at zero.

Chapter 10

1. PROM requires 2^{12} or 4096 12-input AND gates; PAL or PLA require only 12 AND gates minimum.

2. $AB\overline{C}D\overline{E}\overline{F}$

3. (a) $\overline{A}BC\overline{D} + A\overline{B}C\overline{D} + \overline{B}CD + \overline{A}\,\overline{B}CD + AB\overline{C}\overline{D}$

4. (a)

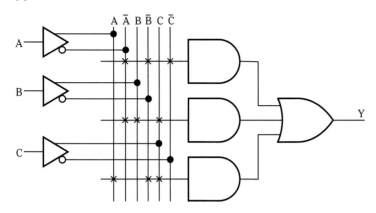

5. (a)

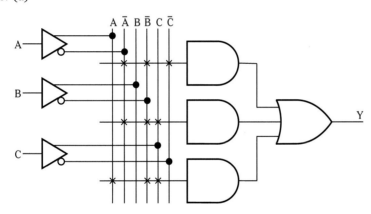

6. (a) 16 inputs, 6 outputs, registered device

7. Eight

11. (c)

9. All eight output pins tied together externally; eight inputs required as EN_1 through EN_8; eight inputs required for data A through H.

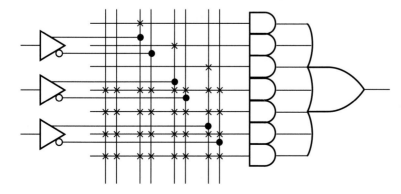

16.

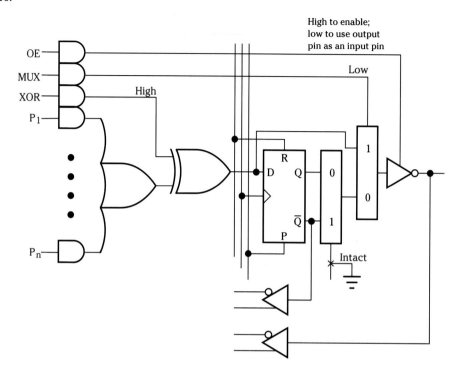

Chapter 11

1.

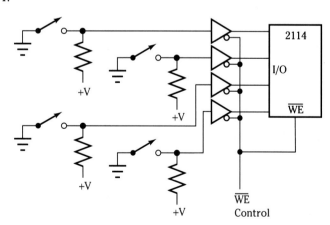

3.

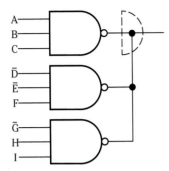

4.

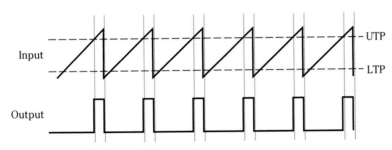

7.

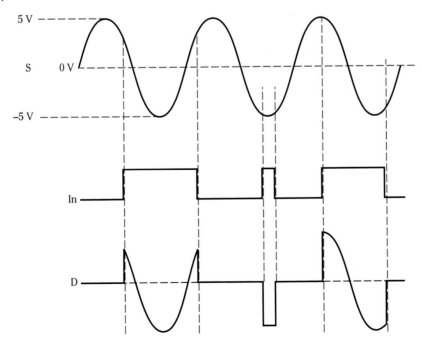

10.

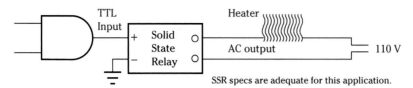

SSR specs are adequate for this application.

11. (a) 64

12. (a) 0.1875 V

13. (a) 11.8125 V

14. (a) 0.0390625 V

16. (a) 1,000,000 con/sec

18. MSB is stuck at one.

19. Assume 1 V comparator = C_1; 2 V comparator = C_2; 3 V comparator = C_3. Binary LSB = $C_3 + \overline{C_2}C_1$; Binary MSB = C_2.

21.

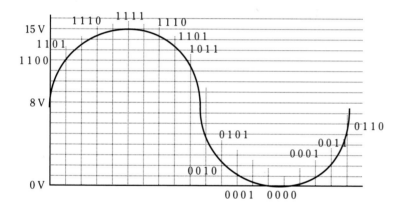

Chapter 12

1. (a) Logic 0   (f) Invalid voltage

2. (a) Logic 0 = 0.9 V; Logic 1 = 2.1 V

3. (b) Logic 0

6. (a) 30 pj

7.

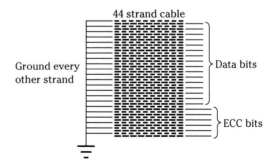

10. (a) 16 mA

11. 17.7 → 17

13. (b) Ground bounce

15. .009 μfd

17.

Chapter 13

1.

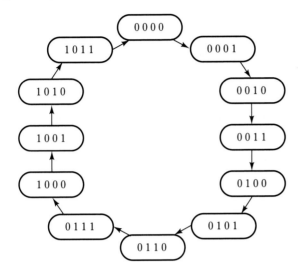

2.

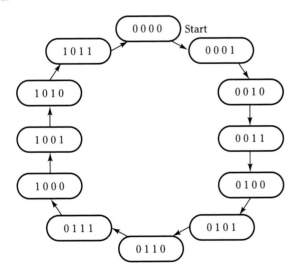

5.

PRESENT STATE	INPUT	NEXT STATE	FLIP-FLOPS
Q_3 Q_2 Q_1 Q_0	Start	Q_3 Q_2 Q_1 Q_0	D_3 D_2 D_1 D_0
0 0 0 0	0	0 0 0 0	0 0 0 0
0 0 0 0	1	0 0 0 1	0 0 0 1
0 0 0 1	d	0 0 1 0	0 0 1 0
0 0 1 0	d	0 0 1 1	0 0 1 1
0 0 1 1	d	0 1 0 0	0 1 0 0
0 1 0 0	d	0 1 0 1	0 1 0 1
0 1 0 1	d	0 1 1 0	0 1 1 0
0 1 1 0	d	0 1 1 1	0 1 1 1
0 1 1 1	d	1 0 0 0	1 0 0 0
1 0 0 0	d	1 0 0 1	1 0 0 1
1 0 0 1	d	1 0 1 0	1 0 1 0
1 0 1 0	d	1 0 1 1	1 0 1 1
1 0 1 1	d	0 0 0 0	0 0 0 0

12. $J_3 = Q_2 Q_1 Q_0$, $K_3 = Q_0$, $J_2 = Q_0 + Q_3 Q_1$, $K_2 = Q_1 Q_0$, $J_1 = 1$, $K_1 = \overline{Q_1} + \overline{Q_3} Q_2 + Q_3 \overline{Q_0}$, $J_0 = \overline{Q_3} + Q_2$, $K_0 = Q_2$

17.

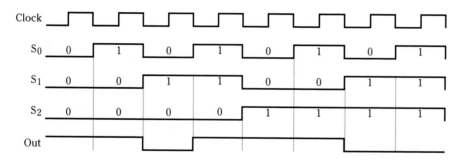

18.

PRESENT STATE		NEXT STATE		FLIP-FLOPS		MULTIPLEXERS		
Q_3 Q_2 Q_1		Q_3 Q_2 Q_1		D_3 D_2 D_1		MUX3	MUX2	MUX1
0 0 0		0 0 1		0 0 1		0	0	1
0 0 1		0 1 0		0 1 0		0	1	0
0 1 0		0 1 1		0 1 1		0	1	1
0 1 1		1 0 0		1 0 0		1	0	0
1 0 0		1 0 1		1 0 1		1	0	1
1 0 1		1 1 0		1 1 0		1	1	0
1 1 0		1 1 1		1 1 1		1	1	1
1 1 1		0 0 0		0 0 0		0	0	0

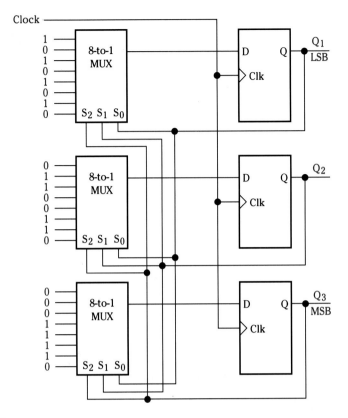

21.

PRESENT STATE			NEXT STATE			FLIP-FLOPS			PROM OUTPUTS		
Q_3	Q_2	Q_1	Q_3	Q_2	Q_1	D_3	D_2	D_1	O_3	O_2	O_1
0	0	0	0	0	1	0	0	1	0	0	1
0	0	1	0	1	0	0	1	0	0	1	0
0	1	0	0	1	1	0	1	1	0	1	1
0	1	1	1	0	0	1	0	0	1	0	0
1	0	0	1	0	1	1	0	1	1	0	1
1	0	1	1	1	0	1	1	0	1	1	0
1	1	0	1	1	1	1	1	1	1	1	1
1	1	1	0	0	0	0	0	0	0	0	0

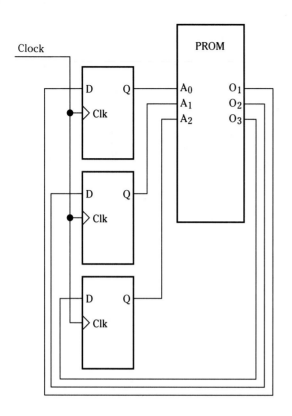

PROM ADDRESS			PROM DATA		
A_2	A_1	A_0	O_3	O_2	O_1
0	0	0	0	0	1
0	0	1	0	1	0
0	1	0	0	1	1
0	1	1	1	0	0
1	0	0	1	0	1
1	0	1	1	1	0
1	1	0	1	1	1
1	1	1	0	0	0

Appendix C

2. (a) /Q*R*/S*T + /Q*R*S*T + Q*/R*/S*T
   + Q*R*S*T

4. (a) $\overline{\text{TIMER\_ENABLE}}$=A*B*/C

5. (a) $\overline{\text{TIMER\_ENABLE}}$=/(A*B*/C)

7. (a) Y=/A*/B*/C + /A*B*C + A*/B*/C + A*B*/C

8. Out = $\overline{\overline{X}YZ + X\overline{Y}Z + XYZ}$

9. (a) SETF /W /X Y /Z

11. (a) X:=/Q*R*/S*T + /Q*R*S*T + Q*/R*/S*T
    + Q*R*S*T

active level   The input logic level (high or low) causing a specified action to occur.

ADC   An analog-to-digital converter.

addressing   The mechanism by which information is read or written to a specified location in a memory device.

alphanumerics   A character set consisting of numerals, letters, and special symbols as found on a standard typewriter keyboard.

analog   A term referring to voltages, currents, or other physical measurements that occur in a continuous range of values.

analog switch   A semiconductor switch that is opened or closed by a digital control signal.

analog-to-digital conversion   The process of converting an analog, voltage or current into a corresponding digital number.

AND   A logic function. The AND output is high when all inputs are high; otherwise the output is low.

architecture   The structure or arrangement of elements in complex logic devices.

arithmetic logic unit (ALU)   A circuit capable of performing a variety of arithmetic and logical operations.

ASCII code   American Standard Code for Information Interchange. A 7-bit code used to represent letters, numbers, and special characters.

ASIC   An application specific IC.

asynchronous counter   A counter designed so that clocking of all stages other than the least significant occurs asynchronously with the clock.

asynchronous load   A design where data is loaded into a counter or shift register unsynchronized with the clock.

binary addition   The process by which 2-bits are added together to produce a sum and a carry output.

binary coded decimal (BCD)   A binary code used to represent decimal numbers.

binary numbers   Numbers represented using two symbols, 0 and 1. Utilizing a positional weighting system, an equivalence to decimal numbers can be obtained.

binary subtraction   The process by which two binary bits are subtracted to produce a difference and a borrow.

bistable multivibrator   A transistorized circuit capable of having two distinct stable output voltages (states).

bit   A single binary digit. (BInary digiT)

Boolean expression   A mathematical expression using Boolean algebra properties describing the operation of a logic circuit.

borrow   A borrow occurs during binary subtraction when a 1 is subtracted from a 0. In order to carry out the process, a 1 is borrowed from the next most significant bit.

bubble   A small round symbol used on logic diagrams to indicate inversion. Also known as a berry.

byte   A grouping of eight binary bits.

cache   A portion of memory in a computer system holding duplicate information as found in the computer's main storage. Used to boost system performance.

CAD   Computer aided design.

carry generate   The carry output of a half adder.

carry look ahead   A circuit designed to compute the carries resulting from the addition of two binary numbers. Carry look ahead circuitry is used to speed up the addition process.

carry propagate   The sum output of a half adder.

CAS   A column address strobe.

check bits   Multiple parity bits created in error checking and correction systems.

checksum   A quantity of information formed by adding together data bits. Used in error detection systems.

circuit carry   The carry output of a full adder. This signal is typically the input to the next most significant stage of a ripple carry adder.

clear   A flip-flop asynchronous input that resets the flip-flop when active.

CMOS   Complementary metal oxide semiconductor. A logic family.

complement mathematics   A process by which negative numbers can be added or positive and negative numbers can be subtracted using addition techniques.

computer simplification   A process using computer software to reduce a Boolean expression.

conversion rate   The number of analog-to-digital or digital-to-analog conversions possible per second.

counter   A digital circuit designed to change state in a defined binary progression.

critical path analysis   An analysis of the delay times in a digital system. The critical path(s) are those most likely to impact system performance.

crosstalk   Interference signals existing on a signal line due to the presence of a signal on an adjacent line.

cyclic redundancy check (CRC)   A representative quantity of information formed by mathematical processes for the purposes of error detection.

D flip-flop   A variation of the D latch where the level sensitive enable input is replaced with an edge sensitive clock input.

D latch   A gated latch with a single data input.

DAC   A digital-to-analog converter.

data selector logic   A form of combinatorial logic using multiplexers to implement truth table functions.

debug   Troubleshooting.

decade counter   A MOD 10 counter. A counter with 10 states.

decode and reset   A counter design technique used to force a truncated counting sequence in an asynchronous counter.

decode and steer   A synchronous counter design technique used to obtain a truncated count sequence.

decoder   A digital circuit designed to detect the presence of one or more binary input combinations.

decoupling capacitors   Capacitors placed near integrated circuits to minimize power supply noise problems resulting when logic devices switch state.

demultiplexers (DEMUXs)   "Data distributor" circuits used to route a single input signal to multiple output lines.

digital   A term referring to measurements occurring in discrete quantities. Digital signal information is expressed using binary numbers.

digital-to-analog conversion   The process of converting digital information into a corresponding analog voltage or current.

disable   A condition by which a logic gate output is not allowed to change. Also called disable, inhibit, or degate.

don't care state   A circuit input condition that is unlikely to occur. Therefore, the associated output level is unimportant.

double word   A grouping of 32 binary bits.

down counter   A counter that counts from a higher binary value to a lower binary value.

dual slope ADC   An analog-to-digital conversion technique utilizing the charging and discharging rates of capacitors to carry out a conversion.

duty cycle   The ratio of how long a waveform is high to how long it is low over the duration of one waveform cycle.

dynamic hazard   A dynamic hazard occurs when a signal line unintentionally changes logic level several times.

dynamic RAM   A kind of semiconductor memory characterized by storage cells requiring periodic refresh.

ECL   Emitter coupled logic. A logic family.

EEPROM   Electrically erasable read only memory.

EMP   An electromagnetic pulse.

enable   A condition by which a logic gate output is allowed to change or a logic device is allowed to function.

encoder   A digital circuit used to combine multiple input signals into a concise coded output(s).

end around carry   A portion of the 1's complement subtraction process where the final circuit carry is added to the difference.

EPROM   Erasable programmable read only memory.

erasable programmable logic device (EPLD)   A highly complex programmable device that allows a designer to interconnect predesigned "macrocells" and other logical building blocks.

error checking codes (ECC)   An error checking–error detection system based on Hamming codes. ECC systems can detect errors as well as correct many errors as they occur.

ESD   Electrostatic discharge.

essential prime implicant   A prime implicant required to express correctly a minimum logic expression.

excess-3 code   A variation of the BCD code used to support BCD arithmetic.

Exclusive-NOR   A logic function. The output is low when one of the inputs is high; the output is high when the two inputs are both low or both high.

Exclusive-OR   A logic function. The output is high when one input is high; the output is low when the inputs are both high or both low. The EX-OR also functions as an even/odd checker.

fan-out   The fan-out specification defines the number of logic gates that can be driven reliably by a logic gate output.

flash converter   A parallel ADC used for high speed analog-to-digital conversion.

flip-flop   A digital device with the capability of storing a single binary value. Flip-flop clock inputs are edge sensitive.

floating gate   A conductive area in EEPROMs used to store electric charge.

floating input   An electrically disconnected logic gate input.

flow chart   A graphical technique used to represent the sequence of events in a state machine or computer program.

four-variable K-map   A Karnaugh map formed from a four-variable truth table. The map will contain 16 squares.

full adder   A digital circuit designed to add three binary bits.

function hazard   A hazard created as a result of the function being designed. Function hazards can only be eliminated by redesign.

fundamental product   A product term consisting of several variables ANDed together. A fundamental product is a simple decoder as well as the basis of a sum of products expression.

fuse map   An output file from programmable logic development software indicating which fuses have been blown in a PAL device.

fusible link   A microscopic fuse within the structure of a PROM. The state of the fuse determines the logic level of the PROM cell.

gated latch   A form of a Set–Reset latch containing set, reset, and enable inputs.

glitch   Any unintended change in logic level.

Gray code order   An arrangement of binary bits where only a single bit changes from one code group to the next.

ground bounce   A signal noise problem caused by variations in ground potential.

ground loop   A poorly constructed ground path in a digital circuit's power distribution system characterized as a high impedance electrical path.

half adder   A digital circuit designed to carry two binary bits. The circuit produces a sum and a carry output.

Hamming code   A mathematical treatment of binary data used as the basis of error correcting systems.

hazard   An unintentional change in logic level. Also known as a glitch.

hexadecimal (hex)   A number system based on powers of 16.

hold time   The time interval between a data signal and a clock applied to a flip-flop after the clock transition.

implicants   The ones on a truth table or Karnaugh map. An implicant represents a product term.

initial listing table   A table used in the Quine–McCluskey process to order truth table input combinations.

input combinations   The input high and low levels driving a logic circuit. The number of combinations is related to the number of inputs by powers of 2; that is, three inputs provide $2^3 = 8$ combinations.

integrated circuit   A small device containing many electronic circuits.

INVERT   A logic function. The output is the complement of the input.

IRAM   An integrated RAM.

JEDEC fuse data file   A file created during PAL device development used to control the PAL programmer.

J-K flip-flop   A flip-flop with two controlling inputs (J-K) allowing a variety of possible operations.

Johnson counter   A ring counter devised by routing the last flip-flop complement output as the data input to the first register flip-flop.

Karnaugh map   A graphical depiction of information displayed on a truth table. Also known as a K-map.

latch   A digital device with the capability of storing a single binary value. Latch inputs are level sensitive.

LCD   A liquid crystal display.

LED   A light emitting diode.

line drivers/receivers   Devices used to allow the transmission of digital signals over long wires or transmission lines.

local area network (LAN)   A networking system whereby personal computers are interconnected to communicate with each other.

logic family   A number of logic devices of varying function possessing similar electrical characteristics. TTL, CMOS, and ECL are common logic families.

logic levels   A single bit of digital information may possess one of two states or logic levels. Typical logic levels are high/low, one/zero, true/complement, and true/false.

LSB   A least significant bit. The bit in a binary number having the smallest positional value.

macrocell   The programmable output circuitry of complex PAL devices.

Master–Slave flip-flop   A flip-flop containing two Set–Reset latches used to isolate input changes from output changes.

Mealy model   A state machine configuration characterized by outputs obtained from a combination of state register and input signal values.

microcode   Hardware-specific program instructions that determine the sequence of events in a computer system.

microprocessor   A computer processing unit on a single integrated circuit. Typically the heart of all personal computers.

microstripline   A printed circuit board trace of specific thickness and width used to establish a known characteristic impedance.

minimum-change code   Binary code exhibiting the property where only a single bit changes from one code group to the next.

minimum row set   The fewest rows required from a prime implicant table to implement a logic function correctly. Used in the Quine–McCluskey process.

modem (modulate-demodulate)   A device used to allow the transmission of digital information over telephone lines.

modulus   The number of states through which a counter passes.

monotonicity   A characteristic of an analog-to-digital converter that shows how accurately a conversion takes place.

Moore model   A state machine configuration characterized by outputs obtained from the state register.

MSB   The most significant bit. The bit in a binary number having the greatest positional value.

multiplexer (MUX)   A "data selector" allowing multiple input signals to share a common output line under control of select control lines.

NAND   A logic function. The output is low only when all inputs are high; otherwise the output is high.

nanosecond   A unit of time equal to one billionth of a second.

natural binary count   A binary counting sequence proceeding from zero to a maximum value in a normal counting progression (0, 1, 2, . . .).

netlist   A computer generated listing of the interconnections in a logic circuit.

nibble   A grouping of four binary bits.

noise margin   A specification detailing a logic gate's ability to reject noise signals and maintain reliable operation.

nonunique conditions   The input combinations on a truth table producing output levels of the same logic value.

nonvolatile memory   Memory devices that retain their data when power is removed.

NOR   A logic function. The output is low when any input is high; the output is high only when all inputs are low.

NVRAM   A nonvolatile RAM.

Nyquist sampling theorem   A theorem stating that accurate A/D sampling occurs if the sampling rate is at least twice the highest frequency of the analog signal.

octal   A number system based on powers of 8.

octet   A K-map enclosure of eight ones. An octet leads to the elimination of three variables.

1's complement number   The value obtained by complementing each bit in a binary number.

one-shot   A monostable multivibrator controlled by an RC network to create a timing pulse of specific duration.

open collector   A logic device whose output signal is obtained directly from the collector of the logic device's output transistor.

open drain   A logic device whose output signal is obtained directly from the drain of the logic device's output transistor.

opto-coupling   A circuit connection where signals are passed from one circuit to another using light rather than electrical current flow.

OR   A logic function. The output is high when any input is high; the output is low when all inputs are low.

OR-AND networks   A logic circuit comprised of OR gates feeding into an AND gate. Usually developed from product of sums equations.

oscillator   An electronic circuit used to create a periodic output signal.

overbar   A line drawn over a variable or logic expression to indicate inversion.

overflow   A condition that occurs when the addition of two numbers produces a result greater in magnitude than can be supported by the computational circuitry.

overlapping   A K-map technique used to enclose ones on the map with previously enclosed ones.

pair   A K-map enclosure of two ones. A pair leads to the elimination of one variable.

PAL   Programmable array logic. A logic device with a programmable AND array and a fixed OR array.

parallel counter   A synchronous counter.

parity   An error checking system used to detect whether a binary code group contains an even or odd number of bits at the high level.

PCB   A printed circuit board.

piezoelectric effect   The property by which a crystal subject to pressure will produce a voltage or when subject to an applied voltage will vibrate.

pin out   A diagram indicating the function of each pin on an integrated circuit.

PIPO   Parallel-in, parallel-out.

PISO   Parallel-in, serial-out.

PLA   Programmable logic array. A logic device with programmable AND and OR arrays.

positional weighting   A mathematical convention where the position of a digit in a decimal number or a bit in a binary number determines the place value of the digit/bit.

positive logic   A logic convention where high logic levels consist of voltages greater than the voltages used to represent the low logic level.

power   The amount of electrical energy dissipated by an electronic device. Measured in watts.

preset   A flip-flop asynchronous input that sets the flip-flop when active.

presettable (programmable) counter   A counter that can be preloaded with any initial starting state.

prime implicant table   A table used in the Quine–McCluskey process to determine the prime implicants required for a successful reduction.

prime implicants   A maximum enclosure of ones on a K-map.

product of sums (POS)   A Boolean expression consisting of multiple OR terms ANDed together.

programmable logic device (PLD)   Any programmable logic device.

PROM   Programmable read only memory.

propagation delay time   The amount of time required for a digital device to change its output level in response to a change in input level.

pure binary counters   Counters that count in a straightforward binary sequence with the number of states exactly equal to a power of 2.

quad   A K-map enclosure of four ones. A quad leads to the elimination of two variables.

Quine–McCluskey simplification   A circuit simplification method frequently used to reduce logic designs employing more than four input variables.

race condition   A design problem that occurs when two flip-flop inputs change in a near simultaneous manner.

RAM   Random access memory.

RAS   Row address strobe.

Read   An operation used to retrieve information from memory.

reduction table   A listing of input combinations utilized in the Quine–McCluskey reduction process to isolate combinations with a single variable difference.

refresh cycle   A portion of time during a dynamic RAM's operation where the memory's storage-cells are replenished with electric charge.

registered devices   This term refers to PAL devices with output lines obtained from flip-flops.

resolution   The number of binary code groups by which an analog signal can be represented.

RFI   Radio frequency interference.

ring counter   A shift register modified so that data from the last register flip-flop is used as the input to the first register flip-flop.

ripple carry adder   A digital circuit comprised of full adders used to add together two multiple bit binary numbers.

ripple counter   An asynchronous counter.

rolling   A Karnaugh mapping technique recognizing that ones along the edges of a K-map can be combined for reduction.

ROM   Read only memory.

RS-232-C   A communications standard used to connect peripheral equipment to computer systems. The standard covers many aspects of data communications including the logic voltage requirements.

SAR   A successive approximation register.

schematic capture   The computer software used by a logic designer to draw and create a digital circuit.

Schmitt trigger   A logic device with two differing switching threshold voltages.

scratchpad memory   A small high-speed segment of memory used in computer systems.

security fuse   A fusible link on a programmable logic device. When blown, the programmable device cannot be read by a PAL programmer.

sequential circuits   A class of digital

logic characterized by its ability to store binary values so that output conditions are not directly dependent on input conditions.

Set–Reset latch   A basic storage device formed by cross coupling two NAND gates or two NOR gates.

settling time   The length of time required for a DAC or ADC to sample and convert an input signal.

setup time   The time interval occurring between the application of a signal to a flip-flop and an active clock transition.

seven segment display   A display technology used to display numbers. Both LED and LCD versions are used.

shaft encoder   A transducer affixed to a motor shaft for the purpose of converting rotational speed or position into a binary coded signal.

shift register   A circuit allowing data to be moved 1-bit per clock pulse either to the left or right.

sign bit   A single bit in a binary number used to indicate the sign (positive or negative) of the number.

signed magnitude   A binary numbering system where the MSB is used to represent the sign of the number.

signed numbers   Numbers containing a sign bit indicating whether the number is positive or negative.

simulation   Using computer programs to determine the response of a digital circuit.

sink current   Defined as the flow of current into the output pin of a logic device.

SIPO   Serial-in, parallel-out.

SISO   Serial-in, serial-out.

skew   The delay in signal time experienced by a signal passing through several logic gates.

speed   The rate at which a digital device can operate. Measured in nanoseconds.

speed-power product   A figure of merit determined for a logic family by multiplying propagation delay time and power dissipation specifications.

state   The particular logic level observed at the output of a digital device. For example, a device output equal to logic level one is referred to as being in the one or high state.

state diagram   A pictorial representation of state machine operation.

state machine   A sequential circuit designed to sequence through a predefined number of states following specific operating conditions.

static one hazard   A momentary change in a high logic level to the low state.

static RAM   A kind of semiconductor memory characterized by a storage cell that retains information as long as power is supplied to the chip.

static zero hazard   A momentary change in a low logic level to the high state.

storage cell   The circuitry comprising the storage element in a memory device. Each cell can store 1-bit of information.

stray capacitance   Capacitance associated with wire length, circuit loading, or other circuit interconnections.

strobe signal   A timing signal used to enable portions of digital circuitry.

successive approximation ADC   An analog-to-digital conversion technique based on successive comparisons of the analog signal to an internally generated reference signal.

sum of products   A Boolean expression consisting of multiple AND terms ORed together.

switch bounce   The temporary change in switch output level due to the mechanical vibrations associated with opening and closing a switch.

synchronous counter   A counter designed so that each flip-flop stage receives the clocking signal.

synchronous load   A design where data is loaded into a counter or shift register synchronized with the clock signal.

terminator   A resistive device placed near a digital load to provide a proper match between the load's input impedance and the impedance of the connecting transmission line.

three-variable K-map   A Karnaugh map formed from a three-variable truth table. The map will contain eight squares.

timing diagram   A pictorial representation of the change in level of a logic circuit output for specific input level changes.

toggle   An action describing a flip-flop switching from one state to another.

transfer curve   A graph relating the input and output characteristics of a logic gate.

transient voltage   A sudden, undesired change in voltage level.

transition table   A table indicating all conditions required for a change from one state to another in a state machine design.

transmission line   A connecting wire exhibiting both inductive and capacitive traits at a specified operating frequency.

tristate   A characteristic of certain interfacing logic devices possessing three output conditions—high, low, and high impedance.

truncated count sequence   A counting progression other than a natural binary count.

truth table   A diagram relating all the possible circuit input combinations to the circuit output levels.

TTL   Transistor-transistor logic. A logic family.

2's complement number   The value obtained by adding 1 to a 1's complement number.

two-variable K-map   A Karnaugh map formed from a two-variable truth table. The map will contain four squares.

unique conditions   The input combination on a truth table producing an output level different from all others for the specified logic operation.

unsigned numbers   Any number assumed to be positive.

unweighted code   A binary code bearing no mathematical relationship between the bits and information represented by the bits.

up counter   A counter that counts from a lower binary value to a higher binary value.

volatile memory   Memory devices that lose their data when power is removed.

voltage comparator   A device whose output level is determined by a comparison of two input voltages.

voltage transient   A sudden variation in voltage level caused by a change in electrical load.

weighted code   A binary code where each bit position carries some numerical value.

word   A grouping of 16 binary bits.

word size   The number of bits comprising a binary number. Usually refers to the quantity of data handled by a digital system. Typical word sizes are 4, 8, 16, and 32 bits.

write   An operation used to place information into memory.